Space and Society

The Space and Society series explores a broad range of topics in astronomy and the space sciences from the perspectives of the social sciences, humanities, and the arts. As humankind gains an increasingly sophisticated understanding of the structure and evolution of the universe, critical issues arise about the societal implications of this new knowledge. Similarly, as we conduct ever more ambitious missions into space, questions arise about the meaning and significance of our exploration of the solar system and beyond. These and related issues are addressed in books published in this series. Our authors and contributors include scholars from disciplines including but not limited to anthropology, architecture, art, environmental studies, ethics, history, law, literature, philosophy, psychology, religious studies, and sociology. To foster a constructive dialogue between these researchers and the scientists and engineers who seek to understand and explore humankind cosmic context, the Space and Society series publishes work that is relevant to those engaged in astronomy and the space sciences, while also being of interest to scholars from the author's primary discipline. For example, a book on the anthropology of space exploration in this series benefits individuals and organizations responsible for space missions, while also providing insights of interest to anthropologists. The monographs and edited volumes in the series are academic works that target interdisciplinary professional or scholarly audiences. Space enthusiasts with basic background knowledge will also find works accessible to them.

More information about this series at http://www.springer.com/series/11929

Nikola Schmidt
Editor

Planetary Defense

Global Collaboration for Defending Earth from Asteroids and Comets

Springer

Editor
Nikola Schmidt
Department of Political Science
Institute of Political Studies
Faculty of Social Sciences
Charles University
Prague, Czech Republic

3D Visualizations of the Multipurpose Lunar Base © Martin Vaněk, AD13
www.ad13.cz

ISSN 2199-3882 ISSN 2199-3890 (electronic)
Space and Society
ISBN 978-3-030-00999-1 ISBN 978-3-030-01000-3 (eBook)
https://doi.org/10.1007/978-3-030-01000-3

Library of Congress Control Number: 2018957842

This Springer imprint is published by the registered company Springer Nature Switzerland AG
The registered company address is: Gewerbestrasse 11, 6330 Cham, Switzerland

The artist's depiction of the multipurpose Moon base, particularly the section dedicated to the hydrogen mining. Visualization by Martin Vaněk, AD13. Depiction is based on the general architectural concept by Colorado Mining School

This book is dedicated to the human species living on Earth in the Milky Way.

Foreword

To the general population, defending planet Earth from fast-moving asteroids and comets invokes the genre of science fiction rather than the traditional concerns of national or international security studies. We may be dimly aware of the Tunguska and Chelyabinsk events of 1908 and 2013, respectively, but the former may sooner trigger memories of a 1996 *X-Files* episode than a geostrategic question and the latter grainy YouTube videos shot from the dash cams of perennially vigilant Russian drivers. Alternatively, there are the Hollywood blockbuster films *Armageddon* and *Deep Impact*, both of which deal with imminent comet impacts that in the case of the first are successfully thwarted and, in the second, only partially mitigated, leaving us with a moving popular meditation on human mortality.

The book you hold in your hand (or view on your screen) demonstrates that planetary defense concerns have been the subject of active scientific research and global governance activity for many years, through cooperative international efforts to surveil and map the orbits of "near-Earth objects" (NEOs) and assess their size; the research on methods of orbit change and deflection using spacecraft, lasers, and nuclear weapons; and—just within the last decade—UN-sponsored activity to build international capacity, dialogue, and coordination, resulting in the formation of an International Asteroid Warning Network (IAWN) and a Space Mission Planning Advisory Group (SMPAG). The 12 observatories linked to the IAWN have identified some 18,000 NEOs, including 887 objects with diameters of 1 km or larger, while the SMPAG has recommended that new space missions be tasked with testing both the "kinetic impactor" and "gravity tractor" techniques for deflection. Indeed, NASA is currently preparing a kinetic impactor test with the DART mission to the binary asteroid Didymos in 2021. The testing of nuclear devices in such missions, which some scientists believe to be the most effective for especially large objects, is more controversial, raising a number of legal, ethical, and security concerns that you can read about in the pages below.

With its 27 chapters and about 200,000 words of multidisciplinary scholarship, this book may well be one of the most comprehensive studies of the near-Earth object threat across its scientific, technical, astronomical, governance, and ethical dimensions. You will be able to learn about our technical capacities for finding and monitoring planetary bodies, the scale and likelihood of the potential threat, evolving structures of international cooperation, planetary defense methods, the international law shaping and governing various activities, and futuristic proposals for new kinds of world governance and cooperative space activity. The careful and expert quality of its authors' scholarship, and the integrative analytical power of the book, is an impressive achievement.

For now, the NEO threat appears minimal, but, if the impact does occur and the object is large enough or strikes a populated area, it could have devastating consequences. The 1908 Tunguska event,[1] in which a comet over 100 m wide exploded over a wilderness area with an estimated force of 3–5 megatons of TNT (equivalent to a midsized hydrogen bomb), destroyed two thousand km^2 of forest. Had it exploded over a city, millions would have been killed. As David Morrison explains, objects over 1 km in size could have global effects similar to a nuclear winter, while mass extinction is possible at still larger sizes. The 10 km-wide asteroid that impacted in prehistoric Mexico at the end of the Cretaceous period 66 million years ago triggered the Earth's fifth mass extinction, which wiped out the dinosaurs and 75% of the Earth's species. While no significantly big NEOs have yet been found traveling on an Earth-impact trajectory, tens of thousands remain undiscovered. It is undoubtedly a serious question for the world's security and survival.

Two particular foci of this book are especially valuable and important. The first is its treatment of the issue as a problem for international security and for security studies. I certainly agree with Nikola Schmidt that the nature of this threat cannot be effectively understood or addressed within the framework of national security or alliance politics; it must thus bring us into a space of not merely global but *planetary* security concern. Given that national and even international security architectures are so distorted and compromised—what with ideologically fueled disarray within NATO, Russia, and the Middle East and the UN Security Council ever more paralyzed by great power geopolitics—and given that space is considered a special zone that ideally should not be militarized, existing security concepts and architectures are gravely inadequate. As a number of contributors note, this raises serious fears about the collateral damage that would ensue should one or a number of countries decide to "go it alone" either in long-term or short-term preparations for a potential NEO strike. This raises the question of what moral and ethical frameworks should govern the way governments think about planetary security and create mechanisms to deal with it—whether that be in terms of prevention missions or subsequent response to the damage and disaster that may come from a NEO impact event.

This is the second great strength of this book: its argument that the planetary nature of both the NEO threat and governance challenge necessitates a global and *cosmopolitan* framework for security governance and policy. It has indeed been

---

[1] Whether the Tunguska 1908 event was caused by an asteroid or a comet, and how wide the object was, is a living scientific question. The most recent meeting about the topic was the Tunguska Workshop at NASA Ames Research Center in January 2018. The results will be published in Icarus, March 2019 (note by the editor).

intriguing to see my own theory of "Security Cosmopolitanism" (alongside like thinkers such as Mary Kaldor, Ulrich Beck, and Ken Booth) advanced in these pages as a valuable framework for action and planning. This theory was first published in 2013, around the same time that the UN General Assembly voted to establish the asteroid warning network and the mission planning group. Even if more traditionally globalized insecurities such as poverty, refugees, climate change, nuclear weapons, and conflict had been uppermost in my mind, I can see the strong relevance of the cosmopolitan approach to the problem of planetary defense. One important aspect of this viewpoint that the book recognizes is that it is not merely the entirety of humanity that must be our fundamental security objective but also the myriad life forms and systems that make up the Earth's biosphere. It thus pushes us toward both a planetary and an ecocentric vision of common security.

Security cosmopolitanism was framed as a morally and strategically rigorous answer to a range of globalized systems of insecurity. Its idealism rests on a vision of common humanity and global purpose in which every human being and the ecosystems we depend on are equally worthy of being secured into a long-term future. Its realism rests on an understanding that contemporary insecurities are structural in nature and transcend national borders, often emerging from within our states and systems, and must be addressed through cooperative transnational efforts to transform those systems. I feel that Nikola Schmidt and Frank White are right to point to the importance of a cosmopolitan idealism for thinking about planetary defense, one prompted by the "overview effect" of orbiting the Earth or indeed retreating from it to such a distance that we gain an appreciation, as Carl Sagan reminded us, of the smallness and vulnerability of planet Earth and of our humble place as "a lonely speck in the great enveloping cosmic dark."

Yet, security cosmopolitanism is more than idealist impetus; it is a standard and a warning that also takes its cues from this paean to human humility. One cannot take just any policy or technical decision—however international its genesis—and justify it on the basis of its cosmopolitan idealism or intent, as the liberal internationalists did when arguing for war against Iraq to "rescue" its citizens from Saddam Hussein. Effects, complexity, and consequences also matter. This is why I would not automatically endorse every new proposal for cosmopolitan global governance, a norm of the responsibility to defend Earth, or new space missions such as asteroid mining, without careful consideration and after rich and informed global deliberation that goes well beyond philosophers and experts. The concept of the Planetary Council certainly has those deliberative virtues and is worth exploring with some seriousness. Such creative proposals must be assessed (and if implemented, then reviewed) against the ethical tests proposed by security cosmopolitanism, as should the decisions that are made by any new deliberative body.

Will such proposals advance the security of humanity and the biosphere without discrimination? Will they ensure the security of future generations and the long-term integrity and survival of ecosystems? Are we sure that our actions will have positive consequences that can be borne by the world as a whole? As I argued in

2013, the theory is not opposed to securitization[2] as such, but is framed with the desire to prevent inappropriate and potentially dangerous securitizations and keep all security practices under scrutiny.

We are certainly on the right track when, at the same time that we support global cooperation to track NEOs and develop viable mission options for diverting those that may threaten the Earth, we also question whether such cooperation is inclusive, democratic, accountable, or reflexive enough. In this light, Schmidt asks an important question as to whether the Space Mission Planning Advisory Group will significantly decrease "the undesirable possibility of a unilateral action by a strong space power"—especially if that unilateral action involves the use of nuclear weapons. This is where cosmopolitan security principles can complement the thoughtful discussions in this book about how planetary defense should be considered in relation to international legal and normative frameworks governing space and nuclear nonproliferation. These are chapters everyone working in this area should read; among other things, they raise important questions about whether testing nuclear devices in space would be lawful or consistent with the normative aims of the nuclear nonproliferation regime. There is already serious normative and political conflict placing great internal stress on the nuclear nonproliferation regime, which is manifest in aggressive nuclear modernization by nuclear weapons states, on the one hand, and, on the other, in the adoption of the Treaty on the Prohibition of Nuclear Weapons.

Ongoing research and consideration of the utility of nuclear devices for planetary defense must avoid further stressing the nonproliferation regime or violating its fundamental norms. We should hear the concerns of those who question the normative consequences of nuclear testing in space, even if it is somewhat ambiguous legally, or who express fears about the long-term retention of dual-use devices. In an ideal world, we would choose nonnuclear means of asteroid and comet deflection. Until nuclear disarmament progresses to a point at which weapon stocks are at very low levels, one might suggest that research on nuclear deflection should be brought under the overview of the IAEA, and its benefits and dangers brought into discussion in the NPT review and preparatory conferences and be opened to civil society, however uncomfortable those dialogues may be.

If we are able to progress into a future of nuclear abolition governed by a Nuclear Weapons Convention, the potential retention of a small number of nuclear devices for planetary defense adds to the complexities and risks of the "virtual deterrence" inherent in the long-term persistence of nuclear scientific knowhow. In such a situation, the retention of small numbers of nuclear explosive devices can seem irredeemably threatening, even if they are not weaponized. As Michal Smetana suggests, any such retention would need to be rigorously supervised and controlled by an international agency rather than by any one state or group of states, the kind of

[2] *Securitization* in the international relations discipline is an act in which a securitizing actor makes a point through a speech act to the audience that in reaction implements policy addressing the security threat. The point of securitization is not to study whether the threat is real, but rather to study the process of how we realized that we face a threat and how we react to it. It does not matter whether the threat is real or not, because it is about the process of threat securitization, not about the threat itself. For a broader discussion on the topic, please refer to Chap. 11 (note by the editor).

agency that will be required in a low-number or abolition scenario in any case. Hence, concerns about nuclear deflection and planetary defense need to be folded into broader global discussions about nuclear disarmament and abolition. We must continue to search and survey the Solar System for very large NEOs while giving concerns about nuclear methods their due.

Similar ethical concerns emerge when we highlight the discussion in some chapters about the dilemmas that will emerge should it become clear that a NEO threat cannot be mitigated and will strike the Earth along a corridor that only affects one or a few states. Questions of resilience, disaster relief, equality, and international solidarity force their way into consideration. The world has seen admirable expressions of international solidarity to disasters such as the devastation of Cyclone Nargis and the 2004 Indian Ocean Tsunami, but it has also seen terrible examples of international (and intranational) disregard and indifference. The Rwanda genocide, as well as the appalling US government failures to respond adequately or justly to the devastation of New Orleans and Puerto Rico by hurricanes Katrina and Maria, comes to mind. If there is one area of the planetary defense research and discussion that I would like to see progress, it is these questions of justice, disaster response, and resilience to an impact that cannot be prevented (or is simply not predicted). Given the surprise of Chelyabinsk, and the difficulty of spotting objects emerging from the vicinity of the Sun, this area of research should not be neglected.

Hollywood narratives about meteorite impacts always presume that the event has been seen before it happens, but not within a timeframe that provides us with the opportunity for deliberation and response. That no doubt heightens the drama, but it seems wrong on two counts. The most likely scenario is that either we will have no time and be dealing with an impact that was not foreseen, or that we will have time to spare tens or hundreds of years to meet the threat. The prospect of having no time means that the global community needs to be working harder on systems of disaster response and international aid that are rapid, effective, and just, avoiding the toxic effects of geopolitics, racism, ideological disregard, and disaster capitalism. The prospect of having time means that politically and strategically expedient approaches cannot be justified; we have the flexibility to build the accountable global institutions and approaches that will do their best.

The question of time also imposes on my final comment. The fear of an asteroid or comet impact is one of the very few international security issues (other than nuclear war) that bring with it the specter of mass extinction. Our anthropocentric point of view always tends to make us think and fear for the extinction of humanity in such a scenario. Yet the Chicxulub impact that brought about the fifth mass extinction eliminated 75% of the Earth's species, none of which were *Homo sapiens*. Over the last decade or so, biologists and Earth-system scientists have been desperately pointing out that a sixth mass extinction of species *is actually underway now*, and it is not due to an asteroid impact or a nuclear war but rather to the routine and devastating impacts of human business, settlement, agriculture, resource extraction, and industry on the Earth's biodiversity, forests, and oceans. Unchanged, such activity could lead to the loss of 25% of the Earth's estimated eight million species by 2050, just three decades from now. This threat is not cosmic or mysterious at all; it arises from deep within our own house.

This is where we need to harness the innate solidarity and the "magical essence of space community" described by Nikola Schmidt and Petr Boháček to address *all* the major threats to the survival and flourishing of life on Earth; as they argue, planetary defense must be "defense of Earth's interests as a whole." The Planetary Defense Council is an intriguing model for scientifically informed and reflexive and globally inclusive security governance that could be expanded (or emulated) to address other structures of threat: climate, biodiversity, poverty, conflict, and more. Biodiversity loss and climate change are in fact linked processes where the planetary threat is bearing down on us so quickly that we really do lack time. If our global governance and security cooperation cannot arrest the drivers of the sixth extinction within the next decade, laudable planetary defense efforts to protect the Earth from another Cretaceous-scale event will be deeply ironic, if not increasingly meaningless. As this book reminds us, defending planet Earth needs to be about much more than asteroids and humans.

International and Political Studies
UNSW, Kensington, NSW, Australia

Anthony Burke

Preface

"Realism is an excuse not to attempt and to blame those who do."

This is a book written by people who are convinced that we can shape our future—that we have the power to make the decision whether to wage war or flourish in space, that we are not gridlocked in realist perceptions of the political world, and that visions we imagine can become a reality if we decide to create them.

Most of us are not idealists but rather cosmopolitans, people who believe that thinking as a species provides an advantage for our next great adventures, reaching other planets and stars. During the writing of this book, I have seen how skeptical, rational colleagues started to leave behind their *prisons of self-persuasion* that changing the world is not possible and began to believe that our goals are worth the effort. President Wilson conceived of an unimaginable global world at the end of World War I—a partly formed concept that failed and gave way to World War II. However, the world we built thereafter retained some of his thoughts and led to the era of revised humanism. We might be experiencing some glitches now, but this world is still the most secure and prosperous it has been ever since.

Realists like to follow Helmut Schmidt's famous quotation "whoever has visions should go to the doctor," but my answer to them is clear: "realism is an excuse not to attempt and to blame those who do." Realists think that we should not shape the world because the animosity between states is a naturally given fact. We think that without visions of a better world, we would not have the United Nations. Following this perception, the political scientists in this volume believe that we should think of and propose new steps that will take us further, enabling our visions of colonizing the Solar System and traveling beyond to reach other star systems. But before we proceed, we need to be sure that civilization is not going to end itself because we did not watch our own surroundings. Realists do not have answers. We try to propose some.

The political scientists who were invited to this project had not previously been exposed to the mystique of Solar System astronomy in detail, but they are now leaving this project with tremendously different perspectives about our surroundings. They thought that space is vast and empty, but it is not. The astronomers, space engineers, and veterans in planetary defense, on the other hand, realized that the solutions they are seeking should not only be effective, but should bring us more—something *good*—because the most *effective* solutions are not necessarily *good* solutions for humanity. Technology in our lives has inherently political implications, and we should take responsibility when making decisions about it. Politics should come with ethics, as engineered solutions are based on solid science. Both can balance the effectiveness of solutions with ethical reflection on their societal implications.

The central convictions behind our motivation to write this volume are clear. First, if we know how to detect all Earth-threatening asteroids in the Solar System, then we have the responsibility to detect and follow them all. Second, if we know that we can develop technology to mitigate the threat of an asteroid collision, then we have the responsibility to test as many of them as possible. Third, if we know that we can develop a kind of technology useful for more than just planetary defense, and which is also less politically sensitive and possesses the potential to stimulate humanity and biosphere flourishing, then there is no doubt we have found the better way forward.

Planetary defense has two possible ways to proceed. We can either let a strong spacefaring nation take the lead, involve its military, and save the world if necessary, or we can begin a massive civilian scientific cooperation and develop a super powerful laser that can bring us to the stars and provide planetary defense capabilities at the same time. The following volume follows the conviction that the latter is the better and most feasible way to proceed—that planetary defense is not about a mere absence of threat but also about the opportunity to do things that have the potential to change the world. Planetary defense has the power to help us think of ourselves as one species: we all share spaceship Earth, our Pale Blue Dot, and we have a responsibility to forthcoming generations to stop thinking that we are our own enemies, because otherwise we won't survive long. If the lasers are not feasible in the next decade, then there is no doubt that we need rapid solutions in kinetic impactors or even nuclear deflection methods. Political scientists are trying to tell us that pursuing these massively complicated objectives now is worth the effort, because even the attempt has the potential to stimulate new global political structures.

We are not living in a world of kingdoms and empires, but neither are we living in a world of nation-states per se. We are living in a multilateral world of various actors, including nation-states, nongovernmental organizations, and international organizations. People once imagined the world we live in, so we now imagine a world where our own proposed ideas can come true. The following volume explains how to fulfill these visions of the future.

It has been an amazing three years, during which I have met unbelievably inspiring people in the space community with motivations to do things nobody has ever tried. Today, I see that the challenging mission to talk between disciplines that usually never talk to each other has achieved fruitful and novel results, and I wish the reader to have a comparably inspiring experience.

Prague, Czech Republic Nikola Schmidt

Acknowledgments

This volume could never have emerged without the steadfast confidence and unwavering support of my wife, Markéta Černohorská, who came into my life just days after my return from the Space Studies Program (SSP) organized by the International Space University (ISU) in 2015, where everything began. Thus, I must thank the ISU for organizing this summer program, ESA for supporting my attendance, and NASA for supporting the planetary defense team project. Without these organizations, I would never have met the planetary defense team project chair, Madhu Thangavelu, who gave me indefatigable support, or Jim Burke, whose energy and support from a space veteran like him helped us to proceed and not give up during the development of this project. Madhu and Jim have introduced me to so many inspirational people in the space community during this long process of studying, learning, traveling around the world, and finally writing. I should also thank my alma mater, Charles University in Prague, which provided me with initial funding to attend SSP in 2015. Thanks also to the director of our Institute of Political Studies, Petr Jüptner, who believed in us and supported our follow-up research project, which will lead to national planetary defense strategy in the Czech Republic and maybe even to the building of a laser that has the potential to save humanity. I should thank Philip Lubin for providing us with his concept of a laser array on the Moon, which we used in our visualizations.

I do thank every single author of this volume for putting in an unbelievable amount of work and dedication to make it happen and for all doing something more for the project. We listened to the arguments of others, including those coming from different disciplines who criticized our own perceptions. Sometimes we did not understand each other because of differences in disciplinary jargon. Some passionately read chapters of others and proofread English for nonnative speakers over the nights before submission; others were making amendments in response to the tenth review of their text and did not complain. We made it through, and I deeply thank all the authors that we reached the point of publication. It is hard to make distinctions between particular scientists in the whole fantastic team I have had the opportunity work with, and it has been amazing to see how some veterans opened their figurative doors to those who are just finishing their PhDs.

Besides the authors of the chapters, I must thank the entire graphic studio AD13, who said that "saving the Earth is a responsibility" and decided to make the realistic visualizations you will see throughout this volume and mainly in Chap. 26 free of charge. A special thanks to Martin Vaněk, whose passion for space is imprinted in his artist depictions of possible Moon base structures, and also his boss, David Foldyna, who released him to work for weeks on our project. Some depictions might not meet realistic ends, but no real actions begin without preceding visions.

I would also like to thank all the people I have met with during the book preparation, as without their detailed explanations of topics I did not understand, we would not be here. Thanks are also due to our editors at Springer, for their flexibility and support while we worked to make this book the best possible version it could be. Finally, thank you to all our readers, who have chosen to take this journey with us. We welcome your ideas—do not hesitate to write them to us.

This study was supported by the grant awarded by the Technological Agency of the Czech Republic, project TL01000181: "*A multidisciplinary analysis of planetary defense from asteroids as the key national policy ensuring further flourishing and prosperity of humankind both on Earth and in Space,*" and co-funded by the Institute of Political Studies, Faculty of Social Sciences, Charles University in Prague.

Contents

About the Authors

Nikola Schmidt (born 1982 in Prague, Czech Republic) is currently a leader of a Planetary Defense project with an objective to deliver a policy strategy on planetary defense and asteroid mining for the Czech Ministry of Transport. The project consists of astronomers, space engineers, international lawyers, and political scientists. He finished his PhD in 2016 at Charles University, Faculty of Social Sciences, and Institute of Political Studies with a dissertation on The Birth of Cyber as a National Security Agenda. Nikola has been extensively teaching courses on space and cyber security at various Czech universities. His passion in space issues can be dated long before his master studies of international relations at Metropolitan University. A pivotal moment for Nikola was his participation in the ESA and NASA-funded planetary defense team project at the Space Studies Program, organized by the International Space University in Athens, Ohio, USA, in 2015. Nikola received a bachelor's degree in sociology and studied international security at Sciences Po in Paris. His recent articles were published in *Acta Astronautica* and *New Space*. Nikola founded and led two software development companies before coming back to academia and worked on an NGO-driven food facility project in Afghanistan.

Petr Boháček is a security and foreign policy analyst with a focus on Europe, transatlantic relations, and US politics. Petr received a BA in Political Science and International Relations at St. Norbert College, Wisconsin, and an MA in Security Studies from Charles University, Prague, while studying for a time at Argentina's Universidad de Buenos Aires, Portugal's Universidade Nova de Lisboa and Germany's Leipzig University. As an analyst for various private intelligence companies, Petr examined political, economic, and security risks before founding the *European Security Journal*, a multilingual analytical publication on security. Besides frequently commenting on foreign and security policy issues in media, Petr is a research fellow for the Association for International Affairs (AMO) and an associate fellow at the Institute of International Relations in Prague. Currently, Petr focuses on space policy and global governance and is involved in a research grant on Planetary Defense led by Nikola Schmidt, which has the ambition to shape the global political perceptions on planetary defense.

Mark Boslough received a BS in Physics from Colorado State University in 1977 and a PhD in Applied Physics with additional studies in Geophysics from Caltech in 1983. He worked at Sandia National Laboratories from 1983 until his retirement in 2017. He currently teaches at the University of New Mexico and chairs the Asteroid Day Expert Panel. His research subjects have included laboratory hypervelocity impact experiments, shock metamorphism, spacecraft debris shield testing, modeling the impact of Comet Shoemaker-Levy 9, airburst formation of Libyan Desert Glass, the 1908 Tunguska explosion, the 2008 TC3 airburst over Sudan, and Jupiter impacts of 2010 and 2012. He coauthored the United States National Academies report "Defending Planet Earth" in 2010. His simulations of the 2013 Chelyabinsk airburst were featured in two NOVA documentaries and appeared on the covers of Nature and Physics Today. He has developed scenarios and contributed impact simulations for disaster-planning exercises at conferences and emergency management agency events. He co-created, organized, and convened planetary defense sessions at major scientific meetings and has appeared in dozens of science documentaries, with filming taking place in Siberia, the Sahara Desert, the Nevada Nuclear Security Site, and other locations. Asteroid 73520 Boslough (2003 MB1) is named after him.

Tomáš Bruner studied law and jurisprudence with a specialization on international public law at Charles University, Prague. He also finished a master's program in international security at the Faculty of Social Sciences of the same university, continuing with his PhD studies there. His primary fields of interests are intersections of international law and security, especially in regard to use of armed force in IR, human security, and cyber security. As a senior legal counsel, he leads the legal department of a Czech branch of global corporation (MetLife) in Prague.

Anthony Burke is Professor of International and Political Studies at UNSW, Australia. His books include *Uranium* (Polity 2017), *Ethics and Global Security: A Cosmopolitan Approach*, with Katrina Lee Koo and Matt McDonald (Routledge 2014), and *Ethical Security Studies*, with Jonna Nyman (Routledge 2016).

William Crowe is a PhD candidate at UNSW Sydney and an expert in spacecraft dynamics and systems, with an emphasis on nanosatellites. His PhD research focuses on using swarms of spacecraft to characterize asteroids during flyby. One of the main findings of his research shows that several dozen flyby missions to asteroids are possible every year for a swarm of satellites without leaving Earth's orbit. William has won several awards, including the Space Generation Advisory Council's "Move an Asteroid" Award, which led to him cofounding a nanosatellite business, High Earth Orbit Robotics.

Pavel Dufek is a lecturer in the Department of Political Science, Faculty of Social Studies, Masaryk University, Czech Republic. His interests lie in issues of political theory and international political theory, especially democratic theory, theories of justice, philosophy of human rights, and meta-theoretical issues in normative political theory. His current research deals with the announced crisis of liberal democracy, focusing on topics such as political representation and its embeddedness in normative political theories, the concept of separation of powers, and the promises and pitfalls of public reason as a justificatory device. Dr. Dufek has authored one monograph (*Levels of Justice: Liberalism, Cosmopolitanism, and Human Rights*, 2010, in Czech) and coedited another (*Human Rights in International Politics*, 2014, in Czech, with Hubert Smekal). He has published articles in *International Theory*, *Representation*, *Journal of International Relations and Development*, *Cosmopolis: A Review of Cosmopolitics*, *Filosofický časopis* (*The Philosophical Journal*), *Právník* (*The Lawyer*), *Czech Journal of Political Science*, and others. He is editor of the *Central European Political Studies Review* (journals.muni.cz/cepsr).

Evie Kendal is a bioethicist whose research focuses on the ethical, legal, and social implications of emerging technologies. As a member of the Ethics, Law and Professionalism team at the Deakin University School of Medicine, she currently teaches health law and ethics to medical, science, and business students. Evie is the author of *Equal Opportunity and the Case for State Sponsored Ectogenesis* (Palgrave 2015) and editor of *Teaching Medicine and Medical Ethics Using Popular Culture* (Palgrave 2017). Her current research projects consider the ethical implications of artificial wombs, epigenetics research, defensive urban design, and space colonization. Since 2015, Evie has also taught at Yale University's Interdisciplinary Center for Bioethics, with courses covering emerging genetic technologies, precision and electronic medicine, synthetic biology, reproductive biotechnologies, social robotics, geoengineering, and global consent for space research and the use of celestial objects.

Romana Kofler is Slovenian and works as a program officer in the Committee, Policy and Legal Affairs section of the United Nations Office for Outer Space Affairs (UNOOSA). In this position, Romana is in charge of substantively supporting the intergovernmental meetings of the Committee on the Peaceful Uses of Outer Space, its subcommittees and working groups. In the area of planetary defense, Romana works closely with the International Asteroid Warning Network (IAWN) and the Space Mission Planning Advisory Group (SMPAG), established in 2014 as a result of the UN-endorsed recommendations for an international response to the NEO impact threat. As of 2017, Romana also serves as the permanent secretariat to SMPAG, on behalf of UNOOSA, pursuant to General Assembly resolution 71/90. Romana holds a BA in International Relations and Political Science from the University of Ljubljana, Slovenia, and an MA in Asia-Pacific and International Relations from the Hitotsubashi University, Tokyo, Japan.

Leviticus A. Lewis joined the Federal Emergency Management Agency in July 2007 as director of the National Response Coordination Center. In November 2009, he was assigned as the first FEMA Senior Agency Liaison to the Federal Bureau of Investigation and the FBI National Joint Terrorism Task Force. Mr. Lewis is currently assigned as chief of the National Response Coordination Branch. The National Response Coordination Branch is responsible for executive direction, program management, and leadership for the programs administered by the branch which include the National Response Coordination Center (NRCC), the national evacuation program, planning for future operations, and policy for aviation operations. He also serves as one of FEMA's space hazards subject matter experts.

Mr. Lewis has served on several space hazard-related working groups, including as a co-lead on the US Group on Earth Observations (USGEO), the 2016 Earth Observations Assessment, the Space Weather Societal Benefit Area Team, the National Preparedness Science and Technology Task Force, and the Space Hazard Review Team. He also serves as a member of the White House Office of Science and Technology Policy (OSTP) Interagency Working Group for Detecting and Mitigating the Impact of Earth Bound

Near Earth Objects (DAMIEN). Additionally, he co-chairs the FEMA-NAS Planetary Impact Emergency Response Working Group (PIERWG). Mr. Lewis is a retired naval officer, having served for 20 years in the US Navy. He attended the US Naval Academy, graduated from Savannah State University with a Bachelor of Science degree in Electronic Engineering Technology, and has a Master of Arts in National Security and Strategic Studies from the US Naval War College. Mr. Lewis was awarded a Certificate of Completion for the National Preparedness Leadership Initiative, Harvard University, John F. Kennedy School of Government and Harvard School of Public Health, Executive Education.

Nahum Melamed has been a project leader at The Aerospace Corporation since 2003. He obtained his BS and MS degrees in Aeronautical Engineering from the Technion, Israel Institute of Technology, and a PhD in Aerospace Engineering from Georgia Tech, USA. Dr. Melamed conducts studies on planetary defense from asteroid impact, serves on planetary defense conference and exercise organizing committees, and gives talks at these venues. Jointly with NASA's Jet Propulsion Laboratory (JPL), he leads the development of a web-based asteroid deflection app and instructs a planetary defense class built around the tool.

David Morrison an astrobiologist, planetary scientist, and educator, is senior scientist at NASA Ames Research Center. Morrison received his PhD in astronomy from Harvard University. He is a founder of the field of astrobiology and is also known for research on small bodies in the Solar System, including defense of our planet against asteroid impacts. Before joining the staff at NASA Ames, he spent 17 years on the faculty of the University of Hawaii, where he worked with the telescopes of Mauna Kea Observatory. At NASA, he has been chief of the NASA Ames Space Science Division and founding director of the NASA Lunar Science Institute and has been a member of science teams for the Voyager, Galileo, and Kepler space missions. Among his NASA awards was an Outstanding Leadership Medal for his work dealing with the impact hazard. Morrison is the author of multiple popular books and college texts in astronomy and planetary science. He received the Carl Sagan Medal for communicating science to the public and the Education Prize of the American Astronomical Society. He is a fellow of the American Association for the Advancement of Science and the California Academy of Sciences. Asteroid 2410 Morrison is named for him.

Joseph N. Pelton is on the Executive Board of the International Association for the Advancement of Space Safety (IAASS) and is the award winning author of over 50 books and 300 articles. His book Global Talk won a Pulitzer Prize nomination and the Eugene Emme Literature Award of the International Association of Astronautics. He is one of the founders of the International Space University and the former dean and chairman of the Board of Trustees of the ISU. He is also the director emeritus of the Space and Advanced Communications Research Institute (SACRI) at George Washington University. In earlier positions, he was director of Strategic Planning for Intelsat and held other management positions at Intelsat and Comsat. He also served as professor and director of the Interdisciplinary Telecommunications Program at the University of Colorado-Boulder. He has received the Arthur C. Clarke Foundation's Lifetime Achievement Award, the IAASS "da Vinci Award" for lifetime achievement, the British Space Society "Arthur Award," and most recently the Guardian Award of the Lifeboat Foundation. He is the founder of the Society of Satellite Professionals and the Arthur C. Clarke Foundation. He announced the formation of the Clarke Foundation at the White House in 1983, while he was acting managing director of the National Committee for World Communications Year—a US presidential appointment. He played a role in the formation of the Arthur C. Clarke Center on the Human Imagination at the University of California-San Diego in 2012 as well as a new STEM education exhibit at the National Electronics Museum that opened in September 2018. Dr. Pelton has degrees and training in physics, telecommunications, and international relations from the University of Tulsa, New York University, and Georgetown University.

Clemens M. Rumpf has conducted Planetary Defense research since 2013, first toward his PhD at the University of Southampton (UK) and then at NASA Ames Research Center (USA) since 2018. His main focus is asteroid impact risk quantification, which he explored further as visiting researcher at the European Space Agency (ESA) Advanced Concepts Team (Netherlands) and at the Italian National Research Council IFAC Institute. From 2016 to 2017, he represented the UK at the United Nations in the Space Missions Planning Advisory Group, addressing the international response to the asteroid impact hazard. Numerous times, he has provided expert perspectives on the topics of space engineering and asteroid impacts in mainstream media. Clemens received his aerospace engineering degree at the Braunschweig University of Technology (Germany) in 2012 after a year-long visit at Purdue University (USA). He has extensive space engineering knowledge, having worked at the DLR Institute of Space Systems (Germany) on planetary lander guidance navigation and control and as a systems engineer on the Lunar Lander mission at ESA (Netherlands). Between 2014 and 2017, he co-initiated and was the project leader for the first satellite of the University of Southampton, which has become part of the ESA Fly Your Satellite program.

Jackelynne Silva-Martinez was born in Cusco, Peru. She earned two BAs from Rutgers University, in Mechanical and Aerospace Engineering and Spanish Translation and Interpretation. She earned a Certificate in Lean Six Sigma and a Certificate in Engineering Management from Drexel University. Jackelynne obtained a master's degree in Aeronautical Science with concentration in human factors aviation/aerospace systems from Embry-Riddle Aeronautical University and a second master's degree in Aerospace Engineering with concentration in space systems integration from the Georgia Institute of Technology. Jackelynne is an alumna of the 2015 Space Studies Program from the International Space University, where she was the project manager for the Planetary Defense team project with over 30 international participants. Jackelynne works at NASA Johnson Space Center in the ISS and Orion Mission Planning

Operations within the Flight Operations Directorate. She worked as a mechanical engineer and test operator at NASA Jet Propulsion Laboratory on the Robotic Manipulators and Deployable Booms group, performing verification and validation ground tests for the Mars Science Laboratory rover mission. Prior to that, Jackelynne worked for Lockheed Martin Space Systems Company as an antennas mechanical design engineer and as a systems integration and test engineer for commercial and government satellite programs.

Michal Smetana is research associate and assistant professor at the Faculty of Social Sciences, Charles University, and coordinator of the newly established Peace Research Center Prague. His main research interests include the dynamics of global nuclear politics, arms control and disarmament, proliferation of weapons of mass destruction, deterrence theory, norms and deviance in international affairs, frozen conflicts, and political psychology. His most recent articles were published in *International Affairs*, *Bulletin of the Atomic Scientists*, *The Nonproliferation Review*, and other academic and policy journals. He is the coauthor of *Global Nuclear Disarmament: Strategic, Political, and Regional Perspectives* (Routledge) and *Indirect Coercion: Triangular Strategies and International Conflict* (Charles University Press).

Martin Staško joined the United Nations Office for Outer Space Affairs (UNOOSA) in August 2017 and currently holds the position of an individual contractor. Previously, he conducted his studies and research and graduated with honors and an MA in the program "European Studies in the field of study Political Sciences" at Comenius University in Bratislava, Faculty of Social and Economic Sciences. He wrote his thesis on "Dynamics of Space Programs in Contemporary International Politics: from Competition to Cooperation," focusing on the evolution of space relations between the USSR and the USA since the beginning of the space age until early 2017. He has contributed to several publications in cooperation with colleagues from UNOOSA and participated in the organization of UNISPACE+50, celebrating the 50th anniversary of the First United Nations Conference on the Exploration and Peaceful Uses of Outer Space.

Martin Švec is a PhD candidate at Masaryk University (Czech Republic), research associate at gLAWcal (UK), and a visiting researcher at Tel Aviv University (Israel). His dissertation "Dimensions of International and European Energy Law: State Sovereignty in Ensuring Energy Security" focuses on the limits of energy security and instruments of international law at states' disposal. His expertise covers international energy law, investment law, international environmental law, and international humanitarian law. He also deals with potential conflicts between international investment law and EU law. In 2015, Martin Švec worked as a legal intern at the Energy Charter Secretariat in Brussels. In 2016, he lectured in international economic law and EU law at Mendel University in Brno (Czech Republic).

Carlos Manuel Entrena Utrilla is a physicist-turned-consultant who entered the space industry after being moved by a deep passion for the expansion of humanity into the Solar System and a great excitement for space mining. He started his space career with a Master of Space Studies from the International Space University, where he integrated his technical background with new business perspectives of new space and space commercialization. Since then, he has been a passionate advocate of the commercialization of all activities in space, especially those that create value in space for space. His publications include a COTS-like program for asteroid mining and a framework for the development of the cislunar economy.

Peter Vereš is an astronomer, mainly interested in dynamical and physical properties of near-Earth objects (NEOs), asteroids, comets, meteors, and meteoroids as well as in discovery and photometric characterization of exoplanets. Dr. Vereš obtained a master's in Physics and a PhD in Astronomy and Astrophysics at Comenius University in Bratislava, Slovakia. He attended the Kobe International School of Planetary Sciences in Japan and Vatican Observatory Summer School in Italy. Dr. Vereš worked as a postdoctoral researcher at the University of Hawaii on the Pan-STARRS project and at the Jet Propulsion Laboratory within the Center for NEO Studies. Dr. Vereš also held a faculty position at Comenius University. He currently works as an astronomer at the Minor Planet Center, an advisor to the Slovak Space Policy Association, and a member of the International Astronomical Union. Dr. Vereš co-discovered thousands of asteroids and dozens of comets within Pan-STARRS. Additionally, he is a reviewer of several journals and is active in public outreach in astronomy and space science in his home country, Slovakia. His research deals with simulations of asteroid surveys, such as LSST, improved astrometry by trail-fitting, orbital evolution, and parent body association of meteoroid streams.

Frank White is a magna cum laude graduate of Harvard College, a member of Phi Beta Kappa, and a Rhodes Scholar. He earned his PhM in Politics from Oxford University. He has authored or coauthored numerous books on topics ranging from space exploration to climate change. His best-known work, *The Overview Effect: Space Exploration and Human Evolution*, is a seminal text in the field of space exploration. A film called *Overview*, based largely on his work, has nearly 8 million plays on Vimeo. He is cofounder of the Overview Institute and founder of the Human Space Program project. Frank considers himself a "space philosopher" and has long advocated developing a philosophy of space exploration. He has contributed several articles to the website 2211.world, which is dedicated to developing a new philosophy of space exploration. His book on this topic, *The Cosma Hypothesis*, is scheduled for publication in the fall of 2018. Frank also has a deep interest in artificial intelligence, and in addition to serving as a mentor for the IBM Watson AI XPRIZE, he is working on a multivolume book called *The (Neo) Singularity is Here*!

James J. Wirtz is dean of the School of International Graduate Studies at the Naval Postgraduate School in Monterey, California. Professor Wirtz is the author of *Understanding Intelligence Failure: Warning Response and Deterrence* (Routledge, 2017) and *The Tet Offensive: Intelligence Failure in War* (Cornell University Press, 1991, 1994) and scores of journal articles and book chapters. He also has coedited over 25 volumes. Along with coeditors John Baylis and Colin Gray, he is currently completing the 6th edition of *Strategy in the Contemporary World* (Oxford).

Professor Wirtz earned his degrees in Political Science from Columbia University and the University of Delaware. He was also a John M. Olin predoctoral fellow at the Center for International Affairs, Harvard University, and a visiting professor at the Center for International Security and Cooperation, Stanford University. In 2016, he was honored as a distinguished scholar by the Intelligence Studies Section of the International Studies Association.

Pete Worden (Brig. Gen., USAF, Ret., PhD) (born 1949, in Michigan, USA) is the chairman of the Breakthrough Prize Foundation. He holds a Bachelor of Science degree in Physics and Astronomy from the University of Michigan and a PhD in Astronomy from the University of Arizona. Prior to joining the Breakthrough Prize Foundation, Dr. Worden was director of NASA's Ames Research Center at Moffett Field, California, USA, until his retirement on March 31, 2015. He has held several positions in the United States Air Force and was research professor of astronomy at the University of Arizona, Tucson, USA. He is a recognized expert on space and science issues both civil and military and has been a leader in building partnerships between governments and the private sector internationally. Dr. Worden has authored or coauthored more than 150 scientific papers in astrophysics, space sciences, and strategic studies. He served as a scientific co-investigator for three NASA space science missions, most recently the Interface Region Imaging Spectrograph launched in 2013 to study the Sun. On July 20, 2015, at the Royal Society in London, UK, Yuri Milner and Stephen Hawking launched the Breakthrough Initiatives. On April 12, 2016, in the One World Observatory in New York, NY, Yuri Milner and Stephen Hawking announced the Breakthrough StarShot Initiative to develop and launch Earth's first interstellar probe within a generation. Dr. Worden leads this Initiative as its executive director.

Daniel Garcia Yarnoz is a former Associate Programme Officer at the United Nations Office for Outer Space Affairs, where he was the technical focal point for Space Applications Section initiatives, and Near-Earth Objects (NEO) related activities. He holds an MSc in aerospace engineering from the Technical University of Madrid, Spain, and a PhD in mechanical and aerospace engineering from the University of Strathclyde, Glasgow, UK. His doctoral research was focused on the dynamics of minor bodies and the study of novel ways of manipulating small NEOs, ranging from capture to material processing. Prior to his stay at the UN, Daniel has experience in the space industry, agencies, and academia. He worked for over 9 years for Spain-based company GMV, seven of them seconded to the European Space Agency in their Mission Analysis Section. There he mainly performed feasibility studies for Solar System exploration missions. During the doctoral studies, he was a researcher at the Advanced Space Concepts Laboratory in Strathclyde, followed by a short postdoc opportunity as a Japan Society for the Promotion of Science fellow at the Japan Space Exploration Agency (JAXA). At JAXA, he studied small satellite interplanetary missions to cislunar space and to asteroid Phaethon.

Chapter 1
Introduction: Planetary Defense as the Unique Historical Opportunity to Shape Our Shared Destiny

Nikola Schmidt

We either win as a team or we die as individuals.

—*Al Pacino*

Peace cannot be kept by force. It can only be achieved by understanding.

—*Albert Einstein*

On February 14, 1990, Carl Sagan asked to turn around the Voyager 1 camera and take one last photo of Earth as the spacecraft sped out of our Solar System and into the deep universe. The result was a famous photo of Earth that Sagan called The Pale Blue Dot.[1] Many astronauts have reported strange feelings while in space, a phenomenon lately named the *overview effect*. Twenty-five years later, Ron Garan wrote a book titled *The Orbital Perspective: Lessons in Seeing the Big Picture* (Garan and Yunus 2015). All these moments share the same mind shift undeniably linked to the philosophical question of morality—a mind shift regarding what we might call an undeniable moral holistic perception. This perception is not unique in political science. It began in ancient Greece, continuing during the enlightenment era and through to 20th century efforts to govern political affairs globally.

This book uses an interdisciplinary approach to investigate the idea of protecting the Earth's biosphere from asteroids and comets. A viable Planetary Defense (PD) capability cannot emerge or be constructed without analyzing philosophical perceptions of morality and its consecutive impact on international politics. It is a project

[1] Carl Sagan requested to turn around the Voyager 1 probe in 1990 to take one last photo of Earth and the result is called Pale Blue Dot. Further info here: http://fettss.arc.nasa.gov/collection/details/the-pale-blue-dot/.

N. Schmidt (✉)
Department of Political Science, Institute of Political Studies, Faculty of Social Sciences, Charles University, Prague, Czech Republic
e-mail: nikola.schmidt@fsv.cuni.cz

N. Schmidt (ed.), *Planetary Defense*, Space and Society,
https://doi.org/10.1007/978-3-030-01000-3_1

that should be pursued from a perspective involving the whole of humanity, as it directly impacts the very survival of our species and our planet's biosphere. Rationally speaking, it does, but politically it certainly adds more questions.

Deniers of the planetary defense necessity argument may have political objectives and incentives that are not directly linked to the planetary defense project, but certainly will influence related decision-making on the international stage. In the prevailing nature of the nation state paradigm, politics is not so much about foreign policy as it is about domestic policy and their inevitable interrelation (Fearon 1998). The current nation-based democratic process is constrained by the domination of domestic policy issues—that is, the agenda people vote for first. The actions of any sovereign state on the international level are not motivated only by foreign policy interests of that state, but should be understood as a two-level game where domestic motivations precede international ones (Putnam 1988). We can say that all people tend to want to save the world, but the question is how big is the one – their one – they are saving?[2] Saving humanity can for some be leverage for domestic political objectives.

Moreover, the objective of planetary defense ought to have an apparent rationale since we already have the technical capability to mitigate such a celestial collision. However, the moral call remains unclear, because not all are willing to consider the overall picture of our shared responsibility, which is fully required for survival and flourishing of mankind within the fragile surrounding biosphere that we depend on. We may be at the beginning of engaging a three-level game—*domestic, foreign* and *global*—as we stand clearly in front of global challenges such as a potential asteroid impact, and also environmental issues that do not respect nation state borders but have potential to influence human flourishing with no regard on our nationalities.

There is a list of arguments that impede the project of planetary defense. In a unilateral scenario, where a superpower takes on the task of preventing a collision, some fear that the leading nation would turn the smaller states into followers without ever considering letting them meaningfully contribute; hence it would deepen the political periphery—or the unequal influence and development between certain nations—on a global political scale (Gourevitch 1978). We already observed this tendency during the talks concerning the Code of Conduct in space. Following the perspective of traditional geopolitics, the United States considered more broad Transparency and Confidence Building Measures (TCBMs) as undesirable, as Washington could lose its leading position in space, and because all the situational awareness data concerning the orbital debris was already shared at a bilateral level with nations that actually needed them (Listner 2015).

Another argument points to the low frequency of celestial collisions. Such reasoning tends to amplify the necessity of statistical probability in decision making, while neglecting the fact that a single unpredicted collision can wipe out civilization as we know it. If we take the latter position, we must deal with a moral and rational dimension of such a mission. However, morality is not direct and clear. Its application through technology raises unintended, colossal political dilemmas. The base

[2] I would like to thank here Kristýna Andrlová from the Czech UNHCR office, who gave me this insight when we were talking about the relation between the planetary defense endeavor and the growing nationalist mood in Europe due to the refugee crisis in 2015.

rationale might be clear—the survival of our species and the biosphere—but at its core, this is also derived from the moral foundations of particular philosophical traditions. This debate has recently been sparked in the discipline of International Relations. In this light, social science should have the courage to once again investigate the right and morally desirable path for the future of mankind.

We can either deny the statistics arguing that the last extinction event caused by asteroid impact occurred 67 million years ago (Chicxulub) and think in the traditional geopolitical way of the interstate world system, or we can start thinking as a species with a new awareness of the hazards in our Solar System neighborhood, determined to pass through another threshold to survive and flourish in the universe. We need to take a broader pragmatic perspective and think in terms of the overview effect, perceiving our fate from the orbital perspective.

The fact that humanity already possesses the technology to proceed does not on its own make it an easily defendable political motive. Which state, society, religion, or civilization should proceed? What can the others do justifiably against the will of a nation willing to act if they realize that the project threatens them in some way? Are there others (Merton 1968, p. 287)—meaning other than just one civilization on Earth? Should we all proceed together? Who is "all"? Who leads? These are critical questions that naturally arise when talking about the planetary defense endeavor. By answering or elaborating on them, we slowly walk toward concepts of applicable democratic procedures and global democracy, global governance or even a global government—simply any cosmopolitan political system. This last term can easily cause goosebumps among the majority of current political science scholars, being written off as too idealistic or even aggressive, especially for those who are inclined toward neorealist thinking. The application of cosmopolitan ideas to planetary defense is a gargantuan objective that should respect the foundations of philosophical knowledge in different cultures dating back over many millennia, especially since the intentions are supposed to be defended globally. We aim for a balanced theoretical perspective in this book, and thus cosmopolitanism is not understood here as the only solution, but rather as a procedural exercise for the study of the *highest ideal* on our way to finding a feasible policy for the political and technological infrastructure of planetary defense. We decided to approach the issue of planetary defense from a multidisciplinary perspective to give the audience a wider view of the different topics scientists are thinking about.

Cosmopolitanism generally falls into the idealist school of thought. However, implementing global peace using moralist language can be interpreted as an imperial form of behavior by the state that promotes it (Coates 2000). It might be even self-destructive for the proposing state, as we can observe in any neorealist argument placing attention on national power. The questions of *peace* and its *implementation* have been well studied and criticized, especially since the events of 9/11 and the subsequent state-building agenda in Afghanistan and Iraq.[3] The critical response

[3] For a principal criticism on liberal peace read for example (Chandler 2004). For a more structural argument you can check for example (Goodhand and Sedra 2010). For an argument that peace is not universal, you can read (Paris 2004). Paris later wrote an inspiring discussion on liberal peace (Paris 2010).

to the idea of possible democratic peace implementation within a state or local society has over the last 17 years been strong and broad among almost all universities with international relations departments around the world. This experience shows us how the legitimacy of the global application of universal values can devolve to the benefit of the neorealist argument of interstate anarchy, the principal sign of a systemically inevitable environment of international relations that cannot be changed by international institutions (Mearsheimer 1994). Or, it can be blamed as pure geopolitics by language in critical geopolitical studies (Dalby 2007).

A planetary defense endeavor can be easily blamed by neorealists as hidden power politics, and from critical geopoliticians, as a discursive occupation of the globe. We disagree of course; for us, planetary defense is the best political agenda to overcome international tensions and promote interstellar exploration as a species, as it puts emphasis on the cosmopolitan identity, which has already been a norm in space for 50 years through the promotion of the UN Outer Space Treaty.

If we adhere to cosmopolitanism as the solution for planetary defense and as a feasible and implementable plan on the interstate level, we will certainly face the well-known neorealist counter-argument that the reality of anarchy between states is not solvable (Waltz 1996) and that the argument of universal values is flawed, as the experiment of neoliberal state-building is ethically bankrupt (Richmond 2009). However, neoliberal institutionalists contend that international institutions are built on shared values (Keohane and Martin 1995) and that cooperation under anarchy is not totally impossible (Axelrod and Keohane 1985). The European Union in its decades-long process is certainly based more on the belief in a better future rather than on pragmatic self-centered national power politics of separate European states (Pollack 2008). If the European Union does not disintegrate under the flash of emotional populism we are currently living through, it can provide us guidance on how to think bigger than in national terms. Universal morality is not then a mere idealistic vision; it is the main motivation for deeper political cooperation and it is worth the attempt.

The overview effect shows how politics come from ideas, from the perspective we take, from the position we meaningfully advocate for, stimulated by a view of our home from space. Neorealism received a strong constructivist criticism by Wendt's argument that anarchy is what states make of it (Wendt 1992). The overview effect can also be understood as a proposal for how to think to overcome the anarchy. In the end, for Wendt, the solution is an inevitable cosmopolitan state (Wendt 2003), which will arrive naturally when we overcome our skepticism, suspicion and disbelief as we did at the end of the Cold War (Wendt 1992).

Politics come from ideas. There is a chance for this project on a global scale that depends on our will to proceed with it; it can set up the foundations for unprecedented international cooperation. Global security perspectives are necessary, and a cosmopolitan approach is the only way to overcome tiny struggles between states that prevent us from solving crucially important issues. Thus has Burke introduced his new theory called *security cosmopolitanism*, asking us to imagine the world that we can achieve to solve the issues on which our survival and flourishing fully depend (Burke 2013).

There are already some observable outcomes that show this is not just mere idealism. The International Space Station was founded as a project of international cooperation and succeeded tremendously (Manzione 2002); The European Space Agency with its equal model of project distribution between European nations would be the next example, as would the European Organization for Nuclear Research (CERN). We also have examples of failures and setbacks, such as Russia's recent cancelled membership in the International Science and Technology Center (ISTC), an institution designed to give meaningful civilian nuclear projects to former Soviet nuclear scientists to avoid investment in nuclear weaponry and promote international nuclear civilian research. High-stakes world politics influenced by a single *delusional* politician have continually been a problem (Byman and Pollack 2001), but they are not proof of the inevitable interstate anarchy—in fact, such events explain the opposite. Individuals do influence politics, but the politics of others can mutually influence them as well; the whole system can contain these risk-prone individuals and still be strong enough to contain their negative will—it's not the agent, nor the political structure but both what constitute the political life (Wendt 1987).

Planetary defense as the ultimate global cooperation endeavor can tame narrow-minded politicians with short-term ambitions and pave the road for a future Solar System regime. A planetary defense program can even help to deal with the current low threat, but constantly growing problem of orbital debris.

Dealing with global space threats requires creative, constructive and critical processes that see saving mankind as a higher priority than preserving the current status quo of international relations. Intergovernmentalism seems to be a stable, growing process, but the question is whether this current political model is desirable. With the rising complexity of the system, the system framework will tend to collapse thanks to a rising number of possible vetoes (Jervis 1997). This is not because of the systemic character of international systems as perceived by neorealism, but because of a structural problem of intergovernmental global politics; a problem well discussed in European integration and accentuated by federalists (Burgess 2000). The mission to save the world undoubtedly has a moral dimension. Philosophical traditions vary, and we cannot ignore these varieties if we want to understand and anticipate the consequences of huge policy proposals concerning the defense of the Earth and the preservation of our species and biosphere. Let us now elaborate on the philosophical concept of morality.

Morality has been differently perceived in different ages. For ancient Greeks, moral behavior meant the acceptance of the Greek Gods, as seen in Homer's *Iliad and Odysseus.* Being a good citizen meant participating in public practices. Rejecting them was understood as a behavior on which the others in that society cannot count on. The relations between people were based on mutual reliability rather than on the actual act of belief (Rawls 2000, p. 4). Morality was linked to reliability in civic pieties, social processes and relations, rather than as an obligation to the actual God. This is contrary to Christianity, in particular dating back e.g. to Aquinas, who understood God as the highest authority (Bonnete 1972).

Things changed with Socrates, who rejected the Homeric ideal of a heroic warrior and replaced it with the universal *highest ideal*. In the end, he was sentenced to death not for disbelieving, but for influencing children and thereby causing social unrest. He was greeted with disrespect and disdain by the general public. Still, he argued that the moral focus must be oriented to the *highest ideal* rather than to the constructed noble warrior of Homer, and he stood behind this idea until the end. His concepts formed the very core of individual moral qualities for the forthcoming millennia. After Socrates, morality in ancient Greece was defined as the focus toward the *highest ideal* as a *reasonable choice*, not a *belief*. Both of these Greek approaches are purely based on reason, not on religion. They were free from request and lacked any obligation which was not the case in Christian ages until the Enlightenment.

Morality in the modern age is based on three important historical developments: the reformation of the Church in the 16th century, the establishment of the nation state as the central political and administrative sovereign actor during the Westphalia peace talks, and the development of modern, rational science as it emerged during the Enlightenment. All three historical developments spanned over 200 years, between the 16th and the 17th centuries, oftentimes clashing with the Christian Church, the fourth influential factor in the moral development of the modern age.

During the medieval age, Christianity was an authoritative, doctrinal, expansionist religion of salvation practiced by the sole authority of priests (Rawls 2000, p. 6). During the Enlightenment, sacred texts gave way to reason as the ultimate source of authority for both philosophy and science. Despite this paradigm shift, the Church fought to protect the status quo and its traditional source of truth. It could be argued that the inquisitions did not end by the 12th century, but rather ended in the 19th century (Lea 2015).

Morality has developed from the requirements imposed by the Church up to the present, despite our current or secular society. The uncritical tendency to obey authorities is what we have learnt in churches and it is in many ways still perceived as the *good* strategy to preserve our personal interests. Nevertheless, the way we perceive our duties, obligations and the idea of an imperative dictate of reason (Rawls 2000, p. 7) comes from the longstanding Christian culture, which is detached from the political interests of the Catholic Church. Furthermore, current universal moral values, although based in ancient Greece and the Enlightenment, are linked to the secular 20th century tradition of the human security concept (Thomas and Tow 2002), which challenges national sovereignties (Axworthy 2001) and can be understood as an example of the imperative imposition that was thoroughly criticized as such (Chandler 2002).

This point seriously resonates in current international relations on a daily basis. One may understand any space agenda discussed by the United States in the perspective of its imperative will, exceptionalism and messianism (Edwards 2009). On the other hand, one can argue that having a massive manned space flight program is an undeniable row of achievements that others want to emulate, *a shining city on the hill* that preserves America's leadership, a leading actor in the international arena *worth to be followed.*[4] The latter is certainly driven by a motivation to develop a reasonable policy in space, based on a conviction that the world will follow without hesitation and celebrate the achievements of humanity. However, the former group of people will still see the exceptionalism and messianism, or yearn for power, behind it.

The saying *shining city on the hill* is drawn from the Bible, particularly in Mathew 5:14. In this part, the town is on a hill where it cannot be hidden. In 5:15, Mathew compares the city to the light on a stand rather than in a bowl, as it gives the light to others. Then in 5:16, Mathew impels us to give the light to the others to glorify the Father. This example clearly shows the difference between Christian and Greek foundations of morality. The former builds on a group of believers who follow, in conformity, a stipulated authority as the highest judge of their moral decisions, whereas the latter focuses on the highest ideal based on reason. Keep in mind that Plato and Aristotle were not sentenced for following Socrates. The society faced uncertainty after dropping Homer's warrior ideal, but remained surprisingly calm during discussions of Socrates' message, which in the end gave Plato the opportunity to found a philosophical concept of *dialogue* in a newly created vision of the Academy.

However, our society is influenced by both and we should elaborate on our arguments reflecting both moral traditions to avoid a clash between the proposed policy and its critics. The origins of morality are of course complex and infinitely debatable; however, some norms and values are common in every major religion and have been quite visible in policies concerning global nuclear armament. Here shines the principle of reciprocity: treat others as you wish to be treated.[5] Nations then should approach each other reciprocally (Granoff 2000, p. 1422) as they (should) do under the umbrella of states according to the UN Charter (UN 1945), taking into consideration the fate of individuals. Thus, paving the way towards global planetary defense should not be interpreted as a threat to anyone. Evolving from the League of Nations, the UN is proof that mutual understanding of core cultural moral foundations is possible, and that criticisms of neoliberal development and state building should not be wielded as an argument to deny the idealist character of planetary defense as a stepping stone to a global moral integrity. This should be used as a counter-argument against deniers arguing with philosophical and cultural diversity (Zolo 2000), implying that a (semi)cosmopolitan solution of planetary defense is impossible. It is possible, and it is necessary for the united defense of the planet against an Earth-crossing asteroid and cometary threats or *the means of planetary defense will be the cause of the next global conflict.*

This introduction to the historical development of morality is important when considering the cosmopolitan political architecture as a serious option. Cosmopolitan proponents principally support the need to achieve the *highest ideal* of social

[4]The term is used in (Johnson-Freese 2013, p. 5), but was previously mentioned by president Ronald Reagan, who drew it from John Winthrop, 1630. The original comes from Mathew 5:14.

[5]Check the footnote no. 37 in (Granoff 2000).

establishment—global peace and thriving humanity.[6] The precondition for our survival is a fundamental change in our collective behavior (Barber and Bartlett 2009). When proposed by a state with Christian tradition, the criticism of using the universal values as an imperative based on indisputable norms is inevitable (Coates 2000), but defendable, as is apparent from the criticism and its defense of the state-building agenda (Easterly 2006). Avoiding asteroid collision on Earth and constructing a human rights regime are two quite different goals. Proposing the cosmopolitan political architecture strictly for planetary defense on a new norm of Responsibility to Defend Earth (R2DE) will certainly receive the same criticism as the universal implementation of the human rights regime. However, the argument for why R2DE creates an acceptable universal moral approach—rather than creating a conflict between competing cultural perceptions of morality as discussed above—is crystal clear, as it is about the survival of the entire planet's biosphere (Caldwell 1999).

We proposed this norm for the first time during debates at the Space Studies Program 2015 (Thangavelu et al. 2015) and raised three consecutive arguments on why the criticism applied to Responsibility to Protect (R2P) (Evans and Sahnoun 2001) cannot be similarly applied to Responsibility to Defend Earth. Firstly, the R2DE, in contrast to R2P, does not question state sovereignty, since all states would benefit from the planetary defense action with no difference (or predictable difference). Secondly, while R2P regulates relations between states and forces them to have an appropriate domestic political model respecting universal human rights, R2DE calls for a cooperative global action without placing any such conditions on states' domestic policy and thus treats all the actors equally. Thirdly, R2P justifies an action against a state if it fails to comply to the universal human rights values. This is not the case in R2DE, which simply calls for global, cooperative responsibility against common threats (Thangavelu et al. 2015, p. 29).

We, mankind, face three problems: sharing the planet, sustaining life and developing a rulebook (Rischard 2002). There are plenty of reasons to be convinced that a global society in one or another shape is inevitable sooner or later. Some argue that while democracy could certainly be established across the world on the state level, there is no reason why it could not be established on the global level as well (Held 2010); they even argue that a global democratic system is a precondition to saving our environment and survival (Barber and Bartlett 2009; Hale et al. 2013). The current lack of global identity is not a valid argument. National identities have been socially constructed, so why not global identity (Habermas 2006)? We can start this debate with a general argument that political life is becoming more interconnected with equally sovereign elements. With the rapidly increasing number of elements in the system, the peril of a veto becomes inevitable and unbearable (Jervis 1997, p. 573). The neorealist outcome is anarchy, tensions and even a war, while the idealist outcome is creative thinking of possible governmental models (Held 2006) to tackle these challenges.

It is up to us how we approach our future. Collaboration is inevitable in a world that is more interconnected with each passing day, as we recently saw in the response

---

[6]An interesting reading in globalization studies is summarized in (Held and McGrew 2003).

to Ebola in Africa. As Wendt said, anarchy is what states make of it, so there is no reason to do the opposite. The most common globalist argument claims that communication has become global, despite the tendency of some states to control information flow. The availability of a universal communication method, the Internet, changed the world and made it a bit more *flat* allowing economic capital to freely move around the globe, unhampered by 20th century political conditions (Friedman 2007). The opposing argument states that the globalization process and globalized world is more fragile than we would expect (Ghemawat 2009). However, the pro-globalist argument is not only about communication, but also about the culture that has been slowly merging around the world for the last century.

Environmental problems are global by principle, and the Paris COP21 conference in 2015 showed that states are aware of the need for serious cooperation in this field (UN 2015). Security threats, traditionally perceived and addressed on the state level, are now undeniably global, as terrorism and environmental threats have emerged (Jackson 2006) triggering global coordination in tackling these threats; not to mention that critical studies have shown us how threats such as global terrorism are not only global in fact, but mainly global in discourse (Ditrych 2014). Every day, we see tragedies around the world on our smartphones as push notifications, which are almost the same single sentences shared across the biggest news houses from various continents. These are giving the image of the world to all cultures; a shared reality that we did not experience even a decade ago. We cannot forget the globalization argument in expanding activities in the field of global economic processes, which is prevalently organized by the growing number and sizes of multinational corporations that drive the core of the technology development in sectors crossing borders around the world. Nowadays, meeting a man on a donkey holding a smartphone in Afghanistan is a common situation. There is a growing tendency and success for states organizing functioning collectives in common defense frameworks such as NATO. Not only do super modern weapons consist of components made in different countries, but even defense assets are now shared among nations and deployed jointly among allied states.

According to David Held, the answer to constantly rising requirements on mankind is cosmopolitanism, because all three problems (sharing the planet, sustaining our humanity and developing the rulebook) are not solvable by individual states. Planetary defense is a great and pivotal moment in human history. We currently do not stand on Earth restricted to watching the asteroid fall; we have the opportunity and capability to act. In contrast to any past period in history, we currently possess the technology needed to avoid a disastrous impact that could end our evolution in seconds. Moreover, not only the evolution of mankind, but also the whole biosphere in which we play a role of a species possessing the capability to reverse conceivable cataclysmic event. In that perspective, in which moral aspect of such decision to proceed cannot be omitted, the ethical security studies we use throughout the chapters discussing desirable political approach to the problem of planetary defense are inseparable theoretical approach critically analyzing the whole endeavor and providing an inspiring normative guidance how can we proceed.

Acknowledgement This study was supported by the grant awarded by the Technological Agency of the Czech Republic, project TL01000181: *"A multidisciplinary analysis of planetary defense from asteroids as the key national policy ensuring further flourishing and prosperity of humankind both on Earth and in Space,"* and co-funded by the Institute of Political Studies, Faculty of Social Sciences, Charles University in Prague.

References

Axelrod, R., & Keohane, R. O. (1985). Achieving cooperation under anarchy: Strategies and institutions. *World Politics, 38*(1), 226–254. http://journals.cambridge.org/production/action/cjoGetFulltext?fulltextid=7631716. Accessed 5 March 2014.

Axworthy, L. (2001). Human security and global governance: Putting people first. *Global Governance, 7*(1), 19–23.

Barber, W. F., & Bartlett, R. V. (2009). *Global Democracy and Sustainable Jurisprudence: Deliberative Environmental Law. Journal of Chemical Information and Modeling* (Vol. 53). Cambridge, Massachusetts and London, England: The MIT Press.

Bonnete, D. (1972). *Aquinas' proofs of God's Existence*. Haag: Martinus Nijhoff.

Burgess, M. (2000). *Federalism and European Union: The Building of Europe, 1950-2000*. Taylor & Francis. https://books.google.com/books?id=Ha99UY2cBtUC&pgis=1. Accessed 18 February 2016.

Burke, A. (2013). Security cosmopolitanism. *Critical Studies on Security, 1*(1), 13–28. doi:https://doi.org/10.1080/21624887.2013.790194.

Byman, D., & Pollack, K. (2001). Let Us Now Praise Great Men: Bringing the Statesman Back In. *International Security, 25*(4), 107–146.

Caldwell, L. K. (1999). Is Humanity Destined to Self-Destruct? *Politics and the Life Sciences, 18*(1), 3–14.

Chandler, D. (2002). *Rethinking human rights: critical approaches to international politics*. Palgrave Macmillan.

Chandler, D. (2004). Constructing Global Civil Society. doi:https://doi.org/10.1057/9780230005846.

Coates, T. (2000). Neither cosmopolitanism nor realism: a response to Danilo Zolo. In *Global Democracy, Key Debates* (pp. 87–102). London and New York: Routledge.

Dalby, S. (2007). Regions, Strategies and Empire in the Global War on Terror. *Geopolitics, 12*(4), 586–606. doi:https://doi.org/10.1080/14650040701546079.

Ditrych, O. (2014). *Tracing the Discourses of Terrorism : Identity, Genealogy and State*. Palgrave Macmillan.

Easterly, W. (2006). *The white man's burden: why the West's efforts to aid the rest have done so much ill and so little good*. Penguin Press. http://books.google.cz/books?id=5Iw5IZCTh-kC.

Edwards, M. (2009). "God has chosen Us": Re-Membering Christian Realism, rescuing christendom, and the contest of responsibilities during the cold war. *Diplomatic History, 33*(1), 67–94. doi:https://doi.org/10.1111/j.1467-7709.2008.00747.x.

Evans, G., & Sahnoun, M. (2001). *The Responsibility to Protect. Report of the International Commission on Intervention and State Sovereignty*. Ottawa, Kanada: International Development Research Centre.

Fearon, J. (1998). Domestic politics, foreign policy, and theories of international relations. *Annual Review of Political Science*, 289–313. http://www.annualreviews.org/doi/abs/10.1146/annurev.polisci.1.1.289. Accessed 5 March 2014.

Friedman, T. L. (2007). *The World Is Flat 3.0: A Brief History of the Twenty-first Century*. Picador.

Garan, A. R., & Yunus, M. (2015). *The Orbital Perspective: Lessons in Seeing the Big Picture from a Journey of 71 Million Miles*. Berrett-Koehler Publishers.

Ghemawat, P. (2009). Why the World Isn't Flat. *Foreign Policy*. http://foreignpolicy.com/2009/10/14/why-the-world-isnt-flat/.
Goodhand, J., & Sedra, M. (2010). Who owns the peace? Aid, reconstruction, and peacebuilding in Afghanistan. *Disasters*, *34*, S78–S102. doi:https://doi.org/10.1111/j.1467-7717.2009.01100.x.
Gourevitch, P. (1978). The second image reversed: the international sources of domestic politics. *International Organization*, *32*(4), 881–912. http://journals.cambridge.org/production/action/cjoGetFulltext?fulltextid=4305592. Accessed 5 March 2014.
Granoff, J. (2000). Nuclear Weapons, Ethics, Morals, and Law. *BYU Law Review*, *2000*(4), 1413–1442.
Habermas, J. (2006). *The Divided West* (Vol. 1). Polity Press.
Hale, T., Held, D., & Young, K. (2013). *GRIDLOCK: Why Global Cooperation is Failing when we Need it Most*. Cambridge: Polity.
Held, D. (2006). *Models of Democracy*. Polity.
Held, D. (2010). *Cosmopolitanism: Ideals and Realities*. Polity.
Held, D., & McGrew, A. (Eds.). (2003). *The Global Transformations Reader: An Introduction to the Globalization Debate*. Polity.
Jackson, R. (2006). Genealogy, Ideology, and Counter-Terrorism : Writing wars on terrorism from Ronald Reagan to George W. Bush Jr 1. *Studies in Language and Capitalism*, *1*(1), 163–193. doi:ideologie; terrorismus; reagan; bush; krieg.
Jervis, R. (1997). Complexity and the analysis of political and social life. *Political Science Quarterly*, *112*(4), 569–593.
Johnson-Freese, J. (2013). *Space as a Strategic Asset*. New York: Columbia University Press.
Keohane, R. O., & Martin, L. L. (1995). The promise of institutional theory. *International Security*, *20*(1), 39.
Lea, H. C. (2015). *A History of the Inquisition of Spain, Volume 1*. BiblioLife.
Listner, M. (2015). The International Code of Conduct: Comments on changes in the latest draft and post-mortem thoughts. *The Space Review*. http://www.thespacereview.com/article/2851/1. Accessed 10 January 2016.
Manzione, L. L. (2002). Multinational Investment in the Space Station : An Outer Space Model for International Cooperation ? *American University International Law Review*, *18*(2), 507–535.
Mearsheimer, J. J. (1994). The False Promise of International Institutions. *International Security*. doi:https://doi.org/10.2307/2539078.
Merton, R. K. (1968). *Social Theory and Social Structure*. Simon & Schuster.
Paris, R. (2004). *At war's end: building peace after civil conflict*. Cambridge University Press.
Paris, R. (2010). Saving liberal peacebuilding. *Review of International Studies*, *36*(02), 337–365. doi:https://doi.org/10.1017/S0260210510000057.
Pollack, M. A. (2008). The New Institutionalism and European Integration. *Webpapers on Constitutionalism and Governance beyond the State*, *1*.
Putnam, R. (1988). Diplomacy and domestic politics: the logic of two-level games. *International organization*, *42*(03), 427. doi:https://doi.org/10.1017/S0020818300027697.
Rawls, J. (2000). Lectures on the History of Moral Philosophy, 416. doi:https://doi.org/10.2307/3655762.
Richmond, O. P. (2009). A post-liberal peace: Eirenism and the everyday. *Review of International Studies*, *35*(3), 557–580.
Rischard, J. F. (2002). *High Noon*. New York: Basic Books.
Thangavelu, M., Wilson, T., Hussein, A., Aliaj, B., Entrena, C., Lee, C., et al. (2015). *READI - Roadmap of Earth Defense Initiatives*. Athens (OH). https://isulibrary.isunet.edu/opac/doc_num.php?explnum_id=722.
Thomas, N., & Tow, W. T. (2002). The Utility of Human Security: Sovereignty and Humanitarian Intervention. *Security Dialogue*, *33*(2), 177–192. doi:https://doi.org/10.1177/0967010602033002006.
UN. (1945). Charter of the United Nations. http://search.ebscohost.com/login.aspx?direct=true&db=a9h&AN=21212379&lang=cs&site=ehost-live.

UN. (2015). *Adoption of the Paris agreement* (Vol. FCCC/CP/20). Paris. http://unfccc.int/resource/docs/2015/cop21/eng/l09.pdf.

Waltz, K. (1996). International politics is not foreign policy. *Security Studies*, *6*(1), 54–57. doi:https://doi.org/10.1080/09636419608429298.

Wendt, A. (1987). The agent-structure problem in international relations theory. *International organization*, *41*(3), 335–370. http://journals.cambridge.org/production/action/cjoGetFulltext?fulltextid=4309572. Accessed 5 March 2014.

Wendt, A. (1992). Anarchy is what states make of it: the social construction of power politics. *International Organization*, *46*(02), 391. doi:https://doi.org/10.1017/S0020818300027764.

Wendt, A. (2003). Why a World State is Inevitable. *European Journal of International Relations*, *9*(4), 491–542. doi:https://doi.org/10.1177/135406610394001.

Zolo, D. (2000). The lords of peace: from the Holy Alliance to the new international criminal tribunals. In B. Holden (Ed.), *Global Democracy, Key Debate* (pp. 73–86). London and New York: Routledge.

Part I
Defining the Threat and Current Efforts

Chapter 2
The Cosmic Impact Hazard

David Morrison

2.1 Historical Perspective on Cosmic Impacts

Planetary defense against cosmic impacts is a response to a relatively new hazard. Only in the past century have scientists and the public become aware that our planet is subject to impacts by comets and asteroids, and that these impacts constitute a significant danger to society. Recognition of this hazard required the development of new paradigms in astronomy and geology, together with a long-term perspective on planetary history.

Before the modern era, we did not know that our planet could be struck by cosmic projectiles. Edmund Halley, early in the 18th century, was the first to recognize that comets were permanent members of our Solar System, following highly eccentric, planet-crossing orbits. Because comets were then thought to be large (thousands of km in size), the fear of a comet crashing into the Earth and doing terrible damage seemed plausible. Continuing discovery of comets during the nineteenth century maintained public interest in this threat. The English romantic poet Lord Byron wrote in 1822: "Who knows whether, when a comet shall approach this globe to destroy it … men will not tear rocks from their foundations by means of steam, and hurl mountains, as the giants are said to have done, against the flaming mass?—and then we shall have traditions of Titans again, and of wars with Heaven."

Worry about a comet hazard was greatly reduced in the early twentieth century, when astronomers realized that the visible comets were largely composed of diffuse gas and dust, surrounding a solid nucleus only a few km in size. Meanwhile, the first rocky Near-Earth Asteroids (NEAs) were discovered, and in 1937 asteroid Hermes passed the Earth at only twice the distance to the Moon, creating public concern about an impact at some future date. Since that time, it is the hazard of impacts by

D. Morrison (✉)
NASA Ames Research Center, Mountain View, CA, USA
e-mail: david.morrison@nasa.gov

N. Schmidt (ed.), *Planetary Defense*, Space and Society,
https://doi.org/10.1007/978-3-030-01000-3_2

NEAs that has been the focus of interest in what has come to be called "planetary defense".

If this hazard is real, there should be scars of past impacts on the Earth. Once geologists recognized that the Earth was billions of years old, it was hard to rationalize the apparent absence of impact features on our planet. Still, the thinking dominated by the uniformitarian ideas of Charles Lyell and his followers could not admit of rare or catastrophic events in Earth history. Besides, there were few if any geologic features that looked like impact scars. Many assumed that any asteroids or comets that might occasionally hit the Earth would disintegrate in the atmosphere and leave no geological evidence.

The one geological feature that did suggest an impact was the unique bowl-shaped crater in the deserts of Arizona, called Coon Butte (and later Meteor Crater). In the 1890s, G. K. Gilbert, Chief Scientist of the US Geological Survey, made a careful survey of Meteor Crater and concluded from apparently sound scientific reasons that is was a *maar*, a crater produced by a volcanic steam explosion. This conclusion seemed consistent with its location near the extensive volcanoes of northern Arizona, but it completely discounted the large strewn field of fragments of meteoritic iron that surrounds the crater. Curiously, Gilbert also concluded at about the same time, from a different line of reasoning, that the craters that dominate the surface of the Moon had an impact origin.

It was another 50 years before geologists fully accepted the impact origin of either lunar craters or Meteor Crater. By the mid-century, however, the pioneering work by Richard Dietz and Eugene Shoemaker had demonstrated the roles of impacts in forming Meteor Crater and a dozen or more other craters on Earth, and the first Apollo lunar landings finally settled the nature of lunar craters. The early planetary space missions also revealed craters on the surfaces of Mercury, Mars and the two largest moons of Jupiter. Indeed, impact cratering soon became the accepted paradigm for formation of the surfaces of rocky planets and moons within the Solar System. It was Earth, with its relative lack of craters, that was the exception, because active geology erases such "astroblems" on Earth in geologically short time scales (Oplk 1958; Powell 1998; Shoemaker 1983).

Meanwhile, there were other mysteries related to possible impacts on the Earth. The most dramatic example was the atmospheric explosion in 1908 in the Tunguska region of Siberia, which toppled and burned trees over an area of 2000 km^2, yet did not produce a crater. Russian scientists suggested that the Tunguska impactor was a loosely bound, low density comet, which exploded before it reached the ground. The energies of the Tunguska and Meteor Crater impacts were similar, about 10–20 Mt (million tons of TNT equivalent), but the different consequences reflected their different compositions. It was not until the 1990s that satisfactory models emerged for the production of large airbursts, calibrated by data from nuclear weapons tests with comparable energies (Boslough and Crawford 1997; Chyba et al. 1993; Hills and Goda 1993; Morrison 2018; Sekanina 1998; Wheeler et al. 2017).

There was also the challenge of reconciling the multi-megaton explosions at Tunguska and Meteor Crater with the more familiar science of meteors and meteorites. Most meteors arrive in meteor showers, bursts of meteor activity lasting a few hours to a few days, repeated annually. Measurements of the trajectories of these shower meteors demonstrated that they were mostly associated with comets; none of them produced meteorites. Meteorites, the objects that did reach the surface, had compositions that suggested asteroidal sources. Among the meteorites (called finds) that were on the surface and identified by their unusual appearance and composition, the majority were composed of iron and nickel. In contrast, the meteorites (falls) associated with a visible bright meteor or bolide were more than 90% rocky. While a number of meteorites were single objects, it became increasingly clear that most meteorites fell in large numbers (as meteorite showers, not to be confused with meteor showers), apparently the result of atmospheric breakup of a much larger incoming object, termed a meteoroid. Here again was evidence that the atmosphere provides protection against meteoroids of up to several meters in size. The fragments that reached the surface had lost most of their original kinetic energy and therefore were not a significant hazard. While meteorites are valuable as samples of larger impactors, they were not necessarily representative of the bulk of the object before it broke up in the atmosphere.

The most important change in our thinking about the cosmic impact hazard happened in 1980, when we truly recognized that an impact could have global consequences, biological as well as geological. At that time, the discovery of an asteroid impact was linked to the Cretaceous-Paleogene (K-Pg) mass extinction 66 million years ago, introducing a paradigm shift in thinking. The initial detection of enhanced quantities of trace siderophile elements such as iridium in two widely separated samples from the end-Cretaceous boundary layer was quickly found to be a global marker of an extraterrestrial event (Alvarez et al. 1980; Powell 1998; Shoemaker 1983). The inferred source was a 10-km diameter rocky asteroid, subsequently associated with a large buried impact crater (Chicxulub) in Mexico. This was the first time that a cause was identified with a mass extinction. Just as revolutionary was the idea that such a relatively small impact on a cosmic scale could severely disturb the biosphere by starting a global firestorm and injecting immense quantities of dust into the upper atmosphere, blocking the sunlight for months. The result was that more than 75% of fossil-forming species went extinct, including the dominant animals—the dinosaurs. An asteroid impact of this size is not remarkable on the timescale of tens of millions of years, and at that time, astronomers had not surveyed the nearby asteroids to determine whether there were any NEAs of this size that could hit us. We were flying blind.

In the decade after the discovery of the K-Pg (originally called K-T) impact, scientists confirmed that the impact took place at the same time as the mass extinction, and they discovered the location where it hit in southern Mexico. This decade also saw several high-profile international and interdisciplinary science conferences devoted to cosmic impacts and to the possible effects of such impacts on the Earth. One of these, sponsored by NASA and organized by Gene Shoemaker, specifically addressed the risk of a similar impact today.

By 1990, the concern about this impact hazard reached the United States Congress, leading to a resolution inserted into the 1991 NASA funding bill:

> The House Committee on Science and Technology believes that it is imperative that the detection rate of Earth-orbit-crossing asteroids must be increased substantially, and that the means to destroy or alter the orbits of asteroids when they do threaten collisions should be defined and agreed upon internationally. The chances of the Earth being struck by a large asteroid are extremely small, but because the consequences of such a collision are extremely large, the Committee believes it is only prudent to assess the nature of the threat and prepare to deal with it.

The resolution asked NASA to evaluate the impact risk, to propose a program to survey the population of NEAs, and to investigate technologies that could be used to defend against threatening objects. George Brown, the Chair of the House Committee, added, "If some day in the future we discover well in advance that an asteroid that is big enough to cause a mass extinction is going to hit the Earth, and then we alter the course of that asteroid so that it does not hit us, it will be one of the most important accomplishments in all of history."

The two requested NASA studies were carried out in 1992. I chaired the study of the impact hazard and how to increase the discovery of NEAs, producing the NASA Spaceguard Survey Report (Morrison 1992). The name Spaceguard was taken (with permission) from the fictional future survey with the same goals described by Arthur C. Clarke in his science fiction novel *Rendezvous with Rama*. In 1995 a follow-up NASA workshop, chaired by Gene Shoemaker, further developed the survey plan and provided a cost estimate. NASA formally initiated the Spaceguard Survey in 1998. The defense-oriented Interception Workshop, held in 1992 at Los Alamos National laboratory, was one of several studies concerned primarily with nuclear defense approaches (Canavan et al. 1994).

The first international scientific conference on Near-Earth Asteroids, organized by Clark Chapman and supported by NASA, was held in 1991 at San Juan Capistrano in California. Two years later, a much larger meeting of the rapidly growing community was held for three days in Tucson. The resulting 1300-page book, *Hazards Due to Comets and Asteroids*, edited by Tom Gehrels, contained 47 papers by 120 authors covering all aspects of the impact hazard and plans for planetary defense (Gehrels et al. 1994). Another comprehensive book a decade later was *Mitigation of Hazardous Comets and Asteroids* with 19 authors, edited by Belton et al. (2004).

Other notable meetings that helped develop international support for planetary defense were "Space Protection of the Earth," held at the Russian Federal Nuclear Center near Chelyabinsk (Simonenko 1995); "Planetary Defense Workshop" at Livermore National Lab in California (Nuckolls 1995); the United Nations NEO meeting in New York (Remo 1997) and "Beginning the Spaceguard Survey" in Vulcano, Italy, sponsored by the International Astronomical Union (Alenia Spazio SpA 1995).

An independent government study of the impact hazard was carried out by a UK Task Force (Atkinson et al. 2000). The task force came to essentially the same conclusions as the NASA studies: "Impacts represent a significant risk to human and other forms of life. Means now exist to mitigate the consequences of such impacts…"

The following sections of this chapter discuss the impact hazard and implications for planetary defense from a variety of perspectives.

2.2 Impact Hazard and Risk

Recognizing the reality of dangerous impacts was a major first step, but in order to plan for possible defense or mitigation, it was necessary to quantify the hazard and the risk posed by cosmic impacts. The *hazard* refers to the consequences of an impact, often expressed in terms of human casualties or physical destruction. This hazard from an impact is a function primarily of the size (energy) of the impactor and the location of the strike. The *risk* is the product of the hazard and the frequency of occurrence. The hazard increases with the size of the impactor, while the frequency declines, so the product of the two (the risk) is a complex function of the size. The impact location is critical, given that most of the planet is uninhabited or nearly so, while most of the population and critical infrastructure are located in urban areas that constitute only a few percent of the surface area. A hit on a large city by a Tunguska-size impactor could kill several million, but a hit in a wilderness area (such as the real Tunguska location) or in the ocean would produce few if any casualties (Ahrens and Harris 1992; Chapman and Morrison 1994; Morrison et al. 1994).

The evaluation of risk is central to planning for mitigation. There is some size (energy) below which the atmosphere provides a shield, and a range of sizes in which airbursts are more likely than surface hits. Depending on the size and location of the impact, there can be combinations of effects from airburst, surface strike, earthquakes, radiative heat and possibly tsunami. As the size increases, there is an increasing possibility of global effects, up to mass extinction for an asteroid several kilometers in diameter. Secondary effects are a likely consequence from a large hit, including mass casualties, destruction of infrastructure and population and economic dislocations, which can spread far beyond the radius of the original hit.

The initial evaluations of the impact hazard, which were used to plan the Spaceguard Survey, were made by Chapman and Morrison (1994), Morrison et al. (1994) and Toon et al. (1997) based on rough estimates of casualties. For impacts by NEAs from a few tens of meters up to a few hundred meters in diameter, the impact area of destruction was calculated from the known effects of nuclear explosions (Glasstone and Dolan 1977) and then translated into casualty estimates using the average population density of the Earth. At larger sizes, there is a transition to global effects, especially agricultural losses from dust lofted into the stratosphere and carried around the world, blocking the sunlight. The onset of global effects produces a shift upward of more than an order of magnitude in the casualty estimates, with a substantial fraction of the planet's human population at risk. Note that such global casualties would not correspond to a mass extinction, even if a majority of the human population perished, and therefore would not be detected in the geological record. This transition to global effects was estimated to take place for NEA diameters between 1 and 2 km. A mass extinction would be possible at still larger sizes, but once most humans are dead, the so-called *kill curve* would flatten out. These hazard estimates led the Spaceguard Survey to focus on finding NEAs of 1 km and larger diameter. (The survey's formal goal was 90% completeness at 1 km, to be

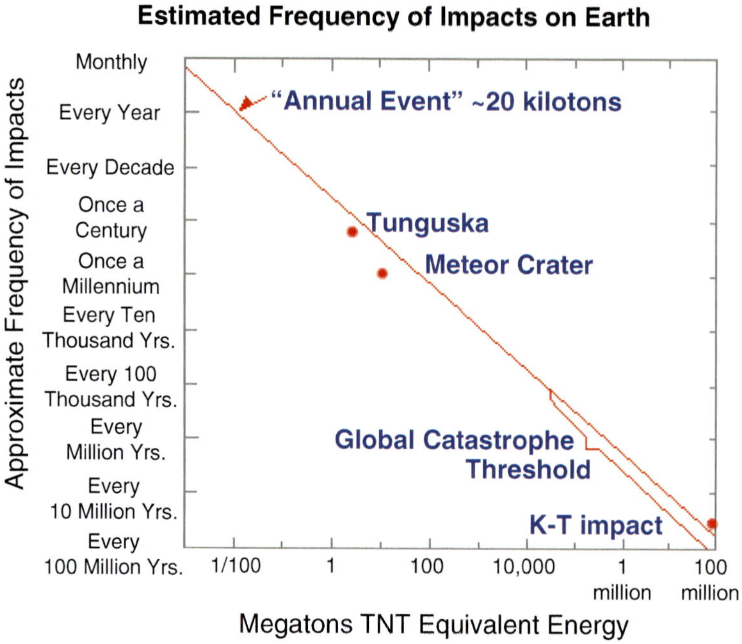

(PSRD graphic, based on figure on page 278 in C.R. Chapman and D. Morrison, 1989, **Cosmic Catastrophes.** Plenum Press, New York, 302 p.)

Fig. 2.1 Frequency of asteroid impacts as a function of energy, using an approximate power law distribution of size. Similar to the original graphs used by Chapman and Morrison (1994)

achieved within a decade.) Because of the nature of the surveys, a nearly complete census of NEAs with diameter greater than 1 km would also imply that the majority would be found down to 0.5 km (Figs. 2.1 and 2.2).

An estimate of the risk required knowledge of the impact frequency as a function of NEA size. Since there have been too few recent impacts on Earth to determine this frequency empirically, an initial estimate was made from the size distribution of lunar craters (Shoemaker 1983). The saturated crater density on the lunar highlands is too great to be useful, but on the lunar maria (lava plains laid down about 4 billion years ago), the observed crater density reflects the average impact rate. Translating from crater diameter to NEA size therefore yields an impact rate for the Earth-Moon system averaged over about 4 billion years. At the small end of the size-frequency distribution, we can derive a current impact rate from bolides, meteoroids that explode in the upper atmosphere and can be detected by down-looking Earth satellites (Harris and D'Abramo 2015). Today, from surveys that have identified many thousands of NEAs, an independent estimate of size-frequency distribution can be made from a census of NEAs that pass close to the Earth, or alternatively from the total NEA population with adjustments for impact rates (Stokes et al. 2017).

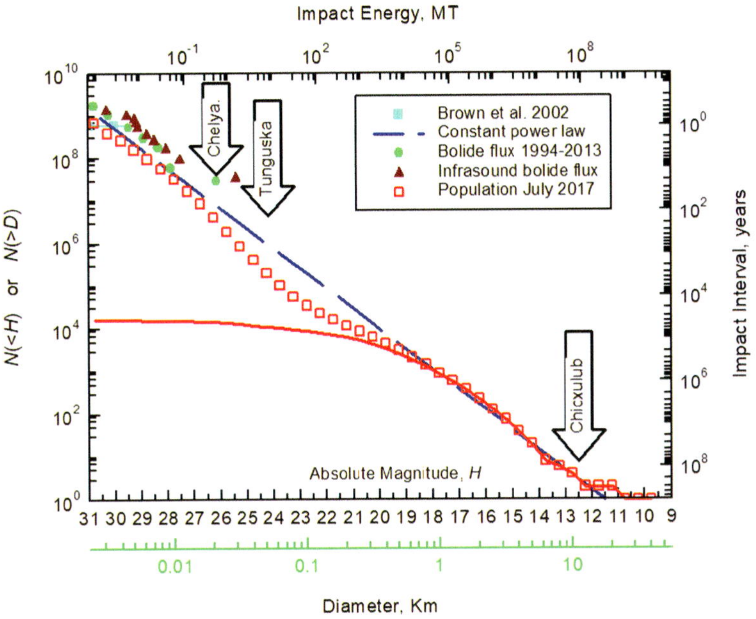

Fig. 2.2 The frequency of impacts as a function of size, as understood at the beginning of 2018. Note departure from the power law, still not fully understood. The data are from asteroid surveys as interpreted by Allan Harris

Over the past two decades, many estimates have been made of the impact risk in terms of expected annual casualties (Chapman and Morrison 1994; Mathias et al. 2017; Rumpf et al. 2017). The distribution of casualties is very different from those more familiar to us, with insurance risks for example (Morrison et al. 1994). The impact casualties are an extreme case of hazards with very high consequences but very low frequency, making it difficult to compare with more familiar risks. The greatest danger is from impacts with global consequences that could kill a billion people, but that happen only once in several million years—much longer than the lifetime of the human species. The fact is that there are no more than a handful of impact casualties that have been recorded in human history. In terms of average annualized deaths, the impact risk (as we understand it) is very low, no more than few hundred per year. This is an average, but decision makers must also consider the terrible consequences from an impact of a large object in the wrong place, which could have severe regional or global consequences.

2.3 Mitigation: Surveys and Predictions

Planetary defense or mitigation is possible only if a specific hazard, from an identified object, has been identified. So far, only two asteroid impacts have been predicted in advance, and they both were from meteoroids only a few meters in diameter, and hence represented no hazard (Figs. 2.1 and 2.2).

The key to any defense effort is early detection. Astronomical surveys are the front line. The requirement is to find hazardous asteroids and calculate their orbits long before they hit. Our concern is focused on NEAs, asteroids that have orbits that come close to or intersect that of the Earth. Comet impacts are also a potential danger, but comet impacts (of any size) are less than 2% as frequent as asteroids (Stokes et al. 2003) and therefore have been generally neglected. This is fortunate, since comets often follow somewhat erratic paths influenced by small non-gravitational forces associated with outgassing, so that prediction of a comet impact more than a few months in advance would be difficult to impossible (Bowell and Muinonen 1994; Morrison et al. 2004; Ostro and Giorgini 2004; Yeomans 2012). However, the surveys do pick up comets as well as asteroids, and their orbits are calculated.

Asteroids are members of the Solar System, and they can be detected telescopically by their reflected sunlight. The first NEAs were discovered early in the twentieth century by photography, when two (or more) images of the same part of the sky were compared to reveal their motion. It was not until the 1970s that several systematic photographic surveys were undertaken, using wide-field telescopes (e.g., the Palomar Leiden survey using the 48-in Schmidt telescope at Palomar Mountain). This was a labor-intensive effort, requiring long nights at the telescope followed by long days of searching the photographs for moving images. Something better was needed to meet the emerging requirements of planetary defense, requirements that were first articulated in the NASA Spaceguard Survey Report of 1992.

Automation of asteroid searches was pioneered by Tom Gehrels at the University of Arizona, who obtained an early version of the now ubiquitous CCD electronic detectors and built a camera for one of the under-utilized telescopes owned by the University. An electronic image is read into a computer, and near-real-time comparison of two images is possible. Gehrels found his first NEA using the CCD camera in 1989. Thereafter, both detector and computer technology rapidly improved, and access to multiple-CCD detector arrays with wide-field cameras was assumed in the recommendations for new surveys.

In the first years of the Spaceguard Survey, which was funded by NASA starting in 1998, the primary search systems were a 1-m-aperture telescope located in New Mexico that had been built for the U.S. Air Force (LINEAR, the Lincoln Laboratory NEA Research Program), the JPL program called NEAT (NEA Tracking), and a program at the University of Arizona called the Catalina Survey. Within three years, more than 500 NEAs had been found, including more than 200 with estimated size larger than 1 km, the size metric for assessing the progress of the Spaceguard Survey. By the end of 2017, nearly 18,000 NEAs of all sizes had been found, including 887 with estimated diameters of 1 km or larger. The primary contributors by this time were the Arizona Catalina Survey (with several telescopes) and the Pan-STARRS (Panoramic Survey Telescope and Rapid Response System) at the University of Hawaii using two specially built survey telescopes. All of these telescopes had apertures of under 2 m, very small by contemporary astronomy standards. Note that rather than organizing a top-down survey with government assets, NASA chose to provide grants to individual investigator teams, who compete with each other. As a result, the efficiency of the survey increased over time as the capability of its individual elements increased (Stokes et al. 2003, 2017) (Fig. 2.3).

The purpose of these observing programs is to discover and track NEAs so that their orbits can be calculated and any future Earth approaches predicted years in advance—it is not to find an incoming object on its final approach to the Earth. For this reason, the absence of a telescope in the southern hemisphere is less of a problem than might be imagined. Any NEA that is missed because it is in the southern sky (or due to telescope problems or cloudy weather) will be picked up on a subsequent pass through the inner Solar System. The important measure of the system is total throughput of discoveries.

An alternative approach is needed if we wish to obtain warnings of smaller NEAs that are at risk of hitting the planet but would likely be missed by current surveys. What is required is a survey that rapidly covers as much of the sky as possible, providing short-term warning. One such system called ATLAS (Asteroid Terrestrial-impact Last Alert System) has been developed at the University of Hawaii with NASA support. However, an incoming meteoroid approaching from the direction of the Sun (as was the case for the Chelyabinsk impact) could not be seen by ATLAS or other similar surveys.

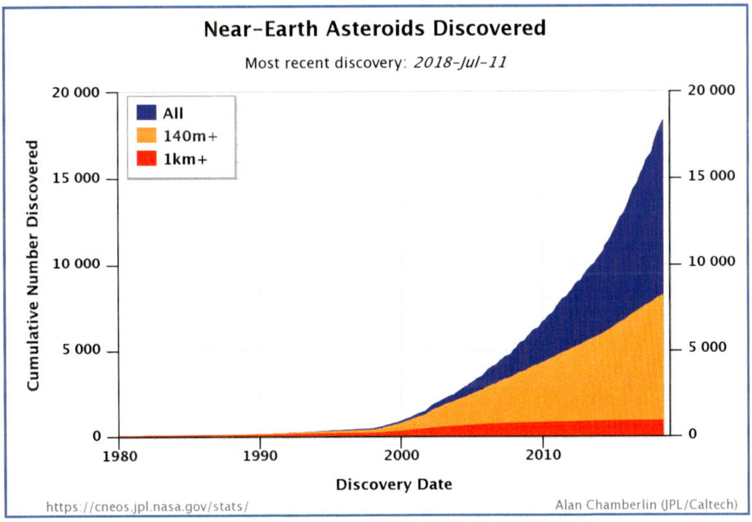

Fig. 2.3 Progress in the discovery of NEOs. The red curve shows the now essentially complete inventory of those larger than 1 km, the Spaceguard goal. From NASA JPL

The most important physical property of an NEA from the perspective of the impact hazard is its impact energy, which requires knowledge of its mass. Mass cannot be measured directly except in unusual cases where it is a binary object, so the next best thing is to estimate the diameter. However, diameters of NEAs are also not easily measured, generally requiring that brightness be measured simultaneously in both reflected sunlight and emitted thermal radiation (Chapman and Gaffey 1979; Morrison 1977). Otherwise, only a crude diameter can be estimated on the assumption that all asteroids have the same reflectivity or albedo. Unfortunately, in this case, nature does not cooperate. As first demonstrated by Chapman et al. (1975), the asteroid albedo distribution is bimodal, with the majority either moderately reflective (about 15%) or very dark (under 5%). We would like to determine the size distribution of NEAs, but for most, the only measured property is the brightness in reflected sunlight, therefore the brightness distribution must serve as a rather poor substitute for a size distribution. If these "sizes" are taken literally, we might overlook (for example) large, dark objects that are faint but still capable of inflicting serious damage if they hit.

An important addition to optical surveys is infrared thermal detection. Since asteroids are mostly dark (low albedo), they are brighter in absorbed and re-emitted thermal radiation than in reflected sunlight (Morrison 1977). In addition, the combination of infrared and optical measurements allows the albedo to be determined and yields a much-improved estimate of the size. A prototype orbital infrared survey has been provided by the NASA WISE (Wide Field Infrared Survey) satellite and its extension as NEOWISE, dedicated to infrared asteroid and comet observations (Mainzer et al. 2011; Stokes et al. 2017).

The size-frequency distribution of the NEAs is an important input to estimates of the risk, which is the product of the size-dependent hazard and the frequency of occurrence. Such curves also allow us to correct for incompleteness of the surveys as a function of brightness and to extrapolate the frequency of occurrence to smaller sizes. A series of papers interpreting the observed size-frequency distribution have been published by Alan Harris (most recently Harris and D'Abramo 2015). One important result from the analyses of the ever-growing observational statistics has been a reduction in the earlier estimates of the risk. As an example, the Tunguska impact of 1908, with estimated energy of 10–20 Mt, now has an estimated frequency of recurrence of millennia, rather than centuries (Morrison 2018).

An essential part of the NEA surveys is accurate calculation of orbits, from which future close approaches to Earth can be predicted. An orbit can be calculated from a series of observed positions extending over several weeks. Initial orbit determinations are made at the Minor Planet Center (MPC) of the Smithsonian Astrophysical Observatory, which is associated with the International Astronomical Union and supported by NASA grants. The MPC posts results immediately, so that other observers (including many amateur astronomers) can follow up on new discoveries and extend the timeline of observations.[1]

[1] https://www.minorplanetcenter.net/iau/MPCORB.html

For each known NEA, the orbit is refined and future positions are calculated at two institutions, which use slightly different software to provide redundancy. One is the NASA Center for Near-Earth Object Studies (CNEOS) at JPL,[2] which posts daily updated orbital elements (ephemeris) and tables of future close approaches, along with other useful information for all known NEAs. The second is the ESA Near-Earth Objects Dynamics Site (NEODyS-2) at the University of Pisa, which posts similar information.[3] These are quite sophisticated calculations, which account for orbit perturbations from gravitational interactions with other Solar System bodies. In addition to the telescopic observations, they utilize any available radar data, which complement the telescopic positions by adding high-precision measurements of distance (range) and velocity (Bowell and Muinonen 1994; Chesley and Spahr 2004).

Radar is a powerful tool for studying NEAs, but it is limited to asteroids that are close to Earth. In addition to improved orbital information, radar can measure the size and rotation of its targets, and if the data have sufficient precision, the shape as well. However, radar is not useful as a survey technique, since it requires a powerful, highly directional transmitter, which can be used only if the position of the target is already known from optical observations (Ostro and Giorgini 2004).

Years of optical observations, and if possible radar data as well, are used to determine an asteroid orbit well enough to make predictions about possible impacts decades in advance. When orbits are less well determined, there are often preliminary indications that an NEA will approach the Earth, but with substantial uncertainties. The JPL and Pisa websites list such future close approaches. In all cases to date, as further observations have improved the orbital precision, the associated impact probability has gone to zero. One example still under study is asteroid Bennu, which was discovered in 1999 and periodically comes very close to Earth. Even with nearly two decades of data, there remains a slight chance that Bennu might impact Earth between 2175 and 2199, with a currently calculated probability of 1 in 2700. Bennu is relatively large (diameter approximately 500 m), and it is the target of the NASA OSIRIS-Rex mission to rendezvous in 2018 and collect surface samples for return to Earth.

2.4 Mitigation: Defense Approaches

If and when an asteroid is determined to be on an impact trajectory, decisions must be made on how to respond (Ahrens and Harris 1992). This is (or should be) an international problem, especially at the time when the location of the predicted impact is not known. In general, there are a range of possible defense responses:

1. Take the hit for small objects, or those likely to hit in an uninhabited area

[2] https://cneos.jpl.nasa.gov

[3] https://newton.dm.unipi.it/neodys/

2. Evacuate the target areas and harden critical infrastructure, using the model of evacuations of cities in the path of major hurricanes
3. Deflect the object by changing its orbit so that it will miss the Earth, or at least miss inhabited areas
4. Disrupt or destroy the object, an option that may be the only course available if the warning time is too short for deflection

The historic Tunguska and Chelyabinsk impacts in Russia illustrate the first option, to take the hit. Neither meteoroid was seen coming, but similar events might be predicted in the future. If we knew today of a pending impact by the 1908 Tunguska object (diameter roughly 70 m) into a nearly uninhabited part of Siberia, we would not need to take any action other than to deploy scientific instruments to record the event. Similarly, the smaller 2013 Chelyabinsk (20 m) impact would not require action unless it was headed for a heavily populated area. Even though it hit within 30 km of a city of a million, there were no deaths. Shelter in place, rather than rushing to windows to see what was happening, would have eliminated even the most minor injuries (Artemieva and Shuvalov 2016; Borovicka et al. 2013; Brown et al. 2013; Kring and Boslough 2014; Morrison 2018; Popova et al. 2013).

Evacuation is the most likely response for larger impacts, especially if they are predicted to take place in populated areas. A warning of a few days or weeks would be sufficient to evacuate populations, but not to mitigate most infrastructure losses. The key requirement is an accurate position for the hit. In one simulation exercise, the general target was known years in advance to be North India and Bangladesh, one of the most densely populated parts of the world. In the final month before the hit, the target was localized to an area with a diameter of about 100 km centered approximately on Dacca. Those carrying out the simulation assumed that with this much warning, the evacuation of even such a large city as Dacca would be possible, although the societal disruption and economic loses would be huge.

Tsunamis generated by impacts in deep water have often been suggested as potentially very dangerous, because the energy of the impact would be propagated for much greater distance in water than the blast wave in air would be. This danger is exacerbated by the concentration of populations in coastal regions. While this is still a concern, recent studies (summarized by Morrison and Venkatapathy 2017) suggest that impact tsunamis dissipate their energy much more quickly than seismic tsunamis, so severe damage would be limited to impacts near the shore.

Before considering the deflection or disruption options, we need to understand the likely pattern of events, which have been simulated in several exercises. When a possible impact is first identified, we will not know the target location, but we will know that the hit will be along a *risk corridor*, a locus no more than a few hundred km wide that wraps part way around the planet. Very likely, several countries will lie within the risk corridor, while many others will be excluded. Hence, the apparent risk is not distributed globally. If deflection is attempted but not fully successful, the target may be moved along the risk corridor from one nation to another, an obvious source for concern or even conflict. It is quite possible that such uncertainty will persist for years, during which there may be competing proposals for how to deal with the hazard. Thus, what would initially be considered an international problem needing collective response will pose unique issues for different nations, depending on their locations.

The most straightforward way to deflect a potential impactor is to change its orbital period, and to do so as far in advance as possible. If the period is changed by as little as a few minutes, the shift in position can be great enough to ensure a miss. The period could be either increased or decreased, depending on whether the force is applied along or against the direction of motion. This impulse would be accomplished by a kinetic impact, striking the asteroid by as heavy a spacecraft as possible. The challenge is to engineer a controlled impact at high velocity between two orbiting objects in deep space. If a single impact would not change the orbit sufficiently, multiple kinetic impactors could be launched.

The alternative approach would be to detonate a nuclear explosive near the surface of the leading or trailing side, using either the vaporization of the surface layer or the formation of a crater to eject material and provide a reaction force against the target. If the nuclear explosive is carried by a fast flyby spacecraft, the orbit would need to be very well determined and the timing of the explosion accurate to a fraction of a second. Since more energy can be imparted by a nuclear explosive than a kinetic impactor, this option might be required to deflect large asteroids.

The technology for deflection by either kinetic impact or nuclear explosive seems straightforward, but neither technology has been tested. A NASA mission called DART (Double Asteroid Redirection Test) is being planned to demonstrate kinetic impact by hitting the 150-m moon of the asteroid Didymos and measuring the change in its orbital period. An alternative ESA/NASA approach is the AIDA (Asteroid Impact and Deflection Assessment) mission, in which the DART impact would be monitored by a second instrumented spacecraft orbiting Didymos. A less expensive variant would use a cubesat rather than a large spacecraft to monitor the event.

At present, international law restricting the use of nuclear explosives in space appears to foreclose the opportunity to test the nuclear technology. While the law could be changed in the case of a real planetary threat, it might be difficult to do so for a test.

The final possibility, to disrupt or destroy the incoming asteroid, is an option of last resort, if the asteroid is too large or the advance notice too short to mount a deflection effort. The most effective approach is probably a direct hit with a large nuclear explosive. This is not an option that would be exercised during the final plunge of the asteroid toward Earth, as is often depicted in Hollywood versions of planetary defense. Rather, the disruption must take place far enough from the Earth that the asteroid fragments (and any radioactive waste) are dispersed, so that that few if any would actually hit our planet.

A related issue called the *deflection dilemma* should be noted (Harris et al. 1994; Sagan and Ostro 1994). New technology that can change the orbits of NEAs could, like any other weapon, be misused. An initial concern was that if we had the capability to deflect an asteroid away from the Earth, we could locate an innocent asteroid passing close to Earth and divert it into our planet, perhaps aimed at an unfriendly nation. More recently, it has become clear from the details of asteroid orbits that this is effectively impossible. However, there are more general concerns about developing and deploying any other defense system, especially if based on nuclear explosives. If we really did have an asteroid headed for the Earth, different nations might have very different opinions about where it should be aimed in a deflection maneuver. These issues are worthy of consideration by those who are advocating for nuclear planetary defense.

2.5 Conclusions

The impact of celestial bodies with Earth, at one time dismissed as an outlandish idea, has become broadly accepted. Indeed, the term "asteroid impact" has entered our vocabulary as a metaphor for any catastrophic and unexpected event, from a stock market crash to a surprise election result. The widely-reported 1995 impact of comet Shoemaker-Levy 9 with Jupiter, and the 2013 atmospheric explosion of a previously unknown asteroid over Chelyabinsk, Russia, provided proof that such events were possible. Two popular 1998 Hollywood films, *Deep Impact* and *Armageddon*, also exposed the public to dramatic fictionalized depictions of such events, and each concluded with an imaginative last-minute nuclear deflection to save the Earth.

Planetary Defense is a global problem, since all parts of the planet are at risk. It is therefore appropriate that solutions be embraced on an international level. Although relatively few nations have the technology to contribute directly, broad international support for planetary defense should be sought. However, the situation is likely to change when a real impact is predicted. It will become apparent that only a few parts of the world, along the track called the impact corridor, are at risk, and as the knowledge of the asteroid's orbit improves, the danger will focus on an area as small as one or a few nations. In these circumstances, the international consensus may break down.

International concern over the asteroid threat led the United Nations General Assembly in 2013 to create an International Asteroid Warning Network (IAWN) to coordinate response to the NEA threat.[4] The tasks of IAWN include a clearing house for receipt and processing of NEA observations, coordinating observing campaigns and developing a database of potential impact consequences. IAWN includes members from Europe, Asia, South and North America. The UN has also established a parallel Space Mission Planning Advisory Group (SMPAG). These UN initiatives are one example of a global awareness of the impact hazard, which, like global warming, represents a challenge to all nations.

In 2016, the United States Government established a Planetary Defense Coordination Office within NASA, directed by Lindley Johnson as Planetary

[4] http://www.unoosa.org/oosa/en/ourwork/topics/neos/iawn.html

Defense Coordination Officer.[5] The office is responsible for tracking and characterizing potentially hazardous NEAs, issuing warnings if required and coordinating US funding for surveys and associated planetary defense activities. NASA has increased its planetary defense annual budget by more than one order of magnitude since the start of the Spaceguard Survey.

Current asteroid surveys are expected to include substantial new capabilities early in the 2020s. The Large Synoptic Survey Telescope (LSST)[6] under construction in northern Chile has in part been justified as an asteroid search engine, repeatedly covering the entire visible sky and generating 20 TB of raw data per night. Meanwhile, NASA is supporting technology development for a dedicated infrared space observatory to carry out surveys detecting NEAs from their thermal emission.[7]

Based on current knowledge of the NEA properties and numbers, we can suggest some general conclusions about future planetary defense requirements.

A large part of the impact risk as defined 20 years ago has been effectively eliminated. When the Spaceguard Survey began in 1998, the great majority of NEAs large enough to risk a global catastrophe were unknown. Now, we know that there are no hazardous NEAs large enough (D > 5 km) on orbits that could pose any risk of a mass extinction. Of the NEAs that risk global effects and might be considered a threat to civilization, more than 90% have been found, and none of these known objects are at risk of hitting the Earth within the foreseeable future (several centuries).

The residual impact risk can be estimated as follows. There are approximately equal hazards from (1) the few remaining undiscovered NEAs with D > 1 km, (2) the much larger undiscovered population of sub-km NEAs peaking at around 100–300-m diameter, which are capable of doing substantial local or regional damage, and (3) the comets, which have received relatively little attention. Quantitatively, the anticipated average annual casualties from these three sources are no more than a few dozen (although clearly this metric takes no account of the possible effects of a very rare, catastrophic hit on a populated area or key infrastructure). It is also worth noting that based on brightness alone, we do not have an accurate measure of the mass or energy of a small NEA, and there may be some dark but relatively massive objects hiding in our surveys.

The most likely impacts are, of course, from the small end of the asteroid size distribution. The Tunguska impact, with energy of 10–20 Mt, provides an example. Such impacts are expected for the whole Earth once in several millennia. If we confine our concerns to the populated parts of the planet (excluding oceans basins, the Antarctic, and areas of low habitation in much of the Sahara, inland Australia, and Siberia), the frequency is about once in 10,000 years. Smaller impacts are much more common but are less likely to cause any great harm.

---

[5] https://www.nasa.gov/planetarydefense/overview

[6] https://www.lsst.org/about

[7] https://neocam.ipac.caltech.edu

Several chapters in this book discuss the role of nuclear explosives in planetary defense. Nuclear approaches are usually proposed in the context of protection against large impacts, taking advantage of their high energy-to-mass ratio. Future progress in NEA surveys will likely reduce the anticipated requirement to deflect or disperse large asteroids, unless of course advanced searches yield a predicted impact in the near future.

If current progress in our surveys continues, there will no longer be a requirement to plan for defense against an asteroid as large as 1 km. Intermediate sizes (a few hundred meter diameter) will continue to be found by existing surveys. We may wish to focus on strategies that can find smaller impactors close to the Earth, within a few weeks or days of impact. This would provide enough time to alert scientists and potential tourists who want to see and study the event, and to evacuate any at-risk populations. It would be gratifying if a risk that has been seen as a threat to the future of civilization should gradually shift toward an emphasis on tourism and entertainment.

References

Ahrens, T. J., & Harris, A. W. (1992). Deflection and fragmentation of near-Earth asteroids. *Nature, 360*(6403), 429–433. doi:https://doi.org/10.1038/360429a0.

Alenia Spazio SpA. (1995). *Beginning the Spaceguard Survey*. Vulcano, Italy. http://spaceguard.rm.iasf.cnr.it/SGF/Vulcano/booklet.ps.

Alvarez, L. W., Alvarez, W., Asaro, F., & Michel, H. V. (1980). Extraterrestrial cause for the cretaceous-tertiary extinction. *Science, 208*(4448), 1095–1108. doi:https://doi.org/10.1126/science.208.4448.1095.

Artemieva, N. A., & Shuvalov, V. V. (2016). From Tunguska to Chelyabinsk via Jupiter. *Annual Review of Earth and Planetary Sciences, 44*(1), 37–56. doi:https://doi.org/10.1146/annurev-earth-060115-012218.

Atkinson, H., Tickell, C., & Williams, D. (2000). *Report of the Task Force on Potentially Hazardous Near Earth Objects*. London.

Belton, M. J. S., Morgan, T. H., Samarasinha, N., & Yeomans, D. K. (Eds.). (2004). *Mitigation of Hazardous Comets and Asteroids*. Cambridge University Press.

Borovicka, J., Spurny, P., Brown, P. G., Wiegert, P., Kalenda, P., Clark, D., & Shrbeny, L. (2013). The trajectory, structure and origin of the Chelyabinsk asteroidal impactor. *Nature*, 1–13. doi:https://doi.org/10.1038/nature12671.

Boslough, M. B., & Crawford, D. A. (1997). Shoemaker-Levy 9 and plume forming collisions on Earth. In *Near-Earth Objects, Annals of the New York Academy of Sciences, Volume 822* (pp. 236–282).

Bowell, E., & Muinonen, K. (1994). Earth-Crossing Asteroids and Comets: Groundbased Search Strategies. In T. Gehlers (Ed.), *Hazards Due to Comets and Asteroids* (pp. 149–198). The University of Arizona Press.

Brown, P. G., Assink, J. D., Astiz, L., Blaauw, R., Boslough, M. B., Borovička, J., et al. (2013). A 500-kiloton airburst over Chelyabinsk and an enhanced hazard from small impactors. *Nature, 503*(7475), 238–241. doi:https://doi.org/10.1038/nature12741.

Canavan, G. H., Solem, J. C., & Rather, J. D. G. (1994). Near-Earth object interception workshop. In T. Gehrels, M. S. Matthews, & A. M. Schumann (Eds.), *Hazards due to Comets and Asteroids* (pp. 93–124). University of Arizona Press.

Chapman, C. R., & Gaffey, M. J. (1979). Reflectance spectra for 277 asteroids. *Asteroids*, 655–687.

Chapman, C. R., & Morrison, D. (1994). Impacts on the Earth by asteroids and comets: assessing the hazard. *Nature, 367*(6458), 33–40. doi:https://doi.org/10.1038/367033a0.
Chapman, C. R., Morrison, D., & Zellner, B. (1975). Surface properties of asteroids: A synthesis of polarimetry, radiometry, and spectrophotometry. *Icarus, 25*(1), 104–130.
Chesley, S. R., & Spahr, T. B. (2004). Earth Impactors : Orbital Characteristics and Warning Times. In M. Belton, T. Morgan, N. Samarsinha, & D. K. Yeomans (Eds.), *Mitigation of Hazardous Comets and Asteroids* (pp. 22–37). Cambridge, UK: University Press Cambridge.
Chyba, C. F., Thomas, P. J., & Zahnle, K. J. (1993). The 1908 Tunguska explosion: atmospheric disruption of a stony asteroid. *Nature, 361*(6407), 40–44. doi:https://doi.org/10.1038/361040a0.
Gehrels, T., Matthews, M. S., & Schumann, A. M. (Eds.). (1994). *Hazards Due to Comets and Asteroids*. University of Arizona Press.
Glasstone, S., & Dolan, P. J. (1977). *The effects of nuclear weapons*. Department of Defense, Washington, DC (USA); Department of Energy, Washington, DC (USA).
Harris, A. W., Canavan, G., Sagan, C., & Ostro, S. (1994). The Deflection Dilemma: Use Vs. Misuse of Technologies for Avoiding Interplanetary Collision Hazards. In T. Gehrels, M. S. Matthews, & A. M. Schumann (Eds.), *Hazards due to Comets and Asteroids* (pp. 1145–1156).
Harris, A. W., & D'Abramo, G. (2015). The population of near-Earth asteroids. *Icarus, 257*(May), 302–312. doi:https://doi.org/10.1016/j.icarus.2015.05.004.
Hills, J. G., & Goda, M. P. (1993). The fragmentation of small asteroids in the atmosphere. *The Astronomical Journal, 105*, 1114–1144.
Kring, D. A., & Boslough, M. (2014). Chelyabinsk: Portrait of an asteroid airburst. *Physics Today, 67*(9).
Mainzer, A., Bauer, J., Grav, T., Masiero, J., Cutri, R. M., Dailey, J., et al. (2011). Preliminary Results from Neowise: an Enhancement to the Wide-field Infrared Survey Explorer for Solar System Science. *The Astrophysical Journal, 731*(1), 53. doi:https://doi.org/10.1088/0004-637X/731/1/53.
Mathias, D. L., Wheeler, L. F., & Dotson, J. L. (2017). A Probabilistic Asteroid Impact Risk Model: Assessment of Sub-300 m Impacts. *Icarus, 289*, 106–119. doi:https://doi.org/10.1016/j.icarus.2017.02.009.
Morrison, D. (1977). Asteroid sizes and albedos. *Icarus, 31*(2), 185–220.
Morrison, D. (1992). *The Spaceguard Survey Report of the NASA International Near-Earth-Object Detection Workshop*. NASA.
Morrison, D. (2018). Tunguska Workshop: Applying New Tools to Understand the 1908 Asteroid Impact. NASA Technical Memorandum.
Morrison, D., Chapman, C. R., & Slovic, P. (1994). The Impact Hazard. In T. Gehrels (Ed.), *Hazards Due to Comets and Asteroids* (pp. 149–198). University of Arizona Press.
Morrison, D., Chapman, C. R., Steele, R. D., & Binzel, R. P. (2004). Impacts and the Public: Communicating the Nature of the Impact Hazard. In M. J. S. Belton, T. H. Morgan, N. H. Samarasinha, & D. K. Yeomans (Eds.), *Mitigation of Hazardous Comets and Asteroids*. Cambridge University Press.
Morrison, D., & Venkatapathy, E. (2017). *Asteroid Generated Tsunami: Summary of NASA/NOAA Workshop*. https://ntrs.nasa.gov/archive/nasa/casi.ntrs.nasa.gov/20170005214.pdf.
Nuckolls, J. (1995). *Proceedings of the Planetary Defense Workshop, Lawrence Livermore National Laboratory, California, May 22-26, 1995*. Lawrence Livermore National Lab., CA (US).
Oplk, E. J. (1958). On the catastrophic effect of collisions with celestial bodies. *Irish Astronomical Journal, 5*, 34.
Ostro, S. J., & Giorgini, J. D. (2004). The role of radar in predicting and preventing asteroid and comet collisions with Earth. *Mitigation of Hazardous Comets and Asteroids, 38*.
Popova, O. P., Jenniskens, P., Emel'yanenko, V., Kartashova, A., Biryukov, E., Khaibrakhmanov, S., et al. (2013). Chelyabinsk Airburst, Damage Assessment, Meteorite Recovery, and Characterization. *Science, 342*(6162), 1069–1073. doi:https://doi.org/10.1126/science.1242642.
Powell, J. L. (1998). *Night Comes to the Cretaceous: Dinosaur Extinction and the Transformation of Modern Geology*. W.H. Freeman. https://books.google.fr/books?id=2RptQgAACAAJ.

Remo, J. L. (1997). Near Earth Objects, the United Nations International Conference. *Annals of the New York Academy of Sciences, 822.*

Rumpf, C. M., Lewis, H. G., & Atkinson, P. M. (2017). Asteroid Impact Effects And Their Immediate Hazards For Human Populations. *Geophysical Research Letters, 44*(March), 3433–3440. doi:https://doi.org/10.1002/2017GL073191.

Sagan, C., & Ostro, S. J. (1994). Dangers of asteroid deflection. *Nature, 368*(6471), 501.

Sekanina, Z. (1998). Evidence for asteroidal origin of the Tunguska object. *Planetary and space science, 46*(2–3), 191–204.

Shoemaker, E. M. (1983). Asteroid and comet bombardment of the Earth. *Annual Review of Earth and Planetary Sciences, 11*(1), 461–494.

Simonenko, V. (1995). *Space Protection of the Earth: Concepts and Approaches to the Development. Institute of Technical Physics, Snezhinsk, Chelyabinsk Region, Russia.*

Stokes, G., Barbee, B. W., Bottke, W. F., Yeomans, D. K., Buie, M. W., Chodas, P. W., et al. (2017). *Report of the Near-Earth Object Science Definition Team: Update to Determine the Feasibility of Enhancing the Search and Characterization of NEOs. NASA, Science Mission Directorate, Planetary Science Division.*

Stokes, G., Yeomans, D., Bottke, W. F., Jewitt, D., Chesley, S. R., Kelso, T., et al. (2003). *Study to determine the feasibility of extending the search for near-Earth objects to smaller limiting diameters. Report of the Near-Earth Object Science Definition Team* (Vol. 32). Washington D.C.

Toon, O. B., Zahnle, K., Morrison, D., Turco, R. P., & Covey, C. (1997). Environmental perturbations caused by the impacts of asteroids and comets. *Reviews of Geophysics, 35*(1), 41–78. doi:https://doi.org/10.1029/96RG03038.

Wheeler, L. F., Register, P. J., & Mathias, D. L. (2017). A fragment-cloud model for asteroid breakup and atmospheric energy deposition. *Icarus, 295*, 149–169.

Yeomans, D. (2012). *Near-Earth Objects Finding Them Before They Find Us*. Princeton University Press.

Chapter 3
What Are NEOs and the Technical Means and Constraints of Solar System Mapping?

William Crowe

3.1 Introduction

The Solar System is filled with many physical bodies in orbit about the Sun. These include the planets, moons, dwarf planets and minor bodies. Minor bodies include what are more commonly known as asteroids and comets, as well as many Kuiper Belt Objects. There are almost a million minor bodies in the Solar System that are known today, with likely two orders of magnitude more that are yet to be discovered (Harris and D'Abramo 2015). A subset of the minor bodies, Near-Earth Objects (NEOs), is of interest because of their proximity to Earth and potential for impact. This chapter explores what NEOs are and how they are detected and characterized.

NEOs are all unique and have undergone a history that has caused them to change dramatically over their lifetimes (Binzel et al. 2015). They are still changing today, which is why non-threatening minor bodies can become threatening and also why it is difficult sometimes to map the properties between objects that have relocated from their original positions and those that have migrated to different parts of the Solar System (DeMeo and Carry 2014). NEOs have had an impact on features of the Earth, including being a source of water throughout the Earth's history (Morbidelli et al. 2000).

More NEOs are now being discovered than ever before, with multiple new NEOs identified every day. Almost all of the NEOs larger than 1 km across have been discovered (Harris and D'Abramo 2015), but there is still a ways to go to identify all of the NEOs that are recognized as being potentially hazardous to life on Earth. Lagging in discovery and orbit determination is the characterization of asteroids. While some efforts have accelerated identification of certain asteroid features by an

W. Crowe (✉)
School of Mechanical and Manufacturing Engineering, UNSW, Sydney, NSW, Australia
e-mail: w.crowe@unsw.edu.au

N. Schmidt (ed.), *Planetary Defense*, Space and Society,
https://doi.org/10.1007/978-3-030-01000-3_3

order of magnitude (Michel et al. 2015), having the rate of characterization increase will take significant cooperation, coordination and investment.

Characterization of asteroids is important to planetary defense, as the ability of some impact mitigation methods to deflect hazardous targets can change greatly between asteroids with different characteristics (Sugimoto et al. 2013). Characterization is also important when predicting if some NEOs will become hazardous over time due to perturbations. There are events that have been identified to change NEO properties over the course of several years, with enough chaos to make future predictions inaccurate (Farnocchia et al. 2013). This means that the process of identifying and then updating details of NEOs should be continued long into the future. Characterization can also assist predictions on the destructiveness of hazardous NEOs, which will be covered further in Chap. 4. This chapter provides an outline of characterization techniques, while specific programs will be described in Chap. 5. Different methods of deflecting NEOs have been outlined in Chap. 7.

This chapter will first provide a brief introduction to NEOs and interesting subsets of this population. It will then cover the different physical and orbital characteristics that NEOs hold, as well as the importance of these characteristics to predict impacts, and how they affect impact severity. These characteristics can change over time, and some guidance is provided as to why this happens. The techniques used to determine these characteristics are then outlined. The final section provides a brief overview of strategies that could be implemented to rapidly expand the current pool of known NEOs and the known characteristics of each.

3.2 Defining NEOs

A Near-Earth Object (NEO) refers to a minor body that orbits the Sun and has an orbital perihelion of <1.3 Astronomical Units (AU), where perihelion refers to the minimum distance that an orbit takes a body to the Sun, and an Astronomical Unit is the average distance from the Sun to the Earth. As of March 2017, there are more than 15,000 known NEOs, with more than 800 larger than a km in diameter and 7,500 larger than 140 m across (Chodas 2017).

NEOs encompass two types of minor bodies: asteroids and comets. The term NEO is often interchanged loosely with NEA (Near-Earth Asteroid), as the vast majority of NEOs are asteroids. The term NEC (Near-Earth Comet) can be used to describe comets that match NEO criteria. There is likely a spectrum that encompasses both asteroids and comets, with many minor bodies showing characteristics of the two (Binzel et al. 2015).

Asteroids are fragments of planets that were left over during the formation of the Solar System and do not have sufficient gravity to become spherical (those asteroids that do become spherical under the forces of gravity are known as dwarf planets). They largely reside in the asteroid belt, situated between the planets Mars and Jupiter, but there are other populations throughout the Solar System, including the Trojan asteroids co-orbiting with Jupiter, as well as NEAs which move through the rocky planets of the inner Solar System. There are numerous asteroids that display some of the same characteristics as comets and are likely extinct comets with orbits that have been greatly perturbed from their original shapes.

Comets are bodies that formed in or beyond the outer reaches of the Solar System. They have water ice as a major component of their mass. They are different from other minor bodies found in the outer Solar System in that they have highly eccentric orbits that take them past the orbits of the planets and release some of their volatiles in a display that is often visible with the naked eye on Earth. Some comets have periods of many thousands of years, but NECs are limited to a 200-year period.

The definition of a Potentially Hazardous Object (PHO) is a minor body that matches the orbital characteristics of a NEO, but additionally has a Minimum Orbital Intersection Distance (MOID) of <0.05 AU and a diameter roughly matching or exceeding 140 m across (often inferred when the NEO has an absolute magnitude, H, of 22 or less). The term PHO is often interchanged with Potentially Hazardous Asteroid (PHA), as there are vastly more asteroids than comets in this category. However, there is discussion over whether the definition should lower the diameter limit, because smaller objects can still inflict significant damage when they hit an inhabited area. Such concerns arose after the Chelyabinsk impact in 2013, where the parent impact was perhaps only 20-m across. Similarly, the Tunguska impact in 1908 was caused by an object of several tens of meters, and the Barringer crater in Arizona originated from an impact of about a 50-m iron asteroid.

There are a number of further categories used to define groups of NEOs, differentiated by orbital characteristics. The defining characteristics are the perihelion distance, q; aphelion distance, Q, which is the maximum distance that an orbit takes a body from the Sun; semi-major axis, a, which is the average distance that an orbit lies from the gravitational center; and orbital period, P, the amount of time that a body takes to complete a revolution about an orbit (Chodas 2017), with some of these illustrated in Fig. 3.1. The characteristics q, Q and a are generally given in AU. The four main NEO groups are the Atiras, which have an orbit entirely within Earth's; the Atens, where the semi-major axis is <1 AU and the aphelion passes Earth's orbit; the Apollos, where the semi-major axis is >1 AU and the perihelion is inside Earth's orbit; and the Amors, where the entire orbit lies outside of Earth's orbit. The number of known NEOs within each of these subgroups varies, and they all have the potential to be PHOs, with a display of the nature of these groups in Figs. 3.2 and 3.3.

It is possible for those small bodies outside of the range of NEO definition to become NEOs over time, due to the dynamic nature of the Solar System. This also applies to objects in some of the orbital groups—they will change to a different group over time.

This movement is due to several perturbations, including gravitational, Solar Radiation Pressure (SRP), the Yarkovsky effect (Vokrouhlický et al. 2015) and collisions with other minor bodies. Comets have a further perturbation due to outgassing. Gravitational perturbations occur when the NEO exchanges energy with a planet or another minor body, causing its orbit to change significantly. The SRP effect sees photons from the Sun hitting the NEO and exchanging momentum, lead-

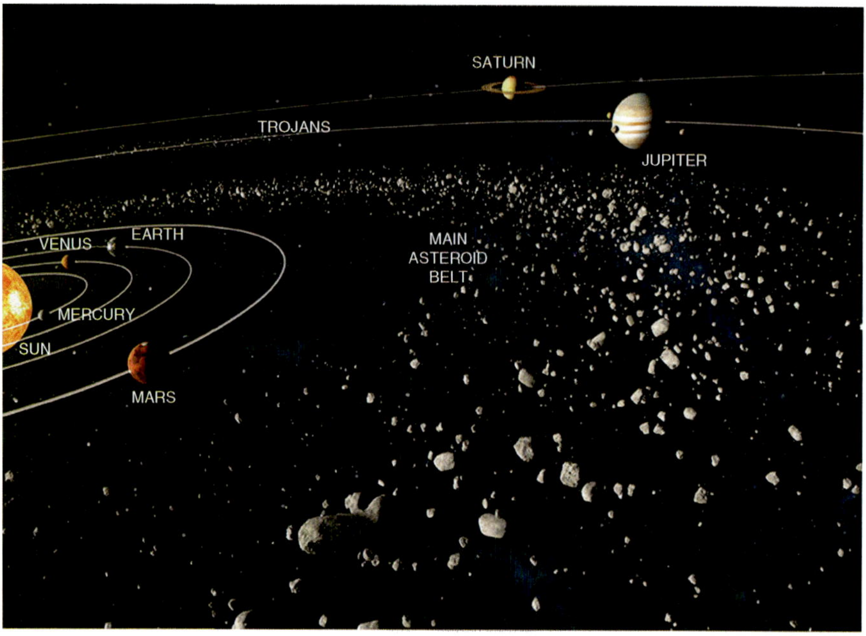

Fig. 3.1 An artist depiction of the Solar System with emphasis on the location of the main asteroid belt (including Trojans on Jupiter's Lagrange point). Not to scale. Visualization by AD13

Fig. 3.2 Diagram of an Apollo-class NEO orbit, where $a = \dfrac{q+Q}{2}$

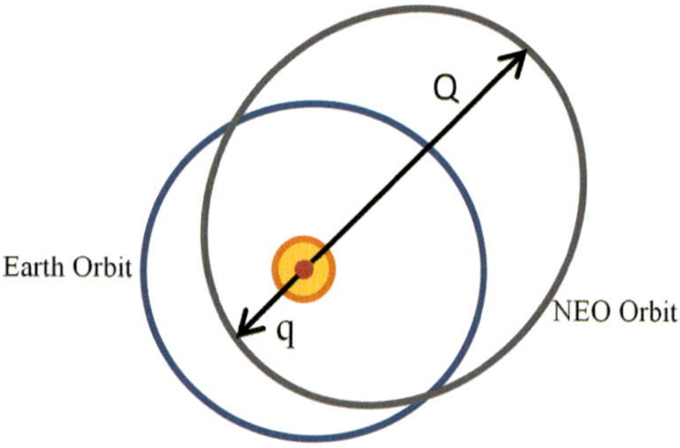

ing to a small force that can produce orbital change over time (this effect is most obvious on small asteroids, especially those <10 m across). The Yarkovsky effect is where the NEO heats up on its Sun-facing side and then re-emits that heat as photons as the NEO rotates, producing a change in momentum that can align with the NEO velocity, producing a significant force over time. Collisions see a rapid change in orbit when two bodies physically connect at speed and then change direction after exchanging kinetic energy.

These perturbation effects mean that observations of NEOs are simply a snapshot in time, with some particularly vulnerable NEOs having noticeable changes in

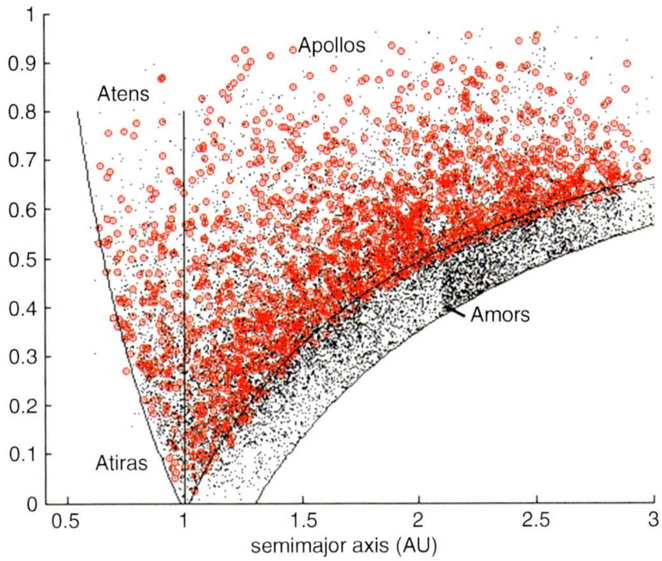

Fig. 3.3 Plot of known NEOs (black dots) and PHOs (circled in red) in various sub-groups, with their semimajor axis a against their eccentricity $\frac{Q}{a} - 1$

orbital parameters over the course of years. Some effects have left certain orbits with few asteroids, especially the Kirkwood Gaps in the asteroid belt, where orbital resonances with the planet Jupiter have left no asteroids with low eccentricity or inclination (Minton and Malhotra 2009). Many of the asteroids that originally resided there have migrated to the inner Solar System and have become NEOs.

3.3 Mapping NEOs

The characteristics of NEOs have been subdivided into two groups in this chapter—those properties of its orbit and those properties of its physical nature. Both have been seen as fundamental in this writing, as they are interrelated and have influence over our ability to prevent them from causing catastrophe on Earth.

3.3.1 NEO Physical Characteristics

There are numerous types of minor bodies that move about the Sun, with NEOs composed of most types. These range from carbonaceous asteroids, to stony and metallic asteroids, to cometary bodies. All of these types have fallen to Earth in the past, and most exist today in meteorite records (Borovička et al. 2015). The physical characteristics of a NEO have implications for the effectiveness of different deflection strategies and so are important to constrain as much as possible (Sugimoto et al. 2013).

Rendezvous missions have been sent to stony asteroids and comets to date, with one mission currently en-route to a carbonaceous asteroid (Tsuda et al. 2013) and a mission in early planning to a metallic asteroid at the time of writing. Past missions include the US Dawn mission, sent to explore the two most massive asteroids, (1) Ceres (now also a dwarf planet) and (2) Vesta, and remain in their orbits for several months each, providing mapping, electromagnetic data and gravimetry (Rayman et al. 2006; Russell et al. 2012). Another past mission was launched by Japan to the asteroid Itokawa, where the relationship between S-type asteroids and ordinary chondrites was identified (Abe et al. 2006; Nakamura et al. 2011), with a small sample sent back to Earth's surface. The Stardust mission returned a sample of cometary dust, (Brownlee 2003; Keller et al. 2006). The European Rosetta mission (Glassmeier et al. 2007) had a rendezvous with a comet and the Near-Shoemaker (Miller et al. 2002; Veverka et al. 2001) mission had a rendezvous with an asteroid. Both of these missions included flybys of other asteroids en-route to their respective targets.

NEOs come in a range of sizes, with the largest, 1036 Ganymed, about 32 km in diameter. The lower limit of NEO size is not defined, although objects smaller than 1 m are called meteoroids rather than asteroids. The largest PHO is about 6.4 km across and ranges down to about 140 m (some objects identified as PHOs could be smaller than this limit, as their diameters are not accurately inferred by their absolute magnitude). The size can be estimated by the brightness of an asteroid and using an assumed albedo (the proportion of the light of the Sun reflected back to observers). Generally a size range is supplied because the albedos are usually unknown. A more accurate size estimate can be made through the use of an infrared brightness, which correlates more closely with asteroid size than visible light brightness, allowing for accurate size estimates for thousands of NEOs and other minor bodies (Mainzer et al. 2015).

Many NEO subgroups have been defined by spectral class, where the spectral signature of a NEO is compared with that of a crushed meteorite specimen (Reddy et al. 2015). Due to the ablation of the outer parts of meteorites when they fall to Earth, as well as poorly defined weathering of the outside of NEOs in space, there is still uncertainty as to how accurate this technique is when telling us about what a NEO is composed of and how similar NEOs are that exist in the same group. This is further confused by the possibility that colliding bodies leave imprints on each other's surfaces, meaning that NEOs contain the material of every other object that has come into physical contact with them (Scott et al. 2015). The identification and spectral analysis of NEO 2008 TC3 before impact meant that the later discovery of its associated meteorites enabled researchers to map the F spectral class of minor bodies to the dark carbon-rich anomalous ureilite meteorites (Jenniskens et al. 2009). This matching was performed similarly for the S spectral class, with the return of the "ground truth" of Itokawa from the Hayabusa rendezvous mission indicating that this was synonymous with the ordinary chondrite meteorite class (Nakamura et al. 2011).

The composition of NEOs is undergoing much study, especially as it is anticipated that they will serve as mineral deposits for exploitation in the near future. It also has some bearing on the way it breaks up in the atmosphere, as some minerals have a higher ablation temperature than others. NEOs originated either from protoplanets that were forming near where the asteroid belt is today, or as small icy objects at the far reaches of the Solar System. The protoplanets were of a more homogenous and larger size than the majority seen today. These early asteroids are known as "parent" bodies, as their compositions are seen in smaller bodies that have since broken away from them. Some of the parent bodies were large enough that they underwent *differentiation*, meaning that in a superheated state, it was massive enough to have a gravity gradient that caused heavier elements to move towards the core of the body and lighter elements to remain at the surface. The Earth and all of the inner planets also underwent this process in the early Solar System.

The mass of a NEO is an important parameter for estimating the damage that it might cause if it were to hit Earth. The energy released by a NEO as it impacts the Earth is close to the kinetic energy contained within the body as it hits the atmosphere, meaning that it is approximately proportional to the mass and the square of the relative velocity.

The mass of NEOs can be estimated in a number of ways. The most accurate is to measure the acceleration caused to another mass due to gravity. This is seen naturally around the largest asteroids, where the effects of their gravity perturb the orbits of other orbiting bodies (Baer and Chesley 2008). Spacecraft that are sent to flyby or rendezvous with minor bodies can be tracked from Earth to accurately detect the change in velocity that occurs due to gravitational acceleration. This has been performed for several minor bodies, including 433 Eros (Yeomans 1999), 21 Lutetia (Patzold et al. 2011) and 25,143 Itokawa (Abe et al. 2006). In binary asteroid systems, observing the rotation of the bodies about one another allows a mass estimate (Margot 2002). The mass can also be estimated by observing the body's reaction to perturbations from the Yarkovsky Effect (Chesley et al. 2014), which requires knowledge of the asteroid size, shape and thermal characteristics. The mass of smaller NEOs can be estimated by applying a standard density to a known or estimated volume. As the density is variable and related to the spectral type, knowledge of the average density of the corresponding meteorite class can yield a more accurate estimate. There are now large sets of data that can be used to make such estimates (Carry 2012), drawn from masses that have been estimated from other techniques.

Density and density distribution within a NEO are important parameters for determining mass, as well as the internal structure and the way the NEO will react to forces that are applied to it. If the bulk (average) mass has been estimated, then the bulk density can be determined with use of an accurate shape model of the same NEO, or with a shape estimate derived from an estimated diameter (Carry 2012). The distribution of density for several asteroids has been estimated using the Yarkovsky-O'Keefe-Radzievskii-Paddack (YORP) effect, which causes an acceleration in rotation of a minor body. As an example, YORP has been observed on the asteroid 25,143 Itokawa to accelerate the rotation in such a way that a change in density is present, likely due to two different original bodies coming together to form one (Lowry et al. 2014). Other methods to estimate density distribution include taking accurate gravity field measurements about an asteroid and then attempting to replicate this distribution through a mathematical model, where hypotheses of the distribution are tested using least squares regression analysis (Scheeres et al. 2000).

Porosity is a measure of the voided space within minor bodies, providing insights into how the body is bound together and the history of collisions and other events. NEOs exhibit a large variation in porosity (Britt et al. 2002). Minor bodies with very high porosity are often termed *rubble piles*, as it is assumed that they are composed of fragments that are held together in a loose fashion. NEOs with low porosity are called boulders, where the entire body originated from the same parent. In the past, it was assumed that NEOs with fast rotation periods must be boulders, as gravity and friction alone is not enough to keep the NEO intact, but this has recently been disproved, with a likelihood that cohesive forces also play a role (Rozitis et al. 2014; Scheeres et al. 2010). There have been observed cases where minor bodies appear to be breaking up due to fast rotation, often spinning up due to YORP (Jewitt et al. 2014), sometimes creating binary minor bodies (Pravec et al. 2010). In all of these cases, porosity can be used to understand how forces combine in a microgravity environment to affect NEO dynamics.

Every NEO has a unique shape, becoming all the more strange due to the lack of gravity and slow process of weathering. This can be due to several different processes, including accretion of two or more bodies (resulting in a *contact binary*) as well as a fission, creating a binary system where a smaller secondary body orbits a larger primary one. The surface can also experience landslides, and regolith lofting changes the appearance (Hartzell and Scheeres 2011; Miyamoto et al. 2007). Impacts from other minor bodies can cause rapid change on a NEO (Richardsonjr et al. 2005), where a large enough impact can cause the entire body to tear apart before re-accreting at a later time. Remotely, NEOs can have their shape explored via detection of light curves, where an ellipsoid bound can be found. This strategy also allows for detection of a secondary body if there is one. The shape can be probed by radar if the NEO flies close enough to Earth and the body is large enough. This provides a precise shape of the parts of the NEO that display during the pass (Ostro et al. 2002). Direct photography is also a possibility, but rarely is an accurate shape obtained except if a telescope is mounted on a spacecraft that makes a close approach. Some of the best shape models have come from rendezvous and flyby missions.

The morphology of a minor body can help describe their porosities and shapes. Regolith is a loosely bound, granular layer, widely believed to be present on most asteroids due to hits from other asteroids and micrometeoroids (Reddy et al. 2015). It has been discovered that regolith is unevenly distributed about a body, as pictured by rendezvous missions. It has even been found to move about the surface at times, especially about more active bodies such as comets. Faults have been discovered on many asteroids and comets that have been visited by spacecraft. They are revealed when a spacecraft maps the surface of an asteroid using photographs. They can often be identified as being caused by a particular impact, as the fault lines emanate from a specific crater (Buczkowski et al. 2012). Craters are found on most asteroids, although the possible sizes are highly varied. They range from giant craters hundreds of km across, as found on the asteroid Vesta, to tiny micrometeoroid craters, as found in the dust brought back from the asteroid Itokawa by the Hayabusa mission (Nakamura et al. 2011).

Rotation is an important secondary characteristic to note, as it allows for inference of the internal structure. The effectiveness of laser ablation methods used to move an incoming asteroid would be impacted by rotation, as it prevents concentration on one part of the surface and dissipates the energy to other parts of space. Most asteroids will eventually rotate about their smallest axis of inertia, although many are known to tumble, where there is motion about two or three axes. Rotation can be found from light curves of asteroids.

The way that an asteroid impacts Earth depends on all of the parameters that help describe an asteroid. Often assumed size, along with knowledge of the asteroid orbit, are used to provide an estimate of the impact energy of an asteroid. The definition of a PHO, with an assumed size bound, limits the NEO population to bodies deemed to have potential to inflict significant damage to the Earth if it were to hit. Smaller bodies have hit in the timeframe of written history, with events such as the Tunguska Explosion (Chyba et al. 1993) and Chelyabinsk Airburst (Popova et al. 2013) causing much damage to the local environment and infrastructure, if not deaths to humans.

3.3.2 NEO Orbital Characteristics

A NEO is also defined by its orbital characteristics, which must be determined with some accuracy before a NEO has officially been discovered. Whether a NEO and the Earth are on a collision trajectory depends on the orbit orientation. When they cross paths, the velocity of a NEO relative to the Earth depends on how closely their orbital shapes match. A simple overview of basic orbital parameters is outlined here, as well as some words on how these can change over time.

Orbits are very complex when considered over a long time, although they are generally simplified when considering the major influences at play. The traditional simplification assumes just two bodies, the Sun and the NEO in this case, and defines an orbit size and shape, orientation of the plane of orbit, as well as the orientation of the orbit and body, using six orbital elements. The six elements can entirely describe an orbit for all time if we assume no perturbations.

The semi-major axis, a, is the average distance of a NEO from the Sun. Together with the fixed gravitational parameter, μ, it defines the period of the orbit. The eccentricity parameter, e, is the second parameter, which is a measure of the ellipticity of an orbit (as previously shown in Figs. 3.2 and 3.3).

Inclination, i, is the angle at which an asteroid orbital plane sits relative to the ecliptic plane. Often, these orbits have better stability characteristics, as they lie further from planets (which largely sit on the ecliptic). Bodies that travel in a retrograde motion, which is opposite to the direction of the majority of orbits, have an inclination >90°. The longitude of the ascending node, Ω, defines the angle at which the NEO passes "upwards" through the ecliptic plane of the reference frame.

The argument of periapsis, ω, is the angle (measured from Ω) at which the periapsis lies on the orbit. The true anomaly at epoch, θ, defines an angle from the argument of periapsis to the point at which the NEO sits on the orbit when given a specific date and time. The epoch can be updated for convenience of the user.

It is generally understood that the majority of NEOs originated in the asteroid belt, where it is likely that there existed about an Earth-mass of material, but where now there exists <1% of this (Bottke et al. 2005). There are three main classes of perturbation causing the orbits to evolve, which include gravitational (Morbidelli et al. 2010), collisional (Bottke et al. 2005) and radiative forces (largely the Yarkovsky Effect) (Bottke et al. 2006). The collisional forces are rare, while the radiative forces are slow-acting and dependent on size. Both of these can lead to a slow change of orbit in the asteroid belt until the minor planet reaches a depletion zone and gravitational perturbations send it to become a NEO or collide with the Sun. The gravitational perturbations of the planets, along with existing data of known NEOs, have allowed statistical models of the orbit locations of populations to be produced (Greenstreet et al. 2012).

Gravitational encounters when a NEO orbit has evolved to the inner Solar System cause its orbit to become chaotic, and it will change shape and orientation over the following several million years. The population of NEOs is in a steady state, with those that collide or escape replaced by new objects supplied from the asteroid belt or outer Solar System. The majority of NEOs are destroyed by colliding with the Sun, with a minority colliding with a planet or leaving the inner Solar System. The orbits of PHOs are more chaotic than those of other NEOs, as the perturbation due to Earth's gravity causes multiple changes over their lifetimes.

Some PHOs have been characterized well enough that they have orbits propagated over a long timeframe. These include (29075) 1950DA, which has had a long observation arc and was imaged with radar, predicted to have a potential Earth-impact in the year 2880 (Giorgini 2002). Uncertainty remains, with some characteristics such as mass still poorly constrained. The change due to perturbation can be projected into the future if the asteroid is well characterized in terms of both physical and orbital features. There are several cases where a large future change in the orbit of the asteroid, for example with a close flyby past Earth, can lead to some uncertainty about if and when an asteroid will collide with Earth (Sokolov et al. 2012).

The uncertainty that grows over time means that there should be regular observations of asteroids to update their orbital characteristics. This is even true of those asteroids with well-known physical characteristics, as collisions may disrupt high fidelity orbital predictions.

3.4 Discovery and Characterization Techniques

Minor bodies are by definition discovered when they have been allocated either a minor planet designation or new comet designation. This is only achieved after there has been an arc of observations long enough from which to make estimates of that body's orbital characteristics. As a minimum, the object must be tracked over a 24-h period before it is assigned a provisional designation. The provisional designation is maintained until the object has been tracked on four or more oppositions, at which point it may be assigned a final number (an opposition is where the object passes on the opposite side of the Earth to the Sun, where the best optical viewing angle exists) (Chodas 2017).

NEOs are primarily discovered and tracked through the use of Earth-based optical telescopes. This means that the discovery of bodies is limited by the upper magnitude of observation telescopes, as well as the rate of repeat coverage over the entire sky. As the apparent magnitude of an object depends on factors such as size, albedo, distance from the observer and percentage illuminated, more regular coverage can allow for the observation of fainter objects that pass close to the Earth for a limited time. Observation can be obscured by bright light sources, especially the Sun and Moon, but also stars and galaxies. Specialized minor body surveys will image the same portion of sky several minutes apart and then look for moving objects against the background stars. If these moving objects do not match previously known minor bodies, details are sent to allow other observatories to make follow-up observations. It is also possible for non-specialized all-sky surveys to make note of objects that do not match known databases. The nature of these surveys makes it difficult for follow-up observations to confirm potential discoveries, however.

Optical telescopes require light to reflect off the object being observed and then become captured by the telescope. As the object observed is at an increasing distance from the Sun and/or the telescope, the ability to pick it up is diminished. Also, interference from other light sources, which include the Moon, Sun and any nearby inhabited areas, can reduce the technique efficacy. Interference from air moisture, dust and other particles can also affect the ability of observatories to make observations. One part of the solution is to have geographical redundancy, with multiple observation sites within the same country or across multiple countries. Geographical redundancy is best served with sites spread across the northern and southern hemispheres, as well as across multiple meridian lines. The southern hemisphere is especially lacking in well-equipped observatories, as are large parts of the northern hemisphere. Another point that could be addressed is long-term observations of known NEOs for characterization of shape, light curves and spectral nature, as well as re-characterization of orbits to check for gravitational perturbations from other NEOs.

Infrared Telescopes require electromagnetic waves from the infrared spectrum to emit from asteroids or other astronomical bodies. It is more difficult to implement, as bands in this part of the electromagnetic spectrum are absorbed through the atmosphere. To account for this, there are infrared telescopes mounted on spacecraft in orbit. The American NEOWISE telescope (Mainzer et al. 2011) is one example in orbit today, although others have been suggested. The Planetary Resources Ceres fleet of spacecraft is another such application of the technology.

Radar requires illumination of an asteroid with radio signals, which then bounce back with distributed strength that reveals aspects of the minor body (Shepard et al. 2017). This strategy often requires two radio telescopes—one for illumination and the other to receive the data. Such telescope pairs exist in several countries, including the US and Australia, although there is good potential to upgrade underutilized telescopes to accommodate these measurements in further locations.

Computation is currently performed well by computers maintained by the MPC and JPL, which both reside in the United States. Further assets distributed around the globe would increase robustness and take pressure off these assets. There is also the possibility of creating new systems to accommodate data and processing from any future characterization programs.

Space missions sent to investigate minor bodies have been sent by several space actors in the past, (namely the US, Europe, Japan and China), which have been the source of ground-truth and confirmation of ground-based observation techniques (Capaccioni et al. 2015; Huang et al. 2013; Miller et al. 2002; Yoshimitsu et al. 2009). There are future missions proposed by private actors as a way of prospecting for minerals on different asteroids. One important point for these missions is that they will all depend on radio downlink from powerful radio antennas to send data, where all of the assets available to do so are currently in national hands.

There is still much to do to pull the rate of characterization of NEOs into line with the rate of discovery and then to have a program of follow-up observations to monitor the changes in properties over time. There are a few programs that have been identified that could help fill these gaps (a great example is the Sentinel project) (Reitsema and Lu 2015). There are also examples of current infrastructure that could be purposed to start looking for and/or characterizing asteroids (for example, disused or underused telescopes). Finally, there are examples of highly successful programs with limited lives that have expired, but could be reopened with a small funding injection (as occurred for the NEOWISE ground segment).

Currently, there is an international process in place to submit discoveries of new asteroids as they are observed. This is through the Minor Planet Center (MPC) operating out of Harvard. All observations have the associated dynamics determined by NASA's Jet Propulsion Laboratory (JPL) (Chodas 2017). These systems allow for several observatories to pick up the same asteroid and create discoveries through multiple-observatory sightings.

Current attempts to discover more asteroids are limited by the nature of Earth's orbit. This means that asteroids that spend a large portion of their time inside of the Earth's orbit are difficult to detect, as they are never in opposition with the Earth. There is also limited ability to make observations of asteroids that pass very close to the Earth, as the angular speed through opposition is inversely proportional to the distance from Earth.

3.5 Conclusion

Understanding the physical and orbital characteristics of NEOs can help us understand which pose a threat in the near future as well as the scale of that threat. All characteristics of NEOs that have been described in this chapter are interrelated and can allow a NEO to develop a potentially hazardous orbit in that NEO's lifetime. The chaotic evolution of NEO orbits means that an ongoing campaign of detection and surveillance is required to update the likelihood and potential severity of impact.

To understand the threat that any particular asteroid poses, larger data sets should be produced. In order to create fuller data sets, there should be investment in a range of different instruments and enabling hardware. This must necessarily be undertaken by multiple international actors for a few reasons, including the necessity for global geographic coverage, where a larger portion of the sky can be covered while avoiding disruption from local weather. Having multiple actors also allows for each to play to their own strengths, for example the US excelling at search programs and Japan with multiple low-cost minor body exploration missions. Significant future investment is required: the benefits of such a program are shared around the planet and so there should be a shared cost.

There is a need to have assets not constrained to Earth orbit in order to achieve a higher rate of discovery of NEOs that largely reside inside Earth's orbit, especially looking at models put forward by the B612 foundation with Sentinel. There also needs to be further characterization of the orbits of those NEOs that will make close approaches with the Earth before they make their flybys.

References

Abe, Shinsuke et al. 2006. "Mass and Local Topography Measurements of Itokawa by Hayabusa." *Science* 312(5778): 1344–47. http://www.sciencemag.org/cgi/doi/10.1126/science.1126272.

Baer, James, and Steven R Chesley. 2008. "Astrometric Masses of 21 Asteroids, and an Integrated Asteroid Ephemeris." *Celestial Mechanics and Dynamical Astronomy* 100(1): 27–42. https://doi.org/10.1007/s10569-007-9103-8.

Binzel, Richard P, Vishnu Reddy, and Tasha Dunn. 2015. "The Near-Earth Object Population: Connections to Comets, Main-Belt Asteroids, and Meteorites." In *Asteroids IV*, University of Arizona Press, 243–56. http://www.jstor.org/stable/j.ctt18gzdvc.19.

Borovička, Jiří, Pavel Spurný, and Peter Brown. 2015. "Small Near-Earth Asteroids as a Source of Meteorites." In *Asteroids IV*, University of Arizona Press, 257–80. http://www.jstor.org/stable/j.ctt18gzdvc.20.

Bottke, W et al. 2005. "Linking the Collisional History of the Main Asteroid Belt to Its Dynamical Excitation and Depletion." *Icarus* 179(1): 63–94. http://www.sciencedirect.com/science/article/pii/S0019103505001958.

Bottke, William F., David Vokrouhlický, David P. Rubincam, and David Nesvorný. 2006. "THE YARKOVSKY AND YORP EFFECTS: Implications for Asteroid Dynamics." *Annual Review of Earth and Planetary Sciences* 34(1): 157–91. http://annualreviews.org/doi/abs/10.1146/annurev.earth.34.031405.125154.

Britt, Dan T., Don Yeomans, K Housen, and G Consolmagno. 2002. "Asteroid Density, Porosity, and Structure." In *Asteroids III*, ed. William Bottke. Arizona, USA: University of Arizona Press, 485–500.

Brownlee, D E. 2003. "Stardust: Comet and Interstellar Dust Sample Return Mission." *Journal of Geophysical Research* 108(E10): 8111. https://doi.org/10.1029/2003JE002087.

Buczkowski, D L et al. 2012. "Large-Scale Troughs on Vesta: A Signature of Planetary Tectonics." *Geophysical Research Letters* 39(18): n/a-n/a. https://doi.org/10.1029/2012GL052959.

Capaccioni, F et al. 2015. "The Organic-Rich Surface of Comet 67P/Churyumov-Gerasimenko as Seen by VIRTIS/Rosetta." *Science* 347(6220): aaa0628-1-aaa0628-4. http://www.sciencemag.org/cgi/doi/10.1126/science.aaa0628.

Carry, Benoit. 2012. "Density of Asteroids." *Planetary and Space Science* 73(1): 98–118. http://linkinghub.elsevier.com/retrieve/pii/S0032063312000773.

Chesley, Steven R et al. 2014. "Orbit and Bulk Density of the OSIRIS-REx Target Asteroid (101955) Bennu." *Icarus* 235: 5–22. http://www.sciencedirect.com/science/article/pii/S0019103514001067.

Chodas, Paul. 2017. "Center for Near Earth Object Studies." 2017(13/04/2017). https://cneos.jpl.nasa.gov/.

Chyba, Christopher F., Paul J. Thomas, and Kevin J. Zahnle. 1993. "The 1908 Tunguska Explosion: Atmospheric Disruption of a Stony Asteroid." *Nature* 361(6407): 40–44.

DeMeo, F E, and B Carry. 2014. "Solar System Evolution from Compositional Mapping of the Asteroid Belt." *Nature* 505(7485): 629–34. https://doi.org/10.1038/nature12908.

Farnocchia, D et al. 2013. "Yarkovsky-Driven Impact Risk Analysis for Asteroid (99942) Apophis." *Icarus* 224(1): 192–200. http://www.sciencedirect.com/science/article/pii/S0019103513000821.

Giorgini, J D. 2002. "Asteroid 1950 DA's Encounter with Earth in 2880: Physical Limits of Collision Probability Prediction." *Science* 296(5565): 132–36. http://www.sciencemag.org/cgi/doi/10.1126/science.1068191.

Glassmeier, Karl-Heinz et al. 2007. "The Rosetta Mission: Flying Towards the Origin of the Solar System." *Space Science Reviews* 128(1–4): 1–21. http://link.springer.com/10.1007/s11214-006-9140-8.

Greenstreet, Sarah, Henry Ngo, and Brett Gladman. 2012. "The Orbital Distribution of Near-Earth Objects inside Earth's Orbit." *Icarus* 217(1): 355–66. http://www.sciencedirect.com/science/article/pii/S0019103511004374.

Harris, Alan W., and Germano D'Abramo. 2015. "The Population of Near-Earth Asteroids." *Icarus* 257(May): 302–12. http://linkinghub.elsevier.com/retrieve/pii/S0019103515002067.

Hartzell, Christine M, and Daniel J Scheeres. 2011. "The Role of Cohesive Forces in Particle Launching on the Moon and Asteroids." *Planetary and Space Science* 59(14): 1758–68. http://www.sciencedirect.com/science/article/pii/S0032063311001644.

Huang, Jiangchuan et al. 2013. "The Ginger-Shaped Asteroid 4179 Toutatis: New Observations from a Successful Flyby of Chang'e-2." *Scientific Reports* 3(1): 3411. https://doi.org/10.1038/srep03411.

Jenniskens, P et al. 2009. "The Impact and Recovery of Asteroid 2008 TC3." *Nature* 458(7237): 485–88. http://www.nature.com/doifinder/10.1038/nature07920.

Jewitt, David et al. 2014. "DISINTEGRATING ASTEROID P/2013 R3." *The Astrophysical Journal* 784(1): L8.

Keller, Lindsay P et al. 2006. "Infrared Spectroscopy of Comet 81P/Wild 2 Samples Returned by Stardust." *Science* 314(5806): 1728–31. http://www.sciencemag.org/cgi/doi/10.1126/science.1135796.

Lowry, S C et al. 2014. "The Internal Structure of Asteroid (25143) Itokawa as Revealed by Detection of YORP Spin-Up." *Astronomy & Astrophysics* 562: A48. https://doi.org/10.1051/0004-6361/201322602.

Mainzer, A et al. 2011. "Preliminary Results from Neowise: An Enhancement to the Wide-Field Infrared Survey Explorer for Solar System Science." *The Astrophysical Journal* 731(1): 53. http://stacks.iop.org/0004-637X/731/i=1/a=53.

Mainzer, Amy K., Fumihiko Usui, and David E Trilling. 2015. "Space-Based Thermal Infrared Studies of Asteroids." In *Asteroids IV*, University of Arizona Press, 89–106. http://www.jstor.org/stable/j.ctt18gzdvc.11.

Margot, J L. 2002. "Binary Asteroids in the Near-Earth Object Population." *Science* 296(5572): 1445–48. http://www.sciencemag.org/cgi/doi/10.1126/science.1072094.

Michel, Patrick et al. 2015. "Asteroids Advances and New Perspectives." In *Asteroids IV*, University of Arizona Press, 3–10. http://www.jstor.org/stable/j.ctt18gzdvc.7.

Miller, J.K. et al. 2002. "Determination of Shape, Gravity, and Rotational State of Asteroid 433 Eros." *Icarus* 155(1): 3–17. http://linkinghub.elsevier.com/retrieve/pii/S0019103501967533.

Minton, David A, and Renu Malhotra. 2009. "A Record of Planet Migration in the Main Asteroid Belt." *Nature* 457(7233): 1109–11. https://doi.org/10.1038/nature07778.

Miyamoto, Hideaki et al. 2007. "Regolith Migration and Sorting on Asteroid Itokawa." *Science* 316(5827): 1011–14. http://www.sciencemag.org/cgi/doi/10.1126/science.1134390.

Morbidelli, Alessandro et al. 2000. "Source Regions and Timescales for the Delivery of Water to the Earth." *Meteoritics & Planetary Science* 35(6): 1309–20. https://doi.org/10.1111/j.1945-5100.2000.tb01518.x.

Morbidelli, Alessandro et al. 2010. "Evidence From the Asteroid Belt for a Violent Past Evolution of Jupiter's Orbit." *The Astronomical Journal* 140(5): 1391–1401. http://stacks.iop.org/1538-3881/140/i=5/a=1391?key=crossref.e262f8e313f49990efff9f23d5938812.

Nakamura, Tomoki et al. 2011. "Itokawa Dust Particles: A Direct Link Between S-Type Asteroids and Ordinary Chondrites." *Science* 333(6046): 1113–16. http://www.sciencemag.org/content/333/6046/1113.abstract.

Ostro, Steven J et al. 2002. "Asteroid Radar Astronomy." *Asteroids III. Univ. of Arizona Press, Tucson*: 151–68.

Patzold, M. et al. 2011. "Asteroid 21 Lutetia: Low Mass, High Density." *Science* 334(6055): 491–92. http://www.sciencemag.org/cgi/doi/10.1126/science.1209389.

Popova, Olga P. et al. 2013. "Chelyabinsk Airburst, Damage Assessment, Meteorite Recovery, and Characterization." *Science* 342(6162): 1069–73. http://www.sciencemag.org/cgi/doi/10.1126/science.1242642.

Pravec, P et al. 2010. "Formation of Asteroid Pairs by Rotational Fission." *Nature* 466(7310): 1085–88. https://doi.org/10.1038/nature09315.

Rayman, Marc D, Thomas C Fraschetti, Carol A Raymond, and Christopher T Russell. 2006. "Dawn: A Mission in Development for Exploration of Main Belt Asteroids Vesta and Ceres." *Acta Astronautica* 58(11): 605–16. http://www.sciencedirect.com/science/article/pii/S0094576506000671.

Reddy, Vishnu et al. 2015. "Mineralogy and Surface Composition of Asteroids." In *Asteroids IV*, University of Arizona Press, 43–64. http://www.jstor.org/stable/j.ctt18gzdvc.9.

Reitsema, Harold J, and Edward T Lu. 2015. "Sentinel: A Space Telescope Program to Create a 100-Year Asteroid Impact Warning." In *Handbook of Cosmic Hazards and Planetary Defense*, eds. Joseph N Pelton and Firooz Allahdadi. Cham: Springer International Publishing, 569–81. http://link.springer.com/10.1007/978-3-319-03952-7_42.

Richardsonjr, J, H Melosh, R Greenberg, and D Obrien. 2005. "The Global Effects of Impact-Induced Seismic Activity on Fractured Asteroid Surface Morphology." *Icarus* 179(2): 325–49. http://www.sciencedirect.com/science/article/pii/S0019103505002472.

Rozitis, Ben, Eric MacLennan, and Joshua P Emery. 2014. "Cohesive Forces Prevent the Rotational Breakup of Rubble-Pile Asteroid (29075) 1950 DA." *Nature* 512(7513): 174–76. https://doi.org/10.1038/nature13632.

Russell, C T et al. 2012. "Dawn at Vesta: Testing the Protoplanetary Paradigm." *Science* 336(6082): 684–86. http://www.sciencemag.org/cgi/doi/10.1126/science.1219381.

Scheeres, D.J., C.M. Hartzell, P Sánchez, and M Swift. 2010. "Scaling Forces to Asteroid Surfaces: The Role of Cohesion." *Icarus* 210(2): 968–84. http://www.sciencedirect.com/science/article/pii/S0019103510002812.

Scheeres, D.J, B Khushalani, and R.A Werner. 2000. "Estimating Asteroid Density Distributions from Shape and Gravity Information." *Planetary and Space Science* 48(10): 965–71. http://www.sciencedirect.com/science/article/pii/S0032063300000647.

Scott, Edward R D et al. 2015. "Early Impact History and Dynamical Origin of Differentiated Meteorites and Asteroids." In *Asteroids IV*, University of Arizona Press, 573–96. http://www.jstor.org/stable/j.ctt18gzdvc.36.

Shepard, Michael K et al. 2017. "Radar Observations and Shape Model of Asteroid 16 Psyche." *Icarus* 281: 388–403. http://www.sciencedirect.com/science/article/pii/S0019103516300288.

Sokolov, L L et al. 2012. "Impact Trajectories of the Asteroid Apophis in the 21st Century." *Solar System Research* 46(4): 291–300. https://doi.org/10.1134/S0038094612040077.

Sugimoto, Y, G Radice, and J.P. Sanchez. 2013. "Effects of NEO Composition on Deflection Methodologies." *Acta Astronautica* 90(1): 14–21. http://www.sciencedirect.com/science/article/pii/S0094576512003402.

Tsuda, Yuichi et al. 2013. "System Design of the Hayabusa 2—Asteroid Sample Return Mission to 1999 JU3." *Acta Astronautica* 91(0): 356–62. http://www.sciencedirect.com/science/article/pii/S009457651300218X.

Veverka, J et al. 2001. "The Landing of the NEAR-Shoemaker Spacecraft on Asteroid 433 Eros." *Nature* 413(6854): 390–93. https://doi.org/10.1038/35096507.

Vokrouhlický, D. et al. 2015. "The Yarkovsky and YORP Effects." In *Asteroids IV*, University of Arizona Press, 509–32. http://www.jstor.org/stable/j.ctt18gzdvc.33.

Yeomans, D K. 1999. "Estimating the Mass of Asteroid 433 Eros During the NEAR Spacecraft Flyby." *Science* 285(5427): 560–61. http://www.sciencemag.org/cgi/doi/10.1126/science.285.5427.560.

Yoshimitsu, Tetsuo et al. 2009. "Hayabusa-Final Autonomous Descent and Landing Based on Target Marker Tracking." *Acta Astronautica* 65(5–6): 657–65. http://www.sciencedirect.com/science/article/pii/S0094576509000368.

Chapter 4
Methods, Means and Governance of NEO Observation

Peter Vereš and Nikola Schmidt

4.1 Introduction

This chapter introduces discovery and observation techniques for NEOs. The current state of all important detection and characterization projects are discussed in detail. In addition to the dominant contribution of the United States, programs in China and Russia and the global collaboration of scientific epistemic communities are explored. The dynamics of relations between these communities and national space programs' policies will be studied, including actors such as the International Asteroid Warning Network (IAWN), the Space Mission Planning Advisory Group (SMPAG), the B612 Foundation, Planetary Resources, and other entities involved or interested in Solar system mapping. The chapter describes three categories of stakeholders involved in the planetary defense effort. First, entities represented by private driven efforts, including epistemic communities of scientists, astronauts, engineers etc. Second, entities represented by national agencies. Third, entities represented by international organizations trying to merge all the efforts into a united endeavor.

P. Vereš (✉)
Harvard-Smithsonian Center for Astrophysics, Cambridge, MA, USA
e-mail: pveres@cfa.harvard.edu

N. Schmidt
Department of Political Science, Institute of Political Studies, Faculty of Social Sciences, Charles University, Prague, Czech Republic

N. Schmidt (ed.), *Planetary Defense*, Space and Society,
https://doi.org/10.1007/978-3-030-01000-3_4

4.2 How Do We Observe Asteroids?

4.2.1 Explanation of Terms for Non-Astronomers

The term *asteroid* was first proposed in 1802 by a British astronomer, William Herschel, only one year after Ceres, the first known asteroid, was discovered. The term asteroid itself comes from ancient Greek and literally means "star-like." Asteroids indeed look like stars in a telescope. The brightest of them, Vesta, is even visible through the naked eye. However, they only reflect the light of the Sun.

While stars are stationary on the celestial sphere, asteroids, like planets, move quickly in front of the static stellar background. The changing brightness and position of the asteroid in the sky is a direct consequence of its motion in the Solar System and the changing geometry between the Sun, Earth, and the asteroid itself. Fast-moving objects are obviously closer to an observer than slow ones. The discovery of Ceres in 1801 brought about a new method for orbit determination derived by Gauss, when only three on-sky positions of an asteroid on three distinct dates were needed for its *orbit determination*. Asteroids, like other objects in the Solar System, are moving around the Sun in elliptical orbits. Unlike planets, which are on almost circular orbits, closely packed near one plane called the ecliptic, asteroid orbits are different. Their orbits are more elliptical (*eccentricity*) and inclined with respect to the ecliptic (*inclination*). Asteroids are mostly found in the Main Belt, between the orbits of Mars and Jupiter. However, Near-Earth Asteroids have orbits that can intersect the orbits of Mercury, Venus, Earth, and Mars, and can reach out to the Main Asteroid Belt—but by definition, their perihelion distance is <1.3 AU. Like Gauss did 200 years ago, the orbit of an asteroid is derived from its multiple on-sky positions. The longer the observing time frame (arc), the better the orbit is constrained. On one night, typically four positions of an asteroid are taken in a so-called tracklet that represents an apparent motion vector of a body on the sky. A single tracklet only gives a hint about where the object is in the Solar System. Typically, at least a few days and brightness measurements are needed for preliminary orbit determination and size assessment. Once an orbit is determined, the on-sky and in-space position of the asteroid and its brightness can be calculated for any given time in the past or future. In astronomy, the collection of predicted positions is referred to as the *ephemeris*.

Magnitude is a logarithmic measure of brightness where a five-magnitude step is proportional to a factor of 100 in brightness. The Sun has an apparent magnitude of −27, the full Moon −13, the brightest stars are around zero, and a typical eye limit is around 6. *Limiting magnitude* represents the faintest detection that a telescope can detect at a given signal-to-noise ratio. *Absolute magnitude* (H) refers to the brightness of an object artificially placed to a unit distance of 1 AU from the Earth and Sun, at zero phase angle (an angle between Earth–object–Sun); on an apparent straight-line Sun–Earth–object. A 1-km object is proportional to H ~ 18, 100-m to H ~ 22 and 10-m to H ~ 28, therefore H is often used as a proxy of asteroid diameter, when albedo is known or assumed. Albedo is a measure describing what fraction of

Sun light is reflected from the asteroid. Dark asteroids have low albedo, bright ones have high albedo. Measuring the apparent brightness of an asteroid in short time intervals for hours or days reveals its *light curve*. This short-term brightness variation is proportional to the shape of the asteroid, and one can infer its spin axis orientation as well.

In the nineteenth century, asteroids were observed by the human eye using optical telescopes. Later, in the same century, photography was invented, and photographic plates soon replaced astronomers' eyes. A photograph can be exposed for hours and collects more light than an eye is capable of collecting in the blink of a second. Photographic plates could thus see much fainter objects, and their usage rapidly increased the number of discovered asteroids.

In 1970, a Charge-Coupled Device (CCD) was developed that could collect light by semiconductors due to photoelectric effect. The following two decades of improvement brought the CCD camera to astronomy. CCDs quickly replaced photographic methods, drastically reducing the time needed for collecting light and allowing the storage of data in an electronic form and data processing by a computer. Conceptually, the same CCD chips are found nowadays in digital cameras or cellphones. Astronomical CCD cameras are extremely sensitive, often able to detect a change of flux on the level of a few photons. The combination of a large telescope and a CCD device allows detection of objects as faint as the 27th magnitude. CCD-equipped telescopes discovered more than 99% of currently known asteroids. CCD detectors denote the brightness of the signal in signal-to-noise ratio (SNR), which can be translated to the magnitude.

Radar observation is the method when a known asteroid is tracked by a Doppler radar, where radio waves reflect from its surface and are detected back. Asteroids need to be close to the radar, typically only a few lunar distances from Earth. Radar observation provides the most precise instantaneous information about the asteroid's distance from the observer, its radial velocity, and can reveal its shape in detail. Except for a direct spacecraft visit to an asteroid, radar images have provided the best shape screening of asteroids to date.

Hundreds of years ago, when the sky was observed visually or by photography, only the optical wavelengths were captured. However, if a specific filter is placed in between the eye or a detector and the telescope, only exact wavelengths are captured. Traditionally, filters are named after a color that is associated with a visual wavelength. For instance, the Johnson-Cousins photometric system has five filters: U (ultraviolet), B (blue), V (visible), R (red) and I (infrared). Newer filter systems, like SDSS, have a better spectral and color separation and are usually denoted by lowercase letters. For example: u (ultraviolet), g (green), r (red), i (near infrared), z (infrared). Asteroids observed through individual filters have different brightness. This depends on the surface properties and composition of the asteroid. Measurement of so-called asteroid colors gives a basic spectral assessment of an object and determines its approximate composition. Modern CCD devices and spectrographs are able to observe asteroids in wavelengths that are not visible to the eye (infrared, ultraviolet, radio). A spectrum can identify specific molecules on the asteroid's surface.

Observation of asteroids in the infrared wavelengths has advantages over observations in visible light. In visible light, reflected sunlight of an asteroid depends on its albedo. Also, in visible light, we see asteroids in phases, similar to the phases of the Moon—depending on viewing geometry and the asteroid's phase angle, we see only a fraction of the asteroid's surface. Lighter asteroids with large albedo are much brighter than dark ones, therefore, for two asteroids of the same size, the darker one is harder to spot. By comparison, infrared light is the heat absorbed by the asteroid and emitted back to space. Higher albedo asteroids absorb less sunlight and are correspondingly fainter in infrared. A combination of optical and infrared observations are used for size determination and albedo of asteroids. If astronomers only have visible wavelength information, they have to assume albedo and then compute the diameter. Figure 4.1 demonstrates how optical and infrared observations differ. In visible light, three asteroids can have the same brightness if they are at the same distance and location with respect to the Earth, even though they have different diameters. The same objects behave differently in infrared, and the flux corresponds to the diameter.

Adaptive optics is a method that allows observers to obtain much higher resolution of observed objects, reducing the distortion of the image due to the atmosphere and allowing ground-based observers to obtain resolution similar to that of space telescopes. Even though adaptive optics is mostly used in astrophysics, recent

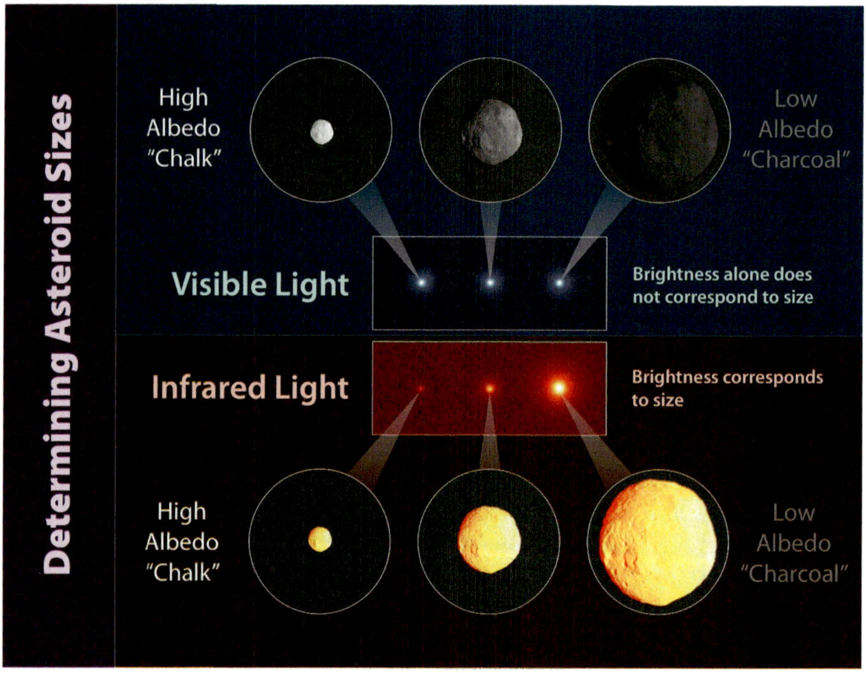

Fig. 4.1 Difference between observing the asteroid in the visible (top) and infrared (bottom) light. Source: Caltech/JPL

observations of asteroids have revealed its usefulness. Adaptive optics allows us to distinguish the shapes of asteroids directly and discover their moons (Hanuš et al. 2013; Marchis et al. 2006), and when it is combined with other optical data, we can learn more about the shapes of asteroids without sending an in-situ spacecraft or utilizing radar.

Space-based telescopes have several advantages in comparison with ground-based observatories. Being on the ground means that half of the sky is not observable for a single observatory because the Earth itself is blocking it, due to the day-night cycle. The Earth obstruction becomes smaller if the space telescope is on a higher, or heliocentric, orbit. Also, telescopes are usually not observing close to the horizon because the atmosphere of the Earth is thicker along the line of sight, thereby decreasing the limiting magnitude. Ground-based observatories lose observing time due to clouds or wind, can only observe at night, and the quality of images or magnitude limits change from night to night. By contrast, space-based telescopes can look almost in any direction (except close to the Sun), are not limited by weather or clouds, and the image quality is orders of magnitude better. However, they are more expensive, they cannot be repaired or upgraded, operation is difficult, and operation lifetime is relatively short. Also, the computing power and data transfer from and to a space-telescope is much slower than what can be achieved on the ground.

4.2.2 General Approaches to Detect an Asteroid

As mentioned previously, asteroids with known orbits are easily found by ephemeris if they are on the night sky and bright enough to be detected by a given telescope. However, undiscovered asteroids can be anywhere in the sky. The general approach of finding new asteroids is to point a telescope to the sky and take multiple images of the same field, preferably by CCD camera, in time intervals, typically of tens of minutes. Images are subsequently stacked together on the same stars. Because the stars do not move but asteroids do, in a stacked image an asteroid would appear like a dot moving across the image along a straight line. Each position of the asteroid is measured with respect to the stars and compared to the catalog of positions computed for all known asteroids. If there is no match by position and rate of motion, the observed asteroid is not known. Additional positions are required to determine its orbit and absolute magnitude.

This *blinking* technique has been used for more than a century, first with photographic plates that were blinked in a so-called blink comparator. Nowadays, use of CCD, computers, and software, makes the entire process significantly faster. The computer identifies all static stars in the image, assigns on-sky positions to the stars, identifies moving targets, and matches them with the catalog. Automated software pipelines are used to produce tracklets of known and unknown asteroids almost in real time at major optical CCD surveys like Pan-STARRS and CSS. All positional observations are sent to Minor Planet Center by email. If the one-night tracklets behave like NEOs, they are placed on the NEO Confirmation Web Site (NEOCP), where observers around the world can follow-up the object and extend its observing arc. If the same object is observed on subsequent nights or weeks, its orbit is determined and its designation and orbit are published by Minor Planet Center.

A novel method of NEO detection is *synthetic tracking*. Described by Shao et al. (2014), synthetic tracking takes very short but many (hundreds) of subsequent exposures of the same field, at a frame rate similar to video cameras. In individual subsecond frames, there is almost no signal and only very bright objects are visible. However, if the images are shifted, rotated, and stacked in the direction of motion of the asteroid and at the same velocity, the signal of moving target becomes visible. Images of static stars will be trailed and faint. This could be a very efficient detection method for small meter-class NEOs that are flying close around or impacting the Earth. Their fast rate of motion makes it very difficult for traditional surveys to detect them, because of the trailing and SNR loss. With synthetic tracking, the time resolution will be more accurate than in the case of traditional surveys. The disadvantage of synthetic tracking is its high demand for computational resources. The direction and rate of motion is unknown; therefore, images of the same field have to be shifted and moved in all possible directions. Parallel computing with top-shelf graphic cards makes this teraFLOP processing possible; for instance, 30-s worth of frames can be processed in 10 s. Shao et al. (2017) proposed that a fleet of five cubesat-based nanosatellites with 10-cm telescopes doing synthetic tracking on a Venus-like orbit, distributed in a manner to cover the entire orbit, could achieve the Congressional mandate of finding 90% of NEOs larger than 140 m in three years, at 10% of the cost of NEOCAM, and much faster than the LSST-class dedicated survey. In 2017, NASA partnered with private companies, and the Small Spacecraft Demonstration Missions Trans Astronautica Corporation was selected. Jointly with Deep Space Industries and the Jet Propulsion Laboratory, they will work on project Theia: Synthetic Tracking Demonstration for Commercial, NASA and Other Government Agency Applications to Space Situation Awareness, Planetary Defense, and Asteroid In Situ Resource Utilization. This will be the first step toward integrating synthetic tracking and NEO observations into cubesats.

4.3 NEO Stakeholders

4.3.1 NASA Planetary Defense Coordination Office (PDCO)

Since its establishment and continuing to the present, PDCO has been led by Lindley Johnson and located at NASA's Headquarters in Washington, D.C. Its mission is to lead national and international efforts to detect any potentially Earth-impacting objects, assess the potential effects of a possible impact, and develop strategies to mitigate impacts or minimize the consequences of the impacts. PDCO is focused on Potentially Hazardous Objects, with orbits <0.05 AU from the Earth's orbit and large enough to reach Earth's surface (30–50 m). PDCO has three major segments:

the NEO Observation Program (NEOO), Interagency and Emergency Response, and Mitigation research. The budget of PDCO has been growing steadily, from 4 million USD in 2002–2010, to 20 million USD since 2012, 40 million USD since 2014, and 50 million USD since 2016. Thanks to the NEOO funding, 98% of currently known NEOs have been discovered and characterized. NEOO's main motivation is reaching the Congressional goal to discover 90% of NEOs larger than 140 m. In addition to discovery facilities (currently including Pan-STARRS, CSS, Spacewatch, SST, ATLAS, NEOWISE and formerly LINEAR, LONEOS, NEAT), the NEOO program funds follow-up networks (Astronomical Research Institute, Magdalena Ridge Observatory, Table Mountain Observatory, Tenagra, David Tholen of the University of Hawaii) and other assets for NEO characterization (radar observations through Goldstone and Arecibo, NASA IRTF, Spitzer space telescope, MANOS project). Within mitigation research, PDCO supports mitigation strategies and the study of asteroid impacts, atmospheric airbursts, and redirect or sample return missions to NEOs. PDCO is involved in international collaboration through the AIDA mission to the binary NEO (65,803 Didymos) and participates in COPUOS. PDCO also funds studies of the future telescope performance to discover NEOs (LSST, NEOCAM), the development of new tools to link NEO orbits (ZTF L-MODE, LSST MOPS), and participates in NEO or asteroid-related missions (Dawn, OSIRIS-REX). PDCO also supports the Asteroid Threat Assessment Project (ATAP) at NASA Ames.

PDCO is currently providing base funding for the Minor Planet Center. Through NASA's Small Body Node and also the JPL Center for NEO Studies, it supports data processing, analysis and management centers, technology projects, and studies for impact mitigation. PDCO manages the Agency's Asteroid Grand Challenge through multidisciplinary collaboration with other agencies home and abroad, between scientists and citizens. PDCO is also involved in a Meteoroid Environment Office of NASA, based at NASA's Marshall Space Flight Center, that provides meteor shower forecasts for spacecraft operations.

4.3.2 The Minor Planet Center (MPC)

MPC is an entity with the mandate of the International Astronomical Union (IAU) to gather observations of asteroids and comets, compute and publish orbits, and coordinate the discovery process. In the past, its role was held by the Astronomisches Rechen-Institut in Berlin since the nineteenth century. In 1947, MPC was independently founded at the Cincinnati Observatory and took over the agenda. In 1978, the MPC was transferred to the Smithsonian Astronomical Observatory in Cambridge, Massachusetts. MPC was recently moved under the management of NASA's Small Body Node, and its operational funding is coming from a NASA NEOO grant. The director of MPC is currently Matthew Holman.

MPC fulfills several important roles for the storage and distribution of asteroid-related data. Its publicly accessible database contains orbital elements of more than

775,000 numbered asteroids and 4015 comets, among them more than 18,000 NEOs, as well as outer irregular satellites of Jupiter, Saturn, Uranus, and Neptune, alongside a small fraction of artificial objects. In addition to orbital elements, MPC provides all historically submitted positional observations for each object—more than 190 million detections. The on-sky location of any object can be computed by using the MPC ephemeris service. MPC also provides daily sky coverage by major asteroid surveys, guides for observers, data files with orbits and observations, orbit diagrams, a database of asteroid light curves, and a database of contribution observatories and statistics. Information is available online. MPC also assigns unique three-digit codes to all contributing observatories. Since the first discovery of an asteroid in 1801, 2500 observatories have reported observation of an asteroid or comet to MPC. More than 130 observatories have discovered and more than 1200 observatories have observed at least one NEO (Fig. 4.2). However, some of the observatories do not operate anymore.

NEO observations are treated with a higher priority. Newly discovered NEO candidates are published at NEOCP. The page contains a table of NEO candidates, which can be sorted by the NEO-probability score (digest score), apparent brightness, on-sky position, designation, information about the length of the observed arc, and its absolute magnitude. Each NEO candidate can have its ephemeris (predicted positions) computed. However, the prediction is rather problematic, especially for short-arc and fast NEOs. If a NEO is confirmed, the Minor Planet Electronic Circular is issued (MPEC) with the orbital elements of its orbit, all positional observations used for the computation, residuals of individual positional astrometry with respect to the computed orbit, as well as the new designation of the object. The discovery process therefore consists of the following steps: initial submission of the tracklet to MPC, follow-up observations on the same and subsequent nights, orbit determination, and issue of MPEC with the new object designation.

There are multiple reasons for a NEO object to be lost. Often, when NEOCP contains too many NEO candidates, the follow-up observers cannot fulfill the needs

Rank	MPC Code	Observatory Name	Discovered	Managing Institution
1	G96	Mt. Lemmon *	5,344	University of Arizona
2	F51	Pan-STARRS1 *	4,088	University of Hawaii
3	703	Catalina *	2,701	University of Arizona
4	704	LINEAR	2,532	MIT
5	691	Spacewatch *	864	University of Arizona
6	E12	Siding Spring	467	University of Arizona
7	644	NEAT	305	Jet Propulsion Laboratory
8	699	LONEOS	296	Lowell Observatory
9	W84	DECAM	290	consortium
10	C51	NEOWISE *	279	Jet Propulsion Laboratory

Fig. 4.2 Most active discoverers of NEOs until June of 2018. First ten observatories discovered 92% of NEOs. One hundred and thirty observatories discovered at least one NEO. Asterisk marks active surveys. Mt. Lemmon and Catalina telescopes are part of one project (Catalina Sky)

of follow-up. Logically, NEO candidates with the highest NEO scores are observed with priority, therefore, lost NEO candidates were most likely non-NEOs. Still, from time to time a highly ranking NEO candidate is lost. Note, that these are all real, observed objects. This happens when the object is too fast and its on-sky uncertainty grows rapidly within hours. Within a day, its location can be anywhere in the sky and the object is lost. Also, flyby geometry plays a role. If the object approaches the Earth from opposition and flies into a day sky, it cannot be observed by telescopes anymore. In addition to fast NEO candidates, the faintest objects are at the highest risk of not being recovered, because the number of large follow-up telescopes is limited. Therefore, without a follow-up, no orbit can be derived and the object becomes lost. Between 2013 and 2016, 5928 NEOs were discovered, however, 877 NEO candidates with digest score equal to 100 (highly likely NEO) were lost, which is 15% of all reported NEOs discovered.

Weather also plays a role in the risks of NEO candidate recovery. For instance, the most productive survey locations (Pan-STARRS; Hawaii and Catalina; Arizona) have seasons where the weather does not allow observations for up to a few weeks. Arizona telescopes are affected mostly during the monsoon season in the summer, while Hawaii cloud coverage is more prominent in the winter months. In this case, the lack of sky coverage by one of the major contributors decreases the follow-up opportunities as well as the number of NEOCP NEO candidates. One of the more useful NEO follow-up services is the NEO Blog, where observers post in real time which targets are being searched for, detected, and submitted to MPC. The blog helps other observers to optimize their time and avoid follow-up of the same object at the same time. Follow-up itself is crucial—if the object is not followed up, it is removed from NEOCP and is no longer accessible for observers. It remains in the internal MPC database. In the future, it can be attributed to an object that will be discovered and will match the lost candidate position in the MPC pipeline. With improved and rapid follow-up coordination, the fraction of lost NEOs will be lower.

One of the most highly anticipated upgrades of MPC will be the new astrometry data exchange standard (ADES). Historically, observers have reported rather limited information about asteroid positions, where all information has to be included in only 80 columns. The historical format is, however, not sufficient for high-precision positions (losing decimal digits) and does not contain observational uncertainties that are needed for accurate orbit determination and other details. ADES is formatted as an XML (table) that can store numerical and text values without a length or decimal digit restriction. ADES development is crucial with the new surveys (LSST) starting in a few years, and high-precision stellar catalogs like GAIA (Lindegren et al. 2016).

4.3.3 Center for NEO Studies (CNEOS)

CNEOS is a group funded by NASA NEOO grant, located at the Jet Propulsion Laboratory, California Institute of Technology, in Pasadena, California, which has been working within the Mission Design and Navigation Section since 2017. Its predecessor was the JPL NEO Program Office. The principal investigator is Paul Chodas. CNEOS products are available online to the public. CNEOS computes orbits of all asteroids and comets by using the MPC-published astrometry independently, and by using the JPL-proprietary orbit determination routines. It is worth mentioning that the independent orbit assessment is not done in competition, but rather as a complement and cross-checking measure. CNEOS offers multiple tools that are not available at MPC. For instance, it lists the close approaching NEOs with the nominal approach distance and time of close approach (Asteroid Watch application). It also provides the automated impact risk web site (Sentry), with impact probability assessed for all asteroids, giving the list of those with the highest impact risk, planetary defense definitions with NEO deflection tool and impact scenarios, a list of NEOs accessible by spacecraft, and NEO discovery statistics. CNEOS is linked to the Solar System Dynamics database of asteroids and comets, which are accessible through database query or HORIZONS, an online data and ephemeris computation service.

In 2017, CNEOS made its SCOUT tool available. SCOUT is a sophisticated alternative to NEOCP. It lists the same NEO candidates, but offers more tools to assess the preliminary ephemeris. It uses systematic ranging (Farnocchia et al. 2015) to compute preliminary orbits and ephemerides of short-arc NEO candidates. SCOUT also provides impact rating, close-approach distance, MOID, NEO and PHA probability scores, score for the large size (1 km), score for geocentric orbit, score for interior Earth orbit, Tisserand invariant score, on-sky predicted location, solar elongation, rate of motion, magnitude, and on-sky uncertainty, which are all—to have a complete perspective—needed integers that help to assess the asteroid as a possible threat.

CNEOS data for close approaches, fireball data, human accessible NEOs, Scout, and Sentry, are also available in machine-readable versions and near-real time updates through provided APIs. CNEOS is thus enlarging the group of programs supporting voluntary global collaboration on asteroid detection. Along with Minor Planet Center and together with other NEO stakeholders, it offers invaluable support for NASA-funded NEO missions and goals.

4.3.4 European Space Agency's (ESA)

The ESA NEO Segment of its Space Situation Awareness (SSA) Programme represents another valuable resource regarding NEOs. The SSA-NEO Coordination Centre is located in Frascati, Italy and led (2017) by Detlef Koschny and Jehn Rüdiger. Its current operations started in 2013 and have since built on the precursor service. The goals of SSA-NEO include prediction of current and future positions of NEOs, estimating the likelihood of impacts, assessing consequences of impacts, and working on NEO deflection methods.

SSA offers databases of asteroids with dynamical and physical properties, observational data, visualization and orbit calculation tools, impact monitoring, and bolide data. It also provides links to its major NEO competitors' activities around the world, as its mission is to offer complete information on NEOs to the public and private sectors, including governments, international entities, defense, space agencies, schools, and the public. NEODyS—a dynamical website managed by the University of Pisa, and its spinoff SpaceDys, is a major asset of ESA SSA-NEO. Independently of MPC or CNEOS, it computes the orbits of NEOs and also the impact probabilities of potentially dangerous NEOs. This independent service is invaluable and offers a unique crosscheck between different tools for computing impact probability, helping to assess the threat of particular asteroids on Earth. Another product of the University of Pisa overseen by SSA-NEO is AstDys, a dynamical database of all asteroids. NEOSHIELD-2, a European Union Horizon 2020-funded international project, is also a major contributor, studying Earth impact scenarios and asteroid characteristics. It has been funded by the EU based on the interests of the EU in planetary defense in general. NEOSHIELD has been useful, particularly in putting emphasis on kinetic impactors as an appropriate response to asteroid threats. It promotes high quality observations, giving us time to deflect them decades in advance and avoid questionable technologies such as nuclear deflection. NEOSHIELD-2 also provided spectral characterization of almost 400 NEOs.

Along with the ESA Optical Ground Station, SSA has its own collaborative network of observatories, which are used for the discovery, follow-up, and characterization of NEOs. Even though SSA's discovery statistics are far less impressive than those managed and funded by NASA, SSA's assets provide coverage of the Eastern Hemisphere. One of the useful tools for NEOs is the "Priority list" that updates regularly and notes NEOs that have a need for follow-up, with information about the current on-sky location, magnitude, visibility windows, and sky uncertainty, sorted by urgency. Some of the NEOs have such a large uncertainty due to their short arc and lack of follow-up that those are most likely lost and left for future incidental recovery. SSA's observatories will help with the follow-up of asteroids discovered by ESA's GAIA missions (GAIA FUN-SSO). GAIA is ESA's space telescope, which precisely measures the positions and motions of the stars and will provide the largest catalog of stellar positions. These positions serve as a reference frame for measuring asteroid on-sky locations. SSA is also funding the development of a Wide Survey network of telescopes, based on novel Flyeye 1-m-class optical devices, which will be able to cover the entire visible sky per night to a relatively faint limit of about 21.5 magnitude. This is not as deep the most productive NEO surveys; however, this is compensated for by the all-sky coverage, which the top surveys do not have.

4.3.5 Japan Aerospace Exploration Agency (JAXA)

The Japan Aerospace Exploration Agency (JAXA) supported NEO research through its two space missions. The first one, Hayabusa (2003–2010), flew to a NEO named Itokawa (Fujiwara et al. 2006), which is a small and irregularly shaped 500-m asteroid. Hayabusa reached Itokawa, successfully orbited the asteroid, retrieved samples from its surface, and delivered the samples back to the Earth. Hayabusa also attempted to land a small experimental lander, Minerva, on Itokawa. Hayabusa became the first spacecraft to bring an asteroid sample to Earth. The second mission, named Hayabusa-2, was launched in 2014 and is currently on the way to a 1-km NEO called Ryugu. Hayabusa-2 is an improved asteroid sample return mission, and the first mission that will visit a carbonaceous asteroid (C-type). C-type NEOs could have brought organic molecules to Earth billions of years ago, contributing to the origin and evolution of life. Hayabusa-2 carries a small explosive impactor to create an artificial impact crater on Ryugu, as well as a small lander.

4.3.6 China

Chinese efforts in NEO research are led by the Purple Mountain Observatory, one of the most active astrometric surveys. The China National Space Administration (CNSA) also sent a spacecraft named Chang'e 2 within its Lunar Exploration Program. After leaving the lunar orbit, Chang'e-2 flew by the NEO Toutatis. CNSA is planning a sample return mission to a binary NEO with a designation 1996 FG3 before the year 2025. The same asteroid was already selected for ESA's Marco Polo mission, which was canceled.

4.3.7 Russia

The Russian Roscosmos State Corporation for Space Activities (ROSCOSMOS) is co-funding a series of new telescopes that will search for NEOs. Its first telescope AZT-33VM, built in the Sayan Mountains of Siberia, was activated in 2016. Within the Russian Federal Space Program 2016–2025, ROSCOSMOS plans to build two space telescopes aimed at NEO discovery and warning: EKOZONT, consisting of 75- and 40-cm telescopes on low-Earth orbit, and a large 1.5-m NEBOSVOD telescope. It also funds two radio telescopes. ROSCOSMOS has a long-term plan for a space mission to the potentially hazardous NEO Apophis, which had a slight chance of collision with the Earth in 2036 that was ruled out in 2013. ROSCOSMOS is also developing a nuclear explosion mitigation method for NEOs on an impact trajectory. However, the nuclear solution of asteroid mitigation is highly disputed internationally and not supported by most stakeholders.

4.3.8 Universities and Research Institutions

NEO research is also widely done around the world at public and private universities and research institutes. In addition to facilities funded directly by NASA or ESA grants, researchers can rely on funding through the National Science Foundation

(NSF, USA), European Research Council (ERC), and other national and international funding sources and grants. Many observatories are operated by universities. Advantages of having universities involved in NEO research include a strong focus on education and the opportunity to include graduate students and young postdoc researchers. The increased volume of discovered NEO and funding sources also supports the enlargement of many astronomy departments (for instance astronomy undergraduate programs at Northern Arizona University in Flagstaff, USA or at University of Hawaii, USA). Other institutes involved in NEO research are government facilities (NASA centers, ESA), or private laboratories (e.g. MIT Lincoln Laboratory, Southwest Research Institute, SETI, Planetary Science Institute).

4.3.9 Public Involvement

The connection of NEOs and the extinction of many life forms 66 million years ago, including the dinosaurs, increased the popularity of NEO research among the public. Amateur astronomers have been useful to the community of professional astronomers, providing routine positions and light-curves of asteroids. Most active amateur astronomers have acquired large astronomical telescopes and are being funded by NASA to observe and characterize NEOs. Some of the most active amateur NEO observers are Robert Holmes (Astronomical Research Institute, Illinois, USA), Clay Sherrod (Arkansas Sky Observatories, Arkansas, USA), and Michael Schwartz (Tenagra Observatories, Ltd., New Mexico, USA).

The public can often help with archival data. Large surveys are using software to detect asteroids in astronomical images. However, each software program works at a given efficiency, and a small fraction of the objects are missed. This often happens for faster objects (NEOs) or objects that are very faint and at threshold detection limits. The human eye can distinguish and identify objects that are even fainter than the detection threshold. Archives of several defunct surveys are accessible to the public (for instance SDSS, NEAT). Other professional observatories, like Pan-STARRS and Catalina Sky Surveys, teamed up with the International Asteroid Search Collaboration (IASC) that connects professionals, amateurs, volunteers, and elementary and high schools around the world, allowing the community to access archival data from surveys and find asteroids that were missed by software routines. This activity introduces astronomy and NEOs to young people, educates the community, and gives a chance to regular citizens to discover an asteroid and become a discoverer. Additionally, the public can search through astronomical images obtained for other science projects and find incidental asteroid observations in archival data. The Canadian Astronomy Data Center (CADC) offers such data (Gwyn et al. 2012). Catalina Sky Survey also offers its archival images for the Asteroid Zoo segment of Zooniverse, a citizen science web portal. Asteroid Zoo is supported by Planetary Resources, which focuses on asteroid mining.

An efficient model of connecting professional and amateur astronomers with the public is the Las Cumbres Observatory Global Telescope Network (LCOGT). This private organization, with headquarters in Goleta, California (USA), operates a network of robotic telescopes around the world (Hawaii, Chile, Canary Islands, Australia, South Africa, Texas) and is expanding quickly (China, Israel). Access to the telescope is provided either to professionals, for a fee and via science proposals, or to the public. A large fraction of the observing time, managed by Tim Lister, is dedicated to NEO follow-up. LCOGT also has research postdocs working on a campus of the University of California in Santa Barbara doing NEO-related research. In addition to research, LCOGT is strongly oriented toward citizen science, automatization, and the future of astronomical research. Another example of a for-hire robotic telescope network is Slooh, which is also oriented toward amateur observers.

B612 foundation is a nonprofit organization established by Rusty Schweickart, Clark Chapman, Piet Huet, and Edward Lu in 2001 in Silicon Valley, California, USA. B612 works toward protecting the Earth from asteroid impacts and dealing with decision-making regarding planetary defense. B612 is currently led by a former NASA astronaut, Edward Lu. B612 represents a unique connection of scientists, collaborators, advisors, and entrepreneurs. The foundation had a plan to raise funds to build and operate a space telescope, Sentinel, that would search for NEOs. However, the mission was canceled, as NASA terminated its funding and B612 did not raise enough money. B612 organizes Asteroid Day, a global event for the public to raise awareness about asteroids and Earth protection. Asteroid Day is held annually on June 30, the anniversary of the Tunguska asteroid impact that occurred in Siberia in 1908. Asteroid Day has achieved unprecedented coverage of media and involves national agencies (NASA, ESA, JAXA). In 2017, B612 established a virtual Asteroid Institute that will collaborate with major universities and research centers worldwide. The Asteroid Institute will fund research postdocs at the University of Washington in Seattle, who will work on planetary defense.

4.3.10 Private Industry

The interest of for-profit private companies in space is growing steadily. The success of private aerospace manufacturers toward launching rockets and delivering cargo and probably men to space in the near future (SpaceX, Orbital, Blue Origin, Bigelow Aerospace), or preparing for commercial suborbital flights (SpaceDev, Virgin Galactic), has led to plans that involve NEOs, particularly asteroid mining. Asteroids, unlike Earth's crust, have atom abundance similar to the ratio present in the Solar Nebula when the Solar System was formed. In the case of Earth, most of the heavy elements fell to the core due to gravity, hence why chemical elements like gold, silver, platinum, and palladium are rare and precious. Depending on the total mass of an asteroid and its composition, one may estimate the total value of useful heavy elements in an asteroid if it were to be entirely harvested. A small, 1-km NEO could have a value of about 85 billion USD. There are currently two private companies involved in prepared asteroid mining—Planetary Resources and Deep Space Industries—and several more emerging companies.

Planetary Resources (Redmond, Washington, USA) has a step-by-step plan, including testing small cubesat-class satellites in low Earth orbit for technology demonstration, small telescopes on orbit to characterize chemical composition of asteroids, spacecraft for detailed in-site mapping of desired targets, and asteroid mining. Planetary Resources manages a website that estimates the value and profit from asteroid mining on currently known data (asterank.com). The company is also involved in the citizen-science search for asteroids in Catalina Sky Survey archives (Asteroid Zoo). Planetary Resources aimed to crowdfund Akryd100, a fleet of small space-telescopes, which eventually failed. Its current focus is on development of the low-cost Arkyd-6, a cubesat-derived telescope.

Deep Space Industries (Mountain View, California, USA) plans the development of small, cost-effective space probes that would be able to reach an asteroid and survey it in-situ. Its technology plans also include refueling by hydrogen and oxygen mined from asteroids, 3D printing in space, and the processing and delivery of precious metals to Earth. Deep Space Industries expects its mining operations to start in the early 2020s. The company developed a new electrothermal thruster micropropulsion, two-camera optical navigation, and cost-effective modular deep space avionics. The new technology is intended for Prospector-X, the first mission to test the technology, and Prospector-1, a low-cost private mission to survey and land on a NEO.

4.4 International Cooperation in NEO Surveys

4.4.1 Rationale for Coordination

Discovery of asteroids is not possible without the coordination of observers around the world. At the time of first detection of a newly found NEO candidate, its orbit is highly uncertain and is practically guessed, for instance, based on statistical ranges. Rapid follow-up is needed for the confirmation of an object and its initial orbit determination. NEOCP is a central node that publishes NEO candidates, gathers follow-up observations from observers, updates the new orbits, and offers ephemerides needed for further observations. More details for NEOCP candidates, mostly focused on Earth-impacting parameters, are published by JPL's SCOUT. Some major surveys have their own follow-up facilities (for instance, Pan-STARRS, Catalina, and Spacewatch), however, the rest is left for the community. Also, follow-up facilities tend to be geographically close to the primary sites. Follow-up on different continents is desirable to cover 24 h of the date. Most of the discovery and follow-up facilities are located in the United States and Europe (Fig. 4.3). Non-representative Earth coverage is apparent, and facilities are needed in continental Asia, Australia, and Africa, and generally in the Southern hemisphere.

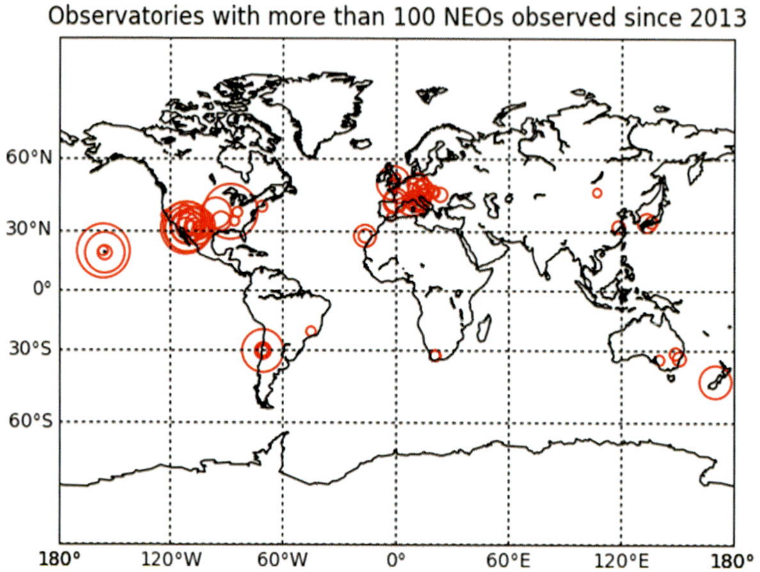

Fig. 4.3 Locations of NEO observers around the world. Circles are proportional to logarithmic number of observed NEOs

Cooperation, as a required prerequisite for coordination, is also a desirable state between various nation states that do not necessarily have good relations. As has been proved by various extremely demanding technical projects, including the International Space Station, cooperation on such Large Technical Systems stimulates good relations in so-called science diplomacy (Ruffini 2017). For Vaughan Turkenian, science diplomacy is "the use and application of science cooperation to help build bridges and enhance relationships between and amongst societies, with a particular interest in working in areas where there might not be other mechanisms for engagement at an official level."[1]

Our general idea in this book is that the observation of the Solar System can help build bridges from one nation to another, no matter if the decision body is a nation state or a multinational state. We can clearly remember the post-World War II era, known for various science races and crowned by the Space Race to the Moon between the United States and Soviet Union. The period was triggered by Vannevar Bush, who was tasked with developing a policy strategy for postwar science and technology development in the United States by President Eisenhower (Bush 1945). Bush proposed to focus on science and technology for military and civilian purposes. This was seen as a critical moment of opportunity for the United States to become a global hegemon as the winner of the war on whose territory nobody fought. He saw a great chance to switch the war economy for extremely effective

[1] http://cordis.europa.eu/news/rcn/30532_en.html.

science and technology, transforming it into a civilian world power. The United States could thus become the source for modernization, and newly adopted technology was seen as a tool to reshape the societies destroyed by war. This technological determinism was quite common in policy decision-making in this age.

The result of the policy was much more about legitimizing the political systems of the winners. It also contained a clear rationale in the continuation of the war economy towards neoliberal application of invented technologies on the open markets. The neoliberal setting has been important in the defense of public spending, because each governmental research project was, and today still is, questioned regarding its practical applications: the case of applied research, its marketization, and monetization. In the case of national defense projects, the effect of providing national security has to be proved. However, this has become more questioned in the Western world since the establishment of NATO as the major security provider, and with the switch from realist thinking of national security to constructivist thinking of the construction of insecurities. The latter questions reactions on various security issues as unnecessary, because the threat to national security is exaggerated by the social construction of its severity. Technology and war technology was no longer understood to be an unbeatable peace creator in this blunt, technologically deterministic way. This cognitive switch, the core theoretical conflict in nuclear technology—whether such weapons bring peace or conflict—sparked decades-long debates about whether technology has politics (Winner 2003), and divided the deterministic ideas into two branches: the one that understood nuclear weapons as peace creators, and the other that understood them as quite the opposite (Moltz 2011).

Today, science diplomacy only rises in importance with the globalist tempers around the world, despite the recent tendency of some to reinvent nationalism. With growing global challenges from climate change, preservation of the ocean biosphere, and planetary defense, nation states have realized that mutual cooperation is inevitable for the preservation and stimulation of human wellbeing. This 180° turn in foreign policy, in comparison to the post-WWII era, is visible globally, and probably started in times when neoliberal logic of international economic relations led to a massive transfer of various productions to China or other countries with cheaper workforces. The technology for NEO or Solar System observation is a good opportunity to produce knowledge based on the shared science and technology capacities of various nation states, which will help to establish a political basis for further cooperation. Observation is not conflictual; it will probably become more a commercial asset when asteroids become feasible resources of various precious materials.

In the case of planetary defense, one can raise the question of whether such a spinoff of delicate technologies is desirable or not, despite the non-conflictual character of observation technology. In the case of observation of the Solar System and surveying asteroids, these questions seem to be irrelevant, because the technology does not have any interventionist character—it is just a survey. The only national security concern can be linked with secret spy satellites, however, even these are usually observed by less sensitive technologies, and their purpose cannot be deduced from the higher sensitivity of the detectors. By this perspective, observation efforts can form and institutionalize a well-interlinked cooperation and help to develop tight relations based on science diplomacy. This is part of our further policy proposal discussed in detail in Chap. 21, in which we argue that deflection methods cannot be easily discussed without a well-established international network of cooperating institutes on the observation of the Solar System that would stimulate international confidence in the endeavor that planetary scientists and astronomers passionately pursue.

In fact, it would be great to see scientists, astronomers, and engineers of their observation systems as pioneers of a global network that will one day provide crucial information related to our survival. Comets, particularly those from the Oort cloud, are unobservable until they get to the inner Solar System. The orbit of a comet is influenced by other forces, not just gravity. The non-gravitational forces, although orders of magnitude weaker than gravity, originate from sublimation of the nucleus volatiles and ices and push the comet slightly off its purely gravitational orbit. The common awareness of NEOs is therefore also about constant computation of changing orbits and knowledge of physical properties, such as cometary non-gravitational parameters, that are unique for each comet. Long-term prediction of cometary orbits is still a challenge. Some comets are even lost and some are disrupted. Even though impact frequency of a comet with the Earth is orders of magnitude lower than that of an asteroid, those coming from the Oort cloud are only visible when incoming and bring devastating kinetic energy due to large encounter velocities. In addition to posing threats, comets represent a rare source of pristine water and organic compounds that can help us understand our prospects as a species in much longer time frames with achievable Solar System colonization. Whether we can make such a cognitive shift remains a question, but one that is already changing the minds of astronauts who have visited space (see Chap. 18). Colonization can be a tough challenge, but the cognitive shift will certainly help us save the only home we have ever known.

4.4.2 *Coordination of Asteroid Observers*

The NEOCP follow-up is coordinated through the NEOCP blog, currently being transformed to a real-time service. This service should help the observers avoid duplicate observation of the same NEO candidate at the same time, thus optimizing follow-up efforts. Some observations and follow-ups are triggered by elevated interests. For instance, NASA is interested in human-accessible NEO targets, which are pointed out to the community through the small-bodies-observations mailing list by NASA's Paul Abell, and to involved astronomical observatories by Brent Barbee. Another mailing list is managed by JPL's SCOUT, which automatically sends emails about NEO candidates to involved observers. JPL's radar group is significantly increasing the number of observed NEO targets and is also requiring improvement of orbits on demand. David Tholen of the University of Hawaii is a world-leading NEO astrometry observer who actively observes the faintest and most difficult NEO candidates that are out of reach of most of the observers, including the virtual Earth impactors (objects with non-zero impact probability). He utilizes telescopes on Mauna Kea. The Priority list of NEOs that have large orbital uncertainties is provided by ESA's SST-NEO. ESA observes through its own facilities. The program is led by Detlef Koschny and Marco Micheli. Last but not least, the amateur community with a non-negligible fraction of professionals is connected through the Minor Planet mailing list (MPML).

The abovementioned coordination of observers is loose. There is no authority to force individual observers to follow-up an individual object. The observation and characterization of targets are left to the discretion of surveys and observers, who are mostly motivated by new discoveries or whether their name appears on MPEC. Some observers are motivated by science, while some are following urgent NEO cases posted either on mailing lists (MPML, Scout, small-bodies-observations) or on other online lists of objects of interest (human accessible NEO, ESA-SSA-NEO list). Sometimes, interest is triggered for objects that are selected for a space mission. Very fast community-driven follow-up was achieved in the case of the first two impacting asteroids that were discovered <24 h before impact—NEOs 2008 TC3 (Jenniskens et al. 2009) and 2014 AA (Farnocchia et al. 2016). Both were only a few meters in diameter. The leftovers from the first one, in the form of meteorites, were found in Sudan within the calculated strewn field. Meanwhile, 2014 AA most likely fell into the Atlantic Ocean. The only coordination that is expected in the case of a discovery of a potentially hazardous asteroid on a direct impact trajectory would be from PDCO, which is funding most of the NEO surveys and has authority to demand the action of their facilities.

Amateur astronomers are also capable of contributing meaningful data. For example, Bob Holmes made 36,000 observations in 2015 alone, using his four home telescopes with primary mirrors between 0.6 and 1.3 m. Results with such magnitude place him in the hall of fame of amateur astronomers. Given the disorganization in prioritization of NEO follow-up, even passion by amateurs can seriously contribute to observation efforts.

To improve worldwide coordination and involvement, in 2013 the UN committee on Peaceful Uses of Outer Space (COPUOS) oversaw the creation of the International Asteroid Warning Network (IAWN). IAWN is an organization that gathers NEO observers, theoretical orbit modelers, astronomers, scientists, amateur astronomers, and policy makers interested in NEOs. Anybody interested in NEO discovery can be involved, as the policy of IAWN centers on openness, inviting both individuals and large institutions to join. Up to the present, NEODYS and the USA have indicated their roles and capabilities within IAWN, and several others signed the statement of intent (NASA, ESA, ESO, JAXA, INASAN, KASI, INAOE). The IAWN website informs its readers about close approaching NEOs, follow-up, meetings, news from surveys, and links related to NEO search and characterization. IAWN organizes conferences and networking among the policyholders.

Another UN-mandated organization is the Space Missions Planning Advisory Group (SMPAG). Members of SMPAG are currently AEM, ASI, BELSPO, CNES,

DLS, ESA, IAWN, ISA, JAXA, KAXI, NASA, SUPRACO, ROSA, ROSCOSMOS, SSAU, and the UK Space Agency. SMPAG has a similar goal to IAWN, to prepare an international response to a potential NEO impact through access to information, improvement of collaborative research and mission planning, and study of NEO mitigation strategies. SMPAG is directly responsible for advising national agencies to build a space mission for Earth impact mitigation, if such an event is predicted. The SMPAG website is hosted by ESA.

The physical properties of NEOs are being acquired either by individual researchers, or mostly through the Mission Accessible Near-Earth Objects Survey (MANOS), led by Nicholas Moskovitz of Lowell Observatory. MANOS involves more than 20 scientists from many institutions. Characterization includes observations in visible and infrared wavelengths, light curve studies to obtain rotation periods, amplitudes and spin-axis orientations, spectra, compositions, diameters, and albedos (Thirouin et al. 2016). Spectroscopy is also provided by the PDCO-funded IRTF telescope in Hawaii, through the MIT-UH-IRTF Joint Campaign for NEO Spectral Reconnaissance, a project led by Francesca DeMeo and Richard Binzel.

4.4.3 Current Obstacles for International Coordination

International coordination is mostly driven by science, not by planetary defense. In the case of observation of asteroids, comets, and orbit issuance, MPC is constantly improving observation submission, which at present is relatively accessible in the form of emails and online submission (in comparison to historically regular mail or literature publication). One obstacle is the lack of funding dedicated for planetary defense, notably in countries outside of the USA and EU. Other observers usually rely on national grants and funding based on scientific research, driven by scientific results and technological innovations. The national grant agencies and strategies differ widely from country to country, and the topic of planetary defense is still not well understood or widespread around the world. This is not an obstacle for amateur astronomers and citizen scientists, who have access to their own telescopes and do not rely on external funding. Citizen scientists are already contributing to planetary defense efforts (e.g. LCOGT, IASC) and may become an even more important contributor in the near future.

Planetary defense-related research will slowly move to space, which is again governed by established players like national space agencies (NASA, ESA, CNSA, ROSCOSMOS), but may be subsidized by emerging private companies with a capability to launch technology to space (e.g. SpaceX, Orbital ATK, Blue Origin) and apply such technology to planetary defense or commercial use (Deep Space Industries, Planetary Resources). The high cost of space-related technologies, like telescopes or space programs, does not allow many countries to participate in the planetary defense endeavor. Perhaps, commercialization will lower the associated costs and will bring in more stakeholders from emerging economies.

4.5 Conclusion

We began by going over the basics of asteroid observations, from terminology to methodology. The chapter then moved on to discussions of key players and stakeholders in Solar System surveys, including government agencies, public universities, amateur astronomers, and private industry. At present, NEO-related research is dominated by the USA and NASA. While international collaboration concerning observing data and orbits is freely governed by the MPC under the IAU mandate, and informal collaboration was established by IAWN under UN COPUOS, the search for Solar System objects and investigations into impact mitigation continue to be pursued by only a select few nations and programs. Because planetary defense is notably a worldwide problem, inclusion of more stakeholders, particularly from non-participating countries, and increasing awareness among policymakers around the world, should be a priority for the next decade.

Acknowledgements This study was supported by the grant awarded by the Technological Agency of the Czech Republic, project TL01000181: *"A multidisciplinary analysis of planetary defense from asteroids as the key national policy ensuring further flourishing and prosperity of humankind both on Earth and in Space,"* and co-funded by the Institute of Political Studies, Faculty of Social Sciences, Charles University in Prague.

References

Bush, Vannevar. 1945. *Science - The Endless Frontier*. Washington D.C.: National Science Foundation.

Farnocchia, D, S.R. Chesley, and M Micheli. 2015. "Systematic Ranging and Late Warning Asteroid Impacts." *Icarus* 258 (September): 18–27. doi:10.1016/j.icarus.2015.05.032.

Farnocchia, Davide, Steven R. Chesley, Peter G. Brown, and Paul W. Chodas. 2016. "The Trajectory and Atmospheric Impact of Asteroid 2014 AA." *Icarus* 274 (August): 327–33. doi:10.1016/j.icarus.2016.02.056.

Fujiwara, A, J Kawaguchi, D. K. Yeomans, M Abe, T Mukai, T Okada, J Saito, et al. 2006. "The Rubble-Pile Asteroid Itokawa as Observed by Hayabusa." *Science* 312 (5778): 1330–34. doi:10.1126/science.1125841.

Gwyn, S, N Hill, and J Kavelaars. 2012. "SSOS: A Moving Object Image Search Tool for Asteroid Precovery at the CADC." In *Astronomical Data Analysis Software and Systems XXI*, edited by P Ballester, D Egret, and N.P.F. Lorente, 461:789. Astronomical Society of the Pacific Conference Series.

Hanuš, J, F Marchis, and J. Ďurech. 2013. "Sizes of Main-Belt Asteroids by Combining Shape Models and Keck Adaptive Optics Observations." *Icarus* 226 (1): 1045–57. doi:10.1016/j.icarus.2013.07.023.

Jenniskens, P, M. H. Shaddad, D Numan, S Elsir, A. M. Kudoda, M. E. Zolensky, L Le, et al. 2009. "The Impact and Recovery of Asteroid 2008 TC3." *Nature* 458 (7237): 485–88. doi:10.1038/nature07920.

Lindegren, L, U Lammers, U Bastian, J Hernández, S Klioner, D Hobbs, A Bombrun, et al. 2016. "Gaia Data Release 1." *Astronomy & Astrophysics* 595 (November): A4. doi:10.1051/0004-6361/201628714.

Marchis, F., M. Kaasalainen, E. F Y Hom, J. Berthier, J. Enriquez, D. Hestroffer, D. Le Mignant, and I. de Pater. 2006. "Shape, Size and Multiplicity of Main-Belt Asteroids. I. Keck Adaptive Optics Survey." *Icarus* 185 (1): 39–63. doi:10.1016/j.icarus.2006.06.001.

Moltz, James Clay. 2011. "The Dynamics of Space Security." In *The Politics of Space Security*, 11–41. Stanford, California: Stanford University Press.

Ruffini, Pierre-Bruno. 2017. *Science and Diplomacy*. Cham: Springer International Publishing. doi:10.1007/978-3-319-55104-3.

Shao, Michael, Bijan Nemati, Chengxing Zhai, Slava G. Turyshev, Jagmit Sandhu, Gregg Hallinan, and Leon K. Harding. 2014. "Finding Very Small Near-Earth Asteroids Using Synthetic Tracking." *The Astrophysical Journal* 782 (1): 1. doi:10.1088/0004-637X/782/1/1.

Shao, Michael, Slava G. Turyshev, Sara Spangelo, Thomas Werne, and Chengxing Zhai. 2017. "A Constellation of SmallSats with Synthetic Tracking Cameras to Search for 90% of Potentially Hazardous near-Earth Objects." *Astronomy & Astrophysics* 603: A126. doi:10.1051/0004-6361/201629809.

Thirouin, A, N Moskovitz, R. P. Binzel, E Christensen, F. E. Demeo, M. J. Person, D Polishook, et al. 2016. "The Mission Accessible Near-Earth Objects Survey (Manos): First Photometric Results." *The Astronomical Journal* 152 (6): 163. doi:10.3847/0004-6256/152/6/163.

Winner, Langdon. 2003. "Do Artifacts Have Politics?" *Technology and the Future* 109 (1): 148–64. doi:10.2307/20024652.

Online Resources

Asteroid Day: https://asteroidday.org/
Asteroid ZOO: https://www.asteroidzoo.org/
B612: https://b612foundation.org/
CADC archive search: http://www.cadc.hia.nrc.gc.ca/ssos
CNEOS: https://cneos.jpl.nasa.gov
Deep Space Industries: http://deepspaceindustries.com/
ESA: http://neo.ssa.esa.int
Hayabusa2: http://global.jaxa.jp/projects/sat/hayabusa2/
IAWN: http://iawn.net
LCOGT: https://lco.global/
MANOS: https://manosobs.wordpress.com/
MPC: http://www.minorplanetcenter.net/
NASA Planetary Defense Coordination Office: https://www.nasa.gov/planetarydefense/
NEAT SkyMorph archive: https://skyview.gsfc.nasa.gov/skymorph/
NEODYS: http://newton.dm.unipi.it/neodys/
Planetary Resources: http://www.planetaryresources.com/
SLOOH: https://www.slooh.com/
SMPAG: https://www.cosmos.esa.int/web/smpag
Spacedys: http://www.spacedys.com/

Chapter 5
Technical Architecture to Deepen Our Solar System Awareness

Peter Vereš

5.1 Introduction

Ground-based telescopes are our eyes for mapping the Solar System and discovery of Near-Earth Objects (NEOs). Almost all NEOs were discovered by a telescope equipped with a charge-coupled device (CCD) camera. Further developments in the CCD camera, software and image processing, along with an increase in funding, have allowed the construction and operation of sophisticated survey telescopes that are currently leading the NEO discovery statistics. In 2013, the Wide-field Infrared Survey Explorer (WISE) mission was extended and renamed NEOWISE, with the purpose of being a dedicated NEO space telescope. The near future counts on using even larger ground-based survey telescopes (LSST) that will drastically increase the number of discovered Solar System objects. This chapter deals with technical details and description of the pioneer optical surveys, current leading surveys and future plans, including the space-based segment and radar characterization.

5.2 First Generation of NEO Surveys

The first asteroid classified since 2006 as a dwarf planet, Ceres, was originally discovered in 1801. In 1900, only 545 asteroids where known. The first NEO, 433 Eros, was discovered 2 years earlier, in 1898. In 1992, recognition of NEOs as a potential threat to Earth led the US Congress to mandate NASA to discover 90% of NEOs larger than 1 km within the next 10 years. The number of NEOs increased from 649 in 1998 to over 5000 by 2008. As a byproduct, other small bodies of the

P. Vereš (✉)
Harvard-Smithsonian Center for Astrophysics, Cambridge, MA, USA
e-mail: pveres@cfa.harvard.edu

N. Schmidt (ed.), *Planetary Defense*, Space and Society,
https://doi.org/10.1007/978-3-030-01000-3_5

Solar System were discovered as well, ranging from main belt asteroids to Jupiter's Trojans and Centaurs.

The increase of known population was a direct result of the Spaceguard Survey initiative aimed at the discovery and study of NEOs. Additionally, CCDs became a dominant detector in astronomy, replacing the photographic method, increasing the detection threshold and sky coverage area, decreasing required exposure time and allowing use of software for data reduction and registration. CCD identification of asteroids uses a method similar to the classical photographic blinking, when two images of the same star field are taken in two different times, typically within tens of minutes. The asteroid moves with respect to a static star background and can be spotted by blinking of the same fields stacked on the static stars. If more than two images are taken, asteroid motion can be approximated by a straight line within a few hours, except for extreme cases of very close Earth approaches. In such cases, there are still limitations, and it is worth noting that we still do not have perfect equipment or methods how to use them. Detection is not a given human or astronomers' skill, as some may expect.

The first generation of dedicated asteroid surveys was built within Spaceguard Survey, led by David Morrison, after a report by 24 civilian scientists from various countries. Surveys work in a different way when compared to classical astronomical observations. Historically, in astronomy, goals were achieved by observing a single object for a very long time, typically through exposure on a photographic plate for hours to gather as much information as possible. The larger the telescope, the more light was collected. Large telescope mirrors yielded to long focal length of the optical assembly and led to very small fields of view. On the contrary, surveys capture large areas of sky in a short, single-exposure visit and observe many targets at once.

Short exposure time, fast CCD readout, software support and sophisticated scheduling led to rapid increase of discoveries of minor planets. Methods matter. Surveys typically use much smaller telescopes than the best astronomical facilities that are of 8–10-m class apertures nowadays. The first asteroid surveys utilized sub-meter or meter-class wide-field telescopes, like Schmidt, with CCD places in the primary focus. Further improvements in optical design led to construction of long focal length but still wide-field optical systems such as Pan-STARRS (corrected Ritchey-Chretien), ATLAS (custom Wright-Schmidt) or LSST (anastigmatic Paul-Baker/Mersene-Schmidt). The Large Synoptic Survey Telescope (LSST)[1] will be also the first wide-field, all-sky survey telescope of a size (8 m) comparable to the largest telescopes in the world. LSST is planned to be fully operational at the end of 2020.

5.2.1 The Spacewatch

The Spacewatch (McMillan and Spacewatch Team 2006) survey was the first dedicated telescope aimed at NEO discovery. The survey started its operation on a 0.9-m Newtonian telescope located at Kitt Peak in Arizona, managed by University of

[1] https://www.lsst.org/.

Arizona's Lunar and Planetary Laboratory and led by Professor Tom Gehrels and Dr. Robert S. McMillan. Its first observations with CCD started in 1984. Until 1989, it only used a small 320 × 512-pixel CCD camera. In 1989, the 0.9-m's CCD was replaced by 2 k × 2 k camera with better quantum efficiency that increased the limiting magnitude from 19.6 to 20.5. Further upgrades of the camera and software improved the limiting magnitude to 21.8. This camera utilized 24 μm pixels and yielded a 37′ × 33′ field of view.

Spacewatch was initially operating in a so-called drift-scanning mode, when the tracking was turned off and CCD readout was continuous when the shutter was open. Originally, this caused the stars to be trailed. However, drift scan shifted the CCD rows in the direction of star motion in order to stack images on the moving target. The advantages included efficient use of CCD, consistent astrometry within a large field, astrometric solution linear in pixel coordinates and precision unaffected by telescope tracking errors. It is difficult to drift scan far from the equator, as it causes some distortion of PSF, its images must be read and written at the same time and variable sky conditions recorded spatially. Since 2002, the 0.9-m has been using a mosaic of four CCD cameras, each of 4608 × 2048, and smaller 13.5 μm pixels that increased the size of visible field of view to 2.9 square degrees. Also, the drift scan was replaced by the point & shoot method, where the telescope precisely tracks stars and CCD is read out after the shutter is closed.

In 2002, the second telescope was built at Kitt Peak. The 1.8-m was using a 2 k × 2 k CCD with 24 μm pixels until 2011, reaching down to 22nd magnitude. The camera was upgraded to a same-sized CCD with 15 μm pixels. Smaller pixels slightly decreased the field of view from 34′ × 34′ to 21′ × 21′, but higher quantum efficiency improved the limiting magnitude, aiding the discovery of more NEOs. The 0.9-m was primarily used for NEO discovery. The 1.8-m has been utilized mostly for follow-up of newly discovered objects. Both Spacewatch telescopes have discovered almost 1000 NEOs.

5.2.2 *The Near-Earth Asteroid Tracking*

The Near-Earth Asteroid Tracking (NEAT) (Pravdo et al. 1999) survey was operated by the Jet Propulsion Laboratory in cooperation with the US Air Force. The survey was led by Eleanor Helin and David Rabinovitz. It used two telescopes (1.0-m GEODSS and 1.2-m AMOS) on Haleakala, Maui. The field of view was 1.2 × 1.6°. In 2001, NEAT transitioned to use of a 1.2-m Schmidt telescope on Palomar with three CCD cameras, each of 4096 × 4096 pixels. The NEAT leadership was overtaken by Steve Pravdo and Ray Bambery and then by Mike Brown at Caltech. NEAT finished its operations in 2007, discovering 442 total NEOs.

5.2.3 The Lowell Observatory NEO Survey

The Lowell Observatory NEO Survey (LONEOS) (Koehn and Bowell 1999, p. 699) was a survey running between 1993 and 2008, using a 0.6-m Schmidt telescope located on Anderson Mesa, approximately 20 miles south of Flagstaff in Arizona. The survey was run by Ted Bowell and Larry Wasserman. In 2000, the CCD camera was upgraded to a large 16 megapixel camera, yielding a 2.88° × 2.88° field of view. During the best observing conditions, the single image could reach down to 19.8 magnitude. The survey scanned the sky, returning to the same field four times per night, each frame separated by 15–30 min. Each night, LONEOS could cover up to 1000 square degrees. LONEOS has discovered 289 NEOs.

5.2.4 The Lincoln Near-Earth Asteroid Research

The Lincoln Near-Earth Asteroid Research (LINEAR) (Stokes 2000) survey was managed by MIT Lincoln Lab and funded jointly by NASA and the US Air Force. It operated from 1993 until 2013 and was using two prime focus GEODSS 1.0-m Ritchey-Chretien telescopes located at White Sands Missile Range in New Mexico. In addition to mapping of the Solar System small bodies, the system was mainly used for space surveillance—satellite and space debris tracking. Until 1998, the field of view was a circle with a diameter of 1.05°. From 1998 onward, following upgrades, the field of view was increased to a rectangle of 1.23° × 1.61°, and it reached down to 19th magnitude. LINEAR operated differently from typical asteroid surveys using long exposures. LINEAR used high frequency frame rates, acquiring several images per second, decreasing the read-out noise to zero and greatly increasing the nightly sky coverage. LINEAR discovered 2504 NEOs, and it was the most successful NEO survey between 1998 and 2004, when its dominant role was taken over by the Catalina Sky Survey. LINEAR submitted more than 32 million astrometric positions of asteroids and comets, which represents 53% of all asteroid and comet observations ever seen.

5.2.5 The Catalina Sky Survey

The Catalina Sky Survey (CSS) (Larson et al. 2003) is operated by the University of Arizona's Lunar and Planetary Laboratory and is led by Eric Christensen. CSS originally began in 1993 as the photographic Bigelow Sky Survey in the Catalina Mountains in Arizona. It was using a 0.4-m Schmidt on Mt. Bigelow. In 1997, the telescope was equipped with the CCD camera, and in 2004 the telescope was replaced by a larger 0.68-m Schmidt (since then referred to as the Catalina Sky Survey). In 2004, CSS started using a 1.5-m on Mt. Lemmon (Mt. Lemmon Survey) and a 4 k × 4 k CCD camera that increased the field of view to 1.2 square degrees. Until 2013, CSS was also running a 0.5-m Uppsala Schmidt Telescope in Siding Spring, Australia, with the identical CCD camera (Siding Spring Survey). Also, Mt. Bigelow Schmidt was upgraded with 4 k × 4 k CCD camera, increasing the field of view to 8.2 square degrees. In 2017, both Mt. Bigelow and Mt. Lemmon telescopes were upgraded with monolithic 10 k × 10 k CCD cameras, increasing the sky coverage to 19.6 and 5 square degrees.

CSS discovery and follow-up is fast as a result of in-person observers who inspect newly detected objects using their own software almost in real time. The two sites often work in tandem, using a 1.0-m telescope on Mt. Lemmon for their own follow-up. Catalina Sky Survey was the most prolific NEO discovery site in 2005. Mt. Lemmon Survey was leading the NEO discovery statistics from 2006 to 2012, when its productivity was overcome by Pan-STARRS.

Among the most interesting objects discovered by CSS were two small asteroids (2008 TC3, 2014 AA) that were discovered less than a day before they hit the Earth and became the first impacting asteroids ever observed before the impact. In total, Mt. Lemmon Survey discovered 5359 NEOs, which is the most of all surveys, Catalina Sky Survey discovered 2702 NEOs (ranking third in discovery rate), and Siding Spring Survey 464 (ranking 6th).

Because of continuous upgrades of the hardware and software, CSS stands as the second most active NEO discovery machine and can be attributed to the next generation of NEO surveys as well.

5.3 Next Generation of NEO Surveys

The next generation of Solar System surveys has emerged in recent years. Typically, these surveys have multiple science goals, use new technology and produce large amounts of data (Pan-STARRS, ATLAS). Yet, they might not be primarily optimized for NEO search. Even with their current assets, the surveys cannot achieve the 90% completeness goal for NEO population larger than 140 m. The objective is currently met at 30% level. Therefore, several future projects have been proposed, including the ZTF, NEOCAM and LSST.

5.3.1 The Panoramic Survey Telescope and Rapid Response System: Pan-STARRS

5.3.1.1 Purpose, Objectives and Achievements

The Panoramic Survey Telescope and Rapid Response System (Pan-STARRS) (Hodapp et al. 2004) is a next generation all-sky survey telescope, built on Haleakala, Maui and operated by the Institute for Astronomy (IfA), University of Hawaii. The original idea of Pan-STARRS included construction of a prototype (PS1), a second identical telescope (PS2) and finally a single-mount, 4-telescope system PS4. Due to development delays and inflated funding requirements, PS4 was never realized.

The development and 3-year mission was led by Nick Kaiser and funded by a consortium of Universities, research institutes and the US Air Force. It dealt with 12 key science projects involving asteroids, comets, low-mass stars and brown dwarfs, exoplanets, the structure of the Milky Way, M31, massive stars, cosmology through explosive transients, galaxy properties, active galactic nuclei and high redshift quasars, cosmological lensing and large scale structures. This next generation survey had a goal of operating, processing and providing data in automated mode.

Pan-STARRS has observed almost 90% of all known asteroids and discovered more than 4000 NEOs. Since 2014, Pan-STARRS has been the most productive NEO discovery site, exceeding the Mt. Lemmon site of the Catalina Sky Survey.

Because of the broad ranges in scientific needs and interests, NEO and Solar System-dedicated activities initially used only 5% of the available time in late 2010 when the survey officially started. After November 2012, the Solar System was increased to 11% of the time. At the end of the consortium and its main mission in 2014, the funding source changed almost exclusively to NASA NEOO grant. Since then, Pan-STARRS utilizes almost 100% of time for NEO survey. The new Principal Investigator of Pan-STARRS NEO Survey is Richard Wainscoat. In 2017, the 3-year mission Pan-STARRS data became available to the public through the Space Telescope Science Institute.

Pan-STARRS has access to several follow-up facilities. For about two weeks a month around the new Moon, they use the 3.5-m Canada-France-Hawaii Telescope (CFHT) on Mauna Kea, Hawaii and its wide-field MegaCam 1-square degree CCD composite camera, made of 40 chips with 378 million pixels. Also, Dave Tholen's group at the Institute of Astronomy, University of Hawaii follows up Pan-STARRS discovery through the University of Hawaii's 2.2-m telescope on Mauna Kea.

5.3.1.2 Technical Solution

Pan-STARRS telescope (Fig. 5.1) is a 1.8-m wide-field Ritchey-Chretien with a 3-square degree field of view and the largest composite CCD camera in the world at the time of construction, having 1.4 billion 10-μm pixels, providing resolution of 0.25″/pixel. Because of the mosaic nature of the camera and low quality of some of the chips, almost 25% of the focal plane is either not populated by pixels or the pixels are not suitable for science exposures. Due to the US Air Force restrictions, during the first year of the telescope's operation, the missing area was even larger, because all satellite streaks were masked from images with a significant margin. Pan-STARRS observes in five narrow-band filters (g, r, i, z, y) and in a wide-band w. The deepest single-exposure images are achieved in w-band (22nd magnitude in 40 s).

The Pan-STARRS image processing pipeline (IPP) (Magnier et al. 2006) is an automated and sophisticated software that deals with large-scale CCD images. The

Fig. 5.1 Pan-STARRS1 telescope. Credit: Photo by Rob Ratkowski and the Institute for Astronomy, University of Hawaii at Manoa

CCD image in IPP is calibrated, rotated and stretched to compensate for small irregularities of the focal plane and optics, and the sky background and noise are subtracted. Subsequently, two CCD images of the same field obtained in a short time interval are subtracted from each other, resulting in a difference image, which does not have any static sources, such as stars or galaxies. Only moving targets, such as asteroids and comets, or static variable sources, such as supernovae or variable stars, are visible in difference images.

Difference images are used for Solar System objects detection. Because two frames are subtracted as first minus the second and second minus the first, each difference frame contains positive and negative images of a moving target. Moving targets identified by IPP are submitted to the Moving Object Processing Pipeline (MOPS) (Denneau et al. 2013). MOPS is an automated pipeline that was originally designed to link detections in a single field into tracklets, link tracklets of the same object to the inter-night tracks and link tracks into orbits. MOPS has a component that also identifies known asteroids with the tracklets, based on velocity and position information. Even though the MOPS pipeline was intensely tested, due to the unexpectedly high volume of false detections resulting from the optical design of the camera as well as IPP flaws, MOPS was not able to create tracks. In addition, the observing cadence was not ideal for inter-night linking and differed from the proposed survey cadence. Thus, instead of automated processing and computing of orbits, human screening of tracklets was needed, and visually inspected tracklets were manually submitted to MPC (see Chap. 4). It takes a few hours from the detection of the object, automated data processing and manual inspection for the object to be submitted to NEOCP. MOPS is currently sensitive for detecting object moving in a velocity range of 0.05–10° per day. In a typical 40-s exposure in the w-band (optical wavelength), Pan-STARRS routinely detects 22-magnitude asteroids.

In 2014, identical Pan-STARRS2 (PS2) was built next to the dome of PS1 and was in operation for a few months. After several years of camera upgrades, PS2 has restarted its operation in mid-2018, joining PS1 with a similar survey cadence.

5.3.2 The Large Synoptic Survey Telescope (LSST)

5.3.2.1 Purpose, Objectives and Achievements

The Large Synoptic Survey Telescope (LSST) (Jones et al. 2015) is the next step of automated surveys. Currently being built on Cerro Pachon in Chile, it is funded by the National Science Foundation, the Department of Energy and private donors. Most of the funding comes from US sources, but the LSST Corporation is welcoming new members, including international contributors. Its main scientific goals are to study Dark Matter and Dark Energy, catalog the Solar System and explore the transients and structure of the Milky Way. Its first light is expected early in 2020, and its 10-year mission will start in 2022. Because of the unprecedented quantities of data and LSST's real-time alerting requirements, the detection, processing and data handling has to be fully automated. Even though NEOs are not the main target of LSST, the proposed LSST cadence will allow it to use the vast majority of its survey data for NEO searches. Because LSST will greatly increase the limiting magnitude, the newly discovered NEOs will be much fainter and more out of reach than existing follow-up capabilities. LSST will utilize its own image processing pipeline and archiving, and for Solar System objects, an improved version of MOPS. The studies have shown that even with the false detection rate, which is expected to be much lower than that of Pan-STARRS, the linking (automated connection of newly discovered objects between nights) would work (Vereš and Chesley 2017). Instead of nightly NEO candidates, LSST will submit its own derived orbits to MPC with the matching detections. Over ten years, LSST is expected to discover more than 100,000 NEOs (the current count is 16,000) and more than 5.5 million other Solar System objects (currently 800,000).

5.3.2.2 Technical Solution

The LSST wide-field optical and near-infrared 3-mirror anastigmatic reflector has a primary mirror of 8.4-m diameter, overall focal length of 10 m and will have the largest CCD camera in the world—a composite 3.2 gigapixel CCD mosaic made of 10 μm pixels (Fig. 5.2). Due to the multiple chips mounted on rafts, the existing gaps will make it so that 10% of the focal plane area will be not visible, a large

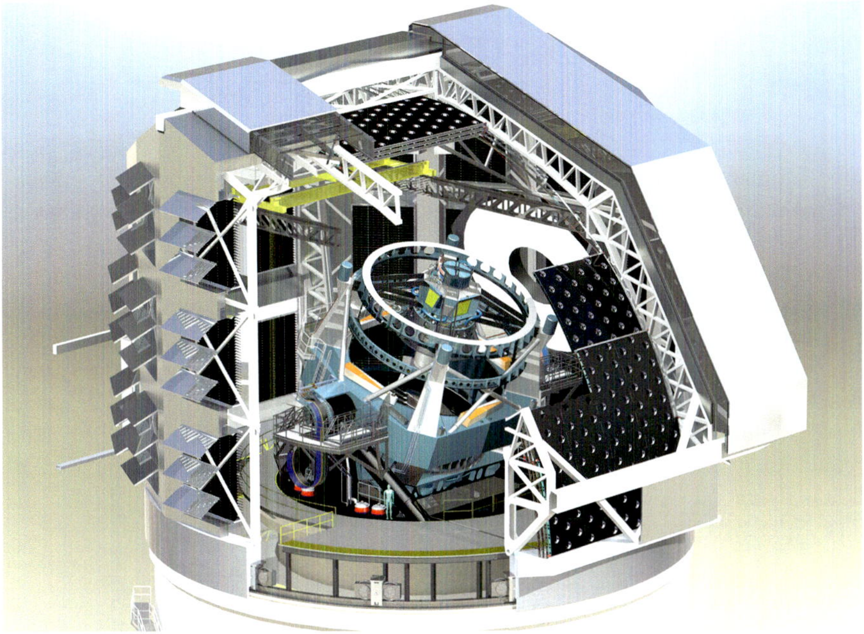

Fig. 5.2 Artist's rendering of the LSST telescope and the cross section of its dome. Source and credit LSST Website Gallery: https://gallery.lsst.org

improvement compared to Pan-STARRS1's 25%. The LSST camera will use six SDSS-like filters, and in a single 30-s exposure will routinely achieve 24.5 magnitude in r-band. In best conditions and in r and g-band, LSST can reach below 25th magnitude. With a 9.6 square degree field of view, LSST will cover around 6000 square degrees per night. The approved budget for building LSST and running its 10-year mission is about $1 billion.

The expected volume of data is on a level of ten million time-domain events per night, and a final catalog containing around 37 billion objects, seven trillion observations, 30 trillion measurements. This will require 100 teraflop of computing power and 200 petabyte of storage, along with real-time alerts requirements. LSST must work in a fully automated routine for identification, linking and orbit determination of Solar System objects. Because of the high demand of computing and data transfer rate, most of its computing services will be distributed and partially done in the cloud. LSST development will steer the future of computing and will be closely related with information technology. Many components, either hardware or software, do not exist at present. During LSST planning, the hardware and software specifications have often exceeded current off-the-shelf products and have had to be invented. It is expected that new technology and spinoffs derived from LSST will be used by the public.

5.3.3 *The Palomar Transient Factory (PTF) and Zwicky Transient Facility (ZTF)*

5.3.3.1 Purpose, Objectives and Achievements

The Palomar Transient Factory (PTF) (Law et al. 2009, p. 141) and its pipeline-upgraded Intermediate Palomar Transient Factory (iPTF) started its operation in 2008. The survey is run by a consortium (Caltech-IPAC, LBNL, IPAC, Berkeley, LCOGT, Oxford, Columbia, Weizmann Institute), and its Principal Investigator is Shrinivas Kulkarni. The scientific goals are diverse, including supernovae, novae, cataclysmic variables, compact binaries, AGN, transiting exoplanets, microlensing, asteroids and comets. The consortium has been working on upgrading the survey, denoted as the Zwicky Transient Facility (ZTF) (Smith et al. 2014), which had its first light in 2017. While PTF discovered only 20 NEOs, ZTF's sky coverage would provide an excellent monitoring of known NEO orbits as well as detection of small NEOs flying very close around Earth or on a direct impact trajectory. So far, the ZTF NEO discovery count is 11.

5.3.3.2 Technical Solution

PTF is using a 1.2-m Samuel Oschin Schmidt Telescope at Palomar Observatory in California. Its camera uses an array of 12 rectangular CCDs of 12 k × 8 k 15 μm pixels, yielding a field of 7.8 square degrees. In a standard 60-s exposure and excellent observing conditions, PTF can detect objects down to 20.5–21 magnitude in r and g-band. PTF has its own 1.5-m automated follow-up telescope on the Palomar Mountain. Image subtraction and object identification are done by automated pipeline in real time, and transient detections issue alerts to the community. The new array of 16 chips made of 6 k × 6x pixels and with 1 arcsecond/pixel resolution will replaced the current CCD camera on the 1.2-m telescope. ZTF will reach 20.5–21 magnitude in r and g-bands in 30 s and cover 47 square degrees in a single exposure. In 1 h, ZTF will cover 3750 square degrees. When built, it will have the world largest field of view for a telescope larger than 0.5 m, equal to the area of 235 full Moons.

5.3.4 *The Asteroid Terrestrial-Impact Last Alert System (ATLAS)*

5.3.4.1 Purpose, Objectives and Achievements

The Asteroid Terrestrial-Impact Last Alert System (ATLAS) is a wide-field, small-telescope class survey funded by NASA NEOO grant and managed by John Tonry at the Institute for Astronomy (IfA), University of Hawaii at Manoa. ATLAS was built as an early warning system for detecting smaller NEOs on a direct impact trajectory. Its data is used for the detection of supernovae, distant transient events, habitable exoplanets, gravitational lensing, variable stars, active galactic nuclei, space debris and mini-moons—small asteroids temporarily captured by Earth. ATLAS could provide a week's warning time for a 50-m impactor on direct collision course with the Earth and a month for a 140-m impactor. To cover the entire Earth, ATLAS has plans to build identical units in Australia, Chile, the Canary Islands and South Africa. ATLAS has discovered 150 NEOs so far, all very bright (17–19 magnitude) and in locations not usually covered by large surveys, e.g. far from the ecliptic. The project is relatively low cost, with the aim of providing off-the-shelf ATLAS to customers for $2 million. Unfortunately, ATLAS will not see NEOs such as Chelyabinsk, because it has the same constraint as other ground-based telescopes—it only observes at night and does not see objects coming from the direction of the Sun.

5.3.4.2 Technical Solution

ATLAS uses a 0.5-m DFM custom Wright Schmidt telescopes. ATLAS is currently installed on Haleakala, its first element deployed in 2015 and the second built in 2017 at Mauna Loa. In a single 30-s exposure, it achieves a field of view of 30 square degrees with its 10.5 k × 10.5 k monolithic CCD camera. Because the camera is not a composite of multiple CCDs, there is practically no loss of active area. The telescope uses two filters for astrophysical science (cyan and orange). Even though predicted limiting magnitude was near 20, the defects in delivered correction plates lowered the limiting magnitude to 18–19. The new optical elements have been ordered and will improve the limiting magnitude. Even though ATLAS uses a small telescope in comparison to Pan-STARRS and Mt. Lemmon, its large field of view and sky coverage help the discovery of nearby small and fast-moving NEOs that are missed by large surveys.

5.3.5 Space Surveillance Telescope (SST)

5.3.5.1 Purpose, Objectives and Achievements

The Space Surveillance Telescope (SST) is primarily operated by the Defense Advanced Research Projects Agency (DARPA) and MIT's Lincoln Laboratory. It was situated on Atom Peak of White Sands Missile Range in New Mexico, but currently, it is being moved to Western Australia. Its aim has been to observe and track artificial satellites and space debris since late 2011, and it is a direct successor of LINEAR. Even though asteroid observation is a side project of SST, it historically ranks sixth in number of astrometric observations of all asteroids ever observed and since 2015 is the most productive astrometric survey of asteroids. However, it does not have its own follow-up network, and the NEO pipeline is rather slow. Thus, the number of discovered NEOs is only 141, which ranks SST as 12th in NEO discoveries. If the telescope improves the pipeline and can dedicate a larger fraction of its time for NEO survey, it may become the most successful discovery machine.

5.3.5.2 Technical Solution

SST uses a 3.5-m Marsene-Schmidt telescope with a very short focal length and large field of view of about six square degrees. SST's specialty is an array of 12 CCDs, each of 2 k × 4x and 15 μm pixels, fabricated to the curved Schmidt focal plane. The alt-azimuthal mount and compact design of SST as well as its primary purpose makes the pointing and positioning of the telescope very rapid. While observing space debris, SST's pipeline detects asteroids and submits them to MPC.

5.3.6 The Dark Energy Survey (DES) and Its Dark Energy Camera (DECam)

5.3.6.1 Purpose, Objectives and Achievements

The Dark Energy Survey (DES), along with its Dark Energy Camera (DECam), is an optical and near-infrared survey studying the expansion of the Universe and its large structure. It observes supernovae, weak lensing, galactic clusters and dwarf galaxies. DES is run by a consortium of institutions. Because of its location in Chile, DECam covers only the Southern Hemisphere. A small fraction of DECam time is also dedicated for DECam NEO survey, led by Valdes, Herrera, James and Rajagopal. So far, DECam has discovered 300 NEOs, however, many of the faintest NEO candidates have been lost due to a lack of follow-up. With the limiting magnitude, large field of view and 4-m aperture, DECam could theoretically be the most productive NEO survey, if more time is assigned to NEO survey.

5.3.6.2 Technical Solution

DECam is mounted at the prime focus of its 4-m telescope at Cerro Tololo Inter-American Observatory in Chile. DECam is made of an array of 62 CCD cameras, yielding a total of 520 megapixel and sky coverage of 3 square degrees with 0.263 arcsecond/pixel resolution. About 30 nights over three semesters are used to cover 320 square degrees in 40-s exposures in VR filter, with five exposures per field with 5-min cadence and repeated on the second and third night. The self-follow-up is therefore crucial, because DECam can routinely detect NEOs of 23rd and 24th magnitudes that are out of reach of the existing NEO follow-up capabilities. Without its self-follow-up, its NEO candidates are lost. DECam uses MODS, a MOPS derivative, to link detections into tracklets.

5.4 Space Telescopes

5.4.1 *Wide-Field Infrared Survey Explorer (WISE)*

5.4.1.1 Purpose, Objectives and Achievements

The Wide-field Infrared Survey Explorer (WISE) (Wright et al. 2010) is a NASA space telescope, operating since 2009. Its scientific goals are diverse, from mapping the sky in infrared, the study of brown dwarfs and cool stars, young stellar clusters and dust, as well as objects in the Solar System. The main mission, operating in four infrared channels, ended in late 2010, when the hydrogen coolant got depleted. Operation was still possible in two bands without the coolant. At the end of cooled mission, one month of observing time was assigned to project NEOWISE, which was dedicated to NEO survey. Due to its success, the mission was extended by three months. In 2011, the spacecraft was put into hibernation. Impressive discovery results, especially of NEOs, led to the re-commissioning of WISE and funding through NASA NEOO grant. NEOWISE restarted in two bands in 2013 and is ongoing. It has discovered more than 30,000 asteroids, observed 200,000 known asteroids and discovered comets as well. The PI of NEOWISE is Amy Mainzer at the Jet Propulsion Laboratory.

NEOWISE has discovered 266 NEOs and ranks 10th among all NEO surveys. Because NEOWISE is observing from Earth's orbit, it can observe asteroids relatively close to the Sun that are not accessible from ground-based telescopes. This means that some NEO candidates are not accessible by follow-up telescopes. Also, the magnitudes reported to MPC are approximate and mostly fainter than they really are, due to the conversion from infrared to visible wavelength. Overestimated magnitude has caused follow-up problems, when ground-based observers do not find the NEO candidates because they are fainter than their telescope limits. One of the controversial results of NEOWISE is that there are actually fewer small NEOs than previously expected (Mainzer et al. 2011), even though data from optical telescopes do not confirm these findings.

5.4.1.2 Technical Description

WISE orbits the Earth on Sun-synchronous polar orbit above the terminator (day/night boundary), faces opposite the Earth and observes narrow stripes of sky. Its 40-cm infrared telescope operated in four wavelengths for about two years, then the cryogenic coolant was depleted. The telescope has been working in only two wavelengths since then. The field of view is 47 arcmin and images are taken every 11 s. Because NEOWISE's orientation is fixed with respect to the direction of the Sun, its field of view depends on the orientation of the spacecraft. NEOWISE looks much closer to the Sun than ground-based surveys. For asteroid and comet processing, NEOWISE uses a MOPS-derivate pipeline (WMOPS) that creates tracklets that are

Fig. 5.3 Artist rendering of WISE. Credit: NASA/JPL-Caltech

manually previewed by observers and submitted to MPC. The field is observed nine times per night, and if there are at least five detections in a tracklet with SNR > 7, not saturated or non-moving, the tracklet is ready for manual preview. The maximum rate of motion of NEOs detected by NEOWISE is 3.32 degree/day (Fig. 5.3).

5.4.2 The Near-Earth Object Camera (NEOCAM)

5.4.2.1 Purpose, Objectives and Achievements

The Near-Earth Object Camera (NEOCAM) (Nugent et al. 2015) is a proposed space infrared telescope, developed on experience learned from WISE. The Principal Investigator is Amy Mainzer. NEOCAM would be a primarily NEO impact-monitoring project, its aim being the discovery and characterization of NEOs and the search of NEOs suitable for automated and crewed spaceflights. It is projected that in a proposed 4-year mission, NEOCAM will discover two-thirds of NEOs

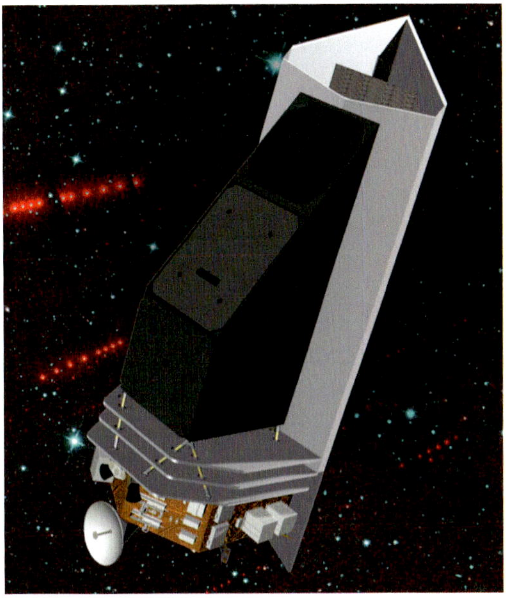

Fig. 5.4 Artist rendering of NEOCAM. Credit NASA/JPL Caltech. Source: https://neocam.ipac.caltech.edu/

larger than 140 m. The proposal was submitted in 2006, selected for technology development to work on detectors by NASA in 2010, and selected among four others for phase two development for NASA's Discovery program in 2015. However, in 2017, NEOCAM was not selected by NASA as a mission, although funding for one more year has been provided. Thus, there is still a slight chance that the project will be funded in the near future, based on the continuous funding flowing to support NEOCAM studies.

5.4.2.2 Technical Description

NEOCAM will have a 50-cm telescope, operating in two infrared wavelengths, without a need of cryogenic coolant. Instead of low-Earth orbit, NEOCAM will be in the Sun-Earth L1 Lagrange point between Sun and Earth, which will allow the telescope to observe asteroids closer to the Sun and the population of interior Earth orbit asteroids (Fig. 5.4).

5.4.3 Sentinel Space Telescope

5.4.3.1 Purpose, Objectives and Achievements

The Sentinel Space Telescope was a proposed space-based infrared NEO mission, designed by the B612 Foundation. Sentinel's planned budget was $450 million in 2013, and its launch was planned for 2018 by SpaceX. B612 intended to secure the

Fig. 5.5 Sentinel telescope render. Credit: B612, CC BY-SA 4.0

funding through private donations. Based on predicted NEO discovery and characterization rates, Sentinel could significantly increase the number of discovered NEOs. However, in September 2015, NASA cancelled the Space Act Agreement with B612, as it did not meet several NASA deadlines, and the future of the mission was withdrawn. Sentinel was also perceived as a direct competition for the infrared NEOCAM, even though the funding did not overlap.

5.4.3.2 Technical Description

The Sentinel's orbit could have been similar to those of Venus, thereby observing NEOs not visible from Earth. Sentinel's primary mirror was to be 0.5-m and could operate in two infrared wavelengths. Its camera required a cryogenic cooling down to 40 K. Sentinel could have a large field of view of $5.5 \times 2°$, and the camera might have been a composite of 16 detectors (Fig. 5.5).

5.4.4 *Other Space Telescopes*

Near-Earth Object Surveillance Satellite (NEOSSAT) was a small Canadian satellite with a 15-cm telescope, funded by the Canadian Space Agency and Defence Research and Development Canada. Launched in 2013, it is located on a Sun-synchronous polar orbit around the Earth, searching narrow regions 45–49° from the Sun. Among observation and discovery of NEOs brighter than 19th magnitude, especially those close to the Sun, NEOSSAT was to track and observe artificial satellites. The primary mission was planned for one year. The satellite was unsuccessful due to software and attitude control issues.

Other space-based telescopes have observed NEOs, mostly incidentally. For instance, the ESA's astrometric missions Hipparcos and currently Gaia had capabilities of observing NEOs. Hipparcos had a 29-cm Schmidt telescope, providing 54 × 94 arcmin field of view and detecting objects to 11th magnitude. Hipparcos was launched in 1989 to a highly eccentric geocentric orbit. The mission ended in 1994. The proper motions and high-precision astrometry from Hipparcos have been used for astrometric measurements of asteroids. Gaia, launched in 2013 to the Sun-Earth Lagrangean point L2, has a NEO pipeline ready and also a follow-up, on-demand network of ground-based observatories (Gaia-FUN-SSO). Gaia has two 1.45 × 0.5-m telescopes that project the light to the focal plane to 106 CCDs, each measuring 4500 × 1966 pixels. Gaia aims at mapping one billion astronomical objects like stars, galaxies, quasars, asteroids and comets, providing proper motions, parallax and distance of stars as well as their astrometric measurements, with a precision an order of magnitude better than before.

In 2016, Gaia published the first data release of its astrometric catalog. Better astrometric cataloging will directly improve the measured positions of asteroids in the future, providing more accurate reference frames for the relative astrometry of Solar System objects. For instance, LSST astrometric quality requirements are expecting post-Gaia precision of star positions. Gaia has not discovered any NEOs yet, but it is expected that as of 2017, the pipeline and network could discover some. The limiting magnitude of Gaia is about 20th magnitude. Another telescope providing incidental discoveries and observations of small Solar System objects is ESA's Solar and Heliospheric Observatory (SOHO), observing the Sun in Earth-Sun Lagrangian L1 point. SOHO discovered more than 3000 comets, most of them colliding with the Sun. Other space-based solar telescopes, the Solar Terrestrial Relations Observatory (STEREO) and Solar Dynamic Observatory (SDO) have been observing sun-grazing comets.

5.5 Radio Telescopes

Radio telescopes working in the radar mode (transmitting and receiving radio waves) are used for observations of known asteroids. The Doppler-delay method is used to determine the instantaneous distance of the object from the observer and

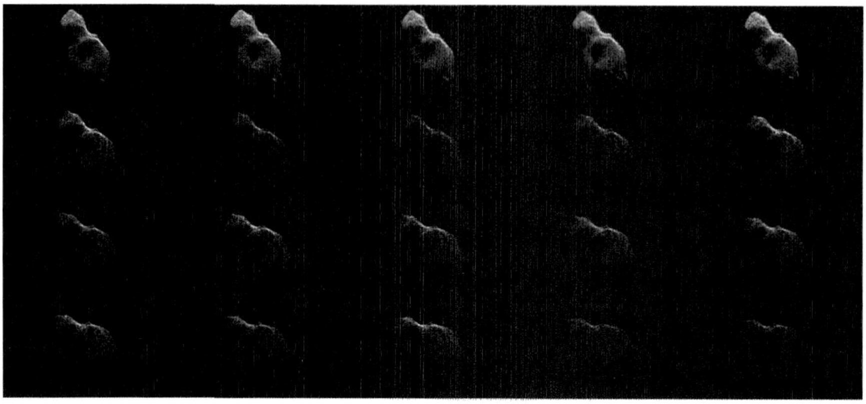

Fig. 5.6 Arecibo and Goldstone radar observations of NEO 2014 HQ124, an irregular asteroid only 370 m across during its close flyby of the Earth on June 8, 2014. Shape, surface features and rotation are apparent. Credit: Marina Brozović and Joseph Jao, Jet Propulsion Laboratory/Caltech/NASA/USRA/Arecibo Observatory/NSF

therefore can highly improve orbit tracking. In comparison to optical and infrared observations, where the accuracy of asteroid's position depends on the detector's resolution, radars can measure the time delay in fractions of a second. Radar observation is often used to resolve and improve the orbit of a potentially hazardous NEO. If objects are close enough to the Earth, typically within several lunar distances, shapes and spin properties can be directly reconstructed from radar data. Radar data has even discovered moons around asteroids and revealed astonishing and distinct shapes of NEOs (Fig. 5.6). Even though radar is a great tool to deliver the physical shape of an NEO and greatly improve its orbit, it does not serve as a useful tool for NEO discovery.

The pioneer of the radar astronomy of asteroid research was Steven Ostro at Jet Propulsion Laboratory, who developed imaging and shape reconstructions of asteroids. Currently, the program at JPL is led by Lance Benner, who utilizes radio telescopes in Arecibo, Puerto Rico and in Goldstone, California. The Goldstone radar is located in the Mojave Desert and is a part of multiple radar dishes that are used within the JPL Deep Space Communication Complex, a communication facility for satellites and space missions. Goldstone radar uses a 70-m moveable dish that is able to track objects in the sky. On the contrary, Arecibo radar, operated by a consortium, is built in a natural depression and measures 305 m across, thus it cannot move. The reflectors near the receiver allow it to cover a 40-degree cone around the zenith, and only objects inside of the field of view can be observed. Until 2016, this was the largest radio telescope in the world, surpassed by a newly built 500-m radio telescope in China. However, the Chinese radio telescope does not have radar capability. Due to funding from NASA and increased interest in NEOs, the number of observed and characterized NEOs is growing rapidly, and known or newly discovered NEOs that are in the reach of Goldstone and Arecibo are regularly scheduled for observation.

5.6 Current Limits

The current NEO discovery limits can be divided into several subcategories. The main restriction is the limiting magnitude of the ground-based telescopes and observational biases, including the weather and location of the observer on Earth. Current dedicated survey telescopes have a limiting magnitude near +22. Deeper limiting magnitude means more discovered NEOs, particularly those smaller and distant, which can be achieved by larger telescopes. In the near future, LSST will increase its limiting magnitude to almost +25 and could therefore provide an order of magnitude more observations and discoveries of NEOs, even though it is not an NEO-dedicated mission. Although there are currently 13 telescopes with a primary mirror larger than eight meters, these are not meant for surveying large areas of the sky, have typically very small fields of view and very specific instruments like spectrographs, so they can rarely be used for asteroid science. Two large telescopes are planned for the next decade: the 30 m Telescope (Mauna Kea, Hawaii) and the European Extremely Large Telescope with a 40-meter primary mirror (Chile), which will change the course of future astronomy. Nevertheless, the two giant telescopes are not expected to contribute to planetary defense or NEO discovery.

The distribution of NEO dedicated telescopes around the world is not uniform. The lack of coverage is obvious in Africa, continental Asia, Australia and southern Oceania. At least on the level of small telescopes, ATLAS has a plan to install its 50-cm telescopes around the world to provide early warning for small Earth impactors at a scale of days to weeks.

Earth itself is a barrier in surveying NEOs. There is always a factor of weather, clouds, duration of night and the phase of the Moon. At any given time, more than half of the celestial sphere is not observable, because half of the Earth is facing the Sun on a dayside. This is particularly a problem for asteroids with orbits mostly or entirely inside the Earth's orbit (IEO). Objects on IEO or Aten orbits could hit the Earth from the dayside without being noticed, similarly to the Chelyabinsk asteroid in 2013. Even though theories do not expect the population of IEO objects to be large, due to observation constraints, only 16 IEO have been discovered. The solution to discovering objects on dubious orbits and mitigating weather conditions is a space telescope, not bound to the Earth and orbiting the Sun closer than Earth, similar to the Sentinel's concept. The disadvantage of such a device is obviously a much higher cost, short life expectancy, difficulty of maintenance, data transfer and analysis. New technology and the growing involvement of private industry could lower the cost and allow the development and deployment of such devices.

An extraordinary class of NEO is a comet. There are 3726 comets or cometary fragments that fulfill the NEO orbit criterion, and 12 comets have PHA-like orbits. Most of these comets are Jupiter Family Comets, with short orbital periods that can be periodically monitored. The rest, however, have long periods (hundreds and thousands of years) and have highly eccentric orbits with large inclinations. Comets are easily distinguished in an image due to their subliming ices and dust, which leave a coma or a tail behind. However, their orbits evolve rapidly, either due to gravitational interactions with Jupiter and/or non-gravitational forces, like jets of the ejecting gas and dust.

Even though the impact risk for Earth from a comet is much lower than that of an asteroid, the consequences could be devastating. The high eccentricity of a cometary orbit increases the comet's kinetic energy, and if such a comet has a very long period, it will be discovered only on a direct impact course, giving humanity only a few years of warning time and almost no means of deflection. Also, the discovery of comets is rather incidental—most of the comets are found among the NEO survey data, and a few are found by amateur astronomers looking close to the Sun. The comets are only active in the inner Solar System, typically closer than Jupiter. If they are further away on their orbit, they are nearly invisible, as comets have very dark surfaces. One of the methods of predicting a distant threat from an unknown comet is to monitor the activity of meteor showers without a known parent body, possibly pointing toward a long-periodic comet crossing the Earth's orbit that has yet to be discovered.

The automation of the discovery process and orbit determination has been in use for some time. Still, the NEO discovery process consists of submitting individual one-night tracklets to MPC and further follow-up observations by either dedicated facilities or available volunteers. Problems may arise with larger telescopes stepping into business, producing too many NEO candidates while follow-up capabilities remain limited. The number of lost NEO candidates was very large when DECAM reported its NEO candidates, as there were no follow-up facilities available that could reach such faint objects. Nevertheless, almost ten years ago, the software for automatic linking of NEO and other Solar System objects (MOPS) was developed, initially for Pan-STARRS, with the aim of helping LSST and others. If MOPS was used in its originally intended configuration, the telescopes would gather data nightly, no follow-up would be needed and objects would be regularly linked to orbits. LSST and other future projects are relying on improvements, both in software and hardware, to make a MOPS-like pipeline work automatically. The future will most likely see the decline of routine manual follow-up, even with much larger numbers of NEOs being discovered. The role of the existing telescopes doing the follow-up will shift to the characterization and detailed studies of individual objects.

Funding and national competence are major limiting factors in NEO and Solar System mapping. NASA is a single federal agency that covers the vast majority of discovery and characterization efforts, followed by ESA. Other countries have rather limited access to funding or resources needed for NEO searches and characterization. Apart from NASA funding, which is dedicated to NEO search and planetary defense, funding applicants from other countries can rely on research grants. However, the call for research grants is either very specific or research-oriented, and discovery itself does not justify these funding sources. Researchers from other countries do not have access to NASA or ESA funds. Beyond individual NEO follow-up by observers and discoverers, cooperation on the political level is lacking. Though COPUOS created IAWN, it lacks cohesiveness, funding and authority.

The global coordination of NEO discovery within planetary defense might need global funding, governance and clearly defined procedures and communication. Thus, researchers and parties from other countries, including citizen scientists, could contribute to the global effort and help in areas that are not covered primarily by NASA or ESA.

5.7 Conclusion

The constantly improving observational techniques, growing number of observing sites and prospects of future space-based telescopes looking for asteroids suggest a fruitful future for NEO discoveries and better characterization of objects threatening the Earth. Better mapping for the Solar System will grant humanity a longer warning time if there is indeed a direct threat from an asteroid or comet. However, there are still some technological limits that prevent us from efficiently mapping particular objects. The smallest NEOs are still discovered only incidentally, and ground-based telescopes cannot not reach asteroids with orbits mostly inside the Earth's orbit. Discovery and follow-up are slowly migrating from manual efforts and individual observers toward an automated procedure, where a large fraction of sky is covered and automatically processed, identifying known objects and linking fresh discoveries into orbits. Still, the discovery and follow-up process, including impact hazard monitoring and mitigation, are mostly concentrated in the United States, leaving space for international collaboration and admission of potential players from countries not yet involved in the global monitoring of the Solar System inventory, including government agencies and private industry.

References

Denneau, Larry, Robert Jedicke, Tommy Grav, Mikael Granvik, Jeremy Kubica, Andrea Milani, Peter Vereš, et al. 2013. "The Pan-STARRS Moving Object Processing System." *Publications of the Astronomical Society of the Pacific* 125 (926): 357–95. doi:https://doi.org/10.1086/670337.

Hodapp, Klaus W., Walter A. Siegmund, Nicholas Kaiser, Kenneth C. Chambers, Uwe Laux, Jeff Morgan, and Ed Mannery. 2004. "Optical Design of the Pan-STARRS Telescopes." In *Ground-Based Telescopes*, edited by Jacobus M. Oschmann, Jr., 5489:667. \procspie. doi:https://doi.org/10.1117/12.550179.

Jones, R. Lynne, Mario Jurić, and Željko Ivezić. 2015. "Asteroid Discovery and Characterization with the Large Synoptic Survey Telescope." Edited by S.~R. Chesley, A Morbidelli, R Jedicke, and D Farnocchia. *Proceedings of the International Astronomical Union*, IAU Symposium, 10 (S318): 282–92. doi:https://doi.org/10.1017/S1743921315008510.

Koehn, B. W., and E Bowell. 1999. "Enhancing the Lowell Observatory Near-Earth-Object Search." In *Bulletin of the American Astronomical Society*, 31:1091. \baas.

Larson, S, E Beshore, R Hill, E Christensen, D McLean, S Kolar, R McNaught, and G Garradd. 2003. "The CSS and SSS NEO Surveys." In *AAS/Division for Planetary Sciences Meeting, Session 36. Asteroid Dynamics II*, 35:982. Bulletin of the American Astronomical Society.

Law, Nicholas M., Shrinivas R. Kulkarni, Richard G. Dekany, Eran O. Ofek, Robert M. Quimby, Peter E. Nugent, Jason Surace, et al. 2009. "The Palomar Transient Factory: System Overview, Performance, and First Results." *Publications of the Astronomical Society of the Pacific* 121 (886): 1395–1408. doi:https://doi.org/10.1086/648598.

Magnier, E, Nick Kaiser, and Ken Chambers. 2006. "The Pan-STARRS PS1 Image Processing Pipeline." In *The Advanced Maui Optical and Space Surveillance Technologies Conference*, E50.

Mainzer, A, T Grav, J Bauer, J Masiero, R. S. McMillan, R. M. Cutri, R Walker, et al. 2011. "Neowise Observations Of Near-Earth Objects: Preliminary Results." *The Astrophysical Journal* 743 (2): 156. doi:https://doi.org/10.1088/0004-637X/743/2/156.

McMillan, R. S., and Spacewatch Team. 2006. "Spacewatch Preparations for the Era of Deep All-Sky Surveys." In *AAS/Division for Planetary Sciences Meeting Abstracts #38*, 38:592. Bulletin of the American Astronomical Society.

Nugent, C. R., A. Mainzer, J. Masiero, J. Bauer, R. M. Cutri, T. Grav, E. Kramer, S. Sonnett, R. Stevenson, and E. L. Wright. 2015. "NEOWISE Reactivation Mission Year One: Preliminary Asteroid Diameters and Albedos." *The Astrophysical Journal* 814 (2): id.117.

Pravdo, Steven H., David L. Rabinowitz, Eleanor F. Helin, Kenneth J. Lawrence, Raymond J. Bambery, Christopher C. Clark, Steven L. Groom, et al. 1999. "The Near-Earth Asteroid Tracking (NEAT) Program: An Automated System for Telescope Control, Wide-Field Imaging, and Object Detection." *The Astronomical Journal* 117 (3): 1616–33. doi:https://doi.org/10.1086/300769.

Smith, Roger M., Richard G. Dekany, Christopher Bebek, Eric Bellm, Khanh Bui, John Cromer, Paul Gardner, et al. 2014. "The Zwicky Transient Facility Observing System." In *Ground-Based and Airborne Instrumentation for Astronomy V*, edited by Suzanne K. Ramsay, Ian S. McLean, and Hideki Takami, 9147:914779. \procspie. doi:https://doi.org/10.1117/12.2070014.

Stokes, Grant. 2000. "Lincoln Near-Earth Asteroid Program (LINEAR)." *Icarus* 148 (1): 21–28. doi:https://doi.org/10.1006/icar.2000.6493.

Vereš, Peter, and Steven R. Chesley. 2017. "Near-Earth Object Orbit Linking with the Large Synoptic Survey Telescope." *The Astronomical Journal* 154 (1): 13. doi:https://doi.org/10.3847/1538-3881/aa73d0.

Wright, Edward L., Peter R. M. Eisenhardt, A Mainzer, Michael E. Ressler, Roc M. Cutri, Thomas Jarrett, J. Davy Kirkpatrick, et al. 2010. "The Wide-Field Infrared Survey Explorer (Wise): Mission Description And Initial On-Orbit Performance." *The Astronomical Journal* 140 (6): 1868–81. doi:https://doi.org/10.1088/0004-6256/140/6/1868.

Online Resources

ATLAS: http://www.fallingstar.com/home.php
Catalina Sky Survey: http://www.lpl.arizona.edu/css/
DECAM: https://www.darkenergysurvey.org/
LINEAR: http://www.ll.mit.edu/mission/space/linear/
LONEOS: http://asteroid.lowell.edu/asteroid/loneos/loneos.html
LSST: https://www.lsst.org/
NEAT: https://neat.jpl.nasa.gov/
NEOCAM: http://neocam.ipac.caltech.edu/
NEOWISE: http://neowise.ipac.caltech.edu/
PTF: http://www.ptf.caltech.edu/
Pan-STARRS: http://pswww.ifa.hawaii.edu/pswww/

SENTINEL: https://b612foundation.org/sentinel/
SST: https://www.ll.mit.edu/mission/electronics/ait/scientific-dod-imagers/space-surveillance-telescope.html
Space Telescope Science Institute: https://panstarrs.stsci.edu/
Spacewatch: http://spacewatch.lpl.arizona.edu/
ZTF: http://www.ptf.caltech.edu/ztf

Chapter 6
Vision of Perfect Observation Capabilities

Peter Vereš

6.1 Discovery and Orbit Determination

Knowledge about the Solar System has been constantly changing throughout history. Before 1781, the known Solar System consisted of the Sun, Mercury, Venus, Earth, Jupiter and Saturn, as well as our Moon, four moons of Jupiter and five moons of Saturn. In the mid-eighteenth century, Sir Edmond Halley deduced that a comet he saw appearing repeatedly over time was the same object orbiting the Sun. He computed its orbit and predicted its return, thereby making the first comet discovery (known thereafter as the famous Halley's comet). The first asteroid recorded, Ceres, was discovered in 1801 by Giuseppe Piazzi. At the time of discovery, Piazzi did not know that he was looking at an asteroid—he in fact believed it was a comet, despite not seeing any coma or tail. A few months before that discovery, astronomer Franz Xaver von Zach actually organized 24 astronomers into the so-called Celestial Police and tasked them with finding a planet that was supposed to be between Mars and Jupiter. The planet's existence was only hypothesized on the basis of the empirical Titius-Bode law. The discovery of Uranus in 1781 increased the importance of the law, as Uranus was discovered at the distance predicted. Piazzi was not among the Celestial Police astronomers, and even though he published his observations in the months following and communicated with other astronomers, nearing the end of 1801, Ceres' orbit was still unknown and some even doubted that the new "planet" existed. Thanks to Carl Friedrich Gauss, who developed a fast method for deriving the elliptical orbit from observations and the predicted position of Ceres, the new object was indeed found at the end of 1801 by von Zach and Olbers near the

P. Vereš (✉)
Harvard-Smithsonian Center for Astrophysics, Cambridge, MA, USA
e-mail: pveres@cfa.harvard.edu

N. Schmidt (ed.), *Planetary Defense*, Space and Society,
https://doi.org/10.1007/978-3-030-01000-3_6

predicted position. A few months later, Olbers discovered a second asteroid between Mars and Jupiter, Pallas. Since then, the inventory of the Solar System has been growing steadily, due to improvements in observational techniques and spaceflights. Nowadays, about 740,000 asteroids, almost 4000 comets, and 181 moons of planets are known, and this number is still growing.

The discovery process consists of the detection of an object, realization that the object's position does not match the orbit of any known objects, its subsequent follow-up and computation of its orbit. Once the orbit is computed, the position of the object in the future or past can be calculated, and the object becomes catalogued. Current availability of computing power and accurate knowledge of mass and the location of planets allow computation of reliable asteroid positions hundreds of years into the future. The length of the observation arc—the time interval between the discovery date and last observation—needed for determining the orbit with the desired precision depends on the type of the orbit, precision of astrometric measurements, observing geometry and orbital period. Currently, NEOs are considered to be discovered only after observations from a few subsequent days since the initial discovery. On the other hand, distant objects, like Centaurs between the orbits of Jupiter and Neptune or Trans-Neptunian objects, need at least months, or even years, of observations for orbit determination.

The derived orbit of a known object is defined by its osculating Keplerian orbital elements (semimajor axis, eccentricity, inclination, argument of perihelion, longitude of ascending node, mean anomaly) and epoch (date) at which the orbital elements are valid. Each Keplerian element is determined with its uncertainty. As noted earlier, osculating orbits are only valid for a given epoch and change constantly under the influence of the gravity of other planets and nearby asteroids. Each orbit can also be denoted by its proper orbital elements, which are averaged over a long interval of time and incorporate minor changes in orbits.

6.1.1 Orbital Uncertainties

The uncertainty in orbital elements is crucial for computing the position of object in space. Generally, NEOs with arcs with years of observations are easily found, and orbits are well determined for centuries. However, extra precision is needed if a spacecraft is visiting the asteroid or a radar observation is planned. Improvement of orbit is often achieved by dedicated observations with large telescopes or through archival searches in survey data. Subsequent radar observations, if possible, nail down the orbit, which is then sufficient for spacecraft navigation. Orbital uncertainty is driven not only by the precision of measurement, but also by the reference frame used for astrometric reduction—the stars (Chesley et al. 2010; Farnocchia et al. 2015). Even if the asteroid orbit is well determined and orbital uncertainties understood, several effects can abruptly change the orbit and seriously undermine the prediction of the object's position in the future, discussed below.

6.1.1.1 Non-Gravitational Effects

Gravity is not the only force that affects the orbit of an asteroid. Even though it is orders of magnitudes weaker than gravity, solar radiation itself can play a role, particularly for asteroids close to the Sun, like NEOs. The Yarkovsky effect (Bottke et al. 2006) changes the semimajor axis when the solar photons irradiating the asteroid are emitted back, mostly in infrared, from the asteroid's surface. Because of the temperature gradient between day- and night-side and the lag in temperature distribution on the surface, the more energetic photons carry more momentum and can change an asteroid's orbit. Only small asteroids (<10 km) are prone to this force. The Yarkovsky effect has two components: the diurnal effect and the seasonal effect. Depending on the spin axis orientation of the asteroid and its surface properties, the object can drift either to or from the Sun. The Yarkovsky effect is very small and was first observed in 2003 on the NEO Golevka (Chesley 2003), and it needs a long time to change the orbit significantly.

The Yarkovsky-O'Keefe-Radzievskii-Paddack effect (YORP) (Rubincam and Paddack 2010) is similar to the Yarkovsky effect, but it changes the spin and obliquity of an object. In principle, YORP is caused by solar radiation irradiating irregular surfaces on an asteroid, which re-radiate and cause torque on the asteroid. Small asteroids are prone to faster changes in spin axis orientation and in slowing down or speeding up their rotation. YORP is believed to be one of the reasons for the creation of binary asteroids that once spun up and broke apart. Although both effects are rather small and extremely hard to observe, even a small change can cause an orbital change that can bring an asteroid close to a perturbing planet or to a mean-motion resonance that will rapidly change its orbit. The non-gravitational effects have to be taken into account for NEOs with large impact probabilities or objects having future nearby flybys of Earth. If we do not know the Yarkovsky effect, an object with a known flyby could actually have a small change in its orbit that will lead the object to a so-called *keyhole* and in fact collide with the Earth in the future. The Yarkovsky effect can be observed directly from changes of asteroid's orbit and by knowing the asteroid's shape, pole orientation, rotation period and thermal properties. For the vast majority of objects, these properties are unknown. Fortunately, the number of hazardous NEOs is currently small, and most of the hazards can be mitigated by improved orbits that reduce the impact risk to zero. Therefore, for the few hazardous NEOs, existing instrumentation, observations and efforts are probably sufficient to study the non-gravitational effects on these targets. NASA's Osiris-REX mission, currently on its way to NEO Bennu, will study the Yarkovsky effect on Bennu directly. Bennu has a non-zero impact probability with Earth between the years 2175 and 2181 and currently ranks as the third most dangerous known asteroid on NASA's Sentry web site, according to the Palermo scale.

The orbits of comets are affected by non-gravitational forces. In contrast to rock-solid asteroids, comets are active due to sublimation of ices, loss of mass, and activity changes. In addition to classical Keplerian motion and gravitational perturbations of the planets, one has to include A1 (radial), A2 (transverse) and A3 (normal) acceleration (Marsden et al. 1973) due to the comet's activity. These parameters are derived individually for every observed comet and could change. Ai acceleration is more significant when compared to the above mentioned non-gravitational forces.

6.1.1.2 Close Flyby of a Planet or Asteroid

Motion of an asteroid in the Solar System is computed by assuming the point mass of the Sun and perturbing planets, neglecting the mass of an asteroid. This approach is generally sufficient, and the accuracy of orbital determination is driven by observational constraints of the asteroid, precision of mass and position of the planets. However, if the asteroid has a close encounter with a planet, the point mass approximation is not sufficient. The shape of the planet and the distribution of the mass inside introduce higher spherical harmonics and complicate the shape of the gravitational field. The high-order gravitational field of Earth or other planets is determined from the motion of the spacecraft. In case of an asteroid flying within a few radii of the planet, its orbit changes dramatically, due to a strong gravitational pull from the planet. After the flyby, uncertainty in orbital elements increases due to the gravitational pull of a planet. Thus, further observations are needed to decrease the orbital uncertainty. Comets are particularly prone to dramatic orbital dynamics in the long term, mostly due to their highly eccentric orbits, which frequently cross the orbits of the planets. The gravity of the most massive planet, Jupiter, has a dominant effect on cometary orbits.

6.1.1.3 Artificially Induced Change of the Orbit

The Deep Impact mission to comet 9P/Tempel has proven that mankind is able to deliver a kinetic impactor to a minor body. Even though the impactor was small and the orbital period of the comet was changed by only 1 s, future plans include studies and attempts at serious orbital changes of an asteroid (e.g. the mission DART). Study of impact dynamics and orbital evolution will be crucial to understand NEO mitigation strategies. The internal structure and physical properties of asteroids are still not well understood; in the case of a kinetic impactor, scientists can only estimate how the asteroid would react. Many asteroids are known for having high porosity—large empty spaces in asteroid interiors or microspaces that can efficiently consume the energy of an impact making the transfer of momentum inefficient. Thus, porous asteroids can behave like a sponge and consume the impact momentum. In such a case, a kinetic impactor may fail. Even a nuclear blast, intended for total destruction of a given target, may not succeed. Therefore, nuclear options have also been proposed for partial evaporation of the asteroid's surface to change the asteroid's orbit through momentum transfer. Future missions to asteroids will be able to bring back samples to study their interiors, such as Osiris-REX, and could significantly improve our mitigation efforts.

Often, uncertain orbits are a source of elevated impact risks of some NEOs with the Earth. The impact probability of an asteroid with Earth is a complex problem.

First, the orbits of Earth and the asteroid should be close enough or even intersect; second, the Earth and asteroid should meet at the intersection at the same time. If these conditions are met, then one can assess how close the asteroid flies around the Earth at a given time, or whether it will hit the Earth. One must remember that each asteroid orbit comes with uncertainties and therefore, instead of a single accurate solution where the asteroid will hit the Earth or miss it, there is always a realm of possible solutions within the orbit uncertainties. The tangent plane to the asteroid's trajectory at the time of impact, or close approach, is called a b-plane. At a given time of a predicted impact, all possible closest distances to the Earth of possible orbits create an area on the tangent plane. If the area contains the Earth, then the impact probability for that epoch is non-zero and in a simple approximation can be denoted as a ratio of an area of Earth cross section and the entire area with possible orbits going through the b-plane. It happens that a newly discovered NEO with a short arc that is coming very close to the Earth has a non-zero impact probability, because its orbit is highly uncertain and the area on the b-plane is very large. Typically, further observations improve the orbit, and the impact risk for a given epoch falls to zero. Some objects, however, have orbits with low orbital uncertainty, but still have non-zero impact probability, such as Bennu. The non-zero impact probability is computed for a given time in the future, but even if the orbit is known very well today, small perturbations from planets and non-gravitational forces increase the uncertainty for future impacts. That is why NASA's Sentry is providing predictions only for the next 100 years.

A close flyby of a spacecraft around an asteroid may improve the asteroid's orbit significantly, however, it does not fully mitigate its impact in the future, due to the presence of keyholes (Chodas 1999)—small areas in space near Earth. Keyholes are specific for asteroids flying very close to the Earth and are rather small, from a few to hundreds of kilometers across. If the keyhole is hit during the NEO flyby, the orbit of the NEO becomes resonant with Earth and the NEO will return to Earth regularly, increasing its impact probability. Thus, in case of a very near Earth flyby, the orbit needs to be known with such precision (~km) that keyhole avoidance is confirmed. NASA has even created the NEO Deflection App,[1] where the public can try to change the orbit of a hypothesized NEO on direct impact trajectory. For Earth impact monitoring, the accuracy of orbits and orbital uncertainties is crucial and deserves more attention.

The future of orbit determination and uncertainty mitigation will depend more and more on sophisticated software that will be able to handle orbital computation in detail; assess uncertainties and errors of measurements; coordinate a list of objects that are crucial for follow-up or orbit improvement, or even automatically point the telescopes in a network to observe those asteroids; measure their positions; and submit the data to MPC. This automated process is more or less implanted by several surveys (CSS, LCOGT) and agencies (ESA, MPC).

[1] https://cneos.jpl.nasa.gov/nda/.

6.2 Small NEOs and Accessibility

The number of discovered NEOs has been increasing yearly, and this trend will continue. The smallest NEOs are the most numerous. However, if the albedo is not known, the diameter is only derived from absolute magnitude H and from assumed albedo. For instance, at $H = 31$, diameter can be in the range of 1–4 m, if albedo is in a range of 0.04 (dark) to 0.3 (bright). According to Harris and D'Abramo (2015), there are more than one billion NEOs larger than about two meters and 60–100 million NEOs larger than about 10 m (Fig. 6.1). Most of these are too small to be detected by current telescopes and are only seen if they come sufficiently close to the Earth. Therefore, only about 350 NEOs smaller than about 10 m have been discovered. One may speculate regarding how many of the smallest NEOs come close to Earth and could be easily accessible. Brown et al. (2002) computed the Earth impact frequency from the bolide data and estimated the annual cumulative number of objects larger than a given size impacting as $log\ N(>D) = c_0 - d_0\ log\ D$, where D is the diameter of NEO in meters and constants are equal to $c_0 = 1.568 \pm 0.03$, $d_0 = 2.7 \pm 0.08$. This equation derives the number of 1 m and larger Earth impactors to be 37 per year. A 1-m NEO is not a threat to Earth—it would burn in Earth's atmosphere and be visible as a very bright bolide in the sky. Only a small fraction of its mass would survive the atmospheric entry in the form of meteorites, perhaps of a total mass of a few kilograms.

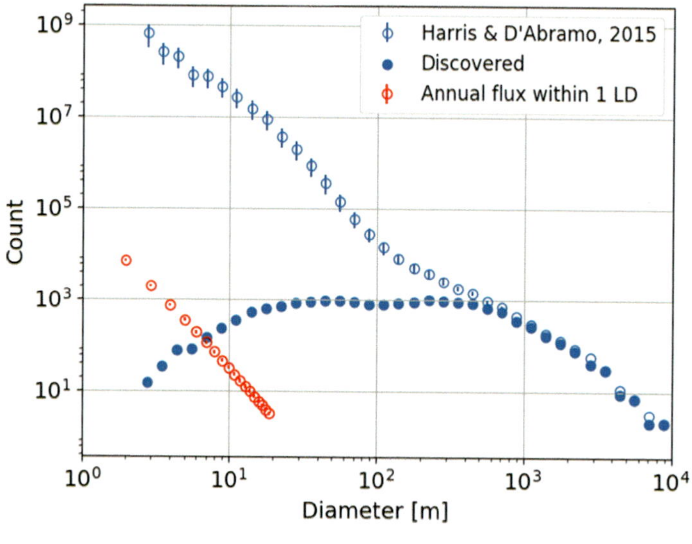

Fig. 6.1 Histograms of size-frequency distribution of NEO population by Harris and D'Abramo (2015) by assuming albedo of 0.14 (Stuart and Binzel 2004), compared to the discovered population of NEOs (mid 2017) and number of small NEOs flying within one lunar distance from Earth annually

We can expand our area and ask how many small NEOs fly around the Earth closer than a given distance annually. To figure out how many NEOs are flying within a given distance, we can expand the search area from the Earth's cross-section to a given distance. We selected one lunar distance (ld = 384,400 km). Earth's capture radius is actually larger than the surface of the Earth and depends on the velocity of an incoming asteroid, as $R_c = R_E\sqrt{1+\left(\frac{v_E}{v_\infty}\right)^2}$, where $R_E = 6378$ km is the Earth's diameter, $v_E = 11.2$ km/s is the Earth's escape velocity and v_∞ is the relative velocity of NEO to a massless Earth. We computed the mean R_c based on mean value v_∞ from CNEOS NEO Earth Close Approach data (Fig. 6.2), for known objects flying closer than one lunar distance from Earth between 1900 and 2200. The mean values are $\overline{v_\infty} = 11.8 \pm 5.7\text{km/s}; R_c = 8810 \pm 193\text{km}$. Expanding the capture cross-section by a factor of $\left(\frac{ld}{R_c}\right)^2$ and multiplying it by the Earth impact frequency N (>D), we derived the number of NEOs as a function of diameter, which annually get closer than one lunar distance to the Earth (Fig. 6.1).

We derived that there are about 10,800 ± 630 NEOs larger than 2 m that come closer than one lunar distance to Earth per year. The smallest NEOs are not yet being systematically searched for and are only found coincidentally, as fast-moving pests in the survey data, often lost without follow-up. The future might favor these, due to their easy accessibility from Earth. We may speculate about future fleets of nanosatellites that will be trailing the Earth in sub-lunar space and could encounter plenty of small NEOs when they get close. The visit might have multiple reasons,

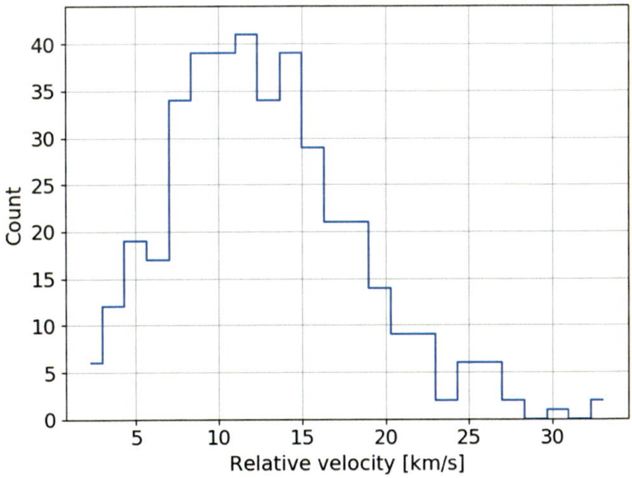

Fig. 6.2 Histogram of relative velocity of NEOs within one lunar distance for known objects between 1900 and 2200

from basic remote or in-situ research to taking and analyzing samples, landing, changing orbits or directly processing the tiny rocks to mine metals and rate elements, which can be delivered to Earth.

6.3 Physical Characterization

Discovery and orbit determination do not provide any information about the physical appearance of NEOs. There are only two parameters associated with basic orbital elements that give a clue: the absolute magnitude (H) and slope parameter (G). Different techniques must be used to obtain the physical characteristics of asteroids. The main difference between the surveys and characterization is that surveys observe many objects at once, while characterization usually focuses on a single object. Light-curve determination means a systematic observation of a single object for hours, or even days. Light-curve provides basic information on the body, including its rotation period and spin axis orientation, and it can even reveal the existence of a moon around the asteroid. Detailed light-curves can be used to generate shape models (Ďurech et al. 2010) of asteroids. The asteroid light-curve database[2] (Warner et al. 2009) stores information about known rotational properties and currently contains 1240 NEOs and 18,000 other asteroids.

In addition to shape, the size can be determined if observation in thermal infrared is provided. Asteroid diameters from NEOWISE missions are available at NASA's Planetary Data System[3] (Mainzer et al. 2016), which contains 951 NEOs and more than 160,000 other asteroids. The mass of an asteroid can be determined either from a spacecraft flyby (in a few cases), if the asteroid is binary, or if two asteroids fly at a close mutual distance and changes in their orbits are noticed. Knowing the mass and volume can directly infer the mean density of an asteroid. The density and volume can serve for calculation of the porosity (void inner space) and internal strength and composition of the asteroid.

Asteroids come in several types. A basic tool for asteroid physical characterization is color photometry, where light from the asteroid is observed in color filters, thereby providing a primitive approximation of its reflectance spectrum. This method is relatively simple and does not require long time periods. It gives the amount of the reflected sunlight in narrow or broad band filters, historically matching the colors that are passing through, e.g. G = green, R = red, B = blue. Some surfaces of asteroids reflect less light, some more light in longer wavelengths toward red and infrared, which can then be used for basic taxonomic classification of asteroids. Spectroscopy requires more time and sophisticated instruments and calibration, but provides accurate measurements in visible, near-infrared, and infrared wavelengths. Spectroscopy can identify specific molecules on the surface (e.g. water, methane, iron and silicate oxides, organic compounds) and contribute to even

[2] http://www.MinorPlanet.info/lightcurvedatabase.html.

[3] https://sbn.psi.edu/pds/resource/neowisediam.html.

Table 6.1 Spectral types of NEOs and the debiased fraction of NEO population based on Stuart and Binzel (2004) with a brief description of the composition

Spectral type	Fraction (%)	Content
X-complex	33.8	Metallic, carbonaceous, stony
S	22.4	Silicates
D	17.5	Organic, silicates, ice
Q	14.1	Silicates with metals
C	9.8	Carbonaceous
V	1.3	Fragments of Vesta/silicates
O	0.5	Silicates with low metal content
P	0.4	Silicates, carbon, ice
A	0.2	Silicates
R	0.2	Silicates

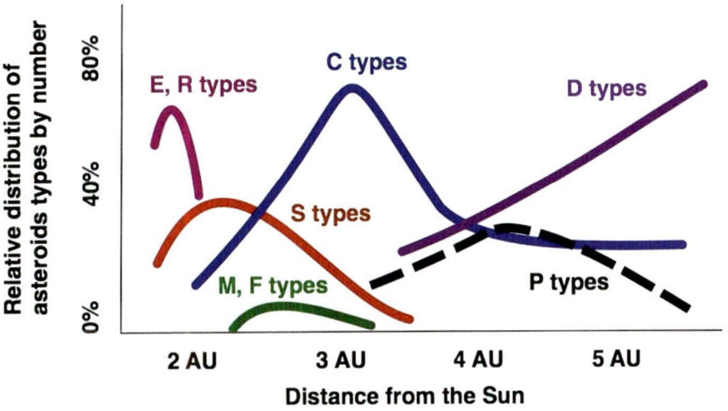

Fig. 6.3 Distribution of asteroid spectral classes in the inner Solar System as a function of distance from the Sun (Gradie and Tedesco 1982)

finer spectroscopic classification of different asteroid types. Until the first sample of asteroid was delivered to Earth aboard the Hayabusa spacecraft, the only information about the composition of asteroids came from spectra, and meteorites believed to be analogs of asteroids and their fragments. Table 6.1 summarizes the basic categories of asteroids by spectral types (DeMeo et al. 2009) in NEO space.

Asteroids of given types are not randomly scattered in the Solar System but generally occupy specific areas. Figure 6.3 shows that the inner part of the Solar System is predominantly filled with S-type (silicate & rocky) objects, the inner Main Belt between Mars and Jupiter has a mixture of M (metallic) and S types, the Outer Main Belt has darker and primitive objects with organic compounds and carbon (C, D, P types) and water in the form of ice. The distribution of given types as a function of distance from the Sun suggest that specific types of asteroids originated at a different distance and temperature, when the Solar System was formed 4.5 billion years ago. NEOs are almost exclusively delivered from the Main Belt and therefore are a mixture of the Main Belt inventory.

Detailed spectroscopy and modeling are currently the only remote way of determining the mineralogy of meteorites. Other methods can help the study of the surface properties, such as polarimetry (polarity of the reflected sunlight) or study of the phase curve and phase effect in visible light, which suggest the surface roughness and presence of dust and regolith. Ideally, a sample of an asteroid, even in the form of a few grams of surface dust, could be sufficient for in-situ mineralogy sampling. Attempts to calculate the value of asteroids based on a few selected minerals and metals was done by the asterank website[4] by Planetary Resources, which based their study on collecting known spectral types and the petrology of those spectral types. Even though they gathered useful information, further studies, including element processing and efficiency, and accurate abundance of metals, are needed to better understand the realistic cost effectiveness of asteroid mining. Up to the present date, the largest database of asteroid spectral types is the SDSS database (Carvano et al. 2010), containing 63,000 asteroids. Only a small fraction of these are NEOs. The SDSS data tables, as well as other physical properties, are accessible through NASA's Planetary Data System Asteroid Data Sets.[5]

6.4 Telescopes in Space

The main advantage of having telescopes in space is the avoidance of Earth's atmosphere. Even though the atmosphere protects life from ultraviolet radiation and keeps the surface habitable, it significantly interferes with the observation of space objects. For instance, some wavelengths are partially (gamma, infrared), or fully (UV, X-rays) obscured by the molecules of the atmosphere and cannot be observed from the surface. On the other hand, visible or radio waves reach the surface. Still, the temperature gradients and motion in the atmosphere, including through clouds and scattering of the light either from light pollution or the Moon, distort the observed image. In the modern era, light pollution decreases the number of ideal dark-sky areas; these have to be located far away from cities and human activity. One way to overcome atmospheric interference is by building the telescope atop a tall mountain. For instance, Mauna Kea observatories are more than 4000 m above sea level, higher than the cloud inversion layer. The atmospheric pressure is 61% of the sea-level value, the humidity is much lower and air is colder, which helps near-infrared astronomy. Yet even the best location for a ground-based telescope does not stop the atmosphere and the motion of the air from causing an effect that disrupts the light of stars and decreases the angular resolution. For instance, the best ground-based visibility is in the range of 0.4–0.6 arcseconds, which is often much larger than the resolution of the optics. Sophisticated methods have been developed, such

[4] http://www.asterank.com/.

[5] https://sbn.psi.edu/pds/archive/asteroids.html.

Fig. 6.4 Laser guide star for the adaptive optics. Credit: ESO/L. Calçada

as adaptive and active optics, to deform the optical elements of telescopes based on analysis of the wave front, with the assistance of the laser guide star (Fig. 6.4). In recent years, further advances in adaptive optics have allowed its implementation for much smaller telescopes and also in automated mode. The first working prototype is ROBO-AO, autonomous laser adaptive optics for few-meter-class telescopes, currently installed in the 2.1-m Kitt Peak Telescope in California (Baranec et al. 2014). ROBO-AO has been successfully tested for asteroid science, particularly in direct imaging and resolution of natural satellites of asteroids, and rapid improvement of asteroid positions. Radio telescopes are also ideally located at locations without any human activities that emit radio waves or electromagnetic fields.

Placing a telescope into space solves issues of the blocking atmosphere, especially in gamma, X-ray, UV and infrared wavelengths (Fig. 6.5). For instance, WISE/NEOWISE or Spitzer would not be able to observe reflected thermal radiation of asteroids from the surface of the Earth. Also, the lack of atmosphere and light pollution allows for much smaller apertures of telescopes that can achieve better resolution and limiting magnitudes than their much larger counterparts on the surface. Space telescopes have a rich history, and space research and astronomy have heavily relied on their data. Nevertheless, there are multiple disadvantages. The cost of any space telescope is usually enormous, incorporating years of development, mission control, launch and operations. Therefore, space telescopes are almost exclusively built and operated by large space agencies. Once the telescope is

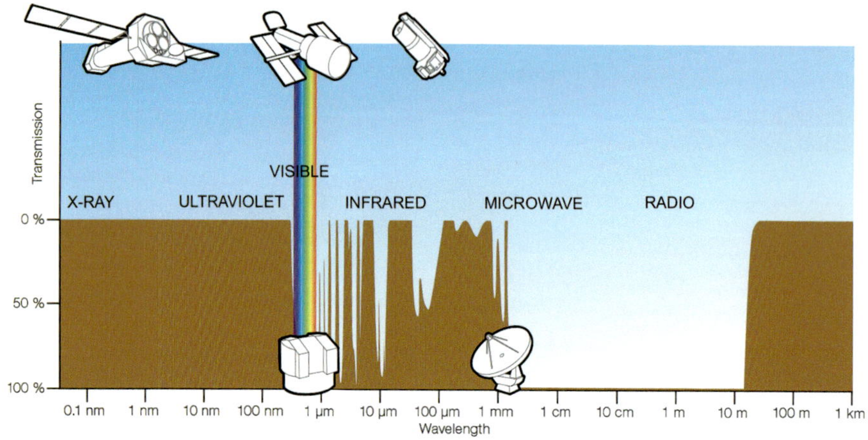

Fig. 6.5 Atmospheric transparency as a function of a wavelength. The visible light, depicted in a rainbow, is only a narrow range of electromagnetic spectrum. Credit: ESA/Hubble (F. Granato)

launched, it is very difficult to implement physical repairs. Only the Hubble Space Telescope (HST) was serviced directly in space through five missions (1993–2009), always by the Space Shuttle and at a very high cost. HST was easily accessible, being in LEO orbit. However, in 2011, the Space Shuttle era has ended. By comparison, the next generation JWST, which will observe in infrared, will be placed around the L2 Lagrange point, far beyond the Moon, beyond the reach of any currently available manned spaceships. Perhaps the need for a future service mission to JWST would also foster manned missions beyond the Moon. Based on the required schedules and plans, this seems to be more feasible through the private sector (SpaceX) rather than government agencies (NASA, ROSCOSMOS).

Communication and data transfer could be a problem: ground-based telescopes are only limited by the speed of already fast optical networks and cables, while space-based telescopes rely on much slower radio communication. Ground-based telescopes often have adjacent computing facilities, while space telescopes must rely on proven but much slower processing power. This is a consequence of their long development, lasting up to decades. By the time of launch, the IT technology onboard is already outdated. Space telescopes are also fragile in the hostile environment of space, facing high-energy particles, large temperature gradients and micrometeorites. Efficient particle shielding is particularly crucial for space telescopes and satellites beyond the Van Allen belts and in outer space. Van Allen radiation belts are zones 500–58,000 km above Earth, containing high energy electrons and protons captured by Earth's magnetic field from the solar wind. Beyond the Van Allen belts, the spacecraft is without any protection of Earth's magnetic field and thus is fully vulnerable to solar wind and high energy particles.

The future may rely on mitigation of rising costs by development of cheaper alternatives by the private sector. This is already happening on the launch vehicle side (SpaceX, Orbital-ATK, BlueOrigin), and may shortly turn in favor of human transportation. Still, private industry is not yet well established in developing telescopes that are meant for basic research. On the other hand, there is plenty of room in technology enhancement and subcontracts for the private sector. Some have already attempted to build low-cost alternatives: Arkyd was a failed project proposed by Planetary Resources, and Sentinel was an expensive proposed project by B612. Lockheed Martin, a well-established private aerospace company, introduced project SPIDER in 2017, a technology that will reduce the weight of a space telescope by 90%.[6] Instead of a large mirror, SPIDER relies on numerous miniature 1-mm lenses that gather and recombine light in a photonic integrated circuit, originally designed for telecommunications. SPIDER is thin and lightweight. If successful, this will represent a significant game changer, since weight is often a limiting factor and major cost when launching a spacecraft. Smaller and lighter space telescopes would allow SPIDER-like technology to be incorporated into much smaller spacecraft, such as cubesats. Perhaps, the rising interest in resources in asteroids and share of private industry in space activities will spur the discovery rate of small and nearby NEOs that can be explored not just for research and hazard mitigation but also for profit.

6.5 The Future of Solar System Exploration

The desired astronomical parameters of future telescopes include increased limiting magnitude, large field of view, instruments covering multiple wavelengths (ideally at the same time), location of the telescopes around the world, improvement of hardware and software and easier access to space for space-based telescopes.

The current limiting magnitude of wide-field asteroid surveys like Pan-STARRS is between 22 and 22.5 magnitudes. In 2020, the next generation LSST, built in Chile, will routinely reach to the 25th magnitude. Still, the survey telescopes are not the largest telescopes available. The larger the mirror and observing time, the fainter the object that could be observed. The largest ground-based telescopes (8–10 m primary mirror) can reach to the 28th magnitude in very long exposures. The faintest object ever observed by the Hubble telescope was 31.5 mag (Illingworth et al. 2013), but it required numerous long exposures of the same field that were stacked and added. Most of the largest telescopes are currently not used for Solar System research, and the observing time, which can be either purchased or awarded in a competitive telescope allocation process, is precious and expensive. The largest telescopes typically focus on other astrophysical targets, and often the existing instruments are not ideal for Solar System observations. The largest planned telescopes (EELT—40-m mirror, TMT—30-m mirror) would not be ideal for Solar System research either.

---

[6] SPIDER, http://www.lockheedmartin.com/us/innovations/011916-webt-spider.html.

One way to estimate the productivity of future survey telescopes is simulation, where synthetic or real orbits are used, numerically propagated to a given epoch, and observed fields of a given telescope are simulated. Often, the simulations could predict the probability of finding objects of interest, for instance the Earth impactors (Vereš et al. 2009) or the future productivity of LSST (Jones et al. 2017). The power of simulations allows for the assessment of desirable optical properties of future telescopes and their location, either on Earth or in space. Myhrvold (2016) simulated the performance of LSST, NEOCAM, Sentinel and a fleet of satellites on heliocentric orbits, to describe the search volumes for NEOs. The future inventory of the Solar System is expected to grow after the commissioning of the LSST. In 10 years, LSST is expected to discover about 100,000 NEOs (currently 16,000), five million Main Belt asteroids (currently 700,000), 300,000 Trojans (currently 6500) and over 40,000 TNOs (currently 1800). The latest theories (Batygin and Brown 2016) even acknowledge the existence of distant and large planets far beyond Neptune's orbit, and years of surveying by dedicated 8–10-m telescopes is underway.

Larger and new telescopes are not the only way to discover more objects. Archival searches are also a viable option. For instance, surveys like NEAT, SDSS, partially Pan-STARRS, and CSS, are providing images for the public. Automated algorithms are typically identifying objects brighter than a given threshold, and signal-to-noise ratio (SNR) that is usually larger than five. Certainly, algorithms can find objects much fainter, however, below SNR = 5 the number of false detections in the images (noise, optical glitches and reflections) is very high. In the recent past and up until the present, it has been humans, or citizen scientists and the public, who have previewed the images and found much fainter objects by eye in the archives. The number of images is enormous. The solution for archival searches and even live image processing is artificial intelligence (AI) and machine learning. The evolving AI could distinguish false detections from the real ones at very low SNR levels and actually find many archival or current objects that are too faint for classical software pipelines. Another example of the usefulness of the IT sector in astronomy is the need for new algorithms and software solutions for dealing with big data and real-time processing. Often, software engineers are hired by research institutes to help scientists develop their instruments and tools. A positive example of collaboration between researchers, the private IT sector and education is NASA's Frontier Development Lab, which connects young professionals from the private IT industry and researchers working on space-related topics.[7] The future will see an increase in the need for a specialist IT workforce for survey and specialized telescopes, on the ground and in space.

The above-mentioned future perspective of larger ground-based telescopes, new space telescopes and development of software and hardware for Solar System observations, require a significant amount of funding. The vast majority of these funds are provided by research funding agencies (NSF, ERC) or space agencies

[7] http://www.frontierdevelopmentlab.org/.

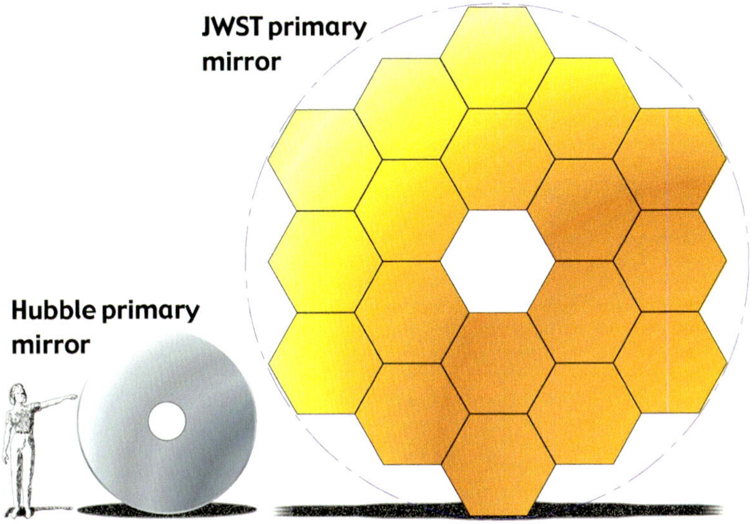

Fig. 6.6 Size comparison of the primary mirror of the Hubble Space Telescope and the James Webb Space Telescope. Credit: NASA

(NASA, ESA), and awardees are mostly universities or research entities. Often, the projects get very expensive, and vast effort, typically over years, is required to get a substantial grant award. For survey telescopes, the cost can be between tens to hundreds of million USD. Space telescopes are even more expensive, ranging from several hundred thousand to billions USD. JWST, the future flagship of space telescopes, is scheduled for launch in 2020. Serving astronomy and cosmology, it will cost around $10 billion (Fig. 6.6). The large cost is usually dealt with by creating consortia of multiple, often international stakeholders. International involvement allows for mobility of the workforce and brings together experts for a given project. Often, access to projects is restricted and data is available to others after the official mission is over, or if a potential stakeholder is able to pay for access to the data. The disadvantage of this form of funding is the soft nature of the funds coming from governments, which can change from year to year based on the priorities of each government. Another disadvantage is the focus of private companies, who are interested in applied research and fast returns on investments. The private sector is not keen on basic research and development that takes a decade. The future may increase the involvement of the private sector in basic research through indirect support of products in IT and consulting. Collaboration in international space would be even more desirable, as well as searching for new sources of funding and involving countries not historically important players, but which may become important in the future. Countries or private companies may search for alternative solutions, such as operation or revival of decommissioned telescopes. A good example is the private LCOGT network, which either purchases new telescopes or takes over spare telescopes not used by professionals. Instead of developing new instruments from the beginning, LCOGT optimizes costs by purchasing the same types and makes of telescopes, with the same instruments—CCD cameras and spectrographs—the same software and same operation procedures across all their assets.

6.6 Spillover Effect

Knowledge of Solar System architecture can serve several purposes. First, the basic research has a goal of discovering unknown objects, surveying their physical characteristics, and computing accurate orbits, thereby allowing past and future predictions of their locations. Generally, basic research enhances the knowledge of our whereabouts and studies the origin and evolution of the Solar System. Second, NEOs get increased attention due to impact hazard, e.g. in 1998, the US congress mandated that NASA complete the so-called Spaceguard Survey to find 90% of NEOs larger than 1 km. The mandate was extended in 2005 as the NEO Survey Act, with the goal to find 90% of NEOs larger than 140 m before 2020. The increased interest led to detailed studies of impact mitigation and the establishment of planetary defense as a growing and important community of experts. Planetary defense, mostly funded and organized by NASA's Planetary Defense Coordination Office, is responsible for funding NEO searches, theoretical studies of impact mitigation and space missions related to NEOs. Third, technological and sociological knowledge concerning space has always had a large economic impact and has brought improvements to our daily lives—the spinoff principle.

Space technology has proven its potential and usefulness for others. Products like LED, chemical sensors, fireproof materials, freeze drying, water purification, cell phone cameras, infrared sensor, memory foam, software, digital imaging and magnetic resonance imaging are used daily by people all around the world. Most of these technologies were derived from devices that were primarily invented for space-related research and applications. Often, restricted science or military products make it to the public domain as well, such as GPS navigation and satellite imagery of Earth.

The rise of private companies dealing with space travel and utilization of resources will most likely continue. The challenges there are even more obvious. Private companies tend to optimize costs, build on existing technology and speed up development to focus on profit. Therefore, technological challenges will certainly focus on lowering costs and development of new technologies, for instance, for space propulsion, autonomous navigation and utilization of lightweight spacecraft, such as cubesats and products that can be quickly used in daily life. The opportunities for private industry in space are the best in history, due to a growing qualified workforce with employment history in space agencies, accessibility of the existing software (such as SPICE[8] for navigation), grant agencies publishing calls for private sector partnerships, and the connection of the space sector to the high-tech industry.

[8] https://naif.jpl.nasa.gov/naif/index.html.

This is a promising opportunity, particularly for non-spacefaring nations, since the aerospace industry enhances the economy of the country, boosts the job market by creating top-paying jobs, stimulates high-level education and mobility of the workforce and attracts qualified labor from abroad. On the other hand, private industry may speed up often long and expensive development, particularly when building on existing and proven technology, such that it does not need to invest in basic research and development.

Undoubtedly, human potential and education related to space research will become even more precious. Space has always created specialized and well-paid, high-value jobs. Currently, there are many more private space agencies or companies than federal or state agencies that are able to launch a rocket to space, deliver cargo to orbit or construct a satellite. Most of the aerospace companies with the most value are in the USA. The total estimate of global space activities in 2014 was $330 billion (The Space Report of Space Foundation). Among the most successful venture-financed space companies is SpaceX, which was valued at $20 billion in 2017. Its future will depend on the demand for space utilization and successful and sustainable business models that will bring profit to private companies.

References

Baranec, Christoph, Reed Riddle, Nicholas M. Law, A. N. Ramaprakash, Shriharsh Tendulkar, Kristina Hogstrom, Khanh Bui, et al. 2014. "HIGH-EFFICIENCY AUTONOMOUS LASER ADAPTIVE OPTICS." *The Astrophysical Journal* 790 (1): L8. doi:https://doi.org/10.1088/2041-8205/790/1/L8.

Batygin, Konstantin, and Michael E. Brown. 2016. "EVIDENCE FOR A DISTANT GIANT PLANET IN THE SOLAR SYSTEM." *The Astronomical Journal* 151 (2): 22. doi:https://doi.org/10.3847/0004-6256/151/2/22.

Bottke, William F., David Vokrouhlický, David P. Rubincam, and David Nesvorný. 2006. "THE YARKOVSKY AND YORP EFFECTS: Implications for Asteroid Dynamics." *Annual Review of Earth and Planetary Sciences* 34 (1): 157–91. doi:https://doi.org/10.1146/annurev.earth.34.031405.125154.

Brown, P., R. E. Spalding, Douglas O. ReVelle, E. Tagliaferri, and S. P. Worden. 2002. "The Flux of Small Near-Earth Objects Colliding with the Earth." *Nature* 420 (6913): 294–96. doi:https://doi.org/10.1038/nature01238.

Carvano, J. M., P. H. Hasselmann, D Lazzaro, and T Mothé-Diniz. 2010. "SDSS-Based Taxonomic Classification and Orbital Distribution of Main Belt Asteroids." *Astronomy and Astrophysics* 510 (February): A43. doi:https://doi.org/10.1051/0004-6361/200913322.

Chesley, S. R. 2003. "Direct Detection of the Yarkovsky Effect by Radar Ranging to Asteroid 6489 Golevka." *Science* 302 (5651): 1739–42. doi:https://doi.org/10.1126/science.1091452.

Chesley, Steven R., James Baer, and David G. Monet. 2010. "Treatment of Star Catalog Biases in Asteroid Astrometric Observations." *Icarus* 210 (1): 158–81. doi:https://doi.org/10.1016/j.icarus.2010.06.003.

Chodas, Paul. 1999. "Orbit Uncertainties, Keyholes, and Collision Probabilities." In *Bulletin of the American Astronomical Society*, 31:1117.

DeMeo, Francesca E., Richard P. Binzel, Stephen M. Slivan, and Schelte J. Bus. 2009. "An Extension of the Bus Asteroid Taxonomy into the Near-Infrared." *Icarus* 202 (1): 160–80. doi:https://doi.org/10.1016/j.icarus.2009.02.005.

Ďurech, J., V Sidorin, and M Kaasalainen. 2010. "DAMIT: A Database of Asteroid Models." *Astronomy and Astrophysics* 513 (April): A46. doi:https://doi.org/10.1051/0004-6361/200912693.

Farnocchia, D, S.~R. Chesley, A.~B. Chamberlin, and D.~J. Tholen. 2015. "Star Catalog Position and Proper Motion Corrections in Asteroid Astrometry." *Icarus* 245 (January): 94–111. doi:https://doi.org/10.1016/j.icarus.2014.07.033.

Gradie, J., and E. Tedesco. 1982. "Compositional Structure of the Asteroid Belt." *Science* 216 (4553): 1405–7. doi:https://doi.org/10.1126/science.216.4553.1405.

Harris, Alan W., and Germano D'Abramo. 2015. "The Population of Near-Earth Asteroids." *Icarus* 257 (May). Elsevier Inc.: 302–12. doi:https://doi.org/10.1016/j.icarus.2015.05.004.

Illingworth, G. D., D. Magee, P. A. Oesch, R. J. Bouwens, I. Labbé, M. Stiavelli, P. G. van Dokkum, et al. 2013. "THE HST EXTREME DEEP FIELD (XDF): COMBINING ALL ACS AND WFC3/IR DATA ON THE HUDF REGION INTO THE DEEPEST FIELD EVER." *The Astrophysical Journal Supplement Series* 209 (1): 6. doi:https://doi.org/10.1088/0067-0049/209/1/6.

Jones, R. Lynne, C. Slater, J. Moeyens, L. Allen, M. Jurić, and Z. Ivezić. 2017. "The Large Synoptic Survey Telescope as a Near-Earth Object Discovery Machine." *Submitted to Icarus*.

Mainzer, A, J.~M. Bauer, R.~M. Cutri, T Grav, E.~A. Kramer, J.~R. Masiero, C.~R. Nugent, S.~M. Sonnett, R.~A. Stevenson, and E.~L. Wright. 2016. "NEOWISE Diameters and Albedos V1.0." *NASA Planetary Data System* 247.

Marsden, Brian G., Z Sekanina, and D. K. Yeomans. 1973. "Comets and Nongravitational Forces. V." *The Astronomical Journal* 78 (March): 211. doi:https://doi.org/10.1086/111402.

Myhrvold, Nathan. 2016. "Comparing NEO Search Telescopes." *Publications of the Astronomical Society of the Pacific* 128 (962): 045004. doi:https://doi.org/10.1088/1538-3873/128/962/045004.

Rubincam, David Parry, and Stephen J. Paddack. 2010. "Zero Secular Torque on Asteroids from Impinging Solar Photons in the YORP Effect: A Simple Proof." *Icarus* 209 (2): 863–65. doi:https://doi.org/10.1016/j.icarus.2010.05.015.

Stuart, J. S., & Binzel, R. P. 2004. "Bias-corrected population, size distribution, and impact hazard for the near-Earth objects." *Icarus* 170(2):295–311. doi:https://doi.org/10.1016/j.icarus.2004.03.018.

Vereš, Peter, Robert Jedicke, Richard Wainscoat, Mikael Granvik, Steve Chesley, Shinsuke Abe, Larry Denneau, and Tommy Grav. 2009. "Detection of Earth-Impacting Asteroids with the next Generation All-Sky Surveys." *Icarus* 203 (2): 472–85. doi:https://doi.org/10.1016/j.icarus.2009.05.010.

Warner, Brian D., Alan W. Harris, and Petr Pravec. 2009. "The Asteroid Lightcurve Database." *Icarus* 202 (1): 134–46. doi:https://doi.org/10.1016/j.icarus.2009.02.003.

Chapter 7
Overview of Active Planetary Defense Methods

David Morrison

7.1 Introduction

The two essential functions of planetary defense are to locate any asteroid on a collision course with Earth and to deflect or destroy it before it hits. Most of the research and funding in planetary defense have been devoted to asteroid surveys and orbit calculations, in order to predict an impact as soon as possible before the actual strike. Once an impact is predicted, it is imperative to have a means of defending ourselves (Ahrens and Harris 1992). For small asteroids, the prediction itself may be sufficient for authorities to issue warnings and allow people living in the target area to evacuate. For larger impacts, however, an active defense is desirable. To date, relatively little has been done to prepare for such an eventuality (Canavan 1994; Chapman et al. 2001; Morrison 2005; Morrison et al. 2002; Morrison and Teller 1994; Simonenko et al. 1994). One European consortium called NEOShield has received funding to survey a broad spectrum of possible responses (Harris et al. 2013).

Once a potentially dangerous impact has been identified, there are several possible approaches to defending the Earth. As noted above, warning and evacuation may be sufficient to protect populations from small asteroids. If active defense is required, we may either deflect the asteroid (change its orbit so that it misses the Earth or strikes in an uninhabited area such as oceans or deserts) or break it up far enough from Earth that the debris are dispersed and miss the planet. There are several ways to accomplish this, as discussed in this chapter (Shapiro et al. 2010).

Many people's ideas about active defense are conditioned to some extent by three Hollywood films that have depicted such scenarios: *Meteor*, *Deep Impact*, and

D. Morrison (✉)
NASA Ames Research Center, Mountain View, CA, USA
e-mail: david.morrison@nasa.gov

N. Schmidt (ed.), *Planetary Defense*, Space and Society,
https://doi.org/10.1007/978-3-030-01000-3_7

Armageddon. All of these films imagine using nuclear explosives to break up the incoming asteroid or comet, and all of these depictions are technically flawed. In *Meteor* (1969), a fleet of American and Russian ICBMs are fired toward the incoming asteroid just hours before impact, and the nuclear explosions cause the object to be vaporized and disappear entirely. In *Deep Impact* (1998), there is also a multi-warhead nuclear attack at the last minute, breaking up the comet and producing a spectacular meteor shower as the fragments enter the atmosphere. In *Armageddon* (1998), a buried nuclear explosion causes the incoming asteroid to break into two pieces, which separate and pass the Earth harmlessly. As we will see, the concept of breaking up an incoming projectile just hours before impact is a poor one, likely to do more harm than good.

Most defense strategies involve deflection, changing the asteroid orbit so that no impact takes place. This is not accomplished by applying a force to the asteroid in any direction, but rather requires that it be accelerated either along or against its orbital velocity vector. The goal is to slightly change the orbital period of the asteroid, a cumulative effect that can eliminate the predicted impact. Such deflections need to be applied a long time (typically a decade) before the impact (Ahrens and Harris 1994; Melosh et al. 1994).

The alternative of disruption must also be applied well in advance, in order for the majority of fragments to disperse and miss the planet. Such an approach could be called *deflection by disruption*. If most of the original mass does hit, the energy input to the atmosphere would be huge, sufficient to cause as much or more damage than would the original single impactor. Rather than a beautiful meteor shower, we would face a firestorm as the impactor energy is deposited in the atmosphere.

We should note that either the original asteroid (in the case of deflection) or some of the fragments (from disruption) have the potential to return to Earth on nearly their original orbits. Most such cases will be benign, but there is often the possibility that they could still strike the Earth. A close pass by the Earth can lead to such an unwelcome return if the asteroid passes through a small region of space called a keyhole (Chodas 1999; Milani et al. 2002; Valsecchi et al. 2003). It is therefore important to apply deflection in a way that misses the keyholes. This may demand additional small adjustments to the orbit, for example by the technique called a gravity tractor, discussed in Sect. 7.2.4.

Finally, the term *mitigation* as used in planetary defense has two different meanings. This chapter discusses efforts to change the orbit of the object so as to eliminate the impact, or at least to move the target point to an unpopulated part of the Earth. Mitigation can also mean the possible evacuation of people from the expected impact area and the recovery that would be needed after a major impact. These topics, which are related to civil defense and responses to natural disasters such as earthquakes and severe storms, are not discussed in this chapter: we concern ourselves here with prevention, not disaster response.

7.2 Deflection and Disruption Technologies

7.2.1 *Introduction*

There are a variety of proposed methods to deflect or disrupt an asteroid that threatens impact with the Earth. We know from existing surveys that there are currently no potentially hazardous Earth-approaching asteroids as large at the 10-km object that led to the K-Pg mass extinction 66 million years ago. The most likely undiscovered asteroids that might cause substantial damage are in the diameter range from approximately 100–300 m. There also remain undiscovered a few near-Earth asteroids larger than 500 m, which are potentially very destructive. Planetary defense technologies are focused on targets in these size ranges. To deal with such impacts, several studies have surveyed the technological possibilities (Ahrens and Harris 1994; Barbee et al. 2018; Chapman et al. 2001; Gritzner and Kahl 2004; Harris et al. 2013; Melosh et al. 1994; Morrison et al. 2002; Shapiro et al. 2010; Simonenko et al. 1994; Solem 2000). The options are summarized in this chapter.

7.2.2 *Kinetic Impactor*

The most straightforward and mature technology for changing the orbit of an asteroid is through kinetic impact (KI), in which the asteroid is struck by one or more spacecraft, moving either along or against its orbital motion. The more massive the impactor and the higher the impact velocity, the greater the momentum transferred to the object and its subsequent change in orbital period will be.

The transfer of momentum to the target is straightforward if the mass of the target is known and the impact is directed toward its center of mass. It does not depend on the physical properties of the target. However, there is an amplification of the effect if the impact forms a crater and ejects surface material. The jet of ejecta itself carries away momentum, and it acts as a simple reaction engine to provide an additional push to the target. The momentum multiplication factor is termed *beta*. If there is no ejection of material, beta equals 1, but beta can have a value as high as 3–4 depending on the impact velocity and the nature of the target surface (Holsapple 2004). The unknown value of beta leads to a substantial uncertainty in predicting the change in momentum produced by a KI. If the asteroid is a loosely bound rubble pile, as might result if it has been partly broken up by impacts and re-accreted, the value of beta is likely to be low.

The change in the asteroid velocity can be estimated in advance of the impact by assuming values for the target mass and the beta factor. In practice, it might be desirable to launch several impactors spaced days or weeks apart, and to evaluate the change in the target's orbit after each impact in order to decide the best strategy. Fortunately, there is no harm if the target is moved farther, once it has been deflected away from an impact on the Earth. It is however important to ensure that the new orbit does not lead to a keyhole, which would offer a free trajectory back to an impact years later. One strategy would be to have a well-instrumented observer spacecraft moving with the target, both to evaluate the change in orbit achieved by each kinetic impact and to provide a fine orbital adjustment, acting as a gravity tractor. However, it is much more difficult to match the asteroid orbit with a rendezvous spacecraft than it is simply to hit the target at high velocity.

There is great interest among the planetary defense community in carrying out a technology demonstration of KI. The NASA Deep Impact spacecraft did successfully impact a comet (9P/Tempel) in 2005; however, the large mass of the target and the absence of an observer spacecraft made it impossible to determine the change of momentum (which was not, in any case, the objective of the experiment). Currently, NASA and ESA are developing a joint mission called Asteroid Impact and Deflection Assessment (AIDA) to test KI technology. The target will be the binary asteroid Didymos, with the deflection experiment to occur in October, 2022. NASA plans to provide the Double Asteroid Redirection Test (DART) spacecraft to perform the KI experiment, while ESA proposes to provide the Hera spacecraft to rendezvous with the target to monitor the deflection. The DART impact on the smaller member of the binary at ~6 km/s will change the binary orbit period, which will be measured by Earth-based observatories as well as Hera. The primary goals of AIDA are to demonstrate KI technology and characterize the deflection caused by the impact (Cheng et al. 2018).

Today's planetary defense planners generally consider that one or more kinetic impactors would be the defense approach of choice in deflecting an asteroid with a diameter up to a few hundred meters, if sufficient lead time is available (Barbee et al. 2018). However, the situation will need to be examined carefully for any specific asteroid, since both physical and orbital parameters are highly variable. In the real world, the asteroid will select us, and we will have no control over these variables.

7.2.3 Nuclear Explosive

Because a nuclear explosive carries far more energy in a given mass than a kinetic impactor, it is often considered to be the better choice in cases where the asteroid is too large, or the warning time too short, to use kinetic impactors. Like KI, nuclear deflection does not require any landing or coupling to the target, and it can be deployed on a fast trajectory with no rendezvous required. The deflection is achieved indirectly, by using a standoff explosion to heat a thin surface layer (primarily by an intense flux of neutrons) and using the escape of the resulting plasma to provide an impulse to the target. The optimum standoff distance is approximately 0.4 times the radius of the object. As with KI, this force needs to be exerted along or against the direction of motion of the asteroid. An alternative way to use a nuclear explosive is to crater the surface, ejecting mass in a direction opposite to the desired force, similar to the enhancement of momentum transfer from a kinetic impact. Options have also been considered where a penetrator emplaces the nuclear charge below the surface for detonation.

An extension of the above option would be to intentionally break up or disrupt an asteroid. Such a breakup can be modeled, but the way the asteroid breaks apart and the velocity distribution of the fragments will depend on its composition and internal structure, which are difficult to determine by remote sensing. Given these uncertainties, and the requirement that the fragments be dispersed so as to miss the Earth, the best strategy might be to use the maximum explosive energy available. This approach might be an option of last resort, after other methods have failed.

While it would be relatively easy to use existing rockets to send a nuclear explosive to an asteroid on a fast orbit, the detonation of the charge as the missile passed the asteroid would have to be achieved with accuracy within a fraction of a second. We also have little empirical experience with the effects of nuclear explosions in space, although they have been extensively modeled.

Nuclear explosives have a high level of technological readiness, although they have not been developed for planetary defense. Unlike the kinetic impactor, it would be politically difficult to test the nuclear technology against an asteroid, and even with a test, the results would probably depend on the specific properties of the target. Currently, there are international agreements that prohibit nuclear explosions in space. While such restrictions could be relaxed in case of a real asteroid impact threat, it may not possible to do so for a demonstration test.

7.2.4 *Gravity Tractor*

The gravity tractor concept has been suggested as the best way to make a controlled change in the orbit of an asteroid. It could be used to trim the orbit in order to miss a gravitational keyhole, which would likely be only a few kilometers in size (Lu and Love 2005). Like the kinetic impactor, the gravity tractor depends on basic physics and does not require any knowledge of the composition or shape of the target. Effectiveness does depend on the mass of the asteroid, which could be derived from the tractor experiment itself.

The gravity tractor uses the gravitational acceleration between the asteroid and the spacecraft to adjust the orbit of the asteroid. Since it requires a relatively massive spacecraft that achieves rendezvous with the asteroid, it demands more time than the KI or nuclear options, which can reach the target using direct high-speed trajectories. The tractor spacecraft is placed in front of or behind the target at a distance of about one asteroid radius. The attraction between them will gradually cause the spacecraft to move closer to the asteroid, and this slow drift must be countered by a slow acceleration of the spacecraft in the direction we wish to pull the target. By maintaining the same distance between the tractor and the asteroid, the two together are gradually deflected into the desired orbit.

The optimum low-thrust propulsion type currently available is solar electric. The ion beam of a solar electric rocket engine must not impinge on the asteroid, or the effects are lost. Therefore, most designs use two tilted ion engines. The goal is to achieve a combination of tilt and distance from the asteroid surface that provides the maximum thrust without hitting the target. The same effects could be achieved with a single thruster, canted away from the asteroid, and spinning slowly around the spacecraft-to-asteroid direction.

Although it provides only a slow shift in orbit, as mentioned before, the gravity tractor requires no knowledge of the physical or chemical properties of the target. It also can work with highly irregular asteroid shapes. Its most likely use would be to make small orbital adjustments to avoid a gravitational keyhole.

The following mitigation approaches are conceptual only, using technologies that are not mature. They also require altering the spin state of the target asteroid, which is something that would require new technology.

7.2.5 Laser and Solar Thermal

A nuclear deflection is based on heating the surface with a stand-off explosion, thereby generating a reaction effect as the hot plasma evaporates. A similar approach could be used if a small part of the surface is heated repeatedly. Two ways to accomplish this have been proposed, using either sunlight or a laser beam to focus energy on the surface. The laser would require its own power source, perhaps a small nuclear reactor. Sunlight is free, but it requires a large reflector to collect the light and focus it on the surface.

In both cases, the reaction force needs to be along or against the orbital vector of the asteroid. The most efficient approach is to remove the asteroid spin and position the spacecraft at an appropriate distance above the surface. Despinning an asteroid, however, is a difficult requirement. It would almost certainly require landing and surface operations, which would present a much greater technical challenge than the previously described deflection methods, made worse by the fact that many small asteroids spin rapidly. An advantage of this type of deflection is that the change in the asteroid orbit could be monitored continually, just as is done with a gravity tractor.

7.2.6 Ion Beam and Direct Push

An alternative way to change the orbit of an asteroid is with a slow push. In the discussion of the gravity tractor, it was noted that one would need to avoid impacting the surface with the spacecraft ion beam. Alternatively, it would be possible to use a more powerful ion beam to directly impact the surface, thereby pushing the asteroid. While this would be most effective if the asteroid were first despun, this might not be necessary for a slowly spinning asteroid.

The second option, a direct push by the spacecraft, would require that the rotation of the asteroid be removed. Once the correct orientation of the asteroid is achieved, the required push could be obtained with a chemical rocket, or more likely an ion beam thruster fastened to the surface. This energy could be provided by sunlight; however, the required direction of push is at 90° from the subsolar point.

7.3 Conclusions

Any planetary defense strategy requires that we consider what actions we can take when an asteroid is discovered on an impact trajectory. If the asteroid is small or is directed toward an uninhabited region (such as deserts, polar regions, or the deep ocean), it may be sufficient to issue a warning and to encourage scientists to set up instruments to record the event. We know from our experience with the 2013 Chelyabinsk airburst that even a short warning time would eliminate most risk of casualties from an impactor with diameter about 20 m. The 1908 Tunguska event teaches us that an asteroid with energy of an order of magnitude greater could severely damage an area of thousands of square kilometers. A hit of that magnitude in an area of low population, such as the Tunguska district itself, would produce few casualties, but it could be disastrous in a populated region. In most cases, however, a warning of a few days would be sufficient to evacuate those in harm's way.

When the diameter of a stony impactor exceeds about 100 m, it may be necessary to attempt to deflect or disrupt it. We can expect an impact by a 100 m or larger asteroid on the land area of Earth only once every few thousand years (Harris 2018), but it could happen at any time. Today, it is unlikely that an impact by an asteroid this large would be predicted long in advance, but the prospects for a timely warning increase as more sensitive telescopes on Earth or in space are added to the NEO surveys.

When an impact is predicted, we calculate that the area of the planet that is at risk is confined to a rather narrow risk corridor. Thus, we will know immediately that most areas are not at risk, even before we are certain that an impact will actually take place. The defense options will depend on the parts of the planet that are within the risk corridor. As more data become available on the orbit of the threatening asteroid, the corridor shrinks. While the obvious goal is to use kinetic impacts or nuclear explosions to deflect the asteroid and move the impact point off the planet, there may also be defense strategy options to move the impact point a smaller distance to an uninhabited region. Schweickart (2004) notes the *Hobson's choice* between doing nothing, thereby suffering the consequences of an impact, or proactively deflecting an asteroid which will, in the process of "protecting the Earth," necessarily place otherwise non-threatened people and property at risk. All such strategies pose difficult and divisive questions for international governance of planetary defense.

The most straightforward way to deflect the threatening asteroid is to change its orbital period. This can be achieved by applying a force along or against the direction of motion. Depending on the orbital details and the time available, one or more kinetic impactors would probably be the preferred option. While the direction of the deflection is easily calculated, its magnitude will be uncertain because of the unknown beta factor, which can add substantially to the momentum transfer from the impactor to the asteroid. Therefore, an observer spacecraft should be in place to monitor the results, and it may be desirable to have the option of multiple KIs spaced several weeks apart. A gravity tractor spacecraft may also be deployed to provide a final trim to the new asteroid orbit, to ensure that it is not heading toward a gravitational keyhole that could lead to a future impact.

As an alternative to the kinetic impact, planners will likely consider the option of a nuclear stand-off explosion to deflect the asteroid. Just as with KI, we will not know in advance how effective this will be, and the asteroid orbit will need to be redetermined to ensure avoidance of any keyholes. There will also be the alternative of using a surface or subsurface explosion to disrupt the asteroid, but we must ensure that no large fragments remain on an impact trajectory. A necessary part of disruption is to disperse the fragments so that few, if any, can strike the Earth.

At present these three options—kinetic impact, gravity tractor, and nuclear explosions—constitute our potential defense technologies. Other suggestions, such as laser or solar heating or various slow-push options, are not technologically mature enough to be considered.

In all such planning, there is an inherent problem in developing defense technologies, or even building defense systems, in advance of a defined threat. While a certain level of planning and technology development is desirable, the impact threats we are discussing are likely to occur on millennial time scales. It is therefore appropriate to remember the caution (sometimes called the *deflection dilemma*) against deploying such technologies in advance of requirements. Their creation may create the possibility of their use as weapon systems, rather than as altruistic means to defend our planet from cosmic threats (Harris et al. 1994; Sagan and Ostro 1994). Absent strong international controls, a misused planetary defense technology could be worse than the threat it is designed to counter.

References

Ahrens, T. J., & Harris, A. W. (1992). Deflection and fragmentation of near-Earth asteroids. *Nature, 360*(6403), 429–433. doi:https://doi.org/10.1038/360429a0

Ahrens, T. J., & Harris, A. W. (1994). Deflection and fragmentation of NEAs. In GehrelsT. (Ed.), *Hazards Due to Comets and Asteroids* (pp. 897–928). Tucson: University of Arizona Press.

Barbee, B. W., Syal, M. B., Dearborn, D., Gisler, G., Greenaugh, K., Howley, K. M., et al. (2018). Options and uncertainties in planetary defense: Mission planning and vehicle design for flexible response. *Acta Astronautica, 143*(August 2017), 37–61. doi:https://doi.org/10.1016/j.actaastro.2017.10.021

Canavan, G. H. (1994). Cost and benefit of near-Earth object detection and interception. In T. Gehrels (Ed.), *Hazards Due to Comets and Asteroids* (pp. 1157–1191). Tucson: University of Arizona Press.

Chapman, C. R., Durda, D. D., & Gold, R. E. (2001). *The comet/asteroid impact hazard, a systems approach*. San Antonio, TX: Southwest Research Institute.

Cheng, A. F., Rivkin, A. S., Michel, P., Atchison, J., Barnouin, O., Benner, L., et al. (2018). AIDA DART asteroid deflection test: Planetary defense and science objectives. *Planetary and Space Science, 157*, 104–115. doi:https://doi.org/10.1016/j.pss.2018.02.015

Chodas, P. (1999). Orbit uncertainties, keyholes, and collision probabilities. In *Bulletin of the American Astronomical Society* (Vol. 31, p. 1117).

Gritzner, C., & Kahl, R. (2004). Mitigation technologies and their requirements. In M. Belton, T. Morgan, N. Samarasinha, & D. Yeomans (Eds.), *Mitigation of Hazardous Comets and Asteroids* (pp. 167–200). Cambridge University Press.

Harris, A. W. (2018). Population and impact frequency of Tunguska-size NEAs (in press). *Icarus*.

Harris, A. W., Canavan, G. H., Sagan, C., & Ostro, S. J. (1994). The deflection dilemma: Use vs. misuse of technologies for avoiding interplanetary hazars. In T. Gehrels (Ed.), *Hazards Due to Comets and Asteroids* (pp. 1145–1156). Tucson: University of Arizona Press.

Harris, A. W., Barucci, M. A., Cano, J. L., Fitzsimmons, A., Fulchignoni, M., Green, S. F., et al. (2013). The European Union funded NEOShield project: A global approach to near-Earth object impact threat mitigation. *Acta Astronautica, 90*(1), 80–84. doi:https://doi.org/10.1016/j.actaastro.2012.08.026

Holsapple, K. (2004). About deflecting asteroids and comets. In M. Belton, T. Morgan, N. Samarasinha, & D. Yeomans (Eds.), *Mitigation of Hazardous Comets and Asteroids* (pp. 113–140). Cambridge University Press.

Lu, E. T., & Love, S. G. (2005). Gravitational tractor for towing asteroids. *Nature, 438*(7065), 177.

Melosh, H. L., Nemenchinov, I. V., & Zetzer, Y. I. (1994). Non-nuclear strategies for deflecting comets and asteroids. In T. Gehrels (Ed.), *Hazards Due to Comets and Asteroids* (pp. 1111–1132). Tucson: University of Arizona Press.

Milani, A., Chesley, S. R., Chodas, P. W., & Valsecchi, G. B. (2002). Asteroid close approaches: Analysis and potential impact detection. In W. F. Bottke (Ed.), *Asteroids III* (pp. 55–70). Tucson: University of Arizona Press.

Morrison, D. (2005). Defending the Earth Against Asteroids: The Case for a Global Response. *Science & Global Security, 13*(1–2), 87–103.

Morrison, D., & Teller, E. (1994). The Impact Hazard: Issues for the Future. In T. Gehrels (Ed.), *Hazards due to Comets and Asteroids* (pp. 1135–1143). Tucson: University of Arizona Press.

Morrison, D., Harris, A. W., Sommer, G., Chapman, C. R., & Carusi, A. (2002). Dealing with the impact hazard. In W. F. Bottke (Ed.), *Asteroids III* (pp. 739–754). Tucson: University of Arizona Press.

Sagan, C., & Ostro, S. J. (1994). Long-range consequences of interplanetary collisions. *Issues in Science and Technology, 10*(4), 67–72.

Schweickart, R. L. (2004). The real deflection dilemma. In *AIAA Planetary Defense Conference* (p. AIAA-2004-1467).

Shapiro, I. I., Vilas, F., A'Hearn, M., Cheng, A. F., Abell, P., Benner, L. a. M., et al. (2010). *Defending Planet Earth*. Washington, D.C.: National Academies Press. doi:https://doi.org/10.17226/12842

Simonenko, V., Nogin, V., Petrov, D., Shubin, O., & Solem, J. C. (1994). Defending the Earth against impacts from large comets and asteroids. In T. Gehrels (Ed.), *Hazards Due to Comets and Asteroids* (pp. 929–954). Tucson: University of Arizona Press.

Solem, J. C. (2000). Deflection and disruption of asteroids on collision course with Earth. *Journal of the British Interplanetary Society, 53*, 180–196.

Valsecchi, G. B., Milani, A., Gronchi, G. F., & Chesley, S. R. (2003). Resonant returns to close approaches: Analytical theory. *Astronomy & Astrophysics, 408*(3), 1179–1196.

Chapter 8
NASA NEO Deflection Application: Current Capabilities and Limitations

Nahum Melamed

8.1 Introduction

The NEO hazard extends from no-warnings to decades-long discoveries, and from insignificantly small objects to those capable of remarkable destruction. There is not a single approach for dealing with the entire range, but when time is short and objects are large, only powerful means may deliver results. The most powerful method known today for counteracting an incoming object is nuclear deflection or destruction, and the technology to launch interplanetary payloads exists. When circumstances allow, nonnuclear methods can be given priority. But exclusion of the stigmatized nuclear option from consideration limits the ability to reduce the risk in demanding situations. Elucidating some of the borderlines between NEO mitigation regimes is the topic of this chapter. A NEO Deflection App (NDA) developed jointly by The Aerospace Corporation and NASA's Jet Propulsion Laboratory (JPL) is used to quantify deflection requirements by the high energy Kinetic Impact method, and highlight limitations of the approach. The NDA was applied at planetary defense conferences and workshops to study hypothetical scenarios, and revealed its utility and limitations in providing realistic estimates. The NDA uses launch vehicle performance and orbital mechanics calculations to deliver deflection payloads to the NEO, but it does that in the nominal sense. Desired updates to the NDA to increase its accuracy are inclusion of NEO orbital uncertainty effects on impact probability, and effect of launch and space systems reliability data on mitigation campaign sizing.

N. Melamed (✉)
The Aerospace Corporation, Culver City, CA, USA
e-mail: Nahum.Melamed@aero.org

N. Schmidt (ed.), *Planetary Defense*, Space and Society,
https://doi.org/10.1007/978-3-030-01000-3_8

8.2 Information Required for Mitigation Decisions

8.2.1 Information Required Before and During Emergency

Priorities and resources are allocated far differently when an impact event is imminent versus if it is theoretical, and these depend on what mitigation capability exists at the time of emergency. Resources that are allocated for building a mitigation capability in advance of an emergency significantly expand response options, especially to short warnings. Conversely, the ability to respond quickly and reliably is compromised in the absence of proactive investment, implying that mitigation capability must be developed when the threat is confirmed and may have a lower likelihood of success. For this reason, decisions that are made before the threat becomes established conclusively reinforce the prospect of a successful and effective mitigation mission.

A key factor affecting decisions is the inherent uncertainty in the early phase of NEO discovery. When an object is discovered, precise knowledge about its orbit, size, composition and other characteristics is typically lacking. As more information is gathered, the initial impact probability assessment may rise or fall over time. In the vast majority of cases, additional observations will rule out an impact in the foreseeable future (Chodas 2018a). This means that because significant impact events are very rare, mitigation systems that are prepared in advance will seldom be used.

The present absence of a confirmed NEO threat routes limited resources to more immediate needs. However, historical impact events tell us that although such events rare, the consequences could be catastrophic. That creates a decision dilemma: should proactive mitigation decisions continue to be based on likelihood alone, or should decisions include both the likelihood and consequences of an impact? Decisions that are informed by taking both likelihood and consequences into account are more likely to achieve reduction in the total NEO risk, when balanced with routine priorities.

The technology required for a system to deflect a threatening NEO is within grasp, but the investment required to build and maintain deflection systems with adequate reliability is substantial, and the return is not clear and definite without a specific threat. In the absence of proactive planning, that expenditure would occur at the onset of a threat. Therefore, low-level investment starting in advance of a confirmed threat would methodically execute the endeavor and would lead over time to a sustainable mitigation capability more ready to defend against a wider range of objects and short warning times than is presently possible. Some of the questions a decision-maker might ask at the onset of a NEO threat may be:

- When is the object predicted to impact?
- What is the probability of impact, and what are the odds the threat will diminish over time?
- What can be done to refine that assessment?

- What is the expected collateral damage and casualty estimate if an impact occurs?
- Has development of mitigation capability started and how far is it from being operational?
- Is the effort internationally agreed upon and well-coordinated?
- What options are possible now?
- Is there time to build and use additional capabilities?
- What is the likelihood of mitigation success?
- How much will it cost?

The answers to some of these questions depend on decisions made well in advance of the threat. For example, the available options depend on what was built and validated proactively and is deployable now. It is therefore constructive to inform today's mitigation decisions in order to create options for tomorrow's real-time decisions. To help answer some of these questions, the physics-based Neo Deflection App (NDA) was jointly developed by The Aerospace Corporation and NASA's Jet Propulsion Laboratory (JPL).

8.3 Description of the NEO Deflection App

8.3.1 The Target Plane, or B-Plane

An important tool for studying planetary encounters relies on the concept of a target plane, or B-plane, which is defined as a geocentric plane oriented normal to the NEO's geocentric velocity vector (Milani et al. 2002). Figure 8.1 depicts the intersection between the target B-plane centered on Earth and the trajectory plane of the approaching NEO (CNEOS n.d.). The point of intersection ŭ35;ŭ35; of the NEO's unperturbed trajectory with the target plane correlates with a reliable estimate of the actual miss distance with Earth and provides considerable insight into the nature of an encounter to a few Earth radii, within which the Linear Encounter Theory is considered applicable. The CNEOS website describes the form by which it is utilized in the web-based NEO Deflection App (NDA), discussed in detail below (NASA 2017).

8.3.2 Purpose of the NEO Deflection App

At the time of writing, no space object capable of inflicting considerable damage in the near future is known to be on a collision course with Earth. Nevertheless, the current rate of discovery, along with new information on atmospheric bolide events created by natural objects, indicates that it is not a matter of if but when a noteworthy impact will occur (Chodas 2018b). Small objects are more abundant, so their

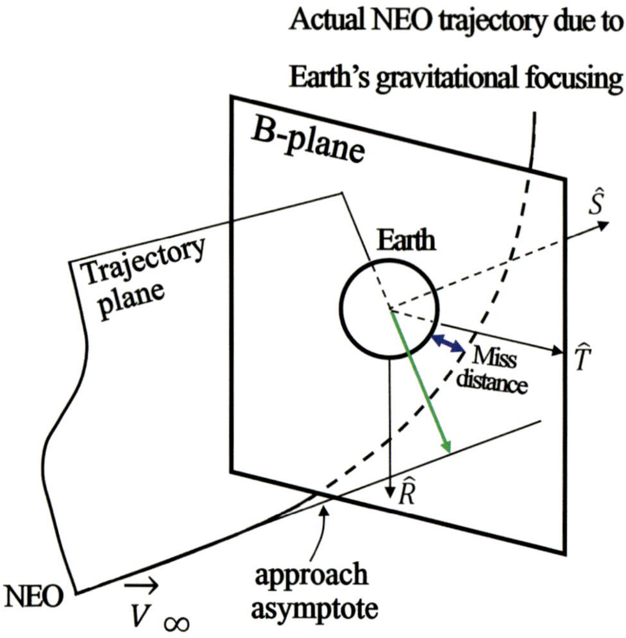

Fig. 8.1 B-plane [Earth Target Plane]

likelihood of collision is higher and they are more likely to be discovered late, but the extent of the damage they can inflict is local. Impacts from less frequent, larger objects are certainly possible, but the true parameters of the encounter are not known until a threat is confirmed, ranging from local, regional and beyond depending on the specifics of the threat. It is therefore helpful to predict beforehand what effort it would take to avert or reduce the risk of disaster, and understand what can gradually be put in place now in order to create a response capability available in times of emergency.

Planetary defense studies and exercises conducted in the 2000s identified the need for a software tool to assist and inform the decision process. Launch of deflection payloads is an essential capability to influence the outcome of an impending encounter with a NEO. How much capability is needed for a given NEO threat scenario can be estimated with the help of NDA, developed as a joint venture between The Aerospace Corporation and NASA's Jet Propulsion Laboratory (JPL). The NDA provides insights into the problem of impulsively deflecting a simulated threatening NEO by a kinetic impactor (KI) mission. The population of real NEOs is represented in the NDA by a set of fictitious, but realistic, impacting NEOs. The app has been pre-loaded with over a dozen JPL-supplied simulated Earth impacting NEOs, including a hypothetical asteroid "2015 PDC" created for the impact scenario discussed at the 2015 IAA Planetary Defense Conference (PDC), and "2017 PDC", for the 2017 PDC.

The NDA enables users to design deflection missions for any of the hypothetical NEOs by applying small velocity changes to the NEO in various directions and at various times before the projected Earth impact (Melamed 2012; Yeomans et al. 2012). These velocity changes are mapped to movement on the Earth impact B-plane of the NEO's approach asymptote intercept point with this plane. To avert an impact with Earth, the intercept point is driven outside the Earth's capture radius superimposed on the Earth impact B-plane. As the NEO gets closer, Earth's gravitational field interacts with it and perturbs its trajectory, so the capture circle is always bigger than Earth to compensate for this "gravitational focusing" effect. In addition, the user can select from several launch vehicles and assess and compare capabilities to launch spacecraft to intercept the NEO and change its velocity. At present, two existing American launch vehicles (Atlas V and Delta IV Heavy), and two planned (Falcon Heavy and NASA's SLS) are featured in the NDA. The mission is considered successful if the user inputs safely move the nominal Earth impact point outside the Earth's capture cross section, as illustrated in Fig. 8.2.

This physics-based tool applies existing and developing launch technology to launch deflection payloads to intercept the oncoming object as early as possible, and to attempt to deflect it to the largest miss distance possible. Because construction and verification of launchers and deflectors is lengthy and is added to the transfer time from launch to NEO intercept, the practicality of this approach is quickly revealed after a brief interaction with the NDA. Large objects require big kinetic

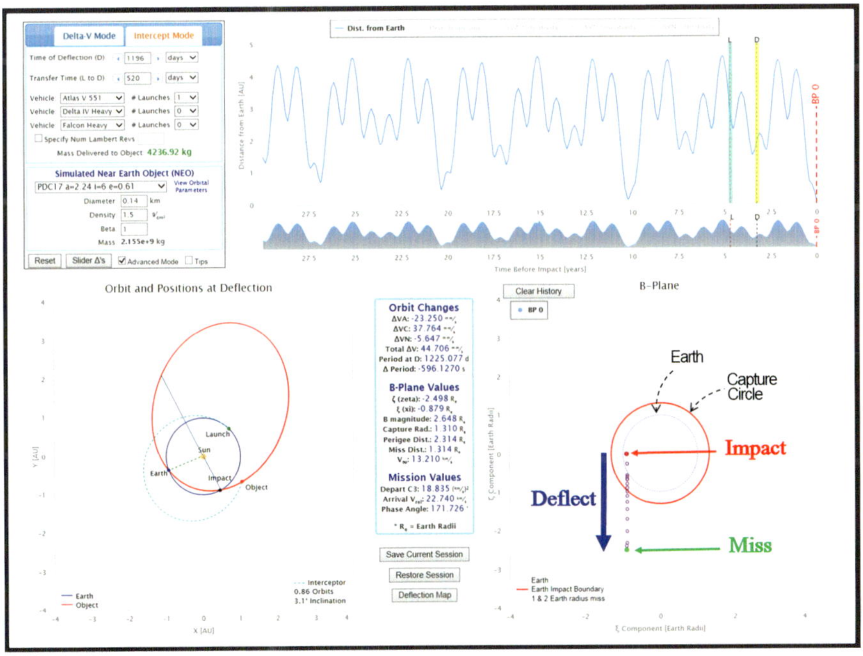

Fig. 8.2 NEO Deflection App [NDA]

deflectors and heavy lift launch vehicles that may or may not exist at the time of need, or alternatively, an assembly of multiple smaller lifters to create a deflection campaign. A campaign involving several deflection systems may also be desired from the perspective of mission assurance due to the imperfect reliability of a single system. Either way, depending on the object size, a substantial amount of hardware must be built and tested to create the necessary deflection capability.

Another key factor affecting mitigation is the amount of warning time available before a predicted impact. Here again, a brief interaction with the NDA quickly reveals the response limit of the kinetic deflection method as the time to impact gets shorter. Transfer time of the kinetic impactor from Earth to the NEO is typically on the order of a year. Consequently, short times to impact impose limitation to the miss distance that can be developed even if the deflection system is ready to launch. The combined times required to build and deliver the systems to the NEO will dictate what range of objects and warning times can be mitigated effectively. The NDA correlates effects of object size, warning time and Earth miss distance, thus providing a first-order assessment of mitigation capability required to negate a range of hypothetical threat scenarios.

The NDA is loaded with a set of hypothetical threatening objects spanning 30 years of heliocentric travel ending with an Earth impact. Because these are not real objects, their discovery time can occur anywhere in the time frame. The amount of warning time is simply assumed by the user to create scenarios of various lengths with the same fictitious object. The physical characteristics of the object such as its mass, composition and velocity are also assumed by the scenario designer, thus expanding the range of cases that can be studied. This ties in with mission and campaign success metrics, which heavily depend on these parameters. For example, the amount of warning time assumed informs whether an in-situ characterization precursor mission for refining the parameters of the threat and optimizing a deflection mission is possible. Given assumed object characteristics, the NDA not only identifies launch windows and deflection opportunities for the kinetic deflectors, but also determines the practical limitations of the approach. It clearly demonstrates that when warning times are short, or objects are massive, kinetic impactors are limited in their capacity to produce adequate deflection, but more powerful means such as a nuclear explosion may still be able to save the day.

The use of nuclear detonation at an optimal orientation and distance from the approaching NEO while it is still far out in interplanetary space has been proposed as an alternative or complementary method to the Kinetic Impact method of deflection in situations where a powerful deflection might be the only solution (Wie 2013). Calculations show that standoff detonation creates considerably larger deflections than is possible with a kinetic impactor, a critical consideration when time is scarce. The nuclear option is not without controversy because of the perceived concern about launch failure with nuclear material onboard or misuse of this capability in space. Another aspect of the controversy is that nuclear weapons were created for the purpose of destruction, are stigmatized because of this origin, and thus are not perceived in the public's mind as a potential means for the mass saving of lives. Nevertheless, these valuable considerations are outside the scope of the NDA in its current form, which is simply to illustrate the potential and limitations of the Kinetic Impact approach, highlighted by the assumed existence or lack of ready-to-launch deflection systems. It is up to the decision-maker to act on this insight and decide what level of proactive investment and what type of protective measures might be rationalized to effectively mitigate a short warning hypothetical NEO threat.

8.3.3 Elements of the NEO Deflection App (NDA)

The NDA uses launch vehicle performance information to approximate the minimum number of launch vehicles required to deliver kinetic energy deflectors to a fictitious NEO designed to impact Earth. Leaders would rely on such information to know what resources should be allocated and when. Although the NDA exclusively deals with fictitious NEOs, it could provide first estimates of feasible solutions for deflection of real objects when the appropriate information becomes available.

The NDA applies a Lambert algorithm to compute the transfer trajectory from Earth launch to NEO intercept by the kinetic impactor. This process determines the Earth excess velocity, which correlates with the minimum number of launch vehicles required to deliver the total payload mass to the NEO. In addition, the relative velocity between the spacecraft and the NEO is computed and used to determine the amount and direction of the applied deflection impulse. In the final step, the NDA maps the modified orbit to a new location on the B-plane based on NEO's mass and density inputs, and computes the resulting Earth miss distance. This process is depicted in the B-plane panel at the lower right corner of Fig. 8.2.

8.4 Application of the NEO Deflection App

8.4.1 Planetary Defense Conferences and Exercises

Natural venues for application of the NDA are scientific planetary defense conferences, workshops and classes. The Aerospace Corporation began organizing biannual Planetary Defense Conferences (PDCs) in 2004 to provide a focus for the international community to discuss and agree on the most promising ways to address the many aspects of planetary defense. Exercises are held at these venues to inform the community, foster collaboration, and formulate solutions to planetary defense-related issues. During these exercises, valuable information about a predicted impact flows between key organizations and emergency managers. Discussions held at the PDCs enable rapid progress within two United Nations-endorsed forums, the International Asteroid Warning Network (IAWN), and the Space Mission Planning Advisory Group (SMPAG), and improve international coordination of activities (Dunbar 2014). One of the many recommendations coming out of the PDCs is that new and improved software tools be developed to design deflection missions using simulated NEOs to assess our ability to deal with real Earth impactors.

The NDA has supported planetary defense exercises, which are an integral part of the PDCs (PDC 2015, 2017) and of NASA-FEMA tabletop workshops. Melamed (2015) examined NEO deflection mission capabilities for the 2015 hypothetical asteroid impact scenario (PDC 2015). The NDA was used in that study to gain insight into deflection requirements, possibilities and limitations in the following categories: NEO discovery time; early vs. late decision to build and launch the deflection device; NEO orbit type, object size and composition; US and international launch vehicles; and the number of launches required in a deflection campaign. The 2015 Planetary Defense's simulated asteroid and hypothetical impact scenario was used to examine launch window availability, attainable relative velocities and kinetic deflector lift capability. Several insights were demonstrated, including: early NEO discovery increases mission options and deflection effectiveness; forthcoming heavy lifters can reach and deflect bigger NEOs and counter the effects of uncertainties; the ability to respond quickly depends on whether launch systems are ready or have to be built when needed; and nuclear detonation remains the only viable option for most short warning situations.

8.4.2 Hypothetical NEO Impact Scenarios

The hypothetical asteroid impact scenario presented at the 2015 IAA Planetary Defense is used as an example (Melamed 2015; PDC 2015). In this case, the asteroid was discovered on April 13, 2015, about 7 years before its potential impact with Earth on September 3, 2022 in the South China Sea (PDC 2015). Its absolute magnitude was estimated at $H = 21.3 \pm 0.4$, which puts its size at roughly 100–500 m. H is the visual magnitude an observer would record if the asteroid were placed one Astronomical Unit (AU) from Earth, and one AU from the Sun and at a zero-phase angle. The large uncertainty in the object's size is due to uncertainties in both albedo (reflectivity) and the absolute magnitude H. Its mean distance from the Sun (semi-major axis) was 1.77 AU, and the orbital eccentricity was 0.49. In mid-June 2015, the probability of Earth impact in 2022 was 1% and continued to rise. The object was observable through the rest of 2015, but in the spring of 2016 the asteroid moved too close to the Sun to be observed, and it remained unobservable for about 7 months.

As the 2015 PDC scenario continued to evolve, additional observations were made before August 2019, adjusting the size range estimate to 150–250 m. Several nations with space launch capabilities joined the effort to deflect the oncoming asteroid. A total of six kinetic impactor spacecraft were scheduled to be launched toward the object in August 2019 and strike the asteroid at very high relative velocities in early March 2020. Images from a spacecraft sent to observe the impacts of the deflection spacecraft confirmed the successful impact of three of the high-speed kinetic impactors, but found that the deflection broke asteroid PDC15 into two pieces. The larger piece was successfully deflected and would not impact the Earth, but a 60- to 100-m fragment received only a partial deflection and would enter the atmosphere on September 3, 2022 in the vicinity of Dhaka, the capital city of Bangladesh. The decision to deflect asteroid 2015 PDC did not anticipate that the object might break in two and create a second threat.

Could better decisions have been made? What resources would be required to support better decisions?

The hypothetical 2015 PDC asteroid impact scenario presumed a campaign of six interceptor missions were more than enough to successfully deflect the original asteroid, without specifying the participating nations and the type of the launch vehicles that would be used. The required size of the deflection campaign is hard to plan due to the asteroid's size and mass uncertainties, so the worst case was assumed. How much deflection is achievable also depends on secondary factors such as the amount and direction of ejecta produced by the impactors, the so-called beta factor.

The worst-case asteroid size at the start of the 2015 PDC asteroid impact scenario was 500 m. The worst-case fragment size following the deflection attempt was 80 m. These values define the range of asteroid sizes considered in the PDC (2015) exercise. The minimum success criterion is deflection of the PDC15 object to just outside the Earth's capture radius.

To do this, the NDA launches a kinetic impactor spacecraft that flies a user-specified amount of time and intercepts with the asteroid at a user-specified time before its impact with Earth. The sum of these times defines when the launch system must be ready. But even if it is ready, launch windows are discrete time segments when the asteroid is accessible from Earth via a heliocentric transfer trajectory obtained through a Lambert solution. For the PDC15 asteroid, there are three deflection campaign opportunities in the decade before impact. These are represented by launching at 3 years before impact with a flight time of 200 days, launching at 5.6 years before impact with a flight time of 350 days, and launching at 7.8 years before impact with a flight time of 250 days (the last segment is considered here for the purpose of illustration, although it is outside the 7-year (PDC 2015) scenario). The results are summarized in the Tables below.

Tables 8.1, 8.3, and 8.5 show the miss distance for deflected asteroid PDC15 at the three campaign opportunities. A negative miss distance corresponds to a perigee distance of less than one Earth radius and implies an impact. Negative miss distances are highlighted in red. It is evident that the total red area is smaller in early campaigns than in late ones. As expected, discovering and responding early is advantageous, assuming a launch system is ready to be utilized. Otherwise, additional time must be factored in for launch systems to be built, potentially causing the early campaign opportunities to be lost.

Also, evident in Tables 8.1, 8.3, and 8.5 is that the total red area is smaller for large launch vehicles. Clearly, larger launch vehicles carry larger payloads that produce larger deflections and are more capable of handling asteroid 2015 PDC, especially at the early phases of the scenario. For example, NASA's SLS 2B is adequate almost throughout the size range on a single launch. A single Delta IV Heavy or

Table 8.1 Earth miss distance (single launch 200 days before deflection, deflect at 900 days before impact)

	80 m	100 m	150 m	250 m	400 m	500 m
Ariane V	3.8	1.6	0.0	-0.5	-0.5	-0.6
Atlas V 551	5.5	2.4	0.2	-0.4	-0.5	-0.5
Delta IV Heavy	9.3	4.4	0.8	-0.3	-0.5	-0.5
Falcon Heavy	9.9	4.7	0.9	-0.3	-0.5	-0.5
H IIA 4S	0.9	0.1	-0.4	-0.5	-0.6	-0.6
H IIA 5S	0.5	0.0	-0.4	-0.5	-0.6	-0.6
Proton Breeze M	4.6	2.0	0.1	-0.4	-0.5	-0.5
NASA SLS 2B	58.2	29.4	8.1	1.2	-0.2	-0.4
Taurus	-0.2	-0.4	-0.5	-0.6	-0.6	-0.6

Table 8.2 Number of launches for Earth miss (launch 200 days before deflection, deflect at 900 days before impact)

	80 m	100 m	150 m	250 m	400 m	500 m
Ariane V	1	1	1	5	20	30
Atlas V 551	1	1	1	5	15	20
Delta IV Heavy	1	1	1	3	10	20
Falcon Heavy	1	1	1	2	10	20
H IIA 4S	1	1	3	15	30	30
H IIA 5S	1	2	5	20	30	30
Proton Breeze M	1	1	1	5	20	30
NASA SLS 2B	1	1	1	1	2	3
Taurus	2	5	15	30	30	30

Falcon Heavy would also be adequate at the latter part of the 2015 PDC scenario when the size estimate is reduced to 150–250 m. Decision-makers might make different mitigation decisions depending on the available launch capability at the threat onset. For example, knowing that a single SLS can accomplish the deflection, they might delay a decision to act and wait for improved orbital and size measurements that confirm or rule out the impact (of course, the possibility of launch failure must be considered as part of the overall success criteria).

Table 8.3 Earth miss distance (single launch 350 days before deflection, deflect at 1700 days before impact)

	80 m	100 m	150 m	250 m	400 m	500 m
Ariane V	12.3	5.9	1.2	**-0.2**	**-0.5**	**-0.5**
Atlas V 551	12.7	6.1	1.3	**-0.2**	**-0.5**	**-0.5**
Delta IV Heavy	20.1	9.9	2.4	0.002	**-0.4**	**-0.5**
Falcon Heavy	23.2	11.5	2.9	0.1	**-0.4**	**-0.5**
H IIA 4S	4.5	1.9	0.1	**-0.4**	**-0.5**	**-0.5**
H IIA 5S	3.8	1.6	0.01	**-0.5**	**-0.5**	**-0.6**
Proton Breeze M	11.5	5.5	1.1	**-0.2**	**-0.2**	**-0.5**
NASA SLS 2B	110.8	56.3	16.1	2.9	0.2	**-0.2**
Taurus	0.2	**-0.2**	**-0.5**	**-0.5**	**-0.6**	**-0.6**

Table 8.4 Number of launches for Earth miss (launch 350 days before deflection, deflect at 1700 days before impact)

	80 m	100 m	150 m	250 m	400 m	500 m
Ariane V	**1**	**1**	**1**	2	7	15
Atlas V 551	**1**	**1**	**1**	2	7	15
Delta IV Heavy	**1**	**1**	**1**	**1**	5	10
Falcon Heavy	**1**	**1**	**1**	**1**	5	7
H IIA 4S	**1**	**1**	**1**	5	20	30
H IIA 5S	**1**	**1**	**1**	5	20	30
Proton Breeze M	**1**	**1**	**1**	2	7	15
NASA SLS 2B	**1**	**1**	**1**	**1**	**1**	2
Taurus	2	5	5	30	30	30

Launch vehicle type is a key factor in deflection campaign sizing. This is evident in Tables 8.2, 8.4, and 8.6, showing how many launches are required to accomplish the smallest deflection that produces a miss. When a single launch is adequate, it is highlighted in green. As expected, the total green area is larger for larger launch vehicles. The total green area is also larger when discovery and action are sooner. Areas not shaded in green estimate the number of launches required if a single launch is not adequate. Tables 8.2, 8.4 and 8.6 show that deflection campaigns

Table 8.5 Earth miss distance (single launch 250 days before deflection, deflect at 2600 days before impact)

	80 m	100 m	150 m	250 m	400 m	500 m
Ariane V	21.6	10.6	2.6	0.04	-0.4	-0.5
Atlas V 551	21.4	10.6	2.6	0.04	-0.4	-0.5
Delta IV Heavy	33.2	16.6	4.4	0.4	-0.4	-0.5
Falcon Heavy	39.2	19.7	5.3	0.6	-0.3	-0.4
H IIA 4S	8.6	4.0	0.7	-0.3	-0.5	-0.5
H IIA 5S	7.5	3.5	0.5	-0.4	-0.5	-0.5
Proton Breeze M	19.7	9.7	2.3	-0.01	-0.4	-0.5
NASA SLS 2B	177.7	90.6	26.3	5.0	0.7	0.04
Taurus	0.8	0.1	-0.4	-0.5	-0.6	-0.6

Table 8.6 Number of launches for Earth miss (launch 250 days before deflection, deflect at 2600 days before impact)

	80 m	100 m	150 m	250 m	400 m	500 m
Ariane V	1	1	1	1	5	10
Atlas V 551	1	1	1	1	5	10
Delta IV Heavy	1	1	1	1	3	5
Falcon Heavy	1	1	1	1	3	5
H IIA 4S	1	1	1	3	10	20
H IIA 5S	1	1	1	3	15	30
Proton Breeze M	1	1	1	2	5	10
NASA SLS 2B	1	1	1	1	1	1
Taurus	1	1	3	15	30	30

quickly become impractical with smaller launch vehicles and larger asteroids. They do become more practical the sooner deflection action is considered. Though an early decision to act would make successful deflection more practical and likely, it is a decision that would have to be made under a greater degree of uncertainty about the asteroid's orbit and size. As noted, it might be advantageous to wait for better data without excessively degrading the capability to respond, and that depends on the types and sizes of available launch vehicles. The NDA provides these mission options for the decision-maker to consider.

The lower end of the size range corresponds to the 80-m fragment still headed toward Earth after the 2019 deflection failed to move the entire object to an Earth miss trajectory. The possibility of a loosely connected binary object was not considered at the outset, so no backup mission was planned to counteract that fragment. Unfortunately, the last opportunity to launch and reach the asteroid with a kinetic impactor is in 2019, about 3 years before impact. A viable alternative in that situation might be to launch a nuclear device with the power to deflect or disrupt the fragment that late in the game. The nuclear option was not considered.

The sizes of the red areas in Tables 8.1, 8.3 and 8.5, and the green areas in Tables 8.2, 8.4 and 8.6 heavily depend on how well the asteroid's composition is known. Superior knowledge of object characterization might justify a larger assumption for beta, the momentum amplification parameter that accounts for asteroidal material ejected backward by the impact of the kinetic impactor. A larger beta value will shrink the red areas and enlarge the green areas in all Tables and is therefore a key mission design parameter. If a larger beta value can be assumed from good object characterization, it would permit reduction of the size of the deflection campaign in terms of the number of launch vehicles required, making it more affordable, more practical to implement and more likely to succeed.

8.5 NDA-Derived Insights

At the 2017 IAA Planetary Defense Conference, a hypothetical threat scenario was exercised on a fictitious 100- to 250-m asteroid termed 2017 PDC. The asteroid was designed to impact Earth about 10 years after its discovery on March 6, 2017. Conference participants used the NDA to assess possible responses to 2017 PDC and exercised mitigation decisions that could be made had the threat been real, and issued recommendations to a group of world leaders who made decisions and shaped the progression of the scenario throughout the conference. At the 2017 PDC and prior exercises, the NDA informed participants of minimum campaign requirements and achievable miss distances, and revealed not only its utility but also its limitations in estimating deflection requirements reliably. These shortcomings helped to identify key desired upgrades that would account for uncertainties in a threatening object's orbit and physical properties, and use launch failure historical data to assess campaign reliability, making NDA's predictions more realistic (Melamed 2017).

At the time of writing, NDA does not account for orbital uncertainty so it can only produce estimates for moving an object on a "nominal" orbit. However, orbital uncertainty is an inherent reality with any newly discovered object before adequate observations or in-situ information are available. Accounting for orbital uncertainty in the NDA will determine how many additional launchers would have to be integrated into the campaign in order to achieve a desired level of deflection assurance. Fig. 8.3 notionally illustrates an uncertainty region on the B-plane resulting from the object's orbital uncertainty, superimposed on the nominal impact point. It is

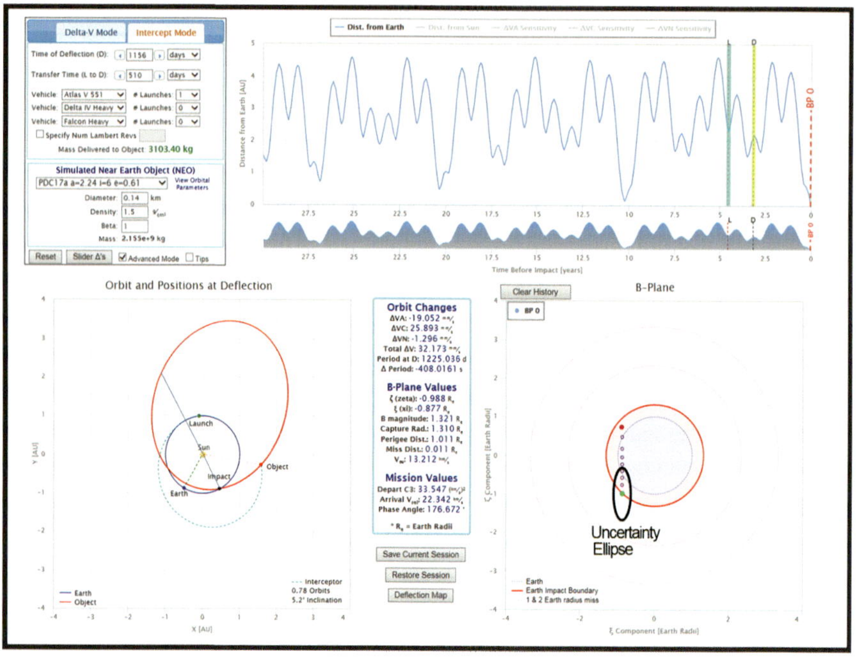

Fig. 8.3 Kinetic deflection with orbital uncertainty (Melamed 2017)

Table 8.7 90% reliability kinetic impact deflection campaign vs. minimum requirement (Iwata 2017)

Launch window	Number of Delta IV Heavy	
	Minimum req.	90% campaign reliability
Soon after discovery	2	5
~7 years before impact	2	5
~3.5 years before impact	3	7

clear that moving the nominal impact location just outside the capture circle leaves an undesired amount of impact likelihood that can only be eliminated by a more powerful deflection that will move the entire uncertainty region outside the capture circle.

In addition to orbital uncertainty of the discovered NEO, reliability data must also be considered to add further realism to NDA deflection campaign estimates. Because the NDA in its current form lacks actual deflection systems performance data, its predictions at the 2017 PDC produced minimum campaigns. Barbee (2017) accounts for all historical launch and space vehicle reliability data to size the PDC 2017 object deflection campaign, assuming a worst-case 250-m object diameter and density of 3.0 g/cm^3. Table 8.7 shows that in order to attain 90% campaign reliability at various launch opportunities in the 2017 PDC 10-year scenario, the number of launch vehicles required for satisfactory deflection has more than doubled. Launch vehicles are more reliable now than they were decades ago, hence, leaving out old data would likely require fewer launches above the minimum requirement than Table 8.7 predicts.

The inclusion of orbit uncertainty and imperfect deflection system reliability in the consideration implies that powerful deflections are required when objects are massive and time is short. In the case of kinetic deflectors, the large number of systems required to deliver the impulse makes a deflection campaign based on this technology less practical. From a technological standpoint, it may be more practical and reliable to apply nuclear deflection or destruction capabilities to achieve the desired deflection in tough situations. The approach applied at the 2017 PDC was to include both kinetic and nuclear spacecraft in the deflection campaign, which markedly expanded the range of options available for deliberation (Barbee 2017).

8.6 Summary and Conclusions

This chapter examined the ability of existing and future domestic and international launch vehicles to deflect an Earthbound asteroid by intercepting it with a Kinetic Impact spacecraft. The 100- to 500-m 2015 IAA Planetary Defense Conference asteroid was the subject object. NDA, a physics-based asteroid deflection app, identified several deflection opportunities to launch kinetic interceptors and move the object to an Earth miss trajectory. Launch windows were not continuous, and there were three distinct time segments in the decade before Earth impact when the asteroid was accessible via a heliocentric transfer trajectory. A similar exercise held at the 2017 IAA Planetary Defense Conference identified key desired upgrades to the asteroid deflection app that would make its predictions more realistic. Lessons learned are:

- The sooner discovery and launch occur, the less deflection effort is required.
- Heavy launch vehicles are advantageous for moving larger objects since they require fewer missions.
- The ability to respond quickly depends on whether launch systems are ready or must be built when needed.
- Nuclear explosion is a powerful method of deflection that may be considered when the time to impact is too short or the object is too large for the Kinetic Impact method to be effective.
- Uncertainties in the threatening asteroid's mass, size and orbit must be included in the design of any deflection campaign.
- Overall launch system and payload reliabilities must be considered as part of any effort to design a campaign for deflecting a threatening asteroid.

References

Barbee, B. (2017). Preliminary 2017 PDC Mission Design Considerations. https://cneos.jpl.nasa.gov/pd/cs/pdc17/2017PDC_Mission_Considerations_Day1.pdf. Accessed 20 November 2017

Chodas, P. (2018a). Sentry: Earth Impact Monitoring. *CNEOS - Center for Near Earth Objects Studies*

Chodas, P. (2018b). NEO Basics, Target Earth. *CNEOS - Center for Near Earth Objects Studies*. https://cneos.jpl.nasa.gov/about/target_earth.html. Accessed 6 July 2018

CNEOS. (n.d.). An Online Educational Tool for Near-Earth Object Mitigation Mission Design. https://cneos.jpl.nasa.gov/nda/nda_handbook_v6.pdf

Dunbar, B. (2014). NASA Experts Continue to Engage United Nations on NASA's Asteroid Initiative. *NASA*. https://www.nasa.gov/content/nasa-experts-continue-to-engage-united-nations-on-nasas-asteroid-initiative/. Accessed 6 July 2018

Iwata, C. (2017). NEO Mitigation Mission Assurance - IAA-PDC-17-05-14. In *Planetary Defence Conference, Tokyo, Japan, 15-19 May 2017*

Melamed, N. (2012). Web Based Mission Design Tool for Planetary Defense from Asteroids. In *Global Space Exploration Conference (GLEX), 22 May 2012, Washington D.C*

Melamed, N. (2015). NEO Deflection Mission Capabilities. In *66th International Astronautical Congress, Jerusalem, Israel*. IAF

Melamed, N. (2017). Simulated Response to Fictitious Asteroid Threat. In *9th IAASS Conference, Toulouse, France, 18-20 October 2017*

Milani, A., Chesley, S. R., Chodas, P. W., & Valsecchi, G. B. (2002). Asteroid close approaches: Analysis and potential impact detection. In W. F. Bottke (Ed.), *Asteroids III* (pp. 55–70). Tucson: University of Arizona Press

NASA. (2017). NASA/JPL NEO Deflection App. *CNEOS - Center for Near Earth Objects Studies*. https://cneos.jpl.nasa.gov/nda/. Accessed 20 November 2017

PDC. (2015). The 2015 PDC Hypothetical Asteroid Impact Scenario. http://neo.jpl.nasa.gov/pdc15/

PDC. (2017). The 2017 PDC Hypothetical Asteroid Impact Scenario. http://neo.jpl.nasa.gov/pdc17/index1.html

Wie, B. (2013). Hypervelocity nuclear interceptors for asteroid disruption. *Acta Astronautica, 90*(1), 146–155. doi:https://doi.org/10.1016/j.actaastro.2012.04.028

Yeomans, D. K., Barbee, B. W., Melamed, N., Anderson, J. P., Chamberlin, A. B., Chesley, S. R., et al. (2012). New On-Line Tools for the Human Exploration and Threat Mitigation of Near-Earth Asteroids. In *Asteroids, Comets, Meteors 2012* (Vol. 1667)

Chapter 9
Near-Earth Objects and the United Nations

Romana Kofler, Daniel Garcia Yarnoz, and Martin Staško

9.1 Introduction

Collisions of near-Earth objects (NEOs) with the Earth can cause significant global damage and disruption to both life and property. Identifying such threats and preventing a potential impact requires a substantive amount of resources. For these reasons, the international community, including national and international space agencies and governments, have over the years recognized the United Nations as one of the main fora for the coordination of NEO mitigation efforts by a diversity of actors, including national and international space agencies and governments. This chapter presents the past, current and future efforts of the United Nations in the field of NEOs and planetary defense.

The United Nations Office for Outer Space Affairs (UNOOSA) and the Committee on the Peaceful Uses of Outer Space (COPUOS), for which UNOOSA serves as the secretariat, have for many years recognized a NEO impact hazard as a global challenge that demands an international response. Impacts of asteroids or comets ranging from a few meters to larger than 1 km in diameter are a regular occurrence in the inner Solar System given the appropriate timeframe (the larger the object, the less frequent) (Ivanov et al. 2001; Strom et al. 2015; Werner 2002). With the current knowledge of the NEO population, impact threats are likely to come from the numerous undiscovered small objects, ranging from a few meters to 100 m in diameter (Shapiro et al. 2010), which could still cause local to regional damage.

R. Kofler (✉) · M. Staško
United Nations Office for Outer Space Affairs (UNOOSA), United Nations Office at Vienna, Vienna International Centre, Vienna, Austria
e-mail: romana.kofler@un.org

D. G. Yarnoz
Formerly United Nations Office for Outer Space Affairs (UNOOSA), United Nations Office at Vienna, Vienna International Centre, Vienna, Austria

N. Schmidt (ed.), *Planetary Defense*, Space and Society,
https://doi.org/10.1007/978-3-030-01000-3_9

9.2 Background and History

In 1995, the United Nations, through UNOOSA, organized the United Nations International Conference on Near-Earth Objects at the UN Headquarters in New York to sensitize Member States to the potential threat of NEOs. There, it proposed to expand existing observation campaigns to detect and track such objects (Remo 1997). It was one of the first gatherings of this magnitude on an international level to discuss NEOs. The conference concluded with several policy perspectives "to increase awareness of the threat from NEOs and to provide guidelines for cooperative observation, research and mitigation programs" (Remo 1997). In particular, it reiterated the importance of estimating probabilities of potential impacts of NEOs, as medium to large-sized asteroids or comets would have severe consequences. Because of this, it was reasoned that there was a need to quantify the probability of a potential NEO impact.

Building on the discussions held in 1995 and consequent debates over the four-year period leading up to 1999, the Third United Nations Conference on the Exploration and Peaceful Uses of Outer Space (UNISPACE III) put forth the recommendation to improve international coordination on activities relating to NEOs (United Nations 1999). Inadequacy of knowledge and the necessity to develop mitigation methods and establish reliable communication with the public were the major issues raised during UNISPACE III. UNISPACE III produced the so-called Vienna Declaration on Space and Human Development (United Nations 1999), containing 33 recommendations for strengthening the use of space for the benefit of mankind. One of the recommendations of the Conference was to encourage increased international collaboration that would improve our understanding of the nature of potential NEOs threats.

In order to implement that particular recommendation, the Action Team on Near-Earth Objects (Action Team 14) (United Nations 2001) was established by COPUOS in 2001. Its mandates were: to structure and organize ongoing efforts in the field of NEO; identify gaps in the ongoing efforts; and propose steps for improvement in the field of international collaboration and coordination between specialized bodies (United Nations 2002). Throughout the years, the Action Team met at NEO-related conferences, on the margins of COPUOS and the Subcommittee's meetings and also held intersessional meetings. UNOOSA provided secretariat functions for these meetings and played a key role in promoting cooperation among all members. At its 42nd session in 2005, the Scientific and Technical Subcommittee of COPUOS considered a new agenda item on the topic of near-Earth objects under a multi-year work plan (United Nations 2005). The Subcommittee noted that the associated activities, including estimating the probability of potential impact and following countermeasures, required a large and coordinated effort at the international level.

9.3 Action Team 14 Results

As one of the outcomes of the Action Team, the General Assembly (United Nations 2008) approved a multi-year work plan on NEOs (United Nations 2007) for the Scientific and Technical Subcommittee (STSC) of COPUOS, to begin in 2008, which would "review progress on international cooperation and collaboration on NEO observations; facilitate, for the purpose of NEO threat detection, a more robust international capability for the exchange, processing, archiving and dissemination of data" and "draft international procedures for handling the NEO threat and seek agreement on those procedures" (Koschny and Drolshagen 2015). A dedicated Working Group on NEOs of the STSC was established under the agenda item "Near-Earth Objects" of the STSC.

In 2013, upon the recommendations of the Action Team 14 for an international response to the near-Earth object impact threat (United Nations 2013a, 2013b) the Working Group recommended:

- "The establishment of an international asteroid warning network (IAWN) by linking together the institutions that are already performing many of the proposed functions, including: discovering, monitoring and physically characterizing the potentially hazardous NEO population; maintaining an internationally recognized clearing house for the receipt, acknowledgment and processing of all NEO observations; recommending policies regarding criteria and thresholds for notification of an emerging impact threat; and developing a strategy using well-defined communication plans and protocols to assist Governments in the analysis of impact consequences and in the planning of mitigation responses." (United Nations 2013a, para. 8, 33)
- "The establishment of a space mission planning advisory group (SMPAG, pronounced "same page") by Member States with space agencies, facilitated by the United Nations on behalf of the international community, with no financial implications for the budget of the United Nations. Its responsibilities should include laying out the framework, timeline and options for initiating and executing response activities, informing the civil-defense community about the nature of impact disasters and incorporating that community into the overall mitigation planning process through an impact disaster planning advisory group." (United Nations 2013a, para. 9, 44)

The recommendations were formally endorsed by COPUOS at its 56th session in June 2013 (United Nations 2013c) and by the General Assembly in its resolution 68/75 in December 2013 (United Nations 2013d). They provide for a coordinated international response to the NEO impact threat, aimed at ensuring information sharing in discovering, monitoring and physically characterizing potentially hazardous NEOs. It aims to ensure that all countries, particularly developing ones with limited abilities to predict and mitigate a NEO impact, are aware of possible threats, and it emphasizes the need for an effective emergency response and disaster management in case of a NEO impact threat (IAWN and SMPAG Chairpersons 2017).

In 2014, the newly established entities, IAWN and SMPAG, held their first meetings with the support of the Action Team on NEOs. To facilitate the interaction of IAWN and SMPAG with the Committee and its Subcommittees, Governments, intergovernmental organizations and non-governmental organizations that were not participating in those NEO bodies, STSC tasked the Action Team on NEOs to support the two bodies in the short term (United Nations 2014). Owing to the successful work of the Action Team in fulfilling its mandate, it was then formally dissolved in 2015. The work of the two groups is facilitated by the United Nations on behalf of the international community (United Nations 2013a). Since 2016, UNOOSA serves as the permanent secretariat to SMPAG, pursuant to the General Assembly resolution 71/90 (United Nations 2016a). UNOOSA closely cooperates with both entities, and besides its mandate as a secretariat to SMPAG, it has an observer status to both entities, while also supporting IAWN in information dissemination.

9.4 International Asteroid Warning Network (IAWN) and the Space Mission Planning Advisory Group (SMPAG)

The International Asteroid Warning Network (IAWN) and the Space Mission Planning Advisory Group (SMPAG) represent key mechanisms at a global level that help strengthen the coordination of planetary defense efforts and prepare for an internationally coordinated response to NEO threats. In that regard, they constitute an example of global space governance, which "encompasses a wide range of instruments, institutions and mechanisms, ranging from international and regional treaties, agreements and regulations, model national laws and regulations to a wide range of international mechanisms utilized in space cooperation" (United Nations 2016b). SMPAG and IAWN participate in the sessions of STSC and report annually to the Subcommittee on the progress of their work on planetary defense.

Figure 9.1 shows a schematic of the relationship of IAWN and SMPAG to the United Nations. Information flow is also possible via their government delegations.

9.4.1 International Asteroid Warning Network (IAWN)

The IAWN[1] is a virtual network linking together institutions involved in the discovery, monitoring and characterization of potentially hazardous NEOs. One of the main functions of IAWN is to maintain an internationally recognized clearing house for the receipt, acknowledgment and processing of all near-Earth object observations. Serving as a global portal and an international focal point for precise and

[1] http://iawn.net.

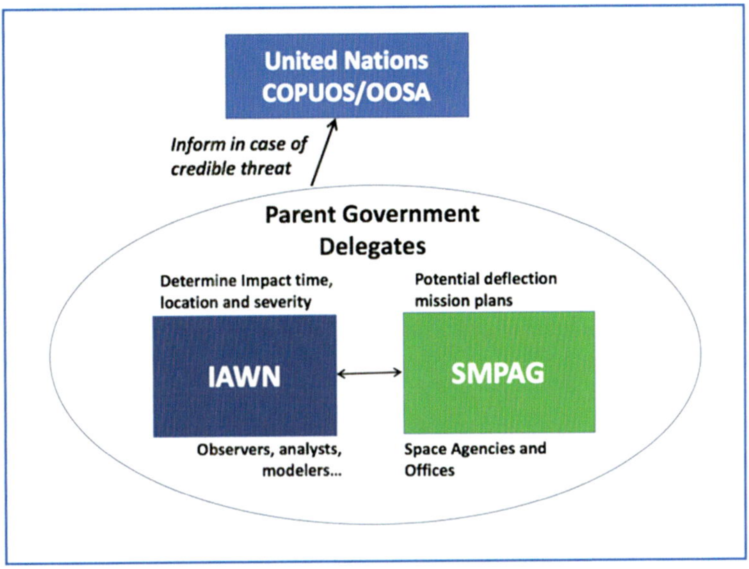

Fig. 9.1 Relationship of IAWN and SMPAG to the United Nations

up-to-date information on the population of NEOs, it is well positioned to recommend criteria and thresholds for notification of an emerging impact threat (its time, location and potential severity). Gathering all types of relevant data enables IAWN to determine hazards and convey the message by using well-defined communication plans and procedures to entities identified by countries responsible for receipt of such notifications. Ultimately, IAWN is mandated to assist Governments in predicting impact consequences and planning a response.

Since its inception in 2014, IAWN has seen continued growth in the worldwide astronomical observation capabilities and its Steering Committee has held a review meeting each year. The IAWN Steering Committee is currently chaired by NASA. As of February 2018, there are 13 official signatories to the IAWN Statement of Intent (IAWN Steering Committee 2014), representing observatories and space institutions from Europe, China, Colombia, Mexico, the Republic of Korea, the Russian Federation, the United States of America, and an amateur observer from the United Kingdom of Great Britain and Northern Ireland. The signatories recognize the benefits of collaborative efforts in collecting and analyzing data and thus contribute with a mixture of different ground and space-based assets equipped for detecting and observing the population of NEOs, together with their expertise in orbit computation and prediction of potential impacts and their consequences. They also use a variety of communication tools and methods to address diverse audiences about NEOs, close approaches and risks of a collision. The possibility to join IAWN is open to groups of observers, analysts and modelers making significant contributions to the study of NEOs or involved in the development of precise and careful communication strategies on NEOs and their impact risks.

9.4.2 Space Mission Planning Advisory Group (SMPAG)

The SMPAG[2] was established as a coordinating body responsible for proposing viable options and implementation plans for a NEO mitigation mission. In case of a NEO on a collision course with Earth, it would be SMPAG's duty to lay out the framework, timeline and options for a deflection mission. In all such cases, the civil-defense community would have to be informed about the nature of impact disasters and be brought into the overarching mitigation planning process through an impact disaster planning advisory group. SMPAG has also an interface role and links Member States with space agencies and other relevant entities.

SMPAG members as of February 2018 include: AEM (Mexico), ASI (Italy), FFG (Austria), Belspo (Belgium), CNES (France), CNSA (China), DLR (Germany), ESA, ISA (Israel), JAXA (Japan), KARI (Republic of Korea), NASA (USA), ROSA (Romania), ROSCOSMOS (Russian Federation), SSAU (Ukraine), SUPARCO (Pakistan), UKSA (UK) and IAWN (ex officio). Present SMPAG observers include: the Association of Space Explorers (ASE), the European Organisation for Astronomical Research in the Southern Hemisphere, also known as the European Southern Observatory (ESO), the International Academy of Astronautics (IAA), the International Astronomical Union (IAU), and the United Nations Office for Outer Space Affairs (UNOOSA).

All national space agencies, as well as other governmental or intergovernmental entities responsible for coordination and funding of space activities that have the capacity to contribute to or carry out a space-based NEO mitigation campaign, are eligible for membership in SMPAG. In the absence of a national space agency, another space-related entity that has been endorsed by a government's internal processes may be designated as a representative of any United Nations Member State.

9.4.3 Building Consensus on Recommendations for Planetary Defense Measures

As per its Terms of Reference (SMPAG Members 2015) adopted in 2015, the objectives of SMPAG are "to develop cooperative activities among its members and to build consensus on recommendations for planetary defense measures." In the event of a credible impact warning by IAWN, SMPAG would suggest mitigation options and implementation plans for consideration by United Nations Member States.

[2] http://smpag.net.

SMPAG addresses four main areas with the following objectives (SMPAG Members 2015):

1. "Reference missions, technology roadmaps and collaborative research:

   (a) Recommend and promote research required for planetary defense. Such investigations can be addressed for example through ground and space-based NEO observations and characterization, computer simulations, laboratory research, technology development, and space missions;
   (b) Develop and agree to a set of reference missions addressing a variety of potential NEO impact scenarios and deflection/disruption possibilities, including evaluations of various scenarios. These reference missions will facilitate realistic technical and resource planning;
   (c) Evaluate the technical maturity and consequences, including failure, of space-based NEO mitigation techniques;
   (d) Recommend, in collaboration with the IAWN, criteria and thresholds for action (e.g., notification of a significant impact risk, initiation of observation and/or mitigation campaigns). Recommend criteria for deflection targeting, such as a minimum acceptable Earth-miss distance;
   (e) Develop decision and event timelines for a variety of potential Earth impactors and trajectories identified in the applicable reference missions.

2. Communication and exchange of information:

   (a) Identify opportunities for international collaboration on research, technologies and techniques for space-based NEO mitigation. This will help avoid costly duplication of effort, and speed the development of effective capabilities;
   (b) Keep Members informed about relevant national activities for planetary defense;
   (c) Communicate its activities to the public following the agreed rules defined in the SMPAG ToRs;
   (d) Offer a yearly briefing to the Scientific and Technical Subcommittee of COPUOS on the status of these activities.

3. Legal and policy aspects: Identify for possible detailed review within appropriate fora any legal and policy issues (e.g., liabilities) that may arise in undertaking NEO mitigation actions or selecting any likely mitigation options.
4. Mitigation planning activities:

   (a) Recommend operational responsibilities for a space-based NEO mitigation campaign;
   (b) Work in coordination with the relevant actors potentially involved in the implementation of the threat response;
   (c) In case of a credible threat, recommend viable concepts for a possible mitigation campaign and directly inform those governments that would coordinate and fund space mission activities and request that they in turn inform COPUOS, via the Office for Outer Space Affairs." (SMPAG Members 2015)

In 2016, SMPAG established an ad-hoc legal working group to discuss and make proposals on the legal issues relevant to the work of SMPAG (SMPAG Members 2017, SMPAG ad-hoc working group on legal issues 2017). The group's work plan, as contained in its Terms of reference, consists of the following tasks:

(a) "Describe the existing legal context, in particular in international law, relevant to the work of SMPAG;
(b) Identify, formulate and prioritize relevant legal questions and issues requiring clarification;
(c) Where necessary, suggest possible ways forward to deal with legal questions and issues;
(d) Provide legal advice to SMPAG as required." (SMPAG Members 2017)

9.5 Work in Progress: First Joint IAWN/SMPAG Results

The work of the two groups has already produced a number of concrete results. In 2016, SMPAG agreed on the following Statement on Asteroid Orbit Deflection Demonstrations:

> Given the degree of international interest in asteroid research and awareness of the impact hazard, advantage should be taken of opportunities to investigate asteroid deflection physics, techniques and effects as a part of science and technology demonstration missions. While general science and technology efforts are vital, SMPAG has identified the need to gain sufficient confidence in the viability of any proposed technique to deflect an asteroid from an impact trajectory. Therefore the performance of the deflection technique to be utilized must have been actively demonstrated in a realistic planetary defense scenario to increase the current level of confidence.
>
> SMPAG encourages actual demonstration of the kinetic impactor technique with a space mission, as it appears at this point in time to be the most technologically mature method of asteroid deflection. SMPAG also supports the investigation of the gravity tractor technique for demonstration as a part of any space mission using ion or other low-thrust propulsion technology planned to visit an asteroid, and encourages the investigation and technology maturation of other potential deflection and impact mitigation techniques to determine their viability, particularly in combination with other missions (SMPAG Members 2016).

9.5.1 IAWN/SMPAG Recommended Criteria and Thresholds for Impact Response Actions

Significant effort has been dedicated to the definition of impact risk thresholds to initiate actions, whether issuing warnings or planning mitigation responses. The two scales most commonly referred to for the determination of the impact threat an asteroid poses are the Torino scale (Morrison et al. 2004) and the Palermo scale (Chesley et al. 2002). Each scale was designed for a different purpose: the Torino scale was designed for communication to the public, while the Palermo scale is a technical scale designed for NEO specialists and engineers and provides a continuous quantitative impact threat level. For the objectives of IAWN and SMPAG members, a more straightforward set of criteria and thresholds were desired, with particular emphasis placed on the warning time until predicted impact.

At the 54th session of STSC of COPUOS in February 2017, IAWN and SMPAG presented to member States of COPUOS a conference room paper (IAWN and SMPAG Chairpersons 2017) containing information on initial agreements achieved on the criteria and thresholds for impact response actions, which has direct relevance for Member States in terms of information sharing on a potential NEO threat:

1. "IAWN shall warn of predicted impacts exceeding a probability of 1% for all objects characterized to be greater than 10 m in size, or roughly equivalent to absolute magnitude of 28 if only brightness data can be collected.
2. Terrestrial preparedness planning is recommended to begin when warned of a possible impact:
   (a) Predicted to be within 20 years;
   (b) Probability of impact is assessed to be greater than 10%; and
   (c) Object is characterized to be greater than 20 m in size, or roughly equivalent to absolute magnitude of 27 if only brightness data can be collected.
3. SMPAG should start mission option(s) planning when warned of a possible impact:
   (a) Predicted to be within 50 years;
   (b) Probability is assessed to be greater than 1%, and;
   (c) Object is characterized to be greater than 50 m in size, or roughly equivalent to absolute magnitude of 26 if only brightness data can be collected (this magnitude threshold would presume an albedo of approximately 3%; i.e., a possible dark object)." (IAWN and SMPAG Chairpersons 2017).

IAWN and SMPAG also agreed to use the distance "above Earth's surface" when reporting flyby distances to Earth, (using an average diameter of the Earth of 6731 km) and on the common use of the "albedo" values.

9.6 Conclusion

The United Nations, through COPUOS and UNOOSA, will continue working on strengthening the collaboration in observation, characterization and information sharing regarding potentially hazardous NEOs, supporting the work of IAWN and SMPAG. In addition to the work carried out by IAWN and SMPAG, COPUOS member States and permanent observers regularly report on their NEO-related activities to the Scientific and Technical Subcommittee of COPUOS under its agenda item

"Near-Earth objects." UNOOSA also maintains a document database containing all the relevant reports on NEO activities.[3]

IAWN's future tasks involve further work as an international body functioning to identify the potentially hazardous NEO population, collect and analyze NEO observation data and communicate the results of its work to the United Nations, including through UNOOSA, and to a wider audience. SMPAG is placing the framework, timeline and possibilities for space mission response activities, as well as promoting international collaboration on research and techniques for NEO deflection. In particular, SMPAG is in favour of actually demonstrating the kinetic impactor technique with a space mission, as it is the most technologically mature asteroid deflection method at the current time.

It is important to further strengthen preparedness in case of a potential NEO impact and raise awareness among Member States, in particular those without capabilities in that area. Under the umbrella of the United Nations, both IAWN and SMPAG continue to work with UNOOSA and Member States to advance awareness both on disaster preparedness and on emergency response in the event that a possible impactor is detected. Existing national and international disaster response agencies would need to prepare for coordinated response activities in the case of both predicted and unexpected NEO impact disasters. As a further step, the global network of the United Nations Platform for Space-Based Information for Disaster Management and Emergency Response (UN-SPIDER), a successful UNOOSA program for various natural disasters, could be used to reach out to disaster risk reduction communities to enhance their understanding and link them to IAWN and SMPAG activities as trusted sources of information. UN-SPIDER would aid in building and testing national and international responses to disasters that might result from NEO impacts, with the final aim of ensuring human security.

** The views expressed herein are those of the author(s) and do not necessarily reflect the views of the United Nations.*

References

Chesley, S. et al. (2002). Quantifying the Risk Posed by Potential Earth Impacts. *Icarus, 159*(2), 423–432. doi:https://doi.org/10.1006/icar.2002.6910

IAWN and SMPAG Chairpersons. (2017). Status report by the International Asteroid Warning Network (IAWN) and the Space Mission Planning Advisory Group (SMPAG). *STSC Reports, STSC 54th Session, A/AC.105/C.* http://www.unoosa.org/res/oosadoc/data/documents/2017/aac_105c_12017crp/aac_105c_12017crp_25_0_html/AC105_C1_2017_CRP25E.pdf

IAWN Steering Committee. (2014). *Statement of Intent for Participation in the International Asteroid Warning Network.*

Ivanov, B. A., Neukum, G., & Wagner, R. (2001). Size-Frequency Distributions of Planetary Impact Craters and Asteroids. In M. Y. Marov & H. Rickman (Eds.), *Collisional Processes in the Solar System* (Vol. 261, pp. 1–34). Dordrecht: Springer Netherlands. doi:https://doi.org/10.1007/978-94-010-0712-2

[3] http://www.unoosa.org/oosa/en/ourwork/topics/neos/docs.html.

Koschny, D., Drolshagen, G. (2015). Activities in Europe related to the mitigation of the threat from near-Earth objects. Advances in Space Research, 56(3), 549-556 doi: https://doi.org/10.1016/j.asr.2015.03.027

Morrison, D., Chapman, C. R., Steele, R. D., & Binzel, R. P. (2004). Impacts and the Public: Communicating the Nature of the Impact Hazard. In M. J. S. Belton, T. H. Morgan, N. H. Samarasinha, & D. K. Yeomans (Eds.), *Mitigation of Hazardous Comets and Asteroids*. Cambridge University Press.

Remo, J. L. (1997). *Near Earth Objects, the United Nations International Conference. Annals of the New York Academy of Sciences* (Vol. 822).

Shapiro, I. I., Vilas, F., A'Hearn, M., Cheng, A. F., Abell, P., Benner, L. a. M., et al. (2010). *Defending Planet Earth*. Washington, D.C.: National Academies Press. doi:https://doi.org/10.17226/12842

SMPAG Members. (2015). *Terms of reference for the near-Earth object threat mitigation Space Mission Planning Advisory Group*.

SMPAG Members. (2016). *Summary of the 6th meeting of the Space Mission Planning Advisory Group, 16-17 February 2016*.

SMPAG Members. (2017) SMPAG Ad-hoc working group on legal issues. (2017). *Terms of reference for the Space Mission Planning Advisory Group's Ad-hoc Working Group on Legal Issues*.

Strom, R. G., Malhotra, R., Xiao, Z.-Y., Ito, T., Yoshida, F., & Ostrach, L. R. (2015). The inner solar system cratering record and the evolution of impactor populations. *Research in Astronomy and Astrophysics*, *15*(3), 407–434. doi:https://doi.org/10.1088/1674-4527/15/3/009

United Nations. (1999). Report of the Third United Nations Conference on the Exploration and Peaceful Uses of Outer Space. *General Assembly Conference Reports, A/CONF.184*. http://undocs.org/A/CONF.184/6

United Nations. (2001). Report of the Committee on the Peaceful Uses of Outer Space. *General Assembly Official Records, General Assembly 56th Session, A/56/20*.

United Nations. (2002). Improve the international coordination of activities related to near-Earth objects. *STSC Conference Room Papers, STSC 39th session, A/AC.105/C*. http://www.unoosa.org/pdf/oosa/Reports/limited/pdf/AC105_C1_2002_CRP13Rev1E.pdf

United Nations. (2005). Report of the Scientific and Technical Subcommittee on its forty-second session. *STSC Reports, STSC 42nd Session, A/AC.105/8*. http://undocs.org/A/AC.105/848

United Nations. (2007). Report of the Scientific and Technical Subcommittee on its forty-fourth session. *STSC Reports, STSC 44th Session, A/AC.105/8*. http://undocs.org/A/AC.105/890

United Nations. (2008). Report of the Committee on the Peaceful Uses of Outer Space. *General Assembly Official Records, General Assembly 63rd Session, A/63/20, p*, 153. https://shop.un.org/books/gaor-63rd-supp20-outer-space-36168

United Nations. (2013a). Near-Earth objects 2011-2012. Recommendations of the Action Team on Near-Earth Objects for an international response to the near-Earth object impact threat. *STSC Reports, STSC 50th Session, A/AC.105/C*. http://undocs.org/A/AC.105/C.1/L.329

United Nations. (2013b). Near-Earth objects 2012-2013. Final Report of the Action Team on Near-Earth Objects. *STSC Reports, STSC 50th Session, A/AC.105/C*. http://undocs.org/A/AC.105/C.1/L.330

United Nations. (2013c). Report of the Committee on the Peaceful Uses of Outer Space. *General Assembly Official Records, General Assembly 68th Session, A/68/20*. http://undocs.org/A/68/20

United Nations. (2013d). International cooperation in the peaceful uses of outer space. *General Assembly Resolutions, General Assembly 68th session, A/RES/68/7*. http://undocs.org/A/RES/68/75

United Nations. (2014). Report of the Scientific and Technical Subcommittee on its fifty-first session. *STSC Reports, STSC 51st Session, A/AC.105/8*. http://undocs.org/A/AC.105/1065

United Nations. (2016a). International cooperation in the peaceful uses of outer space. *General Assembly Resolutions, General Assembly 71th session, A/RES/71/90*. http://undocs.org/A/RES/71/90

United Nations. (2016b). Fiftieth anniversary of the United Nations Conference on the Exploration and Peaceful Uses of Outer Space: the Committee on the Peaceful Uses of Outer Space and global governance of outer space activities. *Committee on the Peaceful Uses of Outer Space, A/AC.105/1137*. http:undocs.org/A/AC.105/1137.

Werner, S. (2002). The Near-Earth Asteroid Size–Frequency Distribution: A Snapshot of the Lunar Impactor Size–Frequency Distribution. *Icarus, 156*(1), 287–290. doi:https://doi.org/10.1006/icar.2001.6789

Chapter 10
A Consideration of a National Approach: The US Planetary Impact Emergency Response Working Group (PIERWG); A Joint Effort Between NASA and FEMA

Leviticus A. Lewis

10.1 The US Plan as a Possible Model

This chapter discusses the overall agency that would be in charge of response and recovery when a space-based defense of the nation is not possible: the Federal Emergency Management Agency (FEMA).

10.2 Why Is this Necessary?

As the resident space "geek/nerd"(or subject matter expert, depends on the day) at FEMA, I am always asked or teased about events happening with NASA or space in general, especially in regard to whether some newsworthy event is scheduled or unscheduled. In this case due to previous briefings we conducted with our NASA colleagues there was an appreciation of possible space hazards and how they could have devastating effects on our citizens. Officials at FEMA are fairly familiar with the space weather hazard, and there is growing awareness of the asteroid impact hazard in particular. About 5 years ago, in 2013, an interesting event took place that hastened interest in this field, around the time when the effort to develop a national policy with regard to asteroid impacts was just starting in earnest.

As required by FEMA's mission, a 24-h national watch covers and monitors all potentially hazardous events nationwide. On February 15, 2013, FEMA's Daily Operations Brief mentioned more than a few articles. Specifically, the flyby of 2012

L. A. Lewis (✉)
National Response Coordination Branch Operations Division, Response Directorate, Department of Homeland Security, Federal Emergency Management Agency, Washington, DC, USA
e-mail: leviticusalewis@gmail.com

N. Schmidt (ed.), *Planetary Defense*, Space and Society,
https://doi.org/10.1007/978-3-030-01000-3_10

DA14 was noted, not as a threat but as an item of interest due to its close approach. Thus, on that day, most of the near-Earth object community was focused on observing the 2012 DA14.

The Chelyabinsk asteroid entered the earth's atmosphere at nearly the same exact time from the opposite direction. As the word and videos of the Chelyabinsk asteroid made social media and news outlets, I was immediately besieged with questions: *Why did you not tell us about this? What do you mean we don't look at all of the sky all the time? What does size and direction of entry have to do with anything? Could we be surprised by something bigger? What could be our possible response?*

That event made many in the international emergency response community realize that this hazard could have severe consequences for any place with a significant population to protect. Many colleagues were surprised that such a small object could cause so many injuries and millions of dollars of damage. In the following months, FEMA participated in a video teleconference with EMERCOM (Russia's FEMA) 2, who detailed its own response to this incident. FEMA and NASA formally chartered a working group to discuss and provide recommendations to respond to this extremely rare but high consequence event, a section of which read:

> Upon notification from NASA of an impending asteroid impact threat to United States territory, the Federal Emergency Management Agency (FEMA) takes the lead to notify appropriate Federal, state and local authorities and emergency response institutions utilizing existing resources and mechanisms. This is analogous to a large re-entering bit space debris and/or hurricane warning procedures. The information to make an informed decision about implementing these procedures will likely be sparse; and the timelines for decision making can vary from years to days or hours."[1]

10.3 The US National Response Framework (NRF)

The National Response Framework (NRF) is a guide to how the nation responds to all types of disasters and emergencies. It is built on scalable, flexible and adaptable concepts identified in the National Incident Management System (NIMS) to align key roles and responsibilities across the United States. This Framework describes specific authorities and best practices for managing incidents that range from the serious but purely local to large-scale terrorist attacks or catastrophic natural disasters. The National Response Framework describes the principles, roles and responsibilities, and coordinating structures for delivering the core capabilities required to respond to an incident and further describes how response efforts integrate with those of the other mission areas (National Response Framework, Third Edition, June 2016, Executive Summary page i).

[1] Planetary Impact Emergency Response Working Group (PIERWG) Charter dated August 2015, Section 5-Background.

10.4 Federal Integrated Operational Plans (FIOP)

The National Preparedness System outlines an organized process for the whole community to achieve the National Preparedness Goal. The National Preparedness System integrates efforts across the five preparedness mission areas—Prevention, Protection, Mitigation, Response, and Recovery—in order to achieve the goal of a secure and resilient nation. One component of the National Preparedness System is a Federal Interagency Operational Plan (FIOP) for each mission area that provides a detailed description of roles and responsibilities, specifies the critical tasks, and identifies federal resourcing requirements for delivering national preparedness core capabilities.[2]

The FIOPs are designed to provide state, local, tribal, territorial and insular area planners an understanding of how the federal government functions in its response, so that they may develop or modify plans to bolster an integrated and whole community effort to make the collective response stronger and more effective.

The Response FIOP builds upon the National Response Framework (NRF), which sets the strategy and doctrine for how the whole community builds, sustains and delivers the Response core capabilities identified in the National Preparedness Goal. The concept of operations and supporting tasks contained in the Response FIOP are scalable, flexible, resilient and adaptable, allowing the FIOP to be used regardless of cause, size, location or complexity. Concepts of operations and/or tasks may be modified, added or deleted depending upon the incident.

In the case of an asteroid impact scenario, it is likely that the framework for coordinating a national response will be based on this core document. At this present time, even with some degree of warning, a response on the ground is most likely the best course of action until space-based mitigation or deflection efforts are feasible. The Response FIOP outlines delivery of federal support to address the impacts of natural or manmade incidents. The Response FIOP is built on the National Incident Management System (NIMS) and utilizes the whole community concept, which incorporates a full range of stakeholders—individuals, families, communities, the private and nonprofit sectors, faith-based organizations, and local, state, tribal, territorial, insular area and federal governments—in national preparedness activities and as full partners in incident response operations. (Response Federal Interagency Operational Plan, Second Edition, August 2016, page 1).

In 2018 efforts began at FEMA to update the National Response Framework and integrate the Response and Recovery Federal Interagency Operational Plans into one document to formalize the relationship between the mission areas. These updates will address the inclusion of other than governmental partners and demonstrate that both response operations and recovery operations can begin simultaneously.

[2]Ministry of the Russian Federation for Civil Defense, Emergencies, and Elimination of Consequences of Natural Disasters. (EMERCOM of Russia).

10.5 An Initial Approach to Planetary Defense: Is this Really an Issue for Emergency Managers?

Current research indicates that a NEO Impact emergency would be an extremely rare event. However, it is recognized that the probability is not zero, and federal officials (NASA and FEMA) have been directed by the US Congress to develop early warning messages and other protocols that may be necessary to respond to such a scenario. The unique nature of an asteroid impact scenario requires that some study and effort be made to prepare the nation as much as practicable. The current approach is to leverage procedures already developed for other large-scale disasters both manmade and natural, such as hurricanes, seismic events, large industrial accidents or acts of terrorism involving weapons of mass destruction. As a result of this directive, which started with a National Space Policy published in June 2010, work began on a national near-Earth object policy. After early education efforts with NASA and other interagency partners, FEMA and NASA formalized the effort to prepare for an asteroid impact scenario. Concurrently, work began to develop a comprehensive National Near-Earth Object Strategy and Action Plan.

10.6 The First Lines of Planetary Defense for the United States: The National Aeronautics and Space Administration (NASA) and the Federal Emergency Management Agency (FEMA)

10.6.1 Planetary Impact Emergency Response Working Group (PIERWG)

Prior to the publication of the national NEO plan, the Planetary Impact Emergency Response Working Group (PIERWG) was established by the partnership of the FEMA and NASA to develop guidance for any potential impact. The PIERWG will coordinate responsibilities and resolve preparedness and operational issues relating to interagency response and recovery activities at the national level in case of a predicted or actual impact of an asteroid or comet in the United States or its territories. One of the primary duties of the PIERWG is to provide guidance and recommendations to senior leadership and to US Federal agencies in preparing for and engaging in ground-based activities to protect and preserve human life and property.

Specific PIERWG roles and responsibilities include:

- Advising senior leadership to ensure a consistent and coordinated Federal response in preparing for, during, and after any NEO impact scenario (the level of senior leadership advised, subject matter experts involved, and roles and responsibilities of PIERWG members will be dependent on the scenario/threat).

- Identifying interagency emergency management issues that cannot be resolved by the Task Teams (see below), especially those identified through ongoing all-hazards planning efforts.
- When necessary, link PIERWG issues, concerns, and recommendations with their respective agency or department program office.
- Informing and providing insights to the development and maintenance of Federal interagency operational plans that may be applied to address the hazards of a NEO impact.
- Providing subject matter expertise and advice through all phases of interagency risk analysis, strategic planning, and capability building related to response and recovery.
- Developing and documenting collaborative interagency courses of action for executing recommendations by the PIERWG.
- Developing, authorizing, and distributing national-level event-focused public information guidelines with key themes, messages, and talking points for use during a NEO impact scenario (NASA and FEMA August 2015).

The current plan is to use the PIERWG as a means to reach out to the federal interagency initially, and as soon as practicable reach out to other state, local and tribal emergency response organizations/officials as well as other organizations and the private sector. The PIERWG is advisory in nature but could be the basis for a "group of experts" to provide advice to senior national leaders should this scenario become a reality. The PIERWG would also interface with any other groups as necessary to ensure that the United States is prepared to respond to this event, should mitigation or deflection efforts fail.

10.6.2 US National NEO Preparedness and Action Plan

NEO impacts are a global issue. Unlike other natural disasters (e.g. hurricanes) and space weather events (e.g. coronal mass ejections from the sun), NEO impacts could be predicted decades in advance and, most importantly in the future, NASA and other international space agencies will develop the protocols and technologies needed to prevent such impacts. Additionally, NEO impacts can cause major environmental, economic and geopolitical consequences for the United States, even if the impact is outside of the country.

This Strategy outlines objectives for enhancing the nation's planetary defense readiness in three key areas: threat assessment, decision-making and response, both in space and on the ground. Some federal departments and agencies have already taken significant steps in these key areas. The challenges posed by a coordinated global response to NEO threats require enhanced research and development to improve observation and modeling capabilities, which are linked directly to preparedness. The goals outlined in this Strategy will leverage these efforts and existing policies, while promoting enhanced coordination and cooperation across the public and private sectors in the United States and abroad.

Even though the United States is currently a global leader in detecting and tracking NEOs, these capabilities depend on international cooperation and coordination. Future capabilities related to deflection and disruption of NEOs will also depend on international cooperation and coordination.

10.6.3 Authority for Creation of the National Planetary Defense Strategy

To address the mandates and challenges above, in 2015, an Interagency Working Group (IWG) for Deflecting and Mitigating the Impact of Earth-bound Near-Earth Objects (DAMIEN) was established by action of the Committee on Homeland and National Security within the National Science and Technology Council (NSTC). The DAMIEN-IWG has developed this National Planetary Defense Strategy as well as the accompanying Action Plan to address the NEO impact risk.

This Strategy ensures that NEO readiness is fully integrated into and builds upon several governmental frameworks already in place. These include: Presidential Policy Directives (PPDs) PPD-8, "National Preparedness" (March 30, 2011) and PPD-21, "Critical Infrastructure Security and Resilience" (February 12, 2013); Section 321 of the NASA Authorization Act of 2005 and Section 804 of the NASA Authorization Act of 2008; the Office of Science and Technology Policy's (OSTP) 2010 response to Congress; and the National Space Policy (Jun 28, 2010).

10.7 International Cooperation for Asteroid Impact Response

When an US National NEO Strategy and Action Plan is published there will be a requirement to foster or continue cooperative efforts with the international community for planetary defense. This level of cooperation has been going on for quite some time via the activities of the United Nations Office for Outer Space Affairs (UNOOSA) and the UN Committee on the Peaceful Uses of Outer Space (UN-COPUOS).

There are two standing activities under the UN COPUOS umbrella. The first is the International Asteroid Warning Network (IAWN), a virtual network linking all institutions performing functions to discover, monitoring and characterizing potentially hazardous near-Earth objects and acting as the clearing house for processing all near earth object observations. The IAWN is working on establishing a network that recommends criteria and thresholds for notification of an emerging impact threat and communications strategies for governments. This network was recently exercised in November 2017 by the NASA Planetary Defense Coordination Office and the European Space Agency, along with other international agencies involved in aspects of planetary defense. This exercise concentrated on tracking a real NEO and coordinating activities to pass along information on the object.

The Space Missions Planning Advisory Group (SMPAG) was established in response to the recommendations for an international response to a NEO impact threat, as endorsed by the UN General Assembly resolution 68/75. SMPAG responsibilities include laying out the framework, timeline and options for initiating and executing space mission response activities as well as promoting opportunities for international collaboration on research and techniques for NEO deflection. SMPAG membership is open to all national space agencies or governmental or intergovernmental entities that coordinate and fund space activities and are capable of contributing to or carrying out a space-based NEO mitigation campaign.[3]

There is no similar group for planning a coordinated international response for an impact that, depending on the size of the object, could possibly have far-reaching international consequences. Consideration should be given to establishing an Impact Disaster Planning Advisory Group (IDPAG) along the lines of the organizations above. The International Academy of Astronautics bi-Annual Planetary Defense Conferences have included an impact tabletop exercise, but participation by emergency response organizations has been minimal. Both NASA and FEMA have conducted three tabletop exercises and plan to continue these exercises in the future. The current plan is to include more state and local officials and private organization in future exercises. The European Space Agency Space Situational Awareness programme and European national disaster response organizations conducted a 2-day exercise in November 2014. Participants included Germany's Federal Office of Civil Protection and Disaster Assistance, the German Space Situational Awareness Centre and the DLR German Aerospace Center. Additional participation was provided by Switzerland's State Secretariat for Education, Research and Innovation and the National Emergency Operations Center/Federal Office for Civil Protection. The previous IAA Planetary Defense Conferences have shown that international response efforts on the ground will be complicated by the uniqueness of an asteroid impact, particularly if a mitigation or deflection effort results in unintended damage a particular nation.[4]

There are six areas to consider with regard to international cooperation in response to an asteroid impact:

- Time to respond
- National Capabilities
- Current International Procedures
- Should an international protocol be developed just for this emergency?
- Obligations if impact is Local (city/state within national border) or International (Regional impact affecting several nations)
- Is the response the same no matter where it impacts?

---

[3] http://www.unoosa.org/documents/pdf/smpag/IAWN_SMPAG_Fact_sheet.pdf.

[4] www.esa.int/Our_Activities/Operations/Space_Situational_Awareness/Preparing_for_an_asteroid_strike/.

10.8 Considerations from Emergency Managers to the Planetary Defense Community

Emergency managers will need to understand and appreciate the science behind what the space/planetary defense community is trying to accomplish with deflection or mitigation efforts. However, for public policy development, it would be beneficial to understand the correlations between the science and public needs it serves. There will be a need to find a credible and authoritative way to communicate with the public in a way that it can understand and absorb that information. Maintaining authority as the source of authentic information may be a challenge. Consider the following questions emergency managers might need to explain or have the space community explain:

- What research, tools and or methods will help you figure out where the NEO will impact?
- How soon will you tell public officials outside of the planetary defense community that there is an impact possibility?
- What determines how much time you can give public officials to react? (remember the CNN factor; this will not be a secret for long)
- How can the current research help explain to public what's going on?
- Can you explain why characterization is important? What risks does that help emergency managers prepare for? Why does it matter if it is a stony or iron or whatever?
- What will be different if the object impacts on the continental shelf? Does depth of water matter? Is the west coast undersea topography different from the east coast? What about blast studies in urban areas vs. rural areas? How will it affect critical infrastructure?
- Your information will help guide public policy—you will not be making it. The response part will be under those officials. However, they will need to know or want to know what your deflection or mitigation plans might be.
- Will whatever you do make it worse? Can your research support doing nothing at all or will it show that the consequences of not doing anything are too high? This will probably be a public policy-driven decision (political), not a science decision.[5]

10.8.1 Is this an Issue for Emergency Managers?

The possibility of an asteroid impact is an important consideration for emergency managers. It is not something that is at the top of every emergency manager's list of concerns, as hurricanes, larges fires and other disasters more easily recognized by

[5] First presented at the First International Workshop Potentially Hazardous Asteroids Characterization, Atmospheric Entry and Risk Assessment, NASA Ames Research Center, Moffett Field, CA July 2015.

citizens all over the world are easier to prepare for. Planetary defense has always had to contend with the so-called "so what" or "big deal" factor—the short-sighted notion that cosmic impacts are so improbable that preparing for them is silly. In the United States, a National NEO Preparedness Strategy and Action Plan was recently published. It is understood that FEMA will have a key role, by statute to be involved in any response, especially if an impact is inevitable. Currently, there are no technological means to mitigate and/or deflect an object in space that threatens impact. Current technology based on spacecraft and ground-based telescopes can hopefully provide some warning, however, there is still a possibility that we will get no warning at all. In that case, FEMA will respond to it as any other no-warning disaster.

10.9 A Hypothetical Response Scenario

Now, let's look at a hypothetical scenario of how these efforts might play out in practice. Upon discovery of a possible asteroid threat on April 29, 2019, the NASA Planetary Defense Coordination Office provides a detailed briefing to the FEMA leadership. The PIERWG is convened to provide additional details for interagency partners. FEMA leadership directs the FEMA Response Directorate to closely monitor the situation and provide monthly updates.

On May 1, 2019, The FEMA-NASA PIERWG meets and establishes a schedule to provide the information as required by senior leadership. The PIERWG assists with monitoring the International Asteroid Warning Network (IAWN) and other information sources to stay updated on the position and latest observations of the object. Response planners at FEMA Headquarters begin crisis action planning with federal agency partners based on the Federal Interagency Operations Plan (FIOP) and applicable annexes. In coordination with the Emergency Support Function Leadership Group (ESFLG),[6] FEMA initiates FIOP Phase 1a, coordinating actions with federal agency partners to include:

- Establishing clearly delineated agency responsibilities, including appropriate lead agencies and notification protocols and standards (The PIERWG was established to provide a group of interagency subject matter experts to assist in this process)
- Review and assess evacuation and mass care strategies
- Identify critical national issues that may require establishment of dedicated federal interagency functional planning teams, such as:

[6] Emergency Support Function Leadership Group (ESFLG). The ESFLG is comprised of the federal departments and agencies that are designated as coordinators for ESFs or coordinating agencies for other NRF annexes. The ESFLG provides a forum for departments and agencies with roles in federal incident response to jointly address topics such as policies, preparedness and training. FEMA leads the ESFLG and is responsible for calling meetings and other administrative functions.

- Support for mass evacuation and potential demands for short-term and long-term housing
- Public information, warning and communication strategies
- Working with states to establish host-state agreements for evacuees
- Managing and providing assistance for large numbers of federal emergency declaration requests
- Continuity of government and continuity of federal response operations
- Mitigation strategies for potential losses and disruptions to critical infrastructure

In July 2019, NASA reports the NEO impact probability to Earth as 2%. FEMA updates Executives with the latest information and explains the meaning of this estimate from NASA. By August 2019, routine reports from NASA have increased the probability of impact from 2% to 35%. NASA announces the estimated date of Earth impact. The FEMA Administrator directs the creation of an Asteroid Impact Working Group to produce a plan with a countdown to impact.

From August–December 2019, FEMA Response Operations and Planning Divisions review previous FEMA/NASA TTX reports and incorporate them into ongoing planning efforts. FEMA Response Planning continues FIOP Phase 1a federal interagency planning activities to cover FEMA Regions II, IV, VI and IX, part of the risk corridor projected by NASA. The FEMA Regions begin dialogue with state partners and crisis action planning with regional federal partners, to include:

- Supporting general public education on potential NEO impact hazards and supporting state, local, tribal and territorial preparedness and public messaging efforts
- Conducting public and private sector outreach regarding NEO threat and preparedness actions
- Conducting pre-impact analyses of likely consequences on the geography, demographics, populations and critical infrastructures of at-risk US interests, including potential long-term effects
- Ongoing FEMA crisis action planning and production of an Information Analysis Brief (IAB) for the ESFLG with input from NASA and scientific communities

On August 1, 2020, FEMA activates the National Response Coordination Center to Level 1.

During this time, US FEMA Response Operations has maintained close coordination with NASA and keeps FEMA Executives informed on a planned NASA mission to deflect the NEO and potentially avert an Earth impact. Between January 2019 and December 2020, FEMA coordinates with the federal interagency, private industry partners, state and local emergency management agencies on all activities with regard to the asteroid. This includes the deflection mission preparations and operations and estimates of the probable success or failure of the deflection mission. Federal planning and preparedness activities intensify, as well as those of state, local, tribal and territorial partners. Federal interagency crisis action planning increases to Phase 1b activities (elevated threat), where FEMA develops courses of action (COAs) in coordination with federal interagency functional planning teams and planning task forces and presents them to leadership for decision-making on time-sensitive, pre-impact mitigation measures and post-impact responses.[7]

In December 2020, NASA announces the deflection mission as a partial success; however a 50-m object still remains an Earth impact threat with 100% probability of impact. With less than 1 year to certain impact, FEMA and federal agency partners increase readiness and crisis action planning to FIOP Phase 1c (Credible Threat). POTUS requests FEMA and NASA to provide a Cabinet Briefing on preparation for mitigation. From December 2019 to December 2020, federal preparations intensify:

- White House issues Executive Orders to federal agencies on readiness measures for an imminent asteroid impact on the US
- FEMA conducts regular coordination calls with State and Territorial Governors, Tribal Executives and International emergency management organizations
- FEMA/NASA team supports Department and White House press conferences and public messages with information to include estimated impact location(s)

10.10 Conclusion

On June 20, 2018, the Trump Administration released a national plan for reducing the risk of near-Earth object impacts: The National Near-Earth Object Preparedness Strategy and Action Plan. This report was an effort conducted by the Interagency Working Group for Detecting and Mitigating The Impact of Earth-Bound Near-Earth Objects (IWG-DAMIEN). This group was chartered in 2016 and has been directed by the White House Office of Science and Technology Policy.[8] The current policy has built on past efforts by the interagency to take a serious look at this low-probability but extremely high-consequence event.

As discussed in this chapter, the Federal Emergency Management Agency (FEMA), a component of the US Department of Homeland Security (DHS), is an all-hazards organization charged with helping people before during and after a disaster. An asteroid impact is one of the many possible scenarios that FEMA must be prepared to respond to. It is recognized that an asteroid impact is a low-probability but high-consequence event (there is no known threat at this time). It is also recognized that an asteroid impact scenario is a unique emergency and may be different enough that some degree of preparedness specific to this threat is necessary. FEMA has been in a partnership with NASA concerning this scenario since 2010 and has formally chartered the PIERWG to make recommendations to senior officials and

[7] Proposed Table-top Exercise (TTX) developed by FEMA and NASA PDCO with the Center for Near-Earth Object Studies (CNEOS). https://cneos.jpl.nasa.gov/.

[8] Interagency Working Group for Detecting and Mitigating The Impact of Earth-Bound Near Earth Objects (NEOS) (DAMIEN) of the National Science and Technology Council, Office of Science and Technology Policy, Executive Office of the President, Chartered in September 2016.

emergency mangers. With the formal announcement of an official US policy, the work of the PIERWG will continue to conduct meetings and exercises and participate where possible on the international level to inform all of our partners and fellow citizens of some of the unique aspects of an asteroid impact emergency.

Further Reading

US National Space Policy, June 28, 2010
FEMA 2018-2022 Strategic Plan
FEMA, National Preparedness Goal, September 2011
FEMA, Overview of the Federal Interagency Operations Plans, August 2016
FEMA-NASA Planetary Impact Emergency Response Working Group Charter, August 2015
National Near-Earth Object Preparedness Strategy, December 2016
National Near-Earth Object Preparedness Strategy and Action Plan, June 2018
NASA Policy Directive, Subject Notification and Communication regarding Potential Near-Earth Object Threats, NPD-8740
European Space Agency, http://ww.esa.intl/Our_Activities?Operations/Space_Situational_Awareness/Preparing_for_anasteroid_strike
Report on the 2013 NASA-FEMA Asteroid–Impact Tabletop Exercise
Report on the 2014 NASA-FEMA Asteroid-Impact Tabletop Exercise
Report on the 2016 NASA-FEMA Asteroid Impact Tabletop Exercise

Independent

"*Planetary Defense Against Asteroid Strikes: Risks, Options, and Costs*" The Aerospace Corporation, Center For Space Policy and Strategy, Melamed, Nahum, The Aerospace Corporation and Melamed Avishai, University of California, San Diego

Websites

NASA: https://www.nasa.gov/planetarydefense
CNEOS: https://neo.jpl.nasa.gov
FEMA: https://www.fem.gov
UNOOSA: http://www.unoosa.org/documents/pdf/smpag/IAWN_SMPAG_Fact_sheet.pdf
ESA: http://www.esa.int/Our_Activities/Operations/Space_Situational_Awareness/Preparing_for_an_asteroid_strike/
IAWN: http://iawn.net

Part II
Thinking About the Risk, Responding to the Threat and the International Dynamics

Chapter 11
Evolution of Security Studies and the Resulting Perspectives of an Asteroid Threat

Nikola Schmidt

11.1 Introduction

How can we translate a threat we are able to identify by scientific means into a policy action, and what dilemmas are raised by following the logic of evidence? This chapter takes a broader perspective on planetary defense as a security studies case (in political science) and aims to reconcile *rational* and *interpretative* thinking. This reconciliation can be achieved by discussing the topic in terms of *pragmatic* thinking, which will be discussed at the end of this chapter.

While rational thinking tends to express the world around us using scientific means only—for example, when we apply math to physics to explain how the world works—interpretative thinking assesses the normative implications of our decisions: "Does this solution lead to a better desired future?" Normative thinking poses questions about whether certain actions and their outcomes can be perceived as good, desirable or permissible, or as bad, undesirable or impermissible. I will use more than one theoretical approach combining rational scientific facts with several critical theoretical approaches enriched by some recent ideas concerning the pragmatic turn in security studies. The pragmatic turn in normative thinking provides us with tools to interpret the world contextually. Thus, normative assessments of certain actions become related to context, not as standalone, unreconcilable statements.

N. Schmidt (✉)
Department of Political Science, Institute of Political Studies, Faculty of Social Sciences, Charles University, Prague, Czech Republic
e-mail: nikola.schmidt@fsv.cuni.cz

N. Schmidt (ed.), *Planetary Defense*, Space and Society,
https://doi.org/10.1007/978-3-030-01000-3_11

11.2 Securitization as a Sincere Intention and a Pragmatic Move

When one talks about a threat, especially of global magnitude, one inherently constructs a threat despite the scientific proofs provided. A *security threat* is not a scientific variable but a political statement, regardless of the obviousness of consequences logically deduced from an asteroid impact. While an asteroid can be perceived by astronomers as an undeniable threat, the point is that an asteroid becomes a threat to our security the moment we realize the threat to our existence and begin talking about it. The process of talking and realizing that an asteroid can pose a threat is a *social construction of the security threat*, again, regardless of whether the asteroid is or is not actually on a terminal trajectory. A security threat in political science is not an objective threat:[1] it has to be related to a referent object that is threatened—a person, society, civilization, city, state, etc. Speaking about the threat—the speech act—is a *securitization move.* It transforms a physical phenomenon into a threat to our security. Although the security threat itself might not be so disputable, the response to it is a sensitive policy question par excellence. Saving the Earth using certain means does not necessarily imply ethical behavior. Such intentions might not necessarily be *normatively desirable.* This chapter discusses the dynamics behind planetary defense efforts in light of current security studies; what challenges the engineering and policy efforts need to overcome; and that the solution focused on rational thinking is not enough and might not lead to desirable outcomes.

We can distinguish between two main critical security studies schools: the Copenhagen school and the Welsh school. Briefly put, the Copenhagen school would like to uncover hidden intentions behind the securitization move—the discursive practice of somebody who wishes to have planetary defense as a central policy priority while having different objectives in mind—for example, taking power. This process is visible in the current rise of populist powers across the transatlantic, as the real deconstruction of the current political world is probably not the intention of the populists. The Welsh school, on the other hand, cares not about the hidden intentions but much more about the normative outcomes of the securitization moves. This author deeply believes—following the Welsh school—that we, political scientists, are here to help steer the technical proposals in the direction that includes not only rational scientific reasons but also the normative outcomes of rationally choosing the most effective solutions.

The Welsh school is based on a conviction that one may have sincere intentions instead of selfish hidden interests in one's securitization move. This is quite a tricky situation, as implementing rational thinking or evidence into policy making may

---

[1] This is valid if we take the perspective of critical security studies that do not accept objective threats. Realists in international relations would argue against saying that states are a threat to each other. For constructivists, threats posed by states are perceived as a state of mind of the political representatives that can be changed and thus, there are no threats that are natural, undeniable or objective.

have two completely different outcomes that are fighting each other in an irreconcilable, theoretical battle. While the rationalist camp thinks that rational thinking will save the world because it solves problems (asteroid threats included), the normative camp will ask how you wish to define what is bad and good before you apply the evidence and propose a solution. In a nutshell, using nuclear weapons against an asteroid might save mankind from extinction, but one should ask in advance whether we would like to see a superpower using nuclear weapons, causing the destigmatization of such weapons, followed by their proliferation, or whether we should prepare other options to avoid this scenario.

Planetary defense is a nice example of a low-probability event with possibly extreme consequences. Probability of impact will never tell us whether we should act now or in a thousand years, since a single asteroid can cause an extinction event. We talk about security threat, an extinction-level threat, a threat to our global civilization, to Earth, to us all, to humanity, to a whole biosphere. One must ask certain questions and think bigger before one proposes a prospective policy without causing global panic or a global disdain: How can we successfully initiate a balanced interest by politicians, the public and the expert community? To whom should we talk first: all or only select groups? Is it a threat to mankind, or should we talk about a threat to the whole biosphere? What is the referent object of our securitization move? What should be the appropriate response, and what kind of security state should this response imply? What is the desirable state of security in case of a successful response by our audience? We need to ask ourselves these questions before discursively formulating the threat—making the speech acts—because the response to such a cataclysmic, exceptional threat will require an equally exceptional political response and might cause exceptional changes in our society, in governance of everything on the globe, and does not necessarily need to lead to the desired state of the affairs.

Thus, the central question should be broad and clear: in what kind of world do we want to live? This question cannot be answered by implementing rational thinking and evidence-driven policy; on the other hand, asteroid threats cannot be conceptualized without particular scientific knowledge, evidence and thus, rational thinking. The current new wave of *ethical security* studies (Nyman and Burke 2016)[2] helps us formulate a policy action based on an *objective basis* (astronomy, physics), with *sincere intentions* (ethics), through an *appropriate response* (normatively desirable), keeping in mind the consequences in case of a successful securitization move (the speech act constructing the security threat).

Planetary defense as an endeavor is an apparent target for critical security studies; however, not too many security studies scholars have approached it. The reason may be twofold: first, the asteroid threat is in a phase in which it waits for its policy action, and political scientists like to study ongoing policy dynamics. As a result, the substance has not yet emerged to be studied. At present, nothing significant has

---

[2] Ethical security studies is a new field in which Anthony Burke and Jonna Nyman have introduced ethical thinking and select philosophical thought to discuss the desirable state of security. They have devoted a whole volume to the topic, as well as articles that will be discussed below.

happened in our political life on a national or global level to yet have a planetary defense-related impact on our lives. Second, political scientists are usually fighting in their dedicated schools of thought over how to study politics, and planetary defense efforts have not touched these schools yet.

Planetary defense is a perfect case for critical security studies and their recent pragmatic turn and ethical assessment, because it calls for an exceptional action to avoid an exceptional extinction event. However, as the planetary-defense case calls for exceptional solutions that can have exceptional consequences, it requires critical study, not disdain or criticism for being too radical or influenced by fiction. Many schools of thought can unite to come up with a solution while keeping critical assessment in mind; however, some of them are irreconcilable, as one school perceives the international system of nation-states as a given (realists), while others perceive it as something we intentionally created (constructivists). This chapter is meant to discuss how we can look for the solution, remaining aware of the fact that the Occam's razor, driven by rationalist logic and aimed at an effective result, would not necessarily produce the answer we want.

11.3 A Brief Introduction to the Evolution of Security Studies

Security studies falls into the discipline of international relations, which is a part of a broader political science discipline. Studying security has been an inseparable part of international relations since its inception about a century ago at the Welsh Aberystwyth University in the United Kingdom. However, the approach to its study has been changing and broadening in the last decades. Different philosophies influence the way we answer a question about what the desired final state of security is. An interesting dilemma that repeats itself in our times, when terrorism is presented as the biggest threat to the global community, is the balance between our security and our privacy. We can perceive it as a balance between our physical security and *security of* our privacy; two securities in competition. Which kind of security is more important, then? To what security the politics incline is influenced by values we wish to cherish or to discursively construct as a necessity; it does not need to be influenced by objective properties, and objective properties do not necessarily bring desirable, normatively good responses. The ideal state of security is thus not defined by the objective state of insecurity, but by our imagination of insecurities, beliefs and what type of security we currently value as the more important one.[3]

[3] Traditional security studies as perceived by realists is much clearer in this way. For realists, the ideal state of security is security of the state—its survival. Other securities, such as security to our privacy or environmental security, are issues for other disciplines. The ethical move holistically includes all these other securities, because decisions on any level can have ethically disputable, and normatively bad implications on our lives. This matters when any individual is included in cosmopolitan thinking.

During the Cold War, political realism was winning in political science as a theory that was able to explain the failure of the League of Nations between the world wars (Carr 1939). In his classical work, Carr defined the foundation of realist thinking and the security dynamics between states. He argued for *real* against *utopia*; subordination is, in his view, a natural behavior of men—real, fact, natural law. This perspective helped him to shed light on saw power as the most important factor in international relations. Carr finished his theory-defining volume during the summer of 1939, just weeks before the spark that set off the most terrible war in history. The book became one of the most influential works after World War II and helped to define the discipline of international relations in realist terms. This perspective might have been better supported due to postwar sentiments; however, it did not stop states from establishing the United Nations, the successor of the League of Nations, which was definitely driven by idealist rather than realist beliefs.

Idealist liberals had called for institutions such as the League of Nations, while realists criticized this approach as utopian. At the same time as such discussions were developing, the discipline of international relations was seeking a research method. Broader social science sought a method that would help develop a positivist methodology capable of studying and maybe also predicting social dynamics through scientific means (rational, quantitative, evidence-based). This movement understandably influenced the discipline of international relations as well as other social-science disciplines that inclined to positivist way of thinking seeking *proofs* for *social facts*.

Realists claimed they were studying politics as-is, because unveiling the *real* behind politics; thus looking for positivist methods was understandable. They criticized liberals for imagining a non-working world of supranational failing organizations, and World War II gave them momentum. Such an appeal, combined with the later criticism of realism by liberals, led to the emergence of neorealism, in which realists added a systemic and relational link between states (Waltz 1979). Claims that the system is anarchic and that power can be measured are in the neorealist tradition. One of the most important contributions was published by Kenneth Waltz at the end of the 1970s. In the Waltzian world, states are natural ends, and anarchy is inevitably the outcome.

Realist and neorealist traditions helped to define security in the so-called four S's: states, strategy, science and the status quo (Williams 2012, p. 3). States were understood as the most important actors in international relations because they wielded (legal) military power in Weberian terms, which is meant to be used in the best strategic way to preserve security of the state, or national security. Science was understood as the key factor to achieve power by improving military capabilities through technological invention and upgrade. This approach was a critical factor in drawing postwar US foreign policy, along with the famous recommendation *Science—the Endless Frontier*, developed by Vannevar Bush for President Franklin Roosevelt (Bush 1945). It was actually broader societal consensus that natural sciences, physics and technology could bring relevant knowledge to the US, and thus also employment and international superiority, as exemplified when the US put the first astronauts on the Moon. Finally, the status quo is understood by realists as the desirable state in international relations—as the only existing natural state of international affairs. Assuming we preserve it, we can avoid revolutions or any other hard-to-control events on the national level, but not necessarily on the international one.

Realists like to measure the absolute power of a state, while neorealists like to put this power in relation to other states. Neorealists introduced the relation between states and the field of anarchic contest—the international system. They would probably focus on the intended technology for planetary defense and argue that developing such game-changing dual-use technologies to mitigate an asteroid threat within borders of one state might lead to a strategic power imbalance between states, despite the global character of the threat. The risk of interstate imbalance is, for any individual state, more important than the global character of the asteroid threat; imbalance will happen immediately.

The biggest problem with (neo)realism in any analysis is the totalization of the state as a historical definite entity and natural political system (Hobson 2000). This perception was carved deeper into the discipline's foundation in the 1970s (Waltz 1979) following the millennia-long development of strategy studies. In this mindset, the state was *objectivized* as a given actor; it has become an objective actor of international interactions. Most present-day social scientists refute this, because the system of states has been intentionally created by men. Some classical proponents of neorealism even argue that other perspectives on security studies that introduce topics beyond the military, such as poverty or environmental security, are not necessary and would destroy disciplinary coherence (Walt 1991). How can we then address threats that would have the same impact on all states? Threats that simply do not emanate from other states but are capable of destroying them all? Threats we can identify by modern science? Are states capable of dealing with the problem, or are they part of it? Is it even a problem, and who decides the shared issues? Realists, the traditionalists in security studies, would say that these questions belong to other disciplines. This author disagrees, as when states claim their power as sovereign, they should take the responsibility that comes with it, and if they are not able to solve issues without bad normative implications, they lose their legitimacy as political actors.

Planetary defense is a quest that requires states to cooperate, because defection is not going to help anyone if and when that the threat materializes. Neorealism, with its focus on the interstate system, does not provide us with appropriate theoretical tools to help us navigate through this global agenda. Neorealism would see the power-laden interests of states and would explain the efforts to build a space-based laser and planetary-defense facility as the development of an ultimate weapon to become a hegemon. That approach does not take into consideration the option that there might be one state with sincere intentions, which are hardly convincing in the neorealist way of thinking. One can ask, how long might these current, sincere intentions last? James Wirtz elaborates on this dilemma in the last chapter of this book.

11.4 Revolution in Theoretical Thinking: The Social Construction of Security and Insecurity

How can we approach a problem of planetary defense in an interstate system and avoid power-laden political moves? Theories in international relations can help us understand how the global political system works, or at least provide some ideas for how it might. Rising economic interdependence gave rise to new liberal thoughts in a masterpiece written by Nye and Keohan (Keohane et al. 1973; Keohane and Nye 1977), regardless of the power-laden perceptions of world affairs. This interdependence logic can be applied to our deepening technological interdependence, situating states in a cooperation mode from which they can hardly defect (Schmidt 2018). First, we must extend our perspective of what security is in order to see how addressing the security issue of asteroids is justifiable.

International relations as a discipline has undergone several great theoretical debates. The last one, the fourth,[4] has developed between rationalists and reflexivists. The former were supporting the rationalists' explanations of the world as discussed above, while the latter were more focused on how to produce knowledge. The reflexivist production of knowledge and its purpose and application were at the root of constructivist approaches in various social scientific disciplines. Since then, the epistemological discussions have grown in substance: how can we know things? How can we say something is self-evident? How we take things for granted?

Barry Buzan, another key author in recent decades, started this reflexivist wave in international relations theory with his publication *People, States and Fear* in 1983; his second edition in 1991 has been more influential. Buzan ignited the discussion over what should or should not be part of security studies inquiry in order to learn something about the political dynamics around us. International relations expert S. M. Walt quickly reacted the same year by calling for the cohesive and simple structure of the international-relations discipline, criticizing the move of social constructivists not only for including more than a military-driven physical security, but also for moving the discipline out of the "naturally evident threats" (Walt 1991). For Walt, a neorealist, the interstate-security nexus is based on natural law of anarchy in the international system, which states try to survive. However, elaborating on the argument of interdependence introduced by Nye and Keohan, Buzan introduced other sectors beyond the military that could cause domestic instability (Buzan 1991, p. 80). In Buzan's view, security can thus be extended from a military perspective to political, economic and environmental ones. He argues that the web of interdependencies between states has made it too difficult even for large and self-contained states such as the Soviet Union or China to insulate themselves from the international system (Buzan 1991, p. 178).

---

[4] Sometimes also addressed as the third by those who refuse the inter-paradigm debate as a theoretical debate. The inter-paradigm debate was between realism, liberalism and radical international relations theories, or between realism, institutionalism and structuralism. To read more about the debates in IR from a critical perspective and how the debates can be fully overridden to perceive the world without the limiting perspectives on one explanation, refer to (Lake 2013).

This was just the beginning. Later, Buzan decided to build his insight into the structural realism introduced by Alexander Wendt (Wendt 1987). Buzan added a new actor, namely international society, and argued that its emergence from the neorealist international system of sovereign states can happen as a natural product of anarchy (Buzan 1993). It is socially constructed by the efforts of various international system actors not limited to states, such as international organizations, intergovernmental organizations, non-governmental organizations, etc. In fact, they do influence states' behavior in the so-called *second image reversed,* as Gourewich famously claims, "The international system is not only a consequence of domestic politics and structures but a cause of them" (Gourevitch 1978) (see Chap. 21). International organizations raise topics, develop norms and push states to follow the stipulated normative good (Barnett and Finnemore 2004). While they are usually focused on a dedicated topic (UNICEF on children, ITU on telecommunications etc.), those actors influence international politics or even states. It is hard to argue against the claim that international affairs are not only about the power between states but also about topics such as the state of the environment—or the state of security of our global environment. Buzan's extremely influential work gave birth to a completely new perspective on how security can be studied, simply as a *social construction of insecurity*, which formed the so-called Copenhagen School on which thousands of studies have since been based (Buzan et al. 1998).

11.5 Copenhagen School

Proponents of the Copenhagen School would first focus on the dynamics of how the threat has been socially constructed through speech acts—as discourse and later subjectively internalized, injected into our minds (to the securitization actor's audience) as something that needs to be addressed, prepared against, solved and implemented as a picture of insecurity into our institutions. This approach is not about problem solving; the Copenhagen School is much more about unveiling the hidden intentions behind the construction of insecurity.

The Copenhagen School triggered an enormous reaction worldwide. As mentioned in the introduction to this chapter, in their perspective, the securitization processes construct security threats through speech acts. These processes need to be unveiled in order to delegitimize the decisions that lead to the adoption of exceptional policies in the so-called desecuritization process (Waever 1993), no matter the character of the threat. They seek speech acts, or discourse, constructing feelings of insecurity with visible impacts on our lives. The Copenhagen School does not focus on the question of whether something is truly or objectively a threat to mankind. Instead, it focuses on representation of insecurities, dangers and threats that make differences in our lives with no regard for any objective—real—threats. As such, the School focuses on security from the negative perspective, with only one normatively good process—to desecuritize the securitized threat. The Copenhagen School is not asking a *normative* question of "what world we want to live in;" rather, it asks such *analytical* questions as "who constructs the dangers we are addressing, and why and for what purpose?" It tries to unveil why we employ exceptional means to enable exceptional politics on issues that, in fact, do not need to threaten us seriously or at all. It uncovers the processes of our real actions based on alleged, fabricated and socially constructed threats. The classical case is the global-terrorist threat against which we employ enormous amounts of real policies that influence our lives, despite the fact that the threat of dying in a car accident is thousands of times higher.

As theorized by the Copenhagen School in security studies just after the era of neorealism, a state actor was an inseparable part of the analysis as the securitizing actor. The theory argues that in the process of securitization, an issue that is not related to security has been securitized (by a state actor) as a threat to the *reference object* (usually state, but also mankind), so *the audience* acts. When linked to a state, it is understood as an issue of national security securitized by the respective state representatives. Finally, this was the answer to the materialist perception of neorealists saying that we live in an anarchic interstate system—we do not: *states make of it* [anarchy] as was argued later; it is a pure fabricated mindset of politicians who simply behave in this way (Wendt 1992).

We can desecuritize a threat by analyzing various factors of the whole process, which is the reason why securitization theory is sometimes called a method rather than a theory. However, it still focuses on states or their principal representatives as the actors. Therefore, the scholars from the Copenhagen School are not so far from neorealists, which is a courageous and provocative statement but not so unusual within the discipline. The state as a central analytical entity is simply limiting us from a broader analysis of what must be done in an ethically desirable way on a global scale, because the security of one state inherently causes insecurity to others.

If a global threat is addressed by one state in defiance of the interest of any other state, shouting to the world that it has right to resist any policy given the right of its sovereignty, what can others seeing an asteroid on a collision trajectory do? One can start thinking about international organizations as an answer, but to this extent the Copenhagen School does not provide us an answer. Much the opposite, says Buzan: scientific evidence of climate change is a social construction and an authoritative statement by a scientific epistemic community that changes its mind time to time (Buzan et al. 1998, p. 72). From this perspective, which is nothing unique in critical studies, it is hard to develop a normatively good policy of planetary defense based on scientific evidence.

11.6 Welsh School

Besides the Copenhagen School, another approach has emerged on the shoulders of Critical Theory, the Welsh School. In this approach, securitization is not limited to a social construction of threats to particular states. The Welsh School understands states as part of the problem, because if the securitization is done by states, the outcome of securing one state causes insecurity to the others, or the state itself becomes the source of insecurity to its inhabitants (Booth 1991). Booth foresaw by the beginning of the 1990s the end of the Westphalian system of sovereign states. For him, the emergence of a borderless world community is inevitable, and the current state of affairs is only an *interregnun* between the Cold War world order and the upcoming new cosmopolitan global-scale entity (see Chap. 21).

For Booth, true security is a state of emancipation, in which emancipation is "freeing people (as individuals and groups) from the physical and human constraints, together with poverty, poor education, political oppression and so on" (Booth 1991). Security, then, brings people the opportunity to fulfill their desires. But one might wonder when the call for more security ends. Extreme pressure on security usually ends in its totalization and, thus, totalitarianism; however, in Booth's terms, security is an emancipatory concept and such a scenario never satisfies. Those who pursue security for others can perceive the concept differently and seek power. One may ask why our desire to seek security must end, especially in a book dealing with security related to the void of our universe, which we still do not understand properly. At present, we do not even have a complex perspective on our own Solar System's asteroid population.

The Welsh School is important in bringing security value, by detaching security from the states and by saying that security is not a mere absence of a threat but an opportunity to fulfill our desires. It brings in the emancipatory argument of security. Booth's term, *security-plus,* refers to the kind of security that gives us opportunities to flourish through security from threats. It provides more than just mere survival, because survival does not eliminate threats (Booth 2007, p. 106).

If we take this normative position, an asteroid threat can be normatively much more tenable than in the previous school briefly described here. This is exactly the *value of security* represented by the differences between various policies, from building underground shelters, to selecting the "right" people to survive, to planetary-scale collision avoidance, to developing multipurpose space infrastructure with more than survival objectives. All of them somehow react on the threat, but each has significantly different ethical implications and normative outcomes.

There needs to be a legitimate process for addressing threats that can be deduced from scientific proofs but which will significantly change current interstate relations by deploying alleged dual-use technologies, perceived by the states as threatening (to their national security) despite their original purpose, such as active space debris removal by a spacecraft capable of catching and de-orbiting a non-functional satellite. On the one hand, we can defend the securitization of an asteroid threat as a normatively desirable move because answering the threat will save humanity from extinction. On the other hand, our reaction to that threat can be disputable due to the proposal of possibly threatening technologies singled out as an exceptional move to secure our survival. The deployment of these dual-use technologies is not limited to their intended purpose, but also includes their politically dedicated purpose by statements, discourse, law, international regime or any other non-objective means that define the political objective of the infrastructure. Simply put, if one party has sincere intentions with dual-use technologies, the other party may not recognize that sincerity due to the lack of transparency and will declare war just because they feel threatened by the deployed technology.

We have experienced several global threats to which humanity was able to react, but which are currently half-forgotten. Recall the battle over lead in automotive fuel, in which Robert A. Kehoe developed false scientific results in corporate-supported research in order to support the car industry, while one single scientist, Clair Cameron Patterson from Caltech, presented research on lead's poisoning effects (Settle and Patterson 1980). It took about two decades to persuade the judges with scientific facts proving the impact of lead on human health. Another example is the reconsideration of gasses in refrigerators depleting the ozone layer in the atmosphere, and the recycling policy around these appliances. In this case, nobody complained too much because this policy did not influence the interests of powerful corporations. Both examples show that securitization of threats based on scientific means is nothing new, and that the implemented policies did not threaten any nation-state worldwide. An asteroid threat is delicate because of the magnitude of the threat and the required response needed if we want to secure not only the survival of our species, but also the survival and flourishing of us all. The deployment of any measures can be easily understood as a threat to any state unless they are properly secured. This author recently argued that the deployment of commercial infrastructure is a way to secure our flourishing through asteroid mining, while the same infrastructure can be used for planetary defense purposes (Schmidt 2017).

The conceptualization of a threat as a global one and its global reaction are critical in bringing about a desirable state of security from a broader cosmic perspective (see Chap. 18). The Welsh School supports our answer, which should not be limited to survival, but should include flourishing—the so-called security-plus state (Booth 2007, p. 102). For example, today we know that an unimpeachable threat comes from invisible radiation. Nevertheless, we lived through epochs during which effects from radiation were ascribed to the gods' powers. Natural science can provide us with a proper tool to unveil explainable roots of objective threats. Yet, as the climate change agenda has shown us, believing in or following scientific evidence or following unless it is refuted or updated is not an obvious habit, and answering the scientific evidence does not need to bring us the desirable solution in all cases. Moreover, key persons in security studies can still perceive scientific evidence as a social construction. Hence, the proper approach involves the plethora of security and scientific practices (the normative and the scientific) we may employ to answer this threat in order to survive as a species and enable the biosphere to flourish. If we care about humanity as a whole and as a species fully dependent on flourishing biosphere, choosing proper means is a delicate ethical question we must address.

11.7 The Pragmatic Move

Rita Floyd argued for a consequentialist evaluation of security in her piece from 2007, in which she tries to reconcile the Copenhagen and Welsh Schools, arguing that both schools have something to add to the discussion. She argued that security is entirely issue-dependent and that the consequences need to be evaluated from the ethical standpoint (Floyd 2007, p. 337). In this perspective, securitization itself is neither good nor bad, and the consequences of securitization matter more. She believes that theories should not be enclosed in their own universes but need to communicate with each other. While the Copenhagen School can be understood as a method to unveil the extreme politicization of securitized threat, the Welsh School can point us toward issues that can spark human flourishing. Planetary defense is the kind of issue that can be exploited for extreme politicization but deserves a more balanced approach, as it can change our perspective of humanity's role on this planet. In order to reach that reflection on planetary defense, we neither need to take the analytical perspective of the Copenhagen School, nor the normative perspective of the Welsh School; we need to reflect on them both and approach the problem more pragmatically, as proposed by Jonna Nyman (2016). Nyman combines the consequentialist theory with the new wave in security studies, focusing on security practices to propose a pragmatic perspective that should focus on ideas that "are also helpful in life's practical struggles;" this perspective is taken from the pragmatic philosophy of William James (1910). As she draws on pragmatism, she asks: what do different security practices do?

Planetary defense is at its beginning, as we currently watch the sky and try to understand the dynamics based on Solar System orbital mechanics. All the practices that will emerge from such complex activities, including observation of the sky and the proposal of various deflection technologies, will produce different contexts and will be somehow a part of the overall asteroid-threat securitization. Here, Nyman argues for overall *normative awareness* in contrast to *normative commitments*; in planetary defense, this argument means that we need to reflect our technological proposal under that normative awareness rather than argue for the immediate necessity of an effective Earth defense policy. From this perspective, security from an asteroid threat is not necessarily good security; more crucial than the asteroid itself are discussions about the outcomes of asteroid-threat securitizations and the broad analysis of possible consequences caused by political decisions that are based on scientific evidence addressing the threat. Scientific evidence proving even a detected asteroid on an impact trajectory does not provide us with an indisputable, normative right to act. Here, one should think twice about the current nuclear-deflection technique. Its immediate and permanent state of preparedness does not bring about a political environment we want to live in, regardless of the fact that nuclear explosive devices may be the only means to save us from an asteroid detected several days before impact; this is the most probable scenario, as we continuously detect asteroids passing along the Earth days or weeks in advance. This is a dilemma for which we do not have a satisfactory answer but feel responsibility to raise it. We should deploy a planetary defense infrastructure capable of bringing us more than just defense—for example, commerce through active asteroid mining (Schmidt 2017)—because its permanent operation would be hardly disputable through national security discourses, and confidence in its purpose would deepen over time.

11.8 Conclusion

This chapter aimed to introduce how differently planetary defense efforts can be perceived through the evolving lens of political scientists. It intended to emphasize various philosophical findings in the discipline's evolution and point out how they can be used in planetary defense policy development.

First, despite the fact that we live in a world consisting of sovereign nation-states, we do not necessarily need to see the same world in a century. International organizations are powerful, focused on particular topics and influences on and between international organizations and states. Perceiving the nexus of global power only through the network of states would be reductionist and shortsighted. Thus, it is not necessarily the state system that would make the planetary-defense endeavor institutionalized.

Second, security, insecurity or a security threat cannot be defined by rational means through objectifiable means. Security is clearly socially constructed, because one has to make a speech act and convince the audience that some phenomenon is a threat to us. Such a move is inherently political and not based on science. However, if the security policy is focused only on negative security—the absence of threat—it would be hard to justify the policy move later when the insecurity does not transpire in a century or so. That means security-plus is not only an ethical move but a necessary one, because then the policy move taken will have been visibly and continuously contributing to human wellbeing.

Taking the security-plus concept, scientifically defined phenomena and an engineered solution is the key approach to develop ethically justifiable policy that addresses the correct problem. Engineers and scientists should accept that social sciences have a strong argument here, even as the acronym STEM (Science, Technology, Engineering, Math) has been slowly moving to the acronym STEAM, which includes Arts.

That said, saving the Earth from asteroids is not self-evidently good policy, as the proposed technology can be applied to different purposes and interests of governing actors. Various schools studying the process of securitization in which threats are elevated to being security threats have been described. In a nutshell, the most common problem is the fact that we live in a system consisting of sovereign states that will most probably elevate these dilemmas, and they may elevate them for their selfish interests rather than for the general good; this is not limited to the normative good, but also to good and right solutions.

The question of what world we want to live in might seem to be naïve, but it is the central guide when discussing how we plan to achieve the desired solution: the normative guidance that should go along with the rational one (discussing scientifically perceivable reality and engineering proper solutions). While Buzan in 1996 approached scientific evidence with harsh criticism for its outcome fluctuations, the pragmatic turn shows us that if we make a mistake in adopting policies based on low-quality science, we should not give up, but rather, try to develop more reliable scientific knowledge that will help us create better policies. Science works on a level of uncertainty, and its core objective is to reduce that uncertainty. The fact that scientific knowledge may have different levels of certainty for possible policies does not mean it is entirely socially constructed. The ability to accept that we are not perfect catches us in the trap of explaining natural processes as social constructions. While social science and constructivism can bring us enormously interesting critical perspectives, this one does not bring us closer to solutions that will make our lives better.

Asteroids may hit us or not, but the adopted policy, the deployed technology and the governing actor(s) will be here regardless of our actions, and regardless of the probability of the asteroids' impact.

Acknowledgments This study was supported by the grant awarded by the Technological Agency of the Czech Republic, project TL01000181: *"A multidisciplinary analysis of planetary defense from asteroids as the key national policy ensuring further flourishing and prosperity of humankind both on Earth and in Space,"* and co-funded by the Institute of Political Studies, Faculty of Social Sciences, Charles University in Prague.

References

Barnett, M., & Finnemore, M. (2004). *Rules of the World: International Organizations in Global Politics*. Ithaca and London: Cornell University Press.

Booth, K. (1991). Security and emancipation. *Review of International Studies, 17*(4), 313–326. doi:10.1017/S0260210500112033

Booth, K. (2007). *Theory of World Security*. Cambridge University Press.

Bush, V. (1945). *Science - The Endless Frontier*. Washington D.C.: National Science Foundation.

Buzan, B. (1991). *People, States and Fear: An Agenda for International Security Studies in the Post-cold War Era*. L. Rienner.

Buzan, B. (1993). From international system to international society: structural realism and regime theory meet the English school. *International Organization, 47*(03), 327. doi:10.1017/S0020818300027983

Buzan, B., Wæver, O., de Wilde, J., & Wilde, J. De. (1998). *Security: A New Framework for Analysis. National Bureau of Economic Research Working Paper Series*. Lynne Rienner Publishers.

Carr, E. (1939). *The Twenty Years' Crisis, 1919-1939*. The Macmillan Press Ltd.

Floyd, R. (2007). Towards a consequentialist evaluation of security: bringing together the Copenhagen and the Welsh Schools of security studies. *Review of International Studies, 33*(2), 327. doi:10.1017/S026021050700753X

Gourevitch, P. (1978). The second image reversed: the international sources of domestic politics. *International Organization, 32*(4), 881–912. http://journals.cambridge.org/production/action/cjoGetFulltext?fulltextid=4305592. Accessed 5 March 2014

Hobson, J. M. (2000). The State and International Relations. doi:10.1017/CBO9780511612442

James, W. (1910). *Pragmatism: A New Name for Some Old Ways of Thinking*. Cambridge University Press.

Keohane, R. O., & Nye, J. S. (1977). *Power and interdependence: world politics in transition*. Little, Brown.

Keohane, R. O., Nye, J. S., & Keohane, R. O. (1973). Power and interdependence. *Survival, 15*(4), 158–165. doi:10.1080/00396337308441409

Lake, D. A. (2013). Theory is dead, long live theory : The end of the Great Debates and the rise of eclecticism in International Relations. *European Journal of International Relations*, *19*(3), 567–587. doi:10.1177/1354066113494330
Nyman, J. (2016). What is the value of security? Contextualising the negative/positive debate. *Review of International Studies*, *42*(05), 821–839. doi:10.1017/S0260210516000140
Nyman, J., & Burke, A. (2016). *Ethical Security Studies: A New Research Agenda*. (J. Nyman & A. Burke, Eds.). Routledge.
Schmidt, N. (2017). Planetary Defense as a Gateway to Space for Commercial and Deep Space Exploration. *New Space*, *5*(4), 219–229. doi:10.1089/space.2017.0005
Schmidt, N. (2018). The political desirability, feasibility and sustainability in planetary defense governance (in review). *Acta Astronautica*.
Settle, D. M., & Patterson, C. C. (1980). Lead in albacore: guide to lead pollution in Americans. *Science (New York, N.Y.)*, *207*(4436), 1167–1176. doi:10.1126/science.6986654
Waever, O. (1993). *Securitization and desecuritization*. Centre for Peace and Conflict Research Copenhagen.
Walt, S. M. (1991). The Renaissance of Security Studies. *International Studies Quarterly*, *35*(2), 211–239. doi:10.2307/2600471
Waltz, K. (1979). *Theory of international politics*. McGraw-Hill.
Wendt, A. (1987). The agent-structure problem in international relations theory. *International organization*, *41*(3), 335–370. http://journals.cambridge.org/production/action/cjoGetFulltext?fulltextid=4309572. Accessed 5 March 2014
Wendt, A. (1992). Anarchy is what states make of it: the social construction of power politics. *International Organization*, *46*(02), 391. doi:10.1017/S0020818300027764
Williams, P. D. (2012). *Security Studies: An Introduction*. New York: Routledge.

Chapter 12
Asteroid Impact Risk Assessment: Rationalizing the Threat

Clemens M. Rumpf

12.1 Introduction

Throughout the evolution of life, asteroid impacts have caused at least one major disruption to the terrestrial ecosystem (Alvarez et al. 1980; Plane 2012). They remain a hazard for the human population today; the record of bolides (NASA 2014b), the recent Chelyabinsk airburst in 2013 (Brown et al. 2013; Popova et al. 2013) and the Tunguska event in 1908 (Chyba et al. 1993) demonstrate this palpably. Due to the sporadic nature of asteroid impacts, this natural disaster is not considered amongst the everyday experiences of the human population. Nonetheless, the potential for destruction by an asteroid is enormous, justifying closer investigation of this hazard. In this chapter, the asteroid hazard and risk is presented from a statistical perspective, providing insight into the frequency and consequences of impact events, along with expected impact conditions such as impact location, speed and angle. In addition, ways of communicating a potential threat with the scientific community or the general public by means of hazard scales are highlighted. The investigation helps to understand this natural disaster, which is characterized by its rarity and potential for global consequences. Smaller impacts can cause damage on a local level and occur more frequent than larger impacts, which can cause destruction on a global scale. The analysis presented in this chapter helps to identify the portion of the impact population that strikes the most hazardous balance of relatively frequent impacts and considerable damage potential.

C. M. Rumpf (✉)
University of Southampton, Southampton, UK
e-mail: drcrumpf@gmail.com

N. Schmidt (ed.), *Planetary Defense*, Space and Society,
https://doi.org/10.1007/978-3-030-01000-3_12

12.2 Near-Earth Asteroid Population

Asteroid population statistics in relation to asteroid size are given in Fig. 12.1 and are presented in (Boslough et al. 2015). As a first order approximation, the asteroid population may be approximated by a constant power law (blue dashed line) as a function of asteroid size. It becomes apparent that smaller bodies are far more numerous in the asteroid population than larger ones. More accurately, the actual estimate of the population varies around the power law (blue circles), and one observation that may be made here is that the population is biased towards small asteroids (<30 m), which are even more abundant than predicted by the constant power law. The abundance of small asteroids explains the far greater impact frequency of small objects as compared to that of larger ones such as the recent Chelyabinsk (Borovicka et al. 2013; Brown et al. 2013; Popova et al. 2013) and Tunguska (Chyba et al. 1993) impact events—both of which are still classified as part of the small asteroid population. Similar impacts can be expected every 15 and 300 years, respectively, as marked in Fig. 12.1. A Tunguska-level event devastates an area comparable to that of Greater London, which houses 8.5 million inhabitants (Rumpf et al. 2017a). This comparison illustrates the damage potential that these objects have, even though they are classified as small objects. The large prehistoric Chicxulub (Morgan et al. 1997; Schulte et al. 2010) impact was likely caused by an

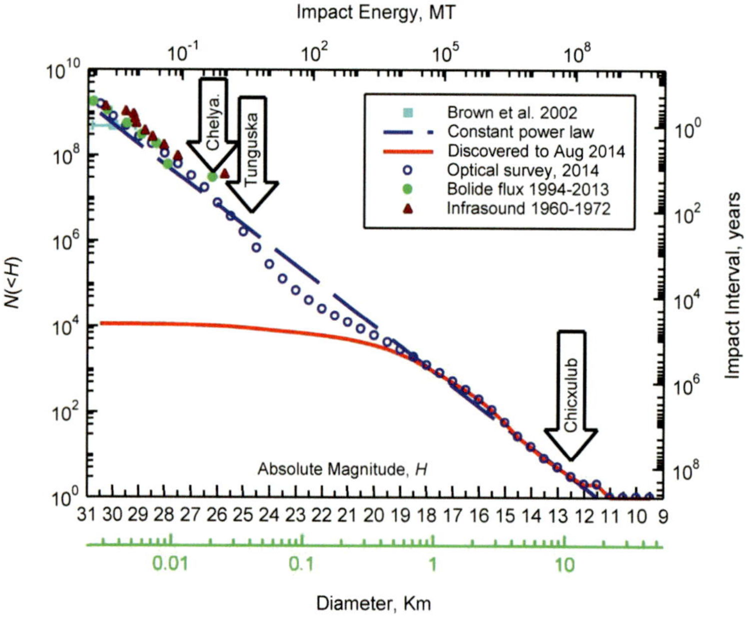

Fig. 12.1 Cumulative population of Near-Earth objects and their impact interval as estimated in 2017. Image provided by Alan Harris

object measuring about 10 km in diameter. An object of this kind has an expected impact frequency of about 100 million years.

It is of great interest to the planetary defense community to discover all objects that could cause harm to the human population. The threat situation posed by asteroids becomes predictable in terms of impact time (European Space Agency 2017; NASA 2014a) and location (Chesley et al. 2014, 2015) once a hazardous object is discovered, and this fact is a significant motivation for the increased asteroid discovery effort. In addition, it is in principle possible to deflect an asteroid and thus to prevent a disaster altogether (Shapiro et al. 2010)· as described in Chap. 7. Prediction and deflection are only possible for objects discovered before they impact.

In addition to the asteroid size and impact frequency distribution, Fig. 12.1 visualizes how complete the catalogue of observed asteroids is in each size regime. The red line provides the number of asteroids of a given minimum size that were known as of August 2014, and this graph may be compared to the expected asteroid population count in that size regime (blue circles). In 1992, the Spaceguard survey (Morrison 1992) was defined and it was subsequently launched in 1998. The Spaceguard Survey stated that 90% of all kilometer-sized asteroids should be detected within 10 years to reduce the risk of catastrophic impact events with global consequences. In 2008, the catalogue was 79% complete for asteroids with a minimum size of 1 km (Harris 2008), and today, the Spaceguard goal has been practically met (Boslough et al. 2015) as the catalogue approaches completeness for asteroids sized 1 km or larger, as described in Chap. 3. However, the vast majority of smaller asteroids remains undiscovered, and this fact, together with the realization that smaller impactors pose a significant threat, has led the United States Congress to mandate the discovery of 90% of asteroids larger than 140 m by the year 2020 (Shapiro et al. 2010). This goal will not be met, as no dedicated funding was appropriated to cover the extended effort. Furthermore, impact events such as the Tunguska and Chelyabinsk airbursts demonstrated that very small asteroids could yield significant destructive force (Boslough and Crawford 2008). Thus, future surveys might have to consider targeting even smaller asteroids than the current aim of 140 m. (Shapiro et al. 2010) recommended to extend the search down to a minimum diameter of 30 m, of which <1% has been observed at the time of writing (2017).

Targeting smaller asteroids in future surveys will incur an increased cost compared to the Spaceguard Survey, because smaller asteroids reflect less light and are thus harder to observe.[1] As a first approximation, the diameter D of an asteroid may be estimated based on its brightness in conjunction with the assumption of spherical shape. Typically, an albedo α (the portion of incoming light reflected by the asteroid's surface) of 0.1–0.154 (NASA 2014a; Universita Di Pisa and European Space Agency 2014) is assumed to calculate the absolute magnitude and size of the asteroid. Absolute magnitude H is the brightness of an asteroid if it was 1 AU from the

[1] The discussion here focuses mainly on optical surveys in the visible spectrum. Asteroids may also be observed using radar or emissions in the infrared. However, the assertion that small asteroids reflect or emit less electromagnetic radiation than larger ones generally holds true.

Sun, as seen by an observer located at the Sun. Hence, asteroid size may be estimated according to (Chesley et al. 2002) with:

$$D = 1329000m \times 10^{-H/5} \alpha^{-0.5} \tag{12.1}$$

In summary, the catalogue of observed asteroids shows a higher ratio of completion in the large asteroid size regime, as demonstrated by the red line being congruent with the population estimate in Fig. 12.1. The smaller asteroid sizes stand out when considering the completion of the asteroid catalogue, as the vast majority of smaller, yet dangerous asteroids between 15-m (local damage) and 500-m (continental damage) diameter remain undiscovered to date.

12.3 Torino and Palermo Asteroid Hazard Scales

Asteroid hazard scales rely on information that is available early after initial discovery of a new asteroid. The trajectory of every asteroid discovered is calculated for about 100 years into the future, and the probability of a collision with the Earth is estimated. The results of this analysis are publicly accessible on the NEO risk webpages maintained by the European Space Agency (ESA) (Universita Di Pisa and European Space Agency 2014) and the National Aeronautics and Space Administration (NASA) (NASA 2014a). The information on these webpages includes the impact probability P, impact date, impact speed v as well as the size estimate for each known object. The kinetic impact energy E of an incoming asteroid of diameter D may be estimated by assuming a bulk density, which typically lies between $\rho = 1000 - 3500\ kg\ m^{-3}$ (Britt et al. 2002; Britt and Consolmagno 2000; Hanus et al. 2016), together with the speed information v available on the risk webpages (European Space Agency 2017; NASA 2014a):

$$E = \frac{\pi}{12} \rho D^3 v^2 \tag{12.2}$$

The Torino and Palermo asteroid impact hazard scales are chiefly in use today to rate the threat of specific asteroids and to relate this information to other experts and the general public. The Torino scale (Binzel 2000) is a simple 11-step scale that assigns a rating of zero if the threat is perceived to be insignificant. or up to 10 for impacts that are certain to happen with global consequences. Figure 12.2 presents the Torino scale. The rating depends on only two parameters: collision probability and kinetic energy expressed in megaton (MT) equivalents of TNT (1MT TNT = 4.3×10^{15} J). Kinetic energy is used as a proxy for impact consequences; larger kinetic energy corresponds to an expectation for more severe consequences for the human population.

The Palermo scale (Chesley et al. 2002) caters to an expert audience. In addition to impact kinetic energy E and collision probability P, it accounts for the time until impact and provides a comparative rating of how likely a similar impact is in the

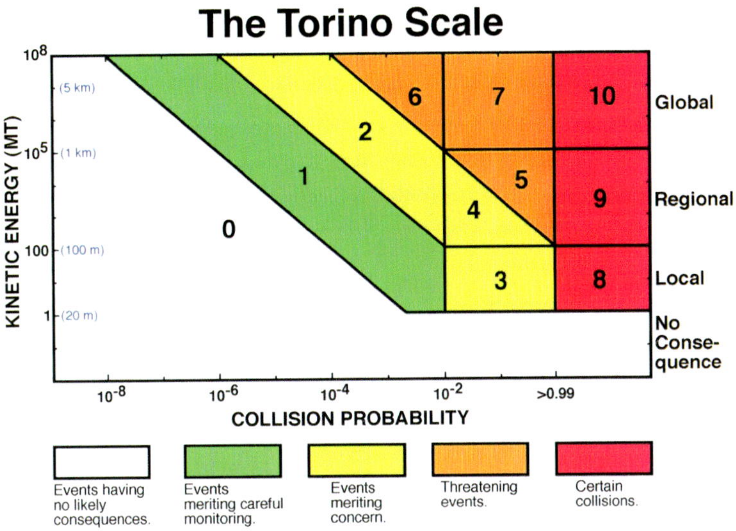

Fig. 12.2 The Torino Asteroid Impact Hazard Scale (Binzel 2000)

same timeframe. A Palermo rating $\mathcal{P}$ is obtained under consideration of time until impact in years ΔT using:

$$\mathcal{P} = \log_{10} \frac{P}{\frac{3}{100} E^{-4/5} \Delta T} \quad (12.3)$$

A negative Palermo rating indicates that another similarly dangerous impact is more likely to happen than the potential impact of the asteroid under consideration. Conversely, a positive Palermo rating indicates that the threat by the asteroid under consideration is more urgent than the abstract background threat of similarly sized objects in the asteroid population.

In general, asteroid impact hazard ratings are useful tools to express the threat level of a given asteroid. However, the Torino and Palermo scales use abstract ratings that do not allow direct comparison to other natural disasters. A new hazard scale is introduced later in this chapter, which uses the unit of expected fatalities and thus remediates this caveat—it is called the Southampton Asteroid Hazard Scale.

12.4 Asteroid Impact Angle, Speed and Location Distributions

A statistical analysis of the possible impact parameter helps us understand the kind of impact scenario we are likely to face. It is therefore useful to have information about the expected impact angle (which is important for the process of atmospheric passage), the expected impact speed (relating to impact energy) and the areas prone to impacts.

12.4.1 *Impact Angle*

An analytical solution to the statistical impact angle distribution problem on planetary bodies is given in (Shoemaker et al. 1962). It was found that a sine-cosine function describes the probability density function for impact angles:

$$dP = 2\sin(i)\cos(i)di \tag{12.4}$$

Where dP is the differential probability of the occurring impact angle i. This function reaches its maximum at 45°, which is the most likely impact angle relative to the planetary surface.

In (Rumpf et al. 2017a), a sample of 10,006 artificial impactors (Chesley and Spahr 2004; Grav et al. 2011) generated in accordance to NEO population statistics was analyzed for individual impact statistics such as impact location, angle and speed characteristics. The impact angle distribution shown in Fig. 12.3c is based on a Beta distribution that yields a correlation coefficient of $R^2 = 0.9998$ with the observed sample. The distribution agrees closely with the analytical solution in Eq. (12.4) yielding a mean impact angle of 42.87°, weakly biased towards shallower impact angles. The minor deviation in mean impact angle could be an artifact in the artificial impactor sample or might hint at a higher order effect not captured in the analytical solution. This could produce a bias towards shallower impact angles and possibly justify further research in this area.

12.4.2 *Impact Speed*

In general, impact speeds between 11.1 km s^{-1} (the escape velocity of the Earth) and over 70 km s^{-1} (depending on the asteroid's apogee) are possible. Analysis of the impact speed distribution showed that a lognormal distribution fitted the sample speeds between 11.1–34.2 km s^{-1} (98.6% of the sample), producing a correlation coefficient of $R^2 = 0.99$ (Rumpf et al. 2017a). The lognormal distribution allows insights into which impact speeds are most likely to be encountered. The mean

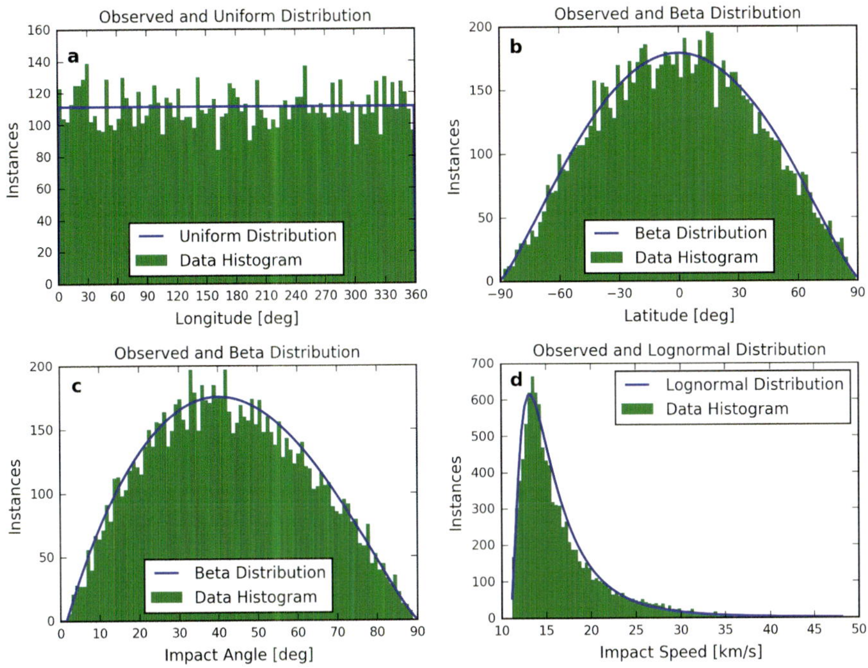

Fig. 12.3 Impact statistics showing impact location distributions for longitude (plot **a**) and Latitude (plot **b**), as well as distributions for impact angle (plot **c**) and speed (plot **d**) based on 10,006 artificial impactors (Rumpf et al. 2017a)

impact speed of the distribution is 16.6 km s^{-1}, which means that 50% of impacts will be slower and 50% will be faster than this value. A mean impact speed of around 17 km s^{-1} matches previous results in the literature, indicating the robustness of this result (Collins et al. 2005). Some research suggests an average impact speed of 20.3 km s^{-1} (Brown et al. 2002) or 20.5 km s^{-1} (Le Feuvre and Wieczorek 2011). However, these higher values might overestimate the average impact speed observed on the Earth, because faster asteroids spend less time in the vicinity of the Earth, therefore degrading the probability (and consequently frequency) of a collision. Slower asteroids spend more time in the volume swept up by the circumference of the Earth along its orbital path, increasing the chance of a collision. Not taking this observation into account biases the average impact speed towards higher values.

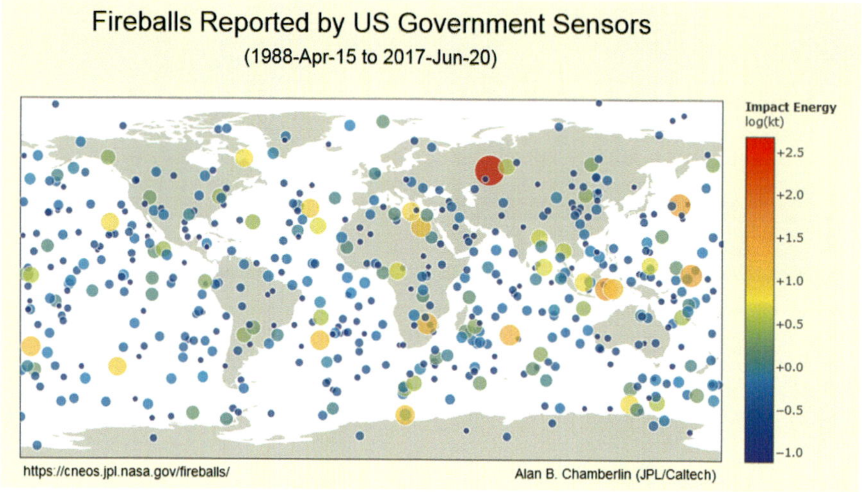

Fig. 12.4 Airburst events recorded by NASA (Chamberlin 2017)

12.4.3 Impact Location

In (Gallant et al. 2009; Rumpf et al. 2016), it was shown that any region on Earth has about an equal chance of being hit by an asteroid. This finding is confirmed empirically by 714 recorded airbursts between 1988 and 2017 (NASA 2014b) (Fig. 12.4).

The artificial impactor sample confirms this finding, producing a uniform impact location density distribution in longitude (Fig. 12.3a) and in latitude, as shown in (Rumpf et al. 2016). This observation is reinforced by a separate study that finds a 1% difference in impactor density between the poles and the equator (Gallant et al. 2009). In Fig. 12.3a, b, the impact location distribution is shown as impact instances per latitude/longitude band. For the longitudinal distribution (Fig. 12.3a), the result remains a uniform distribution, as every longitude band has equal area and thus collects an equal count of impacts. In contrast, the surface area of latitudinal bands shrinks from the equator to the poles. Consequently, fewer impact instances are recorded near the poles, even though the impact location density (and, therefore, impact probability) is near-uniform across the globe. The decreasing impact count towards the poles is suitably expressed by a Beta distribution, as visualized in Fig. 12.3b. In summary, the resulting spatial impact density is uniform in longitude and latitude, as shown in (Rumpf et al. 2016). Any population on Earth can be affected from the asteroid impact hazard.

12.5 Impact Consequences

Predicting the impact consequences of an asteroid and understanding the severity of the situation helps to formulate an adequate response to the asteroid impact threat (Binzel et al. 2010; Drolshagen et al. 2010; Harris et al. 2017; Shustov et al. 2013). The consequences of asteroid impacts in terms of human fatalities have been studied by (Boslough 2013; Chapman and Morrison 1994; Harris 2013; Reinhardt et al. 2016; Stokes et al. 2003), focusing on large asteroids. In general, it is accepted that asteroids larger than 1 km can cause global consequences that result in long-term damage to the terrestrial ecosystem (hence, the focus of Spaceguard). Impactors smaller than 1 km are expected to cause regional (on the scale of countries or continents) or local (on the scale of states or municipalities) damage depending on size, composition and speed. New studies such as (Mathias et al. 2017; Rumpf et al. 2017a) have started focusing on the more frequent, smaller asteroid impacts. Rumpf et al. simulated 50,000 impacts that were spread around the globe according to the distributions identified in Fig. 12.3. Based on the results, we were able to derive the average number of human loss per impact for smaller asteroids of a given size up to 400 m. Using a mean asteroid density value of 3100 kg m^{-3} (Brown et al. 2002), the consequences in terms of expected human loss are shown in Fig. 12.5a.

The green line in Fig. 12.5a is the loss curve for an average impactor anywhere on the globe, and it exhibits a steep rise in the airburst regime that corresponds to asteroids smaller than 50–60 m (Collins et al. 2017). The green dashed-dotted line

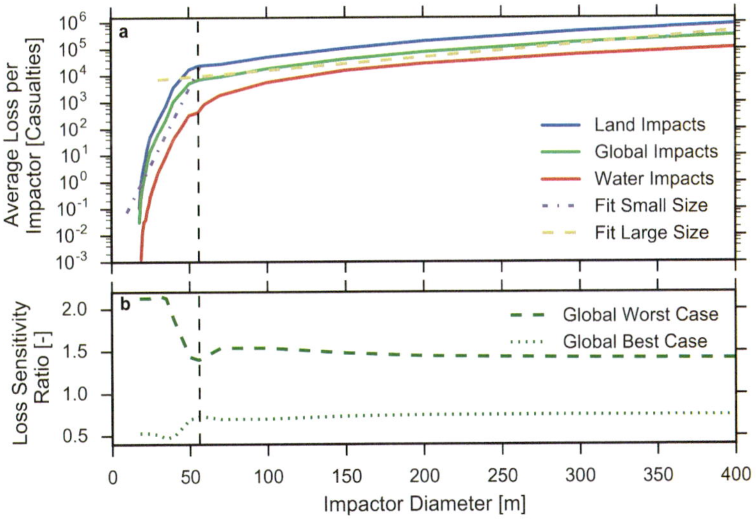

Fig. 12.5 Average loss of human life per asteroid impactor for global, land and water impacts (plot **a**). In plot **b**, the sensitivity of the global impact loss figures is shown as a factor to the loss numbers identified in plot **a**. The vertical dashed line identifies the size threshold where smaller impactors in this study airburst and larger impactors experience first surface impact

in Fig. 12.5a is a fit to the loss curve for smaller asteroids up to 50 m. Its corresponding exponential function is:

$$y_{small} = 0.083 \times 1.14^{1.75x} \tag{12.5}$$

Correspondingly, losses for asteroids larger than 50 m may be approximated with the fitted function (the beige dashed line in Fig. 12.5a) given by:

$$y_{large} = 4374.91 \times 1.01^{0.99x} \tag{12.6}$$

The fatality numbers presented in Fig. 12.5 match the casualty estimates of (Boslough 2013; Chapman and Morrison 1994; Harris 2013; Stokes et al. 2003) for asteroids of hundreds of meters in size but provide finer resolution for smaller sized asteroids. The higher resolution of the results is accomplished through a high fidelity modeling approach. Unlike most previous work, the results here are not based on rough impact consequence averages. Instead, each of the 50,000 impacts was modeled individually, accounting for the asteroid's passage through the atmosphere and the subsequently generated impact effects such as wind and pressure blast, thermal radiation, cratering, seismic shaking, eject deposition, and tsunamis. Additionally, these impact effects were propagated over the local population near the impact site, and vulnerability models were used to estimate the fatality count.

Vulnerability models enable a direct estimation of fatalities based on the impact effect severity that affected human populations are exposed to, as described in (Rumpf et al. 2017b). This reference also provides best and worst-case vulnerability models that aim to capture one-sigma variations in vulnerability for asteroid impact mortality, giving some insight into the uncertainty with which these models produce fatality predictions. Figure 12.5b presents the variability factor to the global loss curve in Fig. 12.5a. It is shown that vulnerability uncertainty is relatively high (factor two) in the airburst regime and decreases to a factor of about 1.4 for larger impactors. This indicates that fatality estimates are generally robust, but higher uncertainty is present for small objects. In addition to the global loss curve and its corresponding fitted functions (Eqs. (12.5) and (12.6)), Fig. 12.5a also presents average loss values for impactors over water or land masses. It was found that the average land impactor of a given size is about an order of magnitude more harmful than a similar water impactor. The main reason for this result is that land impactors are naturally close to population centers, and the impact effects reach such populations before they weaken by dispersing over long distances. In contrast, the average water impactor will strike far away from any inhabited coast and, while a tsunami may travel long distances, it attenuates sufficiently enough to lose much of its destructive energy, thus causing relatively little damage.

12.6 Asteroid Risk Assessment

This section presents three asteroid risk assessments based on separate data sets. The first one focuses on the asteroid population sized 15–400 m in diameter. The second is focused on large asteroid impact events and serves to visualize the risk reduction effect that observational efforts can have. The third covers larger asteroids and uses updated airburst models, which has an effect on the risk estimate originating from the smaller asteroid population. Together, the three risk assessments provide a more rounded picture of the risk situation than any one could provide by itself. Additional analysis is presented that places the asteroid impact risk in context with other natural and modern lifestyle hazards.

12.6.1 Asteroid Impact Risk Distribution up to 400 m

The yearly average asteroid impact risk can be calculated through multiplication of the asteroid impact frequency (Fig. 12.1) and the number of casualties per impact (Fig. 12.5). Current impact frequency data that takes into account the drop in asteroid population numbers between about 50 and 500 m are available in (Boslough et al. 2015) and are reflected in Fig. 12.1 of this chapter. Using the latter frequency

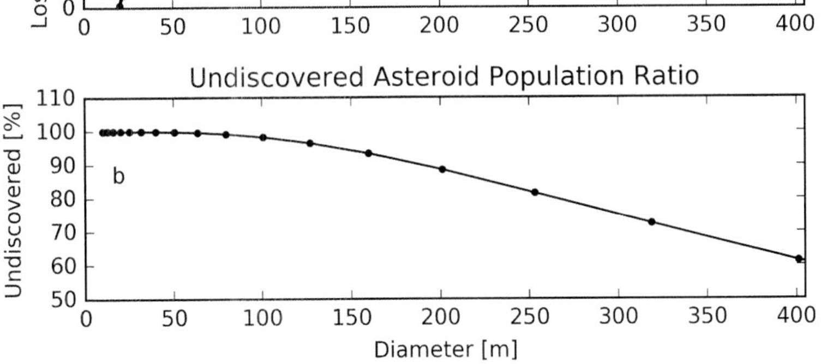

Fig. 12.6 Risk distribution over asteroid size is shown in plot **a**. In plot **b**, the ratio of the asteroid population (0–400 m) that is not yet discovered is shown. The markers indicate the midpoint of each size bin and have a width of half an absolute magnitude (Rumpf 2016)

estimates and the average loss per impactor curves for best, expected and worst case scenarios shown in Fig. 12.5, the risk distribution (averaged over long periods of time) over the size regime of 0–400 m was calculated, as shown in Fig. 12.6a. This figure shows the amount of yearly risk in terms of average fatalities per year and attributes this risk to a specific part of the asteroid population that is sorted by size. The numbers presented here are yearly averages, and it should be considered that asteroid impacts, especially larger ones, occur only every tens of years or even less often. Rather than experiencing this risk on a yearly basis, the risk manifests itself in large-scale events in the rare occasion of an asteroid hit. Interestingly, the risk shows a peak around the 50-m mark, even though the consequences of a larger asteroid are more severe. The peak is the result of the balance between more severe consequences of larger asteroids (about 10,000 per average 50-m impactor) and higher impact frequency for smaller ones (about once every 700 years). The total risk in this size regime (for asteroids up to 400 m in diameter) was found by adding up the risk in each size bin (markers denote the midpoint of each bin); the sum is 32 casualties per year. It should be noted that larger asteroids account for the majority of the risk due to their globally catastrophic impact consequences. These are not considered in Fig. 12.6, as these results only consider risk for objects up to 400 m in diameter.

The risk distribution in Fig. 12.6b is based on the overall asteroid population, considering its size-population numbers and average impact frequencies. However, thousands of asteroids have already been discovered and for these asteroids, the likelihood of an unwarned collision with the Earth has been reduced significantly by propagating their orbits and concluding that they are not on a collision course within the next 100 years. This information needs to be included in the risk analysis to accurately show what residual risk remains today.

Figure 12.6b shows the percentage of the overall asteroid population—sorted by size—that is yet undiscovered (Harris and D'Abramo 2015). The catalogue of discovered asteroids is much more complete for larger asteroids as they are easier to detect while the majority of smaller objects remains undiscovered. As stated before, only about 1% of the small asteroids in the 100-m size range capable of causing regional devastation have been discovered. The situation improves for larger objects, and today <10% of the very large objects (>1 km) that could cause global disaster remain undiscovered, thanks to survey efforts such as Spaceguard. Still, some large objects remain undiscovered and could impact unwarned, even though this chance is extremely small in the foreseeable future. For those asteroids that are discovered, it can be practically ruled out that they will impact the Earth within the next 100 years. The risk distribution was updated taking only into account the undiscovered asteroid population for the remaining (or residual) risk.

Taking into account only the undiscovered asteroid population yields the residual risk distribution shown in Fig. 12.7. The total risk in this impactor size regime (up to 400 m) sums up to about 29 casualties per year, which is a minor improvement over the risk situation of the total asteroid population. The improvement is only minor because the presented analysis focuses on smaller asteroids up to 400 m in diameter, of which only a small fraction has been discovered. The residual risk situ-

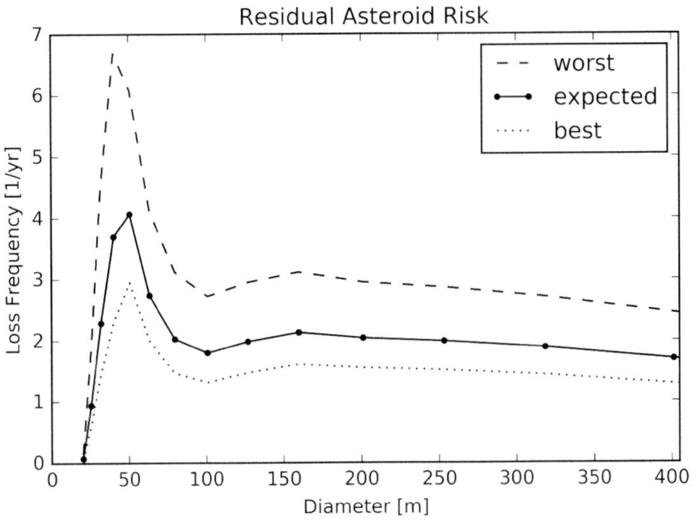

Fig. 12.7 Residual asteroid impact risk distribution over size. The data considers the risk-size distribution and the ratio of undiscovered asteroids for a given size (Rumpf 2016)

ation improves more significantly for larger asteroid sizes, and this is visible in Fig. 12.7, where the risk improves by about 50% for 400-m-sized asteroids when accounting for the already discovered asteroid population.

The reduction in asteroid impact risk based on the discovered asteroid population illustrates the value of NEO surveys in a tangible way. To further improve the risk situation, it becomes necessary to target smaller asteroid sizes, including 50-m-sized objects where the current residual risk peak is located, as visualized in Fig. 12.7. Residual risk will be reduced most effectively if 50-m-sized objects are targeted by new asteroid surveys, providing clarity about whether or not they will collide with the Earth in the next century. Such extended survey efforts would also create a wealth of scientific data about the asteroid population, building upon what has already been delivered by previous campaigns (Harris 2008; Shapiro et al. 2010; Stokes et al. 2003).

12.6.2 Asteroid Impact Risk Distribution up to 10 km

After focusing on the small asteroid size regime and explaining the method of calculating the risk distribution, the next analysis considers the entire asteroid population, including 10-km-sized objects. For very large objects, the benefit of asteroid discovery for risk reduction is even more apparent. The complete risk distribution is first shown without considering the knowledge of non-impacts for observed asteroids, and the resulting risk distribution is shown in Fig. 12.8 for asteroids up to

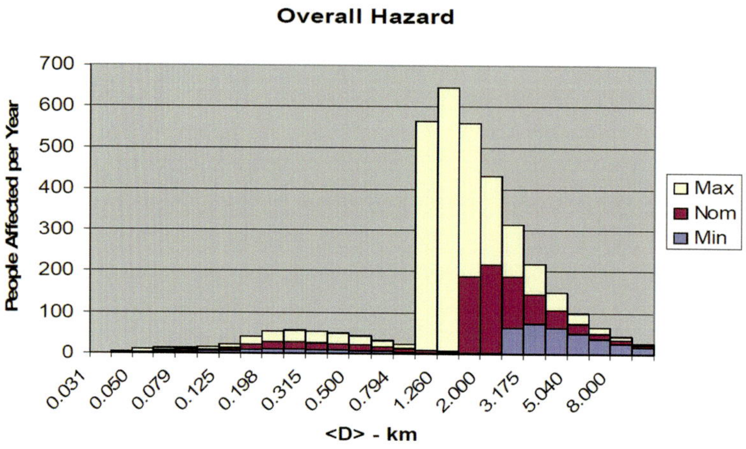

Fig. 12.8 Asteroid impact hazard for water and land impacts for impactors up to 10 km in diameter. Credit: Figs. 3–10 in (Stokes et al. 2003)

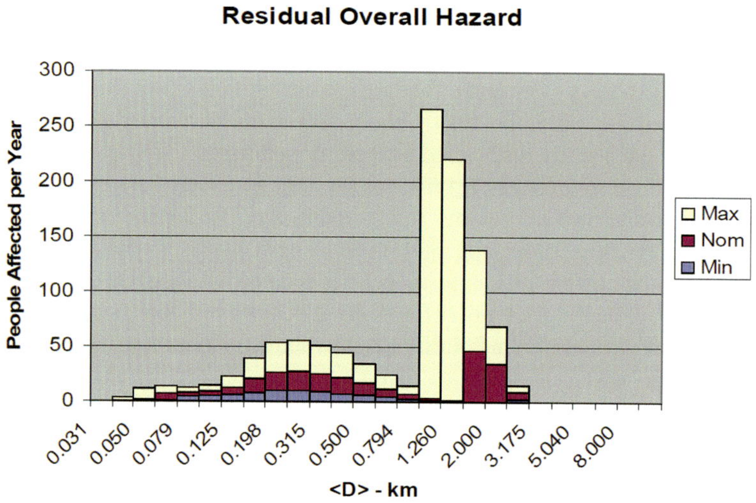

Fig. 12.9 Residual asteroid impact hazard per asteroid size up to 10 km, taking into account that no impact is expected from the known (in 2003) asteroid population. Credit: Figs. 3–11 in (Stokes et al. 2003)

10 km in diameter (Stokes et al. 2003) (An update to this report is available (Stokes et al. 2017)). Kilometer-sized impactors dominate the risk distribution, as impact consequences of such a massive object would affect the entire globe. The total risk as a sum of all size bins is about 1260 casualties per year (Stokes et al. 2003).

The risk situation improves significantly when the already discovered (and deemed harmless) population is excluded from the risk analysis. One should note that these results were based on data available in 2003, 5 years after the official start of the Spaceguard Survey targeting kilometer-sized objects (Stokes et al. 2003). Still, the positive effect of the survey on risk reduction is already apparent and has further improved with additional asteroid discoveries. The corresponding residual risk distribution is shown in Fig. 12.9, and the risk for objects larger than 1 km is reduced by more than a factor of four compared to the uncorrected risk distribution. Risk improvements for smaller asteroids are less apparent, since the discovery gap for these objects could not be closed at a comparative rate. The overall risk situation improved by 77% to 293 casualties per year (Stokes et al. 2003).

12.6.3 Asteroid Impact Risk Distribution up to 2 km with Improved Airburst Model

The risk distributions for asteroids up to 2 km in diameter based on more current data are presented in (Boslough et al. 2015). Figure 12.10 visualizes this data and quantifies the average risk by the total asteroid population as 1427 casualties per year. These numbers are based on research presented in (Boslough and Crawford 2008), which indicates that airburst events may be significantly more dangerous than previously assumed due to a more efficient transfer of destructive energy to the ground in these mid-air explosions, as compared to surface impactors.

The discovered asteroid population is taken into account to obtain the residual risk for impactors up to 2 km in diameter and the resulting risk distribution is shown in Fig. 12.11. As observed previously, major risk reductions are accomplished in the kilometer size regime due to high discovery ratios in this range (Harris and

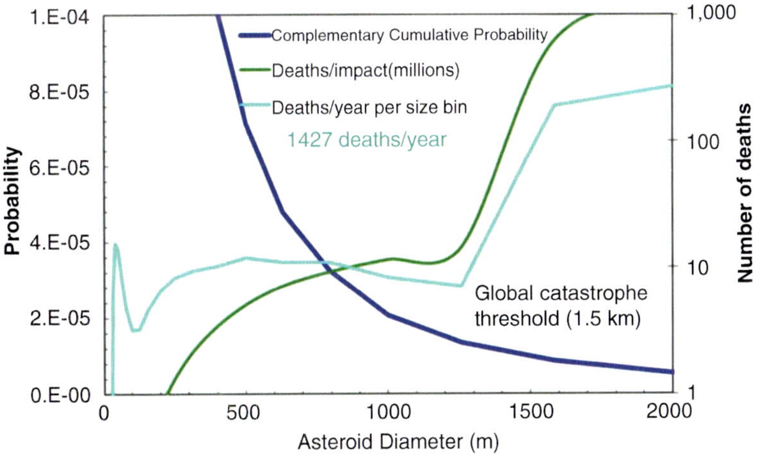

Fig. 12.10 Overall asteroid impact risk up to 2-km-sized impactors based on complete asteroid population. Presented in (Boslough et al. 2015)

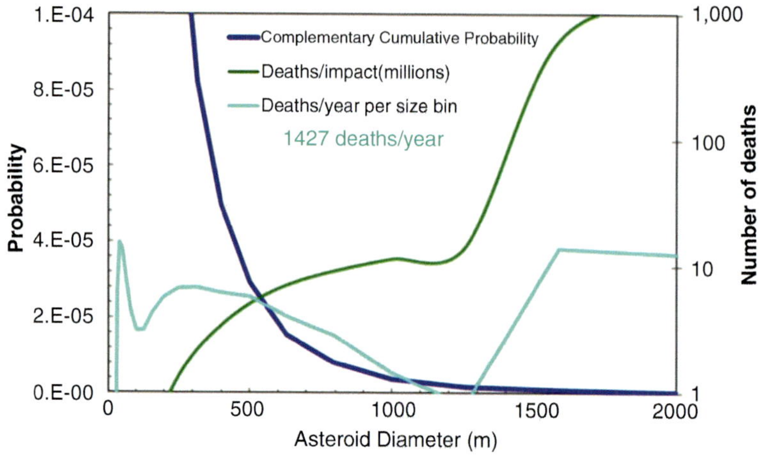

Fig. 12.11 Residual asteroid risk from the undiscovered asteroid population (Boslough et al. 2015)

D'Abramo 2015), taking into account asteroid discovery data up to the year 2014. Consequently, the overall risk was reduced by 90% to 145 casualties per year, compared to the level of 1427 deaths per year prior to asteroid discovery considerations, presented in Fig. 12.10. In fact, the surveys have been successful to such a degree that the peak residual risk of small impactors is comparable to that of kilometer-sized objects. Furthermore, the peak residual risk for small impactors in (Boslough et al. 2015) coincides with the findings presented in Fig. 12.7, indicating that objects of roughly 50 m in diameter should be the main focus for future NEO surveys. This outcome provides confidence in the accuracy of the results, as three different analyses have come to similar conclusions (albeit focusing on differing asteroid population regimes). It is worth mentioning that the size of asteroids in this peak risk regime allows for civil defense measures such as evacuation of the impact area to avoid fatalities. Evacuation could be accomplished in a matter of days after a warning has been issued. If evacuation is the preferred disaster response, one still has to accept significant loss of infrastructure. A deflection mission is feasible for this size regime and would protect infrastructure but would require a significantly longer warning time on the order of years.

The results presented here document an ever-present risk of asteroid impacts to the human population. These results are expressed in terms of expected fatalities per year. However, it needs to be considered that large asteroid impacts do not occur on a yearly basis and that the yearly risk compounds between large-scale impacts. Consequently, a large impact will cause significantly more damage than the yearly average value, and expected damage levels in the event of an actual impact are shown in Fig. 12.5a. Furthermore, the three risk analyses presented here highlight the risk reduction benefit that asteroid discovery efforts such as Spaceguard offer.

Table 12.1 Chances of dying from select causes in the USA based on data from 1994

Hazard	Chance Of dying
Motor vehicle accident	1 in 100
Murder	1 in 300
Fire	1 in 800
Firearms accident	1 in 2500
Asteroid/comet impact (lower limit)	1 in 3000
Electrocution	1 in 5000
Asteroid/comet impact (expected)	1 in 20,000
Passenger aircraft accident	1 in 20,000
Flood	1 in 30,000
Tornado	1 in 60,000
Venomous bite or sting	1 in 100,000
Asteroid/comet impact (upper bound)	1 in 250,000
Fireworks accident	1 in 1,000,000
Food poisoning by botulism	1 in 3,000,000

The table is reproduced from (Chapman and Morrison 1994)

We can reliably predict the trajectories of newly discovered asteroids and rule out an impact in the foreseeable future, which usually allows us to exclude such asteroids from the risk calculation. The remaining peak risk stems from the 50-m-sized asteroid population. Future survey efforts should target 30–50-m sized objects in order to reduce this risk.

12.7 Risk Comparison to Other Hazards

A major advantage of expressing the asteroid impact hazard in terms of the average annual casualty rates is that the asteroid hazard becomes comparable to other hazards. However, it should be noted early on that the nature of the asteroid hazard differs significantly from the other hazards listed below, insofar as it is generally a very low frequency, yet large consequence event. The statistics about annual asteroid casualties are driven by rare events that might be separated by thousands of years but that may produce significantly higher casualty numbers than the other more common hazards. It is important to keep the low frequency, large consequence nature of the asteroid impact hazard in mind when drawing conclusions from risk comparisons.

Before the Spaceguard Survey took effect, the chance of dying from select causes in the USA, including the asteroid impact hazard, was quantified in (Chapman and

Table 12.2 Chances of dying from select causes in the USA based on data from 2008

Hazard	Chance Of dying
Motor vehicle accident	1 in 90
Drowning	1 in 9000
Airplane crash	1 in 30,000
Earthquakes	1 in 130,000
Fireworks accident	1 in 600,000
Asteroid impact (all sizes)	1 in 720,000
Asteroid impact (global effects)	1 in 1,600,00
Asteroid impact (regional level)	1 in 2,800,000
Food poisoning by botulism	1 in 3000,000
Asteroid impact (local level)	1 in 6,000,000
Shark attack	1 in 8,000,000

The table is reproduced from (Harris 2008)

Morrison 1994) with data from 1994; these numbers are reproduced in Table 12.1. Based on these results, an individual from the United States population has an equal chance of dying from a passenger aircraft accident or from an asteroid impact, based on estimates over long periods of time (meaning that no impact might occur in any given lifespan).

Tables 12.1 and 12.2 clearly illustrate the benefit provided by increased asteroid discovery rates. The effect of Spaceguard becomes apparent when data from Table 12.1 is compared with risk estimates after Spaceguard took effect, based on data available in 2008. The chance of dying from an asteroid impact was significantly reduced from 1 in 20,000 before Spaceguard to 1 in 720,000 after Spaceguard, as one could exclude the possibility of most of the larger objects impacting the Earth in the predictable future. In other words, while previously, the fatal asteroid impact risk would have been ranked as comparable to an airplane crash, it was reduced to a similar level as that of a fireworks accident. It should also be mentioned that the differences in risk numbers are not exclusively due to the effect of the Spaceguard Survey but are also due to updated impact consequence estimates and a general change in the external situation (e.g. population size).

12.8 Southampton Asteroid Hazard Scale

The previous sections focused on quantifying the asteroid impact hazard from the overall asteroid population in a statistical sense. However, the concept of risk can also be applied to concrete threat situations to obtain an intuitive understanding of the severity level. Whenever a new asteroid is discovered, the risk of that object in terms of expected fatalities, taking into account impact probability and impact consequences, can be calculated. The risk metric of expected fatalities provides a

Table 12.3 Risk scale comparison for fictitious and historical impact scenarios based on their maximum threat level

Asteroid	Diameter [m]	P	t_{impact} [yr]	Torino [−]	Palermo [−]	Southampton [expected fatalities]
Large urgent	140	0.1	10	4	1.25	2200
Apophis	325	0.027	25	4	1.17	776
Medium urgent	40	0.1	5	3	0.25	299
1950 DA	1100	3.3×10^{-3}	878	2	−0.017	32
2002 NT7	2000	1×10^{-6}	17	1	−1.2	26
Small urgent	19	0.1	5	0	−0.52	3
Normal	140	1×10^{-6}	40	0	−4.35	0.001

Torino and Palermo scale are unitless, while the Southampton scale is expressed in expected fatalities (Rumpf et al. 2017c)

contextual framework that helps humans better comprehend the asteroid hazard. Correspondingly, the Southampton hazard scale is introduced in (Rumpf et al. 2017c), which applies the concept of risk in terms of human casualties to the asteroid hazard. The new hazard scale, expressed in expected loss of life S, is comprised of expected impact consequences C, impact probability P and decision time factor $\tau_{decision}$:

$$\text{Southampton rating} = S = C \times P \times \tau_{decision} \tag{12.7}$$

The consequences C are given by Eqs. (12.5) and (12.6), where the maximum permissible value is the size of the world population. The impact probability P is available on the NEO webpages of the European and US Space Agencies (NASA 2014a; Universita Di Pisa and European Space Agency 2014). The unitless decision time factor $\tau_{decision}$ is based on the premise that impactors that are closer to the present day should receive higher priority, because they leave less time to react and prepare. Impacts in the distant future are less threatening. The time until impact is expressed by t_{impact} in years. The factor considers that a warning time until impact t_{impact} of 10 years provides sufficient time to decide on an appropriate response to the threat, including a deflection mission, and such a lead time produces a decision time factor value of one. More urgent threats may receive a five times larger value, while impacts with longer lead times receive a lower rating. These considerations are expressed in the term:

$$\tau_{decision} = \min\left(\frac{\left(10_{yr}\right)^2}{t_{impact}^2}, 5\right) \tag{12.8}$$

To illustrate example hazard ratings using the Southampton, Torino and Palermo scales, the hazard ratings of select historical impact threats as well as fictitious threats are presented in Table 12.3. The risk order amongst the scales is consistent, except for a reverse order of *2002 NT7* and *small urgent* in the case of the Palermo scale. The major advantage of using the Southampton scale is that it expresses the threat level as a tangible unit (expected fatalities) rather than an abstract number without physical meaning, as provided by the Torino and Palermo scales. It is therefore easier to obtain a natural understanding for the threat situation and to place the threat into context with other hazards to human life. Using this risk scale might help formulate an adequate response to a given threat scenario.

12.9 Conclusions

This chapter provides statistical insights into the nature of asteroids (i.e. population size and distribution as well as impact angle, speed and location distributions), which then lead to expressions of asteroid impact risk. Utilizing extensive simulations, we calculated the average consequences of asteroid impacts up to 400 m in size. This knowledge together with the impact frequencies for asteroids of varying sizes allowed the calculation of the impact risk distribution over size. The resulting risk distribution is valid for asteroids up to 400 m in diameter, providing high resolution in this small size regime. Similar analysis from the literature is presented to cover asteroids up to 2 km and 10 km in size, respectively. The overall risk is about 1400 casualties per year but can be reduced for the predictable future through exclusion of observed asteroids where it is known that they will not collide with the Earth. The value of asteroid surveys for impact risk reduction is illustrated by the overall risk reduction facilitated by the Spaceguard Survey. Spaceguard targeted large objects and has yielded a roughly 90% reduced risk compared to original levels as it became clear that none of the observed objects would impact the Earth. The current risk distributions reveal that the asteroid population in the 50-m regime retains a risk peak through the combination of high impact frequency and considerable damage in case of an impact. It was therefore recommended that future survey efforts should be able to detect objects in this size regime to effectively reduce the residual risk. Such surveys would of course not only improve the asteroid impact risk situation but would also produce a wealth of related scientific data with additional benefits beyond planetary defense, such as prospecting for asteroid mining.

Context for the overall asteroid impact risk situation was provided through direct comparison of the asteroid impact risk to other hazards relevant to human populations. The key to accomplishing risk comparison is a common scale for each hazard, which was found through the unit of expected fatalities. Correspondingly, the Southampton Asteroid Hazard Scale was presented, which employs the unit of expected fatalities to allow better comprehension of a given asteroid threat and its associated urgency.

The asteroid impact risk ranks comparatively low because asteroid impacts are rare and the extreme damage that can occur in the event of an impact is averaged out over long periods of time. The risk numbers in this chapter show that a significant background hazard exists. This background threat is driven by rare impact events, which have the potential for destruction on a global level. The large fatality counts that can be expected in a given event are distributed over the long, intermittent periods between asteroid impacts, thus producing the average yearly risk values seen here. Although these risk values may be misinterpreted as low, they are valuable metrics that provide a tangible way of measuring the danger of asteroid impacts. The illustration of the asteroid impact hazard in this way provides a compelling motivation to prepare for the serious threat of an asteroid impact. Preparation may be in the form of continued survey efforts, development of deflection technology and demonstration, as well as hazard response protocol implementation. These steps will help to increase preparedness when an asteroid is discovered that carries the potential of causing massive damage to the human population.

Acknowledgements The work was supported by the Marie Curie Initial Training Network Stardust, FP7-PEOPLE-2012-ITN, Grant Agreement 317185. The author acknowledges the use of the IRIDIS High Performance Computing Facility at the University of Southampton. Many of the results and concepts in this chapter have been produced as part of a PhD thesis (Rumpf 2016). As such, the author is thankful to his supervisors Hugh Lewis and Peter Atkinson for many fertilizing and fruitful discussions. Thanks to David Morrison for reviewing this chapter and helping improve its quality.

References

Alvarez, L. W., W. Alvarez, F. Asaro, and H. V. Michel. 1980. "Extraterrestrial Cause for the Cretaceous-Tertiary Extinction." *Science* 208(4448): 1095–1108. http://www.sciencemag.org/cgi/doi/10.1126/science.208.4448.1095.

Binzel, Richard P. et al. 2010. "Report of the NASA Advisory Council Ad Hoc Task Force on Planetary Defense." : 1–25.

Binzel, Richard P. 2000. "The Torino Impact Hazard Scale." *Planetary and Space Science* 48(4): 297–303.

Borovicka, Jiri et al. 2013. "The Trajectory, Structure and Origin of the Chelyabinsk Asteroidal Impactor." *Nature*: 1–13.

Boslough, Mark. 2013. "Airburst Warning and Response." *Acta Astronautica* 103: 370–75.

Boslough, Mark, Peter Brown, and Alan Harris. 2015. "Updated Population and Risk Assessment for Airbursts from Near-Earth Objects (NEOs)." In *2015 IEEE Aerospace Conference*, Big Sky, MT: IEEE, 1–12. http://ieeexplore.ieee.org/document/7119288/.

Boslough, Mark, and D.a. Crawford. 2008. "Low-Altitude Airbursts and the Impact Threat." *International Journal of Impact Engineering* 35(12): 1441–48.

Britt, Dan T., and G. Consolmagno. 2000. "The Porosity of Dark Meteorites and the Structure of Low-Albedo Asteroids." *Icarus* 146(1): 213–19.

Britt, Dan T., Don Yeomans, K Housen, and G Consolmagno. 2002. "Asteroid Density, Porosity, and Structure." In *Asteroids III*, ed. William Bottke. Arizona, USA: University of Arizona Press, 485–500.

Brown, P. et al. 2002. "The Flux of Small Near-Earth Objects Colliding with the Earth." *Nature* 420(6913): 294–96. http://www.nature.com/doifinder/10.1038/nature01238.

Brown, P. G. et al. 2013. "A 500-Kiloton Airburst over Chelyabinsk and an Enhanced Hazard from Small Impactors." *Nature* 503(7475): 238–41.

Chamberlin, Alan B. 2017. "Fireballs." *NASA Center for Near Earth Object Studies*.

Chapman, Clark R., and David Morrison. 1994. "Impacts on the Earth by Asteroids and Comets: Assessing the Hazard." *Nature* 367(6458): 33–40. http://www.nature.com/doifinder/10.1038/367033a0.

Chesley, Steven R. et al. 2002. "Quantifying the Risk Posed by Potential Earth Impacts." *Icarus* 159(2): 423–32.

Chesley, Steven R., Paul W. Chodas, and Donald K. Yeomans. 2014. "Asteroid 2008 TC3 Strikes Earth: Predictions and Observations Agree." *NASA Near Earth Object Program*: 1–6.

Chesley, Steven R., Davide Farnocchia, Peter G. Brown, and Paul W. Chodas. 2015. "Orbit Estimation for Late Warning Asteroid Impacts: The Case of 2014 AA." In *IEEE Aerospace Conference*, Big Sky, MT: IEEE, 1–8.

Chesley, Steven R., and Timothy B. Spahr. 2004. "Earth Impactors: Orbital Characteristics and Warning Times." In *Mitigation of Hazardous Comets and Asteroids*, eds. Michael Belton, Thomas Morgan, Nalin Samarsinha, and Donald K. Yeomans. Cambridge, UK: University Press Cambridge, 22–37.

Chyba, Christopher F, Paul J Thomas, and Kevin J Zahnle. 1993. "The 1908 Tunguska Explosion: Atmospheric Disruption of a Stony Asteroid." *Nature* 361(6407): 40–44. http://adsabs.harvard.edu/cgi-bin/nph-data_query?bibcode=1993Natur.361...40C&link_type=ABSTRACT%5Cnpapers://0be24a46-325a-4116-a3c6-fd8a3b614472/Paper/p12890%5Cnhttp://www.nature.com/doifinder/10.1038/361040a0.

Collins, Gareth S., Elliot Lynch, Ronan McAdam, and Thomas M. Davison. 2017. "A Numerical Assessment of Simple Airblast Models of Impact Airbursts." *Meteoritics & Planetary Science* accepted.

Collins, Gareth S., H. Jay Melosh, and Robert a. Marcus. 2005. "Earth Impact Effects Program: A Web-Based Computer Program for Calculating the Regional Environmental Consequences of a Meteoroid Impact on Earth." *Meteoritics & Planetary Science* 40(6): 817–40. http://doi.wiley.com/10.1111/j.1945-5100.2005.tb00157.x.

Drolshagen, Gerhard, Detlef Koschny, and N. Bobrinsky. 2010. "The Near-Earth Objects Segment of the European Space Situational Awareness Program." *Cosmic Research* 48(5): 399–402.

European Space Agency. 2017. "ESA NEO Risk Page." neo.ssa.esa.int.

Le Feuvre, Mathieu, and Mark A. Wieczorek. 2011. "Nonuniform Cratering of the Moon and a Revised Crater Chronology of the Inner Solar System." *Icarus* 214(1): 1–20.

Gallant, J., B. Gladman, and M. Ćuk. 2009. "Current Bombardment of the Earth-Moon System: Emphasis on Cratering Asymmetries." *Icarus* 202(2): 371–82.

Grav, Tommy et al. 2011. "The Pan-STARRS Synthetic Solar System Model: A Tool for Testing and Efficiency Determination of the Moving Object Processing System." *Publications of the Astronomical Society of the Pacific* 123(902): 423–47.

Hanus, J. et al. 2016. "Volumes and Bulk Densities of Forty Asteroids from ADAM Shape Modeling." *Astronomy and Astrophysics* under revi: 1–49.

Harris, Alan W. 2008. "What Spaceguard Did." *Nature* 453(7199): 1178–79. http://www.nature.com/articles/4531178a.

Harris, Alan W. 2013. "The Value of Enhanced NEO Surveys." *Planetary Defense Conference*.

Harris, Alan W., and Germano D'Abramo. 2015. "The Population of Near-Earth Asteroids." *Icarus* 257(May): 302–12. http://linkinghub.elsevier.com/retrieve/pii/S0019103515002067.

Harris, Alan W., Line Drube, and Lindley N. Johnson. 2017. *Roadmap of Relevant Research for Planetary Defense*.

Mathias, Donovan L., Lorien F. Wheeler, and Jessie L. Dotson. 2017. "A Probabilistic Asteroid Impact Risk Model: Assessment of Sub-300 m Impacts." *Icarus* 289: 106–19. https://doi.org/10.1016/j.icarus.2017.02.009.

Morgan, Jo et al. 1997. "Size and Morphology of the Chicxulub Impact Crater." *Nature* 390(December).

Morrison, David. 1992. *The Spaceguard Survey Report of the NASA International Near-Earth-Object Detection Workshop*. NASA.

NASA. 2014a. "NASA NEO Impact Risk."

NASA. 2014b. "World Bolide Events 1994–2013."
Plane, John M. C. 2012. "Cosmic Dust in the Earth's Atmosphere." *Chemical Society Reviews* 41(19): 6507.
Popova, Olga P. et al. 2013. "Chelyabinsk Airburst, Damage Assessment, Meteorite Recovery, and Characterization." *Science* 342(6162): 1069–73. http://www.sciencemag.org/cgi/doi/10.1126/science.1242642.
Reinhardt, Jason C. et al. 2016. "Asteroid Risk Assessment: A Probabilistic Approach." *Risk Analysis* 36(2): 244–61. http://doi.wiley.com/10.1111/risa.12453.
Rumpf, Clemens M. 2016. "Asteroid Impact Risk." University of Southampton.
Rumpf, Clemens M., Hugh G. Lewis, and Peter M. Atkinson. 2016. "The Global Impact Distribution of Near-Earth Objects." *Icarus* 265: 209–17.
Rumpf, Clemens M., Hugh G. Lewis, and Peter M. Atkinson. 2017a. "Asteroid Impact Effects and Their Immediate Hazards for Human Populations." *Geophysical Research Letters* 44(8): 3433–40. http://doi.wiley.com/10.1002/2017GL073191.
Rumpf, Clemens M., Hugh G. Lewis, and Peter M. Atkinson. 2017b. "Population Vulnerability Models for Asteroid Impact Risk Assessment." *Meteoritics and Planetary Science* 52(6): 1082–1102.
Rumpf, Clemens M., Hugh G. Lewis, and Peter M. Atkinson. 2017c. "Southampton Asteroid Impact Hazard Scale." In *5th IAA Planetary Defense Conference*, Tokyo, Japan: IAA.
Schulte, P. et al. 2010. "The Chicxulub Asteroid Impact and Mass Extinction at the Cretaceous-Paleogene Boundary." *Science* 327(5970): 1214–18.
Shapiro, Irwin I. et al. 2010. *Defending Planet Earth : Near-Earth Object Surveys and Hazard Mitigation Strategies*. Washington D.C.
Shoemaker, Eugene M et al. 1962. *PHYSICS and ASTRONOMY of the MOON*. London, UK: Academic Press Inc.
Shustov, B. M. et al. 2013. "A Concept of a Space Hazard Counteraction System: Astronomical Aspects." *Solar System Research* 47(4): 302–14.
Stokes, Grant H. et al. 2003. 32 Report of the Near-Earth Object Science Definition Team *Study to Determine the Feasibility of Extending the Search for Near-Earth Objects to Smaller Limiting Diameters*. Washington D.C.
Stokes, Grant H. et al. 2017. *Update to Determine the Feasibility of Enhancing the Search and Characterization of NEOs*.
Universita Di Pisa, and European Space Agency. 2014. "NEODyS-2 Risk Page."

Chapter 13
Uncertainty and Risk at the Catastrophe Threshold

Mark Boslough

13.1 Introduction

Planetary defense is the multidisciplinary and internationally coordinated effort to protect the Earth and its inhabitants from impacts by near-Earth objects (NEOs). It requires surveys to discover and track NEOs, campaigns to characterize those that are hazardous, modeling efforts to understand and predict impact effects and associated consequences, and mitigation though impact avoidance and/or civil defense. Mitigation requires the development of ways to deflect or disperse an object on a collision course with sufficient warning, as well as emergency response planning for unexpected or unpreventable impacts.

A cost/benefit analysis is required to justify the allocation of resources to any major risk-reduction enterprise. The purpose of this chapter is to make the case that a form of probabilistic risk assessment is the most appropriate method, and to provide simple but concrete examples of uncertainty and risk calculations that include catastrophic events.

This approach was originally developed to objectively quantify and compare risk associated with high-consequence failures of engineered systems like bridges, dams, aircraft, space systems, power plants, and nuclear weapons. As a method of cost/benefit analysis in the face of uncertainty, it can be used to estimate the cost of reducing a risk compared to the benefit of doing so. Others argue that this method is too restrictive when the future of the planet is at stake, and that a cost of preventive measures that exceeds the probability-weighted value of the benefit can be justified.

M. Boslough (✉)
University of New Mexico, Albuquerque, NM, USA
e-mail: mbeb@unm.edu

N. Schmidt (ed.), *Planetary Defense*, Space and Society,
https://doi.org/10.1007/978-3-030-01000-3_13

13.2 The Cost/Benefit Risk Analysis

The cost side of the equation is relatively straightforward. The individual components and capabilities of planetary defense (astronomical surveys, follow-up observations, data maintenance, orbital dynamics calculations, asteroid characterization, impact and explosion modeling, deflection/disruption studies, emergency planning "tabletop" exercises (Boslough et al. 2015a, 2016), conferences, and publishing) all require funding and budgets that can be estimated.

The benefit side requires quantification of the losses that are to be prevented, which is a more difficult task. For impacts from NEOs of a size that falls below the threshold for global catastrophe, proximal or immediate losses can be estimated using physical models to determine the area that is subjected to various conditions such as air blast, thermal radiation, debris, and tsunami inundation. Because the possibilities span a large range of NEO properties (mass, strength, composition, etc.), impact conditions (velocity, entry angle, orientation, etc.), and geographic location, the consequences can vary greatly. The resulting risk must be based on some kind of long-term average of all possible scenarios. In practice, analysts perform probability-weighted ensemble simulations or parameterized sums over discrete NEO size bins.

Monetizing the results of these estimates requires that value be assigned to the losses. In principle, this is straightforward for estimating the replacement cost of physical assets and infrastructure that can be rebuilt. Irreplaceable losses—such as individual human lives, ecosystems, cultural sites, historical buildings and cities, religious treasures, and even pets—have wildly different values to different people and cannot be objectively quantified.

For example, British Astronomer Royal Martin Rees was quoted in the Guardian (2015) saying "The cost of an impact would be colossal, which means—if you calculate an insurance premium in the usual way by multiplying probability by consequences, it turns out to be worth spending $1 [billion] a year to reduce asteroid risk." This was only a ballpark estimate for the Guardian journalist. It assumes a rare catastrophic event every few million years, big enough to kill a billion people. The cost is based on each life being valued at several million dollars, plus a recovery cost amounting to a decade of world GDP (at about $100 trillion per year).[1]

13.3 The Price of Life

The figure of several million dollars per fatality prevention does not reflect what policymakers are actually willing to spend to save lives. Paul Weissmann, in his chapter on Impact Hazards for the 1993 Hazards volume of Gehrels (Weissman 1994), pointed out that antibiotic treatments costing only 25 cents each could be

[1] Based on private email communication with M. J. Rees on 21st June 2015.

preventing most of the 3.6 million pneumonia deaths per year, according to the World Health Organization. The fact that resources were not available for such a vaccination program suggested that 25 cents per life was not what the world was willing to spend. More currently, the European Union reduced funding for migrant rescues in the Mediterranean. In 2014, 100,000 lives were saved with a budget of about $120 million, at a cost of about $1000 per life. This figure reflects actual lives saved, as opposed to theoretical quantification of risk reduction. However, such quantification is complicated by the complex dynamic nature of decision making under circumstances in which the cost of collective risk reduction can encourage more individual risk taking. The optimal cost for an effective rescue program must account for this behavioral feedback, which is not well understood or quantified.

Despite what we may believe is a moral truth, not all lives have equal value in the harsh judgment of economics and policy. Payouts of several million dollars per life tend only to happen in the developed world, and then only when legal liability is involved. Beyond that, only certain kinds of liability involving such things as transportation safety, food and drugs, and nuclear power that put "everyday people" involuntarily at risk (or are those that are perceived to do so) are covered. In the developed world, where people demand it, the governments have regulations that put this kind of value on life. There is also an element of irrationality that leads to a premium on what people fear (e.g. nuclear power) and a discount associated with risks they connect with optional lifestyle choices, personal responsibility, or politically favored activities. Individual tobacco and firearms victims tend not to receive large payouts, for example.

Large-scale disasters can precipitate a cascade of subsequent losses through famine, disease, mass migrations, economic recessions, geopolitical realignment, and/or violent conflict, which transcend national boundaries and are impossible to predict or quantify, even though they may ultimately dominate the bottom line. Because these two elements (irreplaceable losses and cascading losses) are not possible to objectively quantify, we must use a proxy. The planetary defense community has traditionally used the long-run average expected number of fatalities per year as such a proxy. This quantity is both flawed and confusing, as discussed below, but has the advantage of being both relatively easy to estimate and universally applicable to all locations without regard to considerations such as a person's economic value or citizenship.

As one of my colleagues put it (only partly tongue-in-cheek), the US constitutes only 5% of the world's population, so 95% of the cost of NASA's planetary defense program can be understood as "foreign aid". It is true that most of those at risk from an asteroid strike are the same desperately poor and anonymous people that our leaders fail to find the money to help now, so it is hard to argue that saving them from asteroids is more important than saving them from starvation or malaria (which are far more likely to kill them). However, because putting a dollar value on a human life is so fraught, we continue to use an individual human life itself as the basic unit of cost for purposes of objective risk assessment. This is also a matter of simple pragmatism. No matter what unit we use, it must be justified somehow. Using an individual human life as the basic quantum of value requires the minimum amount of justification. As stated in the US Declaration of Independence, "We hold these truths to be self-evident, that all men are created equal." In other words, it is an axiom that requires no justification at all.

Even this proxy begins to break down when it is applied to highly improbable global catastrophes that could lead to the collapse of planetary-scale ecosystems, fisheries, agriculture, and even society itself, possibly ending in human extinction. Such a calculation would require determining the value of the planet, civilization, and the human race. The value of the human race, to the human race, is arguably more than the sum of the value of each individual. One could make the case that the value of our planet and our species is infinite. If the probability of losing these things to a large asteroid or comet impact is non-zero, then the product of probability and consequences is infinity. But is it worth going into debt and breaking the economy to prevent a catastrophe that almost certainly won't ever happen in our lifetimes or the foreseeable future? This "global catastrophe singularity" beyond the threshold makes quantitative arguments difficult.

Stepping back from the catastrophe threshold, we can use methods similar to those that were developed to quantify the risk from climate change. Despite politically motivated objections and ongoing media hype, there is no scientific controversy about the fact that global warming is dominated by anthropogenic greenhouse gas pollution. Nevertheless, the scientific community acknowledges substantial uncertainty about the speed and intensity of future temperature increases. Projections related to other changes in the Earth's climate and ecological systems, such as ocean acidification, extreme weather events, ice sheet melting, sea level rise, habitat loss, droughts, and famines, are even more uncertain. Researchers have created a parameter called climate sensitivity, which is the equilibrium rise in global mean surface temperature that would theoretically be caused by the doubling of atmospheric carbon dioxide. Climate scientists can quantify risk using the tools of probabilistic risk assessment, even when faced with uncertainty about how fast and how much the planet will warm.

Likewise, there is no longer any real scientific debate about the danger posed by near-Earth objects. Similar to climate science, the field of planetary defense is concerned with a major existential threat to the planet and to humanity, but it is dominated by uncertainty. The vast majority of asteroids in Earth-crossing orbits remain to be discovered. We know from many independent lines of evidence that an object is already on a collision course for the next deadly impact, but we don't know where it is, how big it is, when it will hit, where it will strike, or how many people it will kill. Still, we can quantify the overall risk from impacts even when confronted with uncertainty and limitless but unquantifiable consequence, and base our policy on conservative assumptions.

At this point it is worth noting that the term "conservative assumption" has a different meaning in the context of engineering and security than it does in science. The climate change literature tends to be written from a scientific perspective that focuses on the most probable future. The most conservative scientific estimates are those that deviate the least from prior expectations (also known as erring on the side of least drama, or "ESLD" (Brysse et al. 2013)). Scientific conservatism, when applied to climate change or planetary defense, tends to downplay the risk. Safety engineering, on the other hand, is focused on high-risk occurrences. The low-probability but high-consequence set of events must be the primary focus of planning and mitigation efforts. Engineers include significant margins into their designs to account for low probability occurrences. In planetary defense, we tend to use the word "conservative" in the engineering sense.

13.4 Thought Experiment: Russian Roulette[2]

It is useful to define a thought experiment to help explain uncertainty and risk at the catastrophe threshold. An imaginary laboratory test can put these concepts into terms that are both concrete and dramatic. Suppose a professor of risk management is determined to learn how his graduate students, including you, make individual safety decisions under conditions of uncertainty. He puts his subjects in a room with three revolvers. The first gun is loaded with six rounds, one in each of its six chambers. The probability that it will fire when the trigger is pulled is 100%. The second gun has three of its chambers loaded, so the odds it will fire on the first pull is 50%. Gun 3 has only one round out of six chambers, so the probability of firing is 16.7%.

When it is your turn, the researcher informs you that you'll be strapped into a chair and allowed only to push one of three buttons controlling the triggers. The three guns will be aimed at three different objects in the room, but the decision about which button to push is all yours. You must decide whether your choice will depend only on probability, or whether you will also consider the direction in which the guns are pointed. The target for Gun 1 is your expensive smart phone. Gun 2 is pointed at your foot. Gun 3 is aimed squarely at your temple. There is a trade-off between probability and consequences. Are you willing to risk a fatal headshot to prevent damage to your prized mobile telecommunications toy? Would the possibility of a crippling injury be worth it?

What if you were given the option of removing a round from one of the guns before pulling a trigger? Would that affect your decision? Going only by stated probabilities, this would present you with a simple optimization problem. To minimize total risk, you would remove the single cartridge from Gun 3 (the one aimed at your head) and pull the trigger. Risk is defined as probability times consequences. With zero probability of firing a bullet, the objectively calculated risk is now zero, regardless of the magnitude of the consequences (certain death, in this case). Deeper reflection might lead to a different decision.

Even firearms industry groups like the National Rifle Association, not noted for prioritizing gun safety, warns its members to never point a gun at anyone—let alone pull the trigger—unless you intend to shoot them. In the real world, it is not unusual for people to be shot by guns that were thought to be unloaded. Even if you are

[2] This idea was first published in the chapter "Defending the only home that we have ever known" by Mark Boslough in the book *Starmus: Origins of the Cosmos* (Israelian et al. 2016).

virtually certain that a gun is not loaded, the probability that it will fire a bullet is still nonzero. Your economic worth might be a thousand or several million dollars to others, but your life is arguably priceless to you and your loved ones, to whom your death would be catastrophic. The risk (probability multiplied by consequences) is impossible to quantify (nonzero times infinity) but is arguably greater than the price of a replacement iPhone. The results of this experiment would be interesting. Would people rather shoot their phone than take an ill-defined risk, even if it seems small?

In the controlled thought experiment, the statistical probability might calculate to exactly zero. But the uncertainty due to lack of knowledge might not be reducible to zero. How carefully did you watch the experimenter take the bullet out? How do you know he's not just a murderous magician attempting to make your death look like a suicide? Did you spin the cylinder and check every chamber for the presence of a round? How confident are you that you understand how revolvers function? This exemplifies the difference between "aleatory" and "epistemic" uncertainty. Risk that can be thought of as being random (dice rolls, Russian Roulette games, or asteroid impacts) are governed by aleatory uncertainty. Risk associated with lack of knowledge (whether or not dice are honest, whether or not a gun is loaded, or the nature of the size distribution of asteroids) is controlled by epistemic uncertainty.

Adding another layer to the thought experiment shows how to estimate total risk. Actual risk calculations are better demonstrated by an experiment in which the triggers of all three pistols must be simultaneously pulled. We are concurrently exposed to independent risks in the real world and they must be summed over all possibilities, so total risk is each probability multiplied by the associated consequence of all three events. The best way to minimize the risk is a cost-benefit exercise. In this experiment, trigger-pull decisions are not yours. You now have the option of selecting whatever rounds you want to take out, but there is a cost. If you only have enough money to pay for one round, the obvious choice is to remove the one from Gun 3. Eliminating that single round would avoid catastrophe by retiring all risk of death (excluding the small epistemic contribution), even though the chances were only one in six. Logic would dictate that one would choose to guarantee survival before trying to lower the risk of a disabling shot to the foot or avoiding the definite loss of a cherished electronic device.

13.5 The Threshold for Global Catastrophe

The planetary defense community came to a similar conclusion. The NEO population is analogous the numbers of rounds in the revolvers of our pretend laboratory experiment. But the expected consequences of an impact depend on the size of the asteroid. The largest asteroids have the greatest effect—including the possibility of extinction—but the quantification of consequence is also very uncertain. We simply do not know how big an asteroid must be to cause an ecological collapse, to destroy agricultural production and end civilization, or to wipe out the human race. This calculation is not possible because we do not understand all the damage mechanisms associated with an Earth system that is complex and nonlinear.

The asteroid that erased the dinosaurs altered the Earth forever, first by direct impact effects—the generation of an enormous crater and expulsion of ejecta. About 100 million megatons of energy was released in a massive explosion that changed the atmosphere, heating it up by an unknown amount. The air became opaque with dust and debris, leading to an impact winter that lasted years. The composition and radiative properties of the atmosphere were forever altered, and the climate changed. The precise mechanism for the resulting mass extinction is still debated and is unlikely to ever be completely understood. Fortunately, impacts by 10-km asteroids occur only once every 100 million years or so. The current risk is zero, because a 10-km asteroid on a collision course would be large enough to have been discovered already. The same cannot be said for long-period comets, however, the frequency of large comets entering the inner Solar System is low.

A 5-km asteroid almost certainly exceeds the global catastrophe threshold, but at half the diameter of the dinosaur killer. An asteroid's mass governs its impact energy and damage potential, so mass is a better measure of "size" for purposes of consequence estimates. A 5-km asteroid is therefore really only an eighth as big as the dinosaur killer, and its impact would deliver about one-eighth the destructive energy (for a given impact velocity). But there are more of the smaller ones, so the Earth is exposed to more frequent impacts from them (once about every 30 million years). The Earth doesn't experience mass extinctions with that high of a frequency, so it is unlikely that 5-km asteroids exceed the extinction threshold, at least not every time they hit. But if one were to hit the Earth today, the energy released (roughly 10 million megatons) and the amount of debris produced would lead to certain global catastrophe, killing billions of people.

The population of asteroids continues to increase as the size (and consequences) go down. Like the "bullets-in-guns" thought experiment, space is a shooting gallery where most of the shots are relatively harmless, but rare ones are catastrophic. There are sound arguments based on physics and backed by evidence in the geological record that more frequent and smaller impacts can have local, regional, or even continental-scale consequences without causing a major climate disruption or global catastrophe. That suggests the existence of an unknown size threshold for global catastrophe. There is no reason to think that such a threshold even corresponds to a definite size. An impact into one spot might release a large quantity of planet-warming greenhouse gases or cause soot-producing firestorms, resulting in an impact winter. On the other hand, if it landed in a deep ocean basin, there might be little if any global consequences. The threshold for catastrophe is therefore fuzzy in addition to being uncertain.

13.6 Avoiding Catastrophe by Situational Awareness

Chapman and Morrison (1994) published the first comprehensive probabilistic risk assessment for asteroids and comets. They used observations of the effects of nuclear weapons along with physics-based scaling laws to estimate the direct damage caused by an impact of a given size. However, such scaling laws only work well for impacts that are too small to cause indirect global environmental effects such as climate change. They argued that above some threshold size (which they estimated to be around 1.5 km in diameter, with large uncertainty) a comet or asteroid impact would create a global catastrophe that would kill at least a quarter of the world's population, increasing all the way up to extinction for the largest impacts. They spliced the nuclear weapons-based estimates together with the global catastrophe estimates to create a single, but crude "kill curve" that related the number of deaths to the size of an impacting body.

In our Russian Roulette illustration, our three different guns were loaded with three different integer numbers of live rounds (since bullets exist as discrete units). This is a discrete math problem with three different possible consequences, each with its own probability. For the planetary defense risk assessment, the size of the comet or asteroid is a continuous parameter, so the sum becomes an integral. We can solve it by integrating the kill curve (as a function of size) times the probability of an impact of that size, over all possible sizes. In practice, this is done by dividing the curves up into discrete size bins. One can construct a table consisting of the number of expected impacts within some size range in a specified interval of time, and the number of resulting fatalities (averaged over all possible scenarios). According to Chapman and Morrison (1994), the expected long-term number of impact fatalities per year is 3000 if the threshold asteroid diameter for a globally catastrophic impact is 1.5 km (for further discussion of the threshold for global impact effects, see (Toon et al. 1997)).

If our ability to simulate the consequences of an impact were perfect, we could improve on these estimates by running a statistically significant number of computer experiments and determining how many people would be killed, on average, from an impact of a given size. We could simulate random impacts in numbers proportional to the size distribution of the asteroid population, add up the numbers of fatalities, and divide by the number of impacts to generate a better kill curve. Unfortunately, our ability to simulate impact consequences is still far from perfect. The estimates for ocean impacts are particularly uncertain because the efficiency of impact tsunami generation is not well understood. The severity of climate-changing global catastrophes from asteroid impacts are even more uncertain because climate is a nonlinear dynamic system with unknown thresholds and feedbacks. With increased uncertainty comes greater assessed risk. Most of the uncertainty is associated with impact consequences and the "kill curve". Complex geophysical simulations will never be perfect, therefore decisions will always need to be made in the face of this uncertainty. Nevertheless, such calculations are the best way to ensure that such decisions are objective.

The estimated risk of a few thousand fatalities per year is counterintuitive, because there are no examples of unambiguous, confirmed asteroid fatalities. It depends on low-probability, high consequence events—something that only happens every million years or so but could kill hundred million people. The odds of such an event taking place in a given year are only about one in a million, but it would contribute 100 fatalities per year to the total. The expected number of fatalities per year is zero, but the long-term average is much greater. This is not the only possible way to quantify risk, and may not even be the best, yet it has become the de facto metric for the impact risk assessments, for intercomparison of contributing factors, and for performing sensitivity studies in support of cost/benefit analyses for various risk-reduction strategies.

As an example, the Chapman and Morrison (1994) analysis led to an obvious policy recommendation: catastrophe avoidance. This is analogous to removing the single live round from the gun that is pointed at your head in the Russian Roulette example. The optimal risk reduction method is to prevent large impacts. The first step toward avoidance of catastrophic impact is to find all the asteroids in Earth-crossing orbits that are above the global catastrophe threshold. This recommendation led to the establishment of a survey program and the 1998 NASA directive to discover 90% of NEOs greater than 1 km in diameter. This was also the easiest solution, because there are only about 1,000 NEOs of that size. Since they are also the biggest and brightest in the sky, they were the easiest to find. The survey was a success and led to a large reduction in assessed risk.

Using astronomical NEO surveys to eliminate catastrophic risk is based on the same philosophy as looking both ways before crossing the street. The survey is an act of situational awareness that doesn't by itself change the probability of impact. An object in a deterministic orbit will either collide with the Earth on some specified time interval or it won't. Its intrinsic impact probability is either zero or one. The situational awareness provided by looking creates the opportunity to take preventive action to mitigate the risk if something is discovered to be on a collision course. A pedestrian can change his or her own course by waiting until a potentially hazardous vehicle passes. For planetary defense, the preventive option of choice is asteroid deflection. But without a survey to discover the threat, that option is not available.

13.7 Risk "Retirement"

There has been confusion over language used to describe risk reduction attributed to surveys. It is often said that risk is "retired" when an asteroid is discovered and is found to be in a benign orbit. However, risk is (by definition) a human assessment that includes uncertainty. *Assessed* risk is a redundant term, but the adjective reinforces this notion. When uncertainty is reduced through more observation or understanding, the assessed risk can change. The act of discovering an asteroid that is not on a collision course reduces the assessed risk. For a population of NEOs in unknown orbits, the risk is aleatory, because the trajectories can be thought of as random within some range. After they are discovered (and determined to be no threat), they can be "retired" or removed from the random population for purposes of risk assessment. The assessed risk is reduced, but the intrinsic (previously unknown) probability of impact is unchanged. An asteroid is either on a collision course or it isn't, regardless of whether or not it has a name and entry in the Minor Planet Center database. A rational policy and course of action can only be based on our current risk assessment, which incorporates all we know. If our knowledge changes because something is discovered to be on a collision course, we can reduce its contribution to the risk by deflecting it.

NEO surveys have greatly succeeded in contributing to risk reduction because our assessment of impact probability has decreased. The 90% goal has been exceeded, and discovery of smaller objects continues to accelerate. The assessed risk of a global impact apocalypse has been virtually eliminated in our time. The likelihood of a continental-scale catastrophe has been greatly diminished, and the overall risk (measured in average fatalities per year) has been cut by an order of magnitude to a round-number estimate of about 100. More recent assessments (Boslough et al. 2015b; Mathias et al. 2017; Reinhardt et al. 2016; Rumpf et al. 2017; Stokes et al. 2017) make use of large-scale computer simulations and include the Earth's population distribution with better estimates of asteroid populations and physical effects over a wide range of energies and asteroid physical properties. They remain in broad agreement with one another.

Much work is required to eliminate the remaining risk. The size distribution of asteroids follows a power law. There are many more small, dim, and difficult-to-find asteroids than large bright ones that are already in the catalog. We have our work cut out for us.

References

Boslough, M., Barbara Jennings, Carvey, B., & Fogleman, W. (2015a). FEMA asteroid impact tabletop exercise simulations. In *Proceedings of the 13th Hypervelocity Impact Symposium*. https://ac.els-cdn.com/S1877705815006815/1-s2.0-S1877705815006815-main.pdf

Boslough, M., Brown, P., & Harris, A. (2015b). Updated population and risk assessment for airbursts from near-earth objects (NEOs). In *2015 IEEE Aerospace Conference* (Vol. 2015–June, pp. 1–12). Big Sky, MT: IEEE. doi:10.1109/AERO.2015.7119288

Boslough, M., Chodas, P., Ezzedine, S., Jennings, B., & Fogleman, B. (2016). Physical and infrastructure modeling for the 2015 PDC asteroid threat exercise. In *Aerospace Conference, 2016 IEEE* (pp. 1–18). IEEE.

Brysse, K., Oreskes, N., O'Reilly, J., & Oppenheimer, M. (2013). Climate change prediction: Erring on the side of least drama? *Global environmental change*, *23*(1), 327–337.

Chapman, C. R., & Morrison, D. (1994). Impacts on the Earth by asteroids and comets: assessing the hazard. *Nature*, *367*(6458), 33–40. doi:10.1038/367033a0

Israelian, G., Eichler, D., & May, B. (Eds.). (2016). *Starmus: Origins of the Cosmos*. Tenerife, Spain.

Mathias, D. L., Wheeler, L. F., & Dotson, J. L. (2017). A probabilistic asteroid impact risk model: assessment of sub-300 m impacts. *Icarus*, *289*, 106–119. doi:10.1016/j.icarus.2017.02.009

McKie, R. (2015). Asteroid Day tries to save life as we know it | Science | The Guardian. *The Guardian*. https://www.theguardian.com/science/2015/jun/13/asteroid-day-anniversary-tunguska-siberia. Accessed 7 April 2018

Reinhardt, J. C., Chen, X., Liu, W., Manchev, P., & Paté-Cornell, M. E. (2016). Asteroid Risk Assessment: A Probabilistic Approach. *Risk Analysis*, *36*(2), 244–261. doi:10.1111/risa.12453

Rumpf, C. M., Lewis, H. G., & Atkinson, P. M. (2017). Asteroid impact effects and their immediate hazards for human populations. *Geophysical Research Letters*, *44*(8), 3433–3440. doi:10.1002/2017GL073191

Stokes, G. H., Barbee, B. W., Bottke, W. F., Buie, M. W., Chesley, S. R., Chodas, P. W., et al. (2017). Update to determine the feasibility of enhancing the search and characterization of NEOs. In *Technical Report*. Prepared by the Near-Earth Object Science Definition Team at the request of the National Aeronautics and Space Administration. https://www.nasa.gov/sites/default/files/atoms/files/2017_neo_sdt_final_e-version.pdf

Toon, O. B., Zahnle, K., Morrison, D., Turco, R. P., & Covey, C. (1997). Environmental perturbations caused by the impacts of asteroids and comets. *Reviews of Geophysics*, *35*(1), 41–78. doi:10.1029/96RG03038

Weissman, P. (1994). The comet and asteroid impact hazard in perspective. In T. Gehrels (Ed.), *Hazards Due to Comets and Asteroids* (pp. 1191–1212). University of Arizona Press.

Chapter 14
Conceptualizing the Asteroid Threat and Searching for a Balanced Answer Between Effectiveness and Desirability

Nikola Schmidt

14.1 Introduction

Chapter 11 introduced non-political science experts to the core discussion around construction of a security threat in critical security studies or, put bluntly, what it means to make an asteroid threat a political topic, and what consequences it may have on convincing policymakers that the topic requires our attention. This chapter aims to show that even critical security studies are capable of delivering a perspective that is not inherently critical but is constructive instead. Thus, they can contribute to solving this problem in a normatively desirable way, which requires not only an effective approach but also a reflexive one.

N. Schmidt (✉)
Department of Political Science, Institute of Political Studies, Faculty of Social Sciences, Charles University, Prague, Czech Republic
e-mail: nikola.schmidt@fsv.cuni.cz

N. Schmidt (ed.), *Planetary Defense*, Space and Society,
https://doi.org/10.1007/978-3-030-01000-3_14

14.2 Securitization Dilemma of an Asteroid Threat

14.2.1 How an Asteroid in the Solar System Becomes a Threat to Our Security[1]

Planetary defense has not been thoroughly studied through the lens of securitization theory[2] yet despite that, it is a clear example of an exceptional security issue calling for exceptional political decisions. As such, it should be framed by bold critical scrutiny. Some insights into the endeavor have been provided by historians adopting discourse analysis (Mellor 2007) or were studied with a reflection on the nuclear-weapons usage dilemma (Bucknam and Gold 2008), but the classical securitization reflection is missing. The core idea of the original securitization theory presumes that no issue is a menace; it becomes a security threat through *discursive politics* and is constructed through various securitization *speech acts*.[3] A security threat is constructed when the audience accepts the described dynamics as a threat to one's security. The point is that if we call something a security threat, something already has been done (Waever 1993), thus security as a speech act has a *performative quality* and *constitutive function* of the social reality.

Performative quality means that the outcome of a speech act is not dependent on scientifically proven preconditions under which it is performed but rather on facilitating social contexts of the environment and social setting. A play is performed by actors in different ways as their personal emotions, the sensitivity of the audience or the assuredness of their arguments change. The performance has a constitutive function because the certitude of the claims, even supported by solid science, may have different implications on further political decisions and thus the quality of our lives. The fact that scientists prove certain processes does not directly imply a political action—the performative quality of the claim does.

Securitization theory points to successful securitization processes and their outcomes. Common situations are those in which securitizing actors are successful in convincing their audience, usually their electorate, that something out there is a threat to national security. For example, a threat of migration[4] with racist connotations has been persistently emerging in iterations since ever; without states' perspective, migration would not be perceived as a threat. Moreover, migration within one state is usually understood as desirable because it diversifies the available workforce

---

[1] I have elaborated on the securitization of planetary defense in the past. For the link between securitization of the threat and possible commercialization of asteroid mining see (Schmidt 2017); for the broad discussion how to securitize planetary defense written for planetary defense community see (Schmidt 2018).

[2] Securitization theory in a nutshell is a critical security studies approach in the international relations discipline falling under political science. In the securitization theory, a security threat is a social construct—a political statement made by a security actor through speech acts claiming that something is a security threat to a referent object, in order to convince the audience to perform a political act. For more details, see Chap. 11.

[3] See also Balzacq (2010).

[4] For details on securitization of migration see (Huysmans 2006).

in different places where the workforce is needed. In the interstate perspective, migration is primarily perceived as a threat to the cultural integrity of the state into which migrants are coming, rather than as a possible workforce. This perspective of migration as a threat to national security is linked to the sociological perspective, based on the classical sociological process of Mead's *generalized others*—in sociological terms, the construction of the *rogue others* to our own interests and norms. Whether a migration is a real threat remains apparently unclear. At present, it is socially constructed: the phenomenon is securitized and becomes a security threat.

In the Foucauldian means that are problematizing the discourse, securitization theory would approach the debate about planetary defense as a discursive process presenting the scientifically unveiled historical evidence and perceptions, along with depictions of possible consequences. This lifts up the history of the asteroid impacts through a combination of scientific evidence and people's perceptions, from the K/T event that wiped out dinosaurs, to comets bringing misery as perceived in the medieval age, to the current collection of evidence of meteorites and the widely recorded impact on Chelyabinsk, through to Hollywood blockbusters such as *Armageddon*. All of these perspectives are contributing to a general discourse based on imaginations linked to our widely known subcultures and related imaginations (Schmidt 2016) in order to establish political strategies and practices (Neal 2006, p. 44), again, regardless of whether or not the threat to our security actually exists or is scientifically proven.

14.2.2 Both Rational Explanation and Normative Assessment are Needed

Scientists would probably criticize this securitization perspective of the asteroid threat, as they are convinced that the threat can be confirmed by scientific means. The analysis of the securitization process related to migration is the apparent target, as it usually unveils an easy path to increased power of the securitization actors. I would argue that this is not a fundamental problem in the case of natural threats, because securitization of natural threats can stand firmly on scientific evidence while still being politically defensible and normatively desirable, if the normative assessment of the securitization outcome is discussed separately. The Copenhagen approach to securitization (see Chap. 11) fails to analyze a justifiable construction of essential global security threats emanating from natural processes that are inherently and indisputably threats to mankind or even to the whole biosphere. Exceptional politics, as seen through the lens of the Copenhagen School, is understood as inherently bad. The Welsh School, in contrast to the Copenhagen School, argues that securitization can have positive implications if the outcome brings something good. The difference between these two schools lies in their different perspectives of what securitization is about. While Copenhagen School is concerned with the process of securitization itself, the Welsh School is much more concerned about its outcome.

In our case of asteroid-threat conceptualization, the justification needs to have both a rational and a normative basis, which are quite different categories in both phases of planetary defense: the asteroid detection and its deflection. The rational basis is mainly about effectiveness of the proposed solution design for observation and deflection of the asteroid. The normative reflections are about the ethical reflexivity of the outcome that needs to be taken into consideration. The normative assessment should not only focus on the threat itself, but also should be a basis for the pragmatic reflection over all the processes and contexts along the way.

The planetary defenders have not found an asteroid on a collision trajectory yet but assume that one can be found—thus they act exactly as securitization actors. The rational explanation by the PD community of why humanity needs to act is the historical record of asteroid impacts (the explanation sits on scientifically proven facts) and the whole origin of Earth as a planet from the protoplanetary disc, which leads to the conclusion that an asteroid impact is not about "if" but "when." The historical record provides evidence that these events occur, and the magnitude of the impact erases the logic of the timescale of probability since one impact can be decisive. The discussion around the risk probability in the time of K/T event asteroids seems to be a bit unimportant. The probability analysis makes sense in relation to asteroids capable of wiping out cities, and as more people are moving to cities, the importance of this probability will rise; however, the clear and direct argument repeated by planetary defenders is that while we possess the knowledge to realize the threat (astronomy) and the means to defend ourselves (engineering), we have the responsibility to act (see Chap. 21). The Solar System observation (see Chaps. 4 and 5 about asteroid observations) and complete Solar System Situational Awareness (see Chap. 6) would help us to answer the question "when," and the proper answer to the threat (see Chap. 7) would tell us "how."

However, knowing *when* and *how* is not the end of the responsibility discussion. The consequences of our decisions need to be added to the decision-making process, because only through a combination of rational solutions and the continuous normative assessments of their consequences (Floyd 2007) can we claim that we are securitizing asteroids for good.

The delicate situation comes with the following argument. Observation technology is not intrusive, and it does not threaten traditional global actors, i.e. states. The increasing quality of observation is rationally reducing the need for planetary defense efforts, and as such it plays against the argument for the need to expand these efforts. This is a paradoxical situation, because on the one hand planetary defenders need the attention of the political audience, but on the other hand increasing our Solar System awareness weakens the argument that we need to be prepared. If all planetary defense efforts are to become crucial in the globalization and cosmopolitanization of the planet, the intentions of securitization need to both be sincere and focused on uncertainties rather than on bolstering the certainty that no K/T event is going to happen soon. We can observe this ambivalent approach in the community when some scientists point out unpredictable comets as they come from the Oort Cloud, a distance of up to 1.5 light years away, while the asteroids that are simply in our front yard between Mars and Jupiter sometimes fly closer to the Sun and cross Earth's orbit.

The community formulates the problem and develops methods for how to find an asteroid, but before reaching this result, the community already speaks to authorities and argues that action is necessary now. Practices like these are usually understood as the classical example of the social construction of insecurity—the securitization fulfilled by speech acts. The establishment of particular institutions such as IAWN (detection collection) or SMPAG (deflection strategies) to deepen knowledge of the issue and to decide on the most effective reaction to the threat, respectively, is proof that something has been done already, which is not based on evidence but on effective discursive practices—high performative quality. The explanation of the need is based on scientifically proven facts as argued above, but the particular political effect is based on the performative quality of the discursive act of the securitizing actor, as this move has a particular constitutive act. As such, it includes the normative responsibility of the securitizing actor regarding the consequences. And this is the crucial reason why planetary defense needs to include the normative arguments: to make the efforts clearly sincere and demonstrate awareness of the consequences related to particular decisions.

This does not necessarily need to mean that the outcome of securitization is normatively undesirable, until one asks for nuclear armament preparedness against asteroids (Lewis 2013) without a detected asteroid on a collision course. In order to explain this delicate dilemma, it might be useful to distinguish between two levels of securitization: first, a phenomenon is claimed to be a security threat and gets attention, and the phenomena is explained. Second, the extraordinary character of the situation, including the broad expertise needed for a proper reaction, requires exemption from democratic processes; setting up Nuclear Explosive Devices (NEDs) for planetary defense would likely not be subject to the democratic test of referendum, for example. The remaining question is where we stand now when we have realized and even observed the possible consequences of a comet impact (the case of the Shoemaking-Levy comet's impact on Jupiter in 1994), and we are currently thinking about proper reaction in the form of new scientific efforts, decision-making processes, international coordination or required preparedness in case a big asteroid is suddenly detected on a collision course with Earth within days.

As mentioned, besides other phases of the planetary defense program, we can distinguish between two different outcomes addressing the two most important tasks in planetary defense: detection and deflection. If the outcome is not nuclear weapons destigmatization (deflection) but full coverage of the Solar System (detection), the securitization outcome probably has indisputably good implications and can be considered as desirable. However, if the outcome is a proposal for decision-makers that NEDs are the most effective way forward without the normative assessment of the consequences related to such a proposal, the outcome is apparently undesirable, because it does not focus holistically on the broader consequences of the decision, but rather on solving one scientifically proven problem, that has not yet transpired.

A way around this problem would be to develop an infrastructure that can be used for scientific, economic and defensive purposes (see Chap. 26). Infrastructure for asteroid mining would help us to understand where all the threatening asteroids are and focus on them first. In the scenario of massive asteroid mining, no debatable technologies need to be used at all, and all technologies would be used in a transparent way. The infrastructure designed for mining is not necessarily the same as planetary defense, but at least this idea provides a complementary option.

Overall securitization of an asteroid threat can easily swing to intensify weapons of mass destruction deployment; however, including normative assessment of consequences related to the adopted decision-making process could deepen our knowledge and sharpen the preciseness of our asteroids orbit predictions. The ethical assessment of the consequences (referring to the consequentialist reconciliation between the two Schools of securitization) makes a huge difference in the planetary defense securitization outcomes.

14.3 Perspective Shift of the Reference Object: From States to International Society and to the Planet Earth

Realists understand security as a state of current status quo between states, preserved by military power exercised in a proper strategy and supported by strong scientific and technology industrial complex. In history, this status quo was elaborated upon in arguments such as the *balance of power theory*, in which the balance of power between states is preserved, or the *hegemonic stability theory*, in which one single global hegemon shapes world affairs and keeps things calm. However, it seems trivial even in realist terms that the survival of states is not limited to their military power because, for example, in our case, cataclysmic impact of an asteroid would probably destroy everything regardless of our political arrangement. From this perspective, it seems duly justified to elevate security studies to levels of analysis above the system between nation-states.

Current efforts surrounding climate change and large players like the currently defecting United States show how states can behave as an international society—as one actor. At least in the case of climate change, the consensus seems to be quite strong. The climate change movement is powerful enough to shape the international system. Discussions on the topic show us how other actors besides nation-states can significantly influence world politics, as happened at Bonn Conference Cop 23 in 2017, at which US cities acted visibly against their own federal administration. The result, at the time of writing this chapter, is an open reconsideration by President Trump to possibly stay in the Paris Agreement.

Realists argue that conflict is natural. The above-mentioned examples show that cooperation can be natural as well, especially in a case in which the source of the problem is beyond their borders. When it comes to a state's survival (a realist's term) on a global scale, the problem seems to be bigger than a possible threat of state disappearance caused by other states. Without ecological consideration, the perspective of a state's own security is incomplete (Hughes 2006). This consensus has been constructed by both scientific evidence and strong discursive efforts on a global scale, and as such, it has become the norm to protect climate from deleterious human behavior. There will be nothing to secure on a state level should the whole planet suffer from cataclysmic hurricanes, for example.

Under these considerations, I would argue that such a perspective shift can be incorporated in and addressed by realist thinking. States might unite against a global threat by balancing the consequences of the threat and not their own power, which is clearly a realist approach. However, the response to the threat is what matters in this dynamic. Effective, yet threatening, disputable or stigmatized responses will not yield good results, even in a mode of cooperation based on the balance of threat. Effectiveness does not necessarily bring desirability in general or from the perspective of other actors. 'Right behavior' toward the climate has been neither properly nor understandably discussed, because the 'right action' to fix the climate is a delicate question. Science does not matter, because climate change deniers not only deny science but also deny that the action taken will surely bring more good than bad. And they might not be fully wrong, at least during the first attempts, especially if the attempts are not negotiated and taken by the global community and are instead carried out by certain actors. The diversification and integration of all climate-saving practices is critical, because each outcome of any separate climate securitization will bring different but mutually influential. They cannot be easily assessed separately in order to secure the normative/ethical argument of the overall effort.

Beginning with the drafting of the UN Charter and the newly adopted international legal regime after World War II, international society was once again institutionalized after the first significant attempt and failure of the League of Nations. Newly emerging perspectives on human security, with the central point of security of an individual, grew beyond the sovereignty of states and later even exceeded it in the newly adopted norm of Responsibility to Protect (Evans and Sahnoun 2001). This was even before states acted under the flag of 'civilized states' with the aim to do something good on a global level. The Haag Conventions of 1899 used the term civilized states', which acted to support and adopt humanistic norms on a global level regardless of criticism that they acted selfishly without involving much of the world. International society emerged as our desirable end. If the concept itself is legitimized firstly through norms, actions that are above the nation-state are then more smoothly commenced. The point here is that the security-referent object can also be the planet, if we securitize its threat from an asteroid. If the asteroid impact is securitized in a justifiable way, we have a successful securitization not only based on a mere social construction of a cataclysmic event that needs to be answered, but also a strong agenda standing on a solid foundation of legitimate arguments supported by scientific means, surrounded by a proportional response and empowered by sincere intentions imagining the implications of the response, which does not threaten our way of life.

It is apparent that various global actors are willing to act for the common good on a global level. However, the consequences of those actions might differ according the perspective of the involved actor and the consequences following the taken action. As we discuss throughout Part III, the global actor, in whatever form, is for all these reasons inevitable.

14.4 Securitization and Natural Science Facts: Objectivization of Science

Here we discuss how a threat can be securitized justifiably, and what dilemmas this raises when we base a normative argument on a rational scientific foundation in order to support a selected solution reflecting its outcome. Critical security studies work with reflexive constructionist concepts rather than with evidence, experiments or scientific knowledge based on natural science rational methods of knowledge production. Evidence-based policy is a rationalist approach that is hugely criticized by the reflexivist camp. The core criticism says that no policy can be fully disentangled from political values and focused on mere effectiveness. Classical criticism uses clear examples such as the Holocaust, which could be effective in removing unwanted social groups from society but cannot apparently be defended as ethical. However, in certain cases, evidence is crucially needed in order to make the right decision.

I would argue that the application of rational reasoning is arguably more defensible in the case of natural threats, rather than in the case of cultural threats such as terrorist attacks motivated by hatred of a particular society, which can vary and can be ignited by a misinterpreted speech or a social network campaign based on lies and propaganda. The securitization process supported by the scientific explanation of natural dynamics has a lower risk of being normatively bad, even if we for now exclude the assessment of the securitization outcome of chosen means of a successful securitization act. On the contrary, in the sphere of (inter)cultural threats, one has to decide that the others are different, have threatening intentions and are prepared to act in a threatening way. This can be constructed more easily by discursive means because of our inability to use scientific evidence in such a case. In the former example, the natural threat, the mere discursive construction, is less likely because the threat can be described by a scientific, evidence-driven basis; the discourse is complementary here but crucial for the normative assessment.

Not many works have been published to deepen the discussion related to apparently objective threats. The work by Rita Floyd (2011) elevates the argument of an objective threat. However, Floyd does not focus on the objectivization of a threat as much as she focuses on the justification of securitization in general. The biggest shortcoming of her argument is how she treats objective threats. For Floyd, an objective threat could be a terrorist openly expressing intentions to carry out an attack and possessing the ability to do so. Here, it is argued that in fact, any culturally driven threats stand and fall on the problem of contingency—on the uncertainty of one's driving motives and the links of a solid causal chain. Contrarily, when objectivizing the natural threat, contingency is not the same. An asteroid threat that is fully predictable by scientific means and cannot be avoided or delayed just by speaking and promising is independent of our perception. The realization that an asteroid is on a collision course is what makes the asteroid an objective threat to our security with a clearly *binary value*—it will hit us, or it won't. This is a unique situation that can give planetary defense a unique opportunity to shape global governance of global security threats.

Floyd posits that the objective threat is a threat with no regard as to whether or not we are aware that it is a threat to us; thus, to label it as a security threat—a threat to a referent object (humanity)—an actor needs to frame it and securitize it first as a security threat. If Floyd understands the condition of objectivity as a necessary one in order to justifiably securitize a threat, her theory lacks clear focus on natural science methods, despite the fact that she mentions radiation as a clear example of an objective threat. Radiation threatens us no matter if we think it is a threat to our security or not; an asteroid on a collision course does, too. Natural science methods are critical to unveiling the causal chain in planetary defense, minimizing uncertainty and avoiding possible contingencies before making political decisions. In a nutshell, without a perfect Solar System Awareness observation program, we will never be able to prepare for threats we contemplate based on our current knowledge, but we will tend to securitize the threat that may happen, and thus, we will have to think about the normative assessment of our decisions and their consequences as mentioned above. Uncertainty is of course not a condition under which one likes to make decisions. However, politicians are often asked to make decisions under uncertainty. It is the task for both social and natural scientists to develop knowledge for more appropriate decisions. It is desirable to have somebody make effective decisions related to planetary defense before we find an asteroid on a collision course, but that decision must reflect rational effectiveness and normative reflexivity.

Aradau and Munster argue that sovereign decisions by states are taken 'at the limit of knowledge', which may cause debatable implications such as war or preemptive strikes, because limited hard facts lead to decisions related to risk management with no regard to hard scientific facts (Aradau and Van Munster 2007). These moves are usually triggered by a high degree of uncertainty caused by the lack of knowledge. As previously argued, terrorist threats cannot be objectivized, but an asteroid threat has a higher chance of being supported by hard scientific facts, thus avoiding political decisions based on insufficient data. Having nuclear weapons prepared against asteroids, and subsequently causing international turmoil, would be exactly the kind of mistake in which science has failed. However, such preparedness is fully rational from mankind's perspective. This is a dilemma that awaits us, requiring unprecedented transparency, confidence and global-scale consensus. It can be more effectively faced with perfect Solar System observation, which would give us time to develop a long-lasting, economically rational global defense structure, and thus secured sustainability.

There is one more reason why normative assessments of the taken countermeasures against the asteroid are critical. As Berling pointed out (2011), science tends to objectify the object of study and influence what has been said and is not related to our point of interest; thus, it creates hierarchies of importance and constructs the 'true' objects of security. This is a double-edged sword because on the one hand, we have science to show us that the asteroid is coming, but on the other hand, science can detach the problem from politics and thus also from the normative assessment. This is what Bourdieu referred to as 'doxic practice.' Following the Sophist philosophy, 'doxa' means a common sense, an opposite of knowledge, a true truth in which we believe, a truth discursively created, a self-evident mixture of natural and social argument, It is an argument that creates the argument as self-evident. This is the unveiling dynamic behind the planetary defenders' claim that the asteroid is the master threat against which every effort is ethically defensible. Such arguing would produce more threats than solutions we would like to see and would produce literature disparaging current efforts as a doomsday invocation.

14.5 Expertise and Authority in the Production of Knowledge

Sheila Jasanoff is known for her extraordinary contribution to social sciences in the realm of Science and Technology Studies. In one of her papers, she elaborates on the concept of *co-production*. Her central idea behind this concept is that all knowledge, whether social or natural scientific, is mutually co-produced (Jasanoff 2004). In her words, 'Co-production can therefore be seen as a critique of the realist ideology that persistently separates the domains of nature, facts, objectivity, reason and policy from those of culture, values, subjectivity, emotion and politics' (Jasanoff 2004, p. 3).

The combination of very specific scientific expertise with normative reflection is needed and is inevitable in planetary defense. The classical battle between rationalists and reflexivists thus needs to be reconciled. This problem is visible even in other nature-driven cases such as climate change. What makes planetary defense different from climate change is that we do not need to undergo a complicated debate about what is the desirable state of security (an asteroid off the collision course vs. a desirable state of engineered climate), but only what are the desirable means to reach that state of security in a binary problem (on/off collision course). We accept that avoiding an asteroid impact is necessary, that avoiding the collision is hardly a disputable effort; it is common sense despite the fact that one might welcome the extinction of mankind for various reasons. Humans are causing climate change and violent extinctions of other species. Both are relevant arguments from the perspective of species endangered by humans.[5] The question of whether avoiding a collision is normatively desirable is not discussed; we take it for granted—it is self-explanatory common sense.

Nevertheless, the question of whether the hard scientific facts are produced with this dilemma in mind remains unanswered. Sheila Jasanoff differentiates between governmental research driven by risk and scientific research driven by curiosity (Jasanoff 2003). She elaborated on this argument when she studied advisory boards that were serving policymakers and which are in fact policymakers themselves (Jasanoff 1990). As terrorists pose an unpredictable risk, the reactions tend to overemphasize the possible impact. 9/11 has served as an example supporting the argument asking what else terrorists can do if they were able to take down two skyscrapers

[5] I admit that I am making an opinion turn here. It is quite simple to be caught in our own carved perspective that some actions are normatively good when the argument is oriented on biosphere survival. However, the biosphere may flourish if the impact is strong enough to cause civilization collapse only. I thank to Pavel Dufek for this opinion turn as he lectured me in one of our discussion not to be caught in my conviction that the planetary defense endeavor is necessarily a normatively good effort. Evie Kendal discusses this ethical dilemma as well in Chap. 17.

in New York City. The catastrophic language behind planetary defense is what would serve as the inappropriate reaction to asteroid impact if it is not objectivized enough through hard scientific means, but with a normative reflection. The process of objectivization through knowledge is what can close other interventions, and a process of securitization can proceed (Berling 2011).

In recent discussions related to climate change, the political arguments of climate change deniers have become viewed as foolish because, despite some glitches, the scientific community is capable of developing respected knowledge, especially regarding nature-driven issues. Climate change already has been objectivized enough to avoid criticism by scholars, because the collected base of scientific knowledge is credible enough to globally mobilize humanity against this security issue. The discourse of necessity has been successful in securitizing climate change as an undesirable change for mankind. The discussion now is not about whether or not climate change is a problem, but how to solve or alleviate the problem. The hard scientific facts only prove that the processes are on the move. However, since we propose limiting the activity of mankind, the intervention seems to be fully desirable as the cause of climate change is driven by human activity, but we are standing just in front of the next question—to engineer the desirable state of climate—which will have to deconstruct this climate threat discourse, the climate doxa.

One problem lies in the authority of the scientists, who are respected as legitimate voices in deliberation over certain issues. In the planetary defense endeavor, as in any other security-related issues dependent on scientific knowledge, scientists are becoming security experts and are required to provide homogenous bodies of knowledge and support decision-makers with balanced opinions. This comes with a responsibility around shaping the world through their advice, which will never be about hard scientific facts, but instead ethics. Justifiable securitization comes with responsibility, confidence in sincere intentions and overall positive consequences in contributing to human wellbeing (Floyd 2011, p. 429).

14.6 The Pragmatic Practice-Oriented Way

The pragmatic turn in security studies has been developed to overcome the irreconcilable positions between emancipatory and classical securitization schools, Welsh and Copenhagen respectively, through revising the 100-years-old pragmatism. In the pragmatic turn (Nyman 2016), the focus is on practically useful knowledge with the understanding that truth is contingent and context dependent. Without understanding the context, any discussions about what truth is are completely irrelevant—a car can be a transportation vehicle or a killing machine, depending on the context. This context dependency is the perpetual debate surrounding US use and ownership of guns. Do people possess them for self-defense, or to kill? From the same perspective, discussions about what is a good practice are also related to the context in which the practice is deemed to be good and as long as it is deemed to be good. Ethical decisions are thus not universal but context-dependent. This argument can be found in a text defending nuclear weapons as a deflection method, where scientists argue that a

Nuclear Explosive Device is not the same as a Weapon of Mass Destruction (Marks 2015); however, the context is limited to their actual use as weapons, while their destigmatization can change the world as we know it (see Chap. 15). Thus, context cannot be definite but only one of many alternatives; the truth also cannot be universal. It would not be wise to describe this thinking as relativism, because using nuclear weapons as an absolute last resort is definitely a good decision. But as long as we know we have an alternative method, the good decision is to avoid them.

A 100 years ago, James wrote that theories are only 'instruments, not answers to enigmas, in which we can rest' (James 1910). Current scholars are revising this, saying that 'we should embrace experience as it is lived, rather than generate universalisable abstractions about a "real world" deliberately removed from everyday practice' (Cochran 2002). Theories can help us to get oriented, serving as a lighthouse in a complex world of alternatives; however, every single experience and any partial practice can make the object less complex. The desirable planetary defense policy cannot be reduced to one sheet of paper but needs to be normatively assessed in various contexts. In a nutshell, avoiding foundational truths, working with context-dependent ethical assessments and focusing on a practical research agenda of partial practices can help us assess whether all global planetary defense efforts are directed in the *proper* way. To keep efforts moving in an ethical direction, we need to engage in continuous reflexivity and reevaluate both decisions made and our ethical assessments (Nyman 2016).

From the overall perspective of how to approach asteroid-securitization, the focus on security practices at all levels, separately is the way. From discussing the need of asteroid observations to deploying various deflection technologies, through establishing powerful supranational institutions, each practice related to planetary defense should be assessed separately. It would be quite reductionist to deconstruct asteroid-threat securitization as a bad practice in general, based on the argument that the overall implications are visibly a call for exceptional politics. The central ethical driver to save the world is hard to dispute if it is divided and analyzed at each separate level; thus, each level should have its own pragmatic assessment[s]. However, it would be too idealistic to think that any reaction to the monstrous asteroid threat will bring only good. The different means that different actors choose can point to sincere or fully insincere intentions. One party may wish to protect national greatness, while others might be interested in uniting humanity around a common threat, as Immanuel Kant predicted.

Scientific knowledge may be used to objectivize the argument and completely detach the normative layer from the assessment. Practically useful knowledge (Nyman and Burke 2016) is necessary in order to think and make decision about the threat by both rational and normative means. This can be done if the whole assessment is taken from the contextualist perspective. Imagine a situation of a detected asteroid both with and without asteroid mining infrastructure deployed in space; the latter would lead to a call for nuclear weapons in possession of nation-states, while the former would make those weapons unnecessary, and the bureaucracy of managing asteroid mining could just move the one asteroid on a collision course. We need to ask ourselves which scenario is more desirable to mankind's future and I would argue that mining asteroids is quite a better one than having an unarmed nuclear warhead out of control, orbiting the asteroid and Earth along with it, a scenario analyzed at the Planetary Defense Conference simulation in 2017. It is definitely not a desirable scenario to be in a situation in which nuclear deflection will be used as the only available rational method. But unfortunately, as we do not have other comparably effective means, the possession of NEDs is the only rational approach at present. Building up a commercial asteroid mining infrastructure will render NEDs irrational. Mining bots can immediately be used as kinetic impactors or can use smaller asteroids as kinetic impactors.

The pragmatic approach embraces all these perspectives. It combines ethics, effectivity, desirability and critical analysis and as such is a neat, unifying answer to the debate between rationalists and reflexivists. Such a pragmatic approach requires the most effective technical solution with the most desirable normative solution and continuously asks whether there is not a better solution (from the ethical and rational standpoints). We need to remember that no argument, no solution, no policy is inevitably and undoubtedly good—including saving the Earth. However, constantly looking for the good answers to challenges such as planetary defense is crucial for our survival and the flourishing of our species.

14.7 Conclusion

It appears that in general, planetary defenders have sincere intentions. They have the power to unveil the threat objectively. However, objectivization of the science, which detaches it from political thinking, is erasing the ethical and normative debate over community efforts. This removes the most important question about how we implement the solution: what world will we create by implementing particular policies based on particular (most effective) technology? Means we use against asteroids have the power to change the world we live in. They may do so in a good way or in a bad way.

The different phases of the planetary defense endeavor need to be assessed separately. While observation is by design hardly intrusive, proposal of deflection technologies and their development and deployment have the potential to cause more imminent global threats than an asteroid itself. The pragmatic approach provides us with a selection of tools to assess the threat and the means to answer it from various perspectives—from an analytical, normative one to a practical one. An asteroid threat is a delicate case that has the potential to change our global community, but it will be the political decisions that determine which direction we will walk. This is not only a job for astronomers and engineers, but also for social scientists, who can help analyze the motives and consequences of proposed rational solutions.

Acknowledgment This study was supported by the grant awarded by the Technological Agency of the Czech Republic, project TL01000181: *"A multidisciplinary analysis of planetary defense from asteroids as the key national policy ensuring further flourishing and prosperity of humankind both on Earth and in Space,"* and co-funded by the Institute of Political Studies, Faculty of Social Sciences, Charles University in Prague.

References

Aradau, C., & Van Munster, R. (2007). Governing Terrorism Through Risk: Taking Precautions, (un)Knowing the Future. *European Journal of International Relations, 13*(1), 89–115. doi:10.1177/1354066107074290

Balzacq, T. (2010). *Securitization Theory: How security problems emerge and dissolve*. Routledge. doi:10.4324/9780203868508

Berling, T. V. (2011). Science and securitization: Objectivation, the authority of the speaker and mobilization of scientific facts. *Security Dialogue, 42*(4–5), 385–397. doi:10.1177/0967010611418714

Bucknam, M., & Gold, R. (2008). Asteroid Threat? The Problem of Planetary Defence. *Survival, 50*(5), 141–156. doi:10.1080/00396330802456502

Cochran, M. (2002). Deweyan Pragmatism and Post-Positivist Social Science in IR. *Millennium - Journal of International Studies, 31*(3), 525–548. doi:10.1177/03058298020310030801

Evans, G., & Sahnoun, M. (2001). *The Responsibility to Protect. Report of the International Commission on Intervention and State Sovereignty.* Ottawa, Kanada: International Development Research Centre.

Floyd, R. (2007). Towards a consequentialist evaluation of security: bringing together the Copenhagen and the Welsh Schools of security studies. *Review of International Studies, 33*(2), 327. doi:10.1017/S026021050700753X

Floyd, R. (2011). Can Securitization Theory be Used in Normative Analysis? Towards a Just Securitization Theory. *Security Dialogue, 42*(4–5), 427–439. doi:10.1177/0967010611418712

Hughes, J. D. (2006). *What is Environmental History?* Cambridge: Polity Press.

Huysmans, J. (2006). *The Politics of Insecurity: Fear, Migration and Asylum in the EU*. London: Routledge.

James, W. (1910). *Pragmatism: A New Name for Some Old Ways of Thinking*. Cambridge University Press.

Jasanoff, S. (1990). *The Fifth Branch: Science Advisers as Policy- makers*. Cambridge, Massachusetts: Harvard University Press.

Jasanoff, S. (2003). Technologies of humiliation: Citizen participation in governing science. *Minerva, 41*(3), 223–244. doi:10.2307/41821248

Jasanoff, S. (2004). *States of Knowledge: The co-production of science and social order*. Routledge.

Lewis, J. (2013). Armageddon 2. *Foreign Policy*. http://foreignpolicy.com/2013/09/24/armageddon-2/. Accessed 18 December 2015

Marks, J. (2015). Heaven Can't Wait: A Critique of Current Planetary Defence Policy. In J. Galliott (Ed.), *Commercial Space Exploration: Ethics, Policy and Governance* (pp. 71–90). Farnham UK: Ashgate.

Mellor, F. (2007). *Colliding Worlds. Social Studies of Science* (Vol. 37). doi:10.1177/0306312706075336

Neal, A. W. (2006). Foucault in Guantanamo: Towards an Archaeology of the Exception. *Security Dialogue, 37*(1), 31–46. doi:10.1177/0967010606064135

Nyman, J. (2016). Pragmatism, practice and the value of security. In *Ethical security studies: A New Research Agenda* (pp. 200–220).

Nyman, J., & Burke, A. (2016). *Ethical Security Studies: A New Research Agenda*. (J. Nyman & A. Burke, Eds.). Routledge.

Schmidt, N. (2016). *The Birth of Cyber as a National Security Agenda (PhD Thesis)*. Charles University.

Schmidt, N. (2017). Planetary Defense as a Gateway to Space for Commercial and Deep Space Exploration. *New Space*, 5(4), 219–229. doi:10.1089/space.2017.0005

Schmidt, N. (2018). The Political Desirability, Feasibility, and Sustainability of Planetary Defense Governance (in review). *Acta Astronautica*.

Waever, O. (1993). *Securitization and desecuritization*. Centre for Peace and Conflict Research Copenhagen.

Chapter 15
Weapons of Mass Protection? Rogue Asteroids, Nuclear Explosions in Space, and the Norms of Global Nuclear Order

Michal Smetana

15.1 Introduction

Approximately 66 million years ago, the so-called Chicxulub impactor—likely an asteroid 10–15 km in diameter—collided with Earth in the area of today's Mexican state Yucatán. The infamous event led to the extinction of numerous species, including most of the non-avian dinosaurs. Some of the latest scientific findings reveal how the impact led to the catastrophe of global scale, causing widespread firestorms and a significant drop in surface and ocean temperatures (Artemieva and Morgan 2017; Morgan et al. 2013).

As the impact of Chicxulub is by no means an isolated event in the history of our planet, at least since the late 1980s, scientists have been calling for the development of measures to cope with the danger that 'rogue asteroids' and other near-Earth objects (NEOs) pose to the existence of human civilization. Some proposals for planetary defense rely on improved observation and conventional engagement with the object (e.g., the use of kinetic impactors or gravity tractors). Others, however, highlight the unique qualities of nuclear explosive devices (NEDs) in deflecting incoming NEOs.[1]

The argument of this chapter does not counter the claim that NEDs may in principle be a technologically viable option to deal with a certain subcategory of extra-planetary threats. There is, however, an important normative dimension to these proposals that collides with some of the key rules and norms of contemporary world affairs. Beyond the current incompatibility of the use of NEDs with a number of

[1] See Chap. 7 for a comprehensive discussion of NEO-deflection methods.

M. Smetana (✉)
Peace Research Center Prague, Charles University, Prague, Czech Republic
e-mail: michal.smetana@fsv.cuni.cz

N. Schmidt (ed.), *Planetary Defense*, Space and Society,
https://doi.org/10.1007/978-3-030-01000-3_15

international agreements,[2] I particularly focus on the negative implications of the development of any options in this direction for the nuclear non-proliferation regime and the global nuclear order as such. I also critically unpack the seemingly objective scientific rationality of these proposals and discuss how support for NED development legitimizes technology that, similarly to the impact of large asteroids, also has the capacity to destroy mankind with a non-zero probability.

The chapter proceeds as follows. First, it offers a brief review some of the existing proposals for the development of nuclear options in the system of planetary defense against NEOs. Second, it situates the problem of nuclear explosions in space in the context of normative landscape of global nuclear order. Third, it elaborates on the tensions between nuclear deflection proposals and the global norms concerning the conduct of peaceful nuclear explosions, production of new NEDs, and the practice of explosive nuclear testing. Fourth, it discusses the negative consequences of legitimizing nuclear explosions in space for the transformational logic and stability of the NPT-based non-proliferation regime. Fifth, it problematizes the discourse on the 'scientific rationality' of nuclear deflection. Finally, it concludes with some thoughts on the normative reconciliation between nuclear deflection proposals and the norms of global nuclear order.

15.2 Enduring Search for Nuclear Options

The end of the Cold War brought together two major developments that provided a fertile ground for debates over employment of NEDs in space. First, the scientific debate over planetary impacts eventually shifted from 'seeing impacts as significant events in Earth history to seeing them as threatening events in the human future—a move from historical to futurological narratives' (Mellor 2007, 2010). This shift arguably provided an impetus for the gradual securitization of NEOs as a threat to mankind.[3] Second, with the sudden redefinition of the US–Soviet rivalry, the defense communities in both countries sought to find a new rationale for the further sustainment of the remaining nuclear weapons stockpiles (McDonough 2006; Ritchie 2009).

Among those who advocated NEDs as a *necessary* component of planetary defense was Edward Teller, the former director of the Lawrence Livermore National Laboratory and one of the physicists involved in the Manhattan Project, post-WWII development of thermonuclear weapons, and Ronald Raegan's Strategic Defense Initiative (Morrison and Teller 1994; Teller 1995). In response to the widely

[2] For most relevant countries, the employment of such nuclear explosions would be prohibited under the 1963 Limited Test Ban Treaty, the 1967 Outer Space Treaty, the 1974 Threshold Test Ban Treaty, and the 1996 Comprehensive Test Ban Treaty (which was signed by 183 countries but has not yet entered into force). In addition, the 2017 Treaty on the Prohibition of Nuclear Weapons (usually referred to as 'Nuclear Ban Treaty') explicitly prohibits those looking to 'develop, test, produce, manufacture, otherwise acquire, possess or stockpile nuclear weapons or other nuclear explosive devices' (UNGA 2017). However, all the current nuclear-armed countries refused to sign the document and expressed their unwillingness to abide by its terms.

[3] See Chap. 14.

observed impact of Shoemaker-Levy 9 comet on Jupiter, Teller called for the commencement of testing of possible deflection technologies. Moreover, he advocated for closer collaboration between US and Russian laboratories on building a new large NED suitable for NEO deflection.

Ideas for the further development of NED options against NEOs outlived the early post-Cold War period and eventually gained an institutional backing as well as extensive funding in the new century. The 2007 NASA report for the US Congress asserted that '[n]uclear standoff explosions are assessed to be 10–100 times more effective than the non-nuclear alternatives' (NASA 2007). Similarly, the 2010 report by the National Research Council claimed that '[n]uclear explosives constitute a mature technology, with well-characterized outputs [and] the most mass-efficient method of energy transport,' and should be therefore 'considered as an option for NEO mitigation' (National Research Council 2010, p. 76).

In this context, scientists from various US research institutions have been receiving considerable funding to explore options for nuclear deflection. This includes David S. Dearborn from Lawrence Livermore and Robert Weaver from Los Alamos, representing two major US national laboratories involved in the design, production, and maintenance of the US nuclear weapons stockpile (Dearborn et al. 2007). Additionally, there is Bong Wie and Brian Kaplinger from Iowa State University's Asteroid Deflection Research Center, working for NASA on the design for hypervelocity nuclear interceptors (Kaplinger et al. 2012; Wie 2013). Another example is the joint NNSA-NASA development of HAMMER (Hypervelocity Asteroid Mitigation Mission for Emergency Response), a modular spacecraft design that can function either as a kinetic impactor or a carrier for a nuclear device (Barbee et al. 2018).

Whereas some of the technical reports still note that '[o]bvious military and political issues currently preclude research into optimised explosive devices and tests of this method in space' (Harris 2015), others simply assert that '[n]uclear standoff explosions are [...] assessed to be much more effective than any other non-nuclear alternatives,' and 'for the most probable mission scenarios, in which the warning time is shorter than 5 years, the use of high-energy nuclear explosives in space will become inevitable.' (Wie et al. 2014). Based on these recommendations from prominent scientists, defense experts regularly draw conclusions about the need for NED designs in planetary defense planning, or even a resumption of nuclear testing to support this endeavor (Bucknam and Gold 2008).

The quest for the development of nuclear options against incoming NEOs gained an international dimension under the Obama administration. In 2013, the US Department of Energy signed an agreement with its Russian counterpart on enhanced scientific collaboration on nuclear research (Department of Energy 2013). While the document only vaguely mentions 'defense from asteroids' among the potential projects, the employment of nuclear explosions in the planetary defense against NEOs—potentially involving new NED designs specifically tailored to counter this threat—seems like an area of interest among weapon laboratories in both countries (Birch 2013a; Bolton 2016).[4]

[4]Nevertheless, it should be noted that many US–Russian R&D initiatives have been suspended after Russia's annexation of Crimea in March 2014 (see Stone 2016).

15.3 Nuclear Deflection and the Global Nuclear Order

> Ever since the 1945 bombing of Hiroshima ushered in what is often dubbed the "atomic age" or the "nuclear era," people have obsessed about the potential for massive, even civilization-ending, destruction seemingly inherent in the weapon exploded there. Over the decades this obsession has variously focused on an endless array of creative, if consistently unfulfilled, worst-case scenarios deriving from fears about the cold war arms race, nuclear apocalypse, and the proliferation of the weapons to unreliable states (or even to reliable ones) (Mueller 2010).

The quote from the preface to John Mueller's controversial 2010 book is mocking the decades-long obsession of scholars, journalists, and policymakers with the exaggerated dangers of nuclear weapons. Despite Mueller's arguments that such 'nuclear alarmism' is unfounded and largely irrational, the fact is that since 1945, the international community has gone to great lengths to prevent nuclear Armageddon from taking place. In today's world politics, most states are subject to the rules of institutions of global nuclear governance, such as the Nuclear Non-Proliferation Treaty (NPT) regime or the International Atomic Energy Agency (IAEA). The violations of these rules are largely considered some of the most serious threats to international peace and security, as demonstrated by the cases of North Korea and Iran (Huntley 2006; Ogilvie-White 2010; Wunderlich 2017). In the last decades, a complex net of relevant international institutions, norms, rules, and practices has gradually evolved in the realm of world politics—a dynamic techno-political space sometimes labelled a *global nuclear order*.[5]

To a large extent, the logic of contemporary global nuclear order rests on the structural ambivalence of nuclear technology, which distinguishes between military and civilian/peaceful applications of nuclear energy (Abraham 2006, 2010). The military use of nuclear weapons has been largely delegitimized in the decades following the Second World War, with a strong normative 'taboo' attached to a hypothetical nuclear employment in warfare (Paul 2009; Press et al. 2013; Sagan and Valentino 2017a, 2017b; Schelling 2005; Tannenwald 2005, 2007). On the other hand, the (controlled) use of nuclear reactions for civilian applications—primarily the generation of energy—is not only considered a legitimate right of states, but has even been encouraged as an instrument of economic development under the NPT.[6] As discussed in more depth elsewhere (Mellor 2007), many proponents of NED deflection similarly approach the employment of nuclear weapons in these dual-use terms, with NEDs understood as 'ploughshares' protecting mankind from asteroid threat, as opposed to 'swords' deployed on delivery systems with clearly military purpose. In other words, they suggest that there should be a clear distinction between nuclear weapons as 'weapons of mass destruction' and anti-NEO NEDs as 'weapons of mass protection.'

---

[5] For different academic perspectives on the global nuclear order and its dynamics, see for example Walker (2000, 2004, 2007b, 2012), Krause (2007), Biswas (2014), Jasper (2016), or Tannenwald (2018)

[6] Besides the economic argumentation, some scholars and experts (Rislove 2006) also highlight the importance of the global diffusion of nuclear industry in the context of global climate change.

Nevertheless, as noted by Joseph Nye, 'distinction between peaceful- and weapons- use of the atom [is] primarily a question of politics rather than physics' (Nye 1981). What would clearly separate the two—the *intent* of the state—is very difficult to assess even with the most stringent controls. To draw a line where peaceful use stops and military use starts is often a political act rather than purely technical assessment (Acton 2009; Fields and Enia 2009).

In this context, the development or even mere retention of nuclear capabilities for NEO deflection represents a substantive challenge to the normative structure of global nuclear order on two levels: first, the level of specific norms and rules that have gradually emerged with respect to, *inter alia*, the peaceful nuclear explosions (PNEs), vertical proliferation of nuclear warheads and bombs, and their testing; second, the level of the transformative quality of the NPT regime, the core normative building block of the global nuclear order.

15.4 (Dynamic) Rules of Nuclear Governance

The normative structure of global nuclear order is far from being fixed or stable. Norms and rules that the global nuclear order rests upon remain both inherently contestable and regularly contested by individual stakeholders. As a result of these contestations, norms undergo *change*, as do the categories of what constitutes a 'normal' ('good,' 'responsible,' 'orderly,' etc.) or a 'deviant' ('bad,' 'irresponsible,' 'disorderly,' etc.) conduct in nuclear affairs. As a consequence of this dynamic development, there are currently several normative standards within the global nuclear order that represent a serious obstacle to the development of nuclear options against NEOs, among them the norms with respect to the conduct of PNEs, production of new NEDs, and the practice of explosive nuclear testing.

The original 1968 text of the NPT included Article V., which still envisioned an international cooperation on the 'peaceful applications of nuclear explosions,' a right for both nuclear weapon states (NWS) and non-nuclear weapon states (NNWS) parties to the treaty.[7] At the time, the notion of PNEs was primarily tied to the idea of economic development, with possible applications in large-scale civil engineering programs such as artificial creation of dams or canals. During the Cold War, both the United States and Soviet Union launched several programs that would explore and test the feasibility of PNE use for civilian purposes (Nordyke 1998).[8]

[7] The NPT established two categories of states with different rights and obligations according to the treaty: *nuclear weapon states*, defined as the ones that tested nuclear weapons before 1967, i.e. the United States, Soviet Union, United Kingdom, France, and China; and *non-nuclear weapon states*, i.e. all the other signatories.

[8] The two superpowers also adopted a 1976 Peaceful Nuclear Explosions Treaty that would specify the rules of appropriate conduct of PNEs on both sides.

The environmental hazards and other externalities connected with the use of PNEs prevented a broader application of this technology beyond the two superpowers:

> [T]he decisive argument against PNE's was that they would accustom the world to nuclear explosions, undermining the belief that nuclear explosions were inherently evil and reducing the inhibitions on nuclear weapons. The prospect of blasting new river beds in northern Russia or a bypass canal for the waters of the Nile, or harbors in developing countries, generated concern about 'legitimizing' nuclear explosions. (Schelling 2005).

As a result, the NPT Article V. eventually became obsolete.

Furthermore, India's 1974 underground nuclear explosion, labeled a 'peaceful nuclear explosion' by the government in New Delhi, but widely perceived as a nuclear weapon test (Abraham 1998; Jaipal 1977), highlighted the structural ambivalence towards NEDs and the related risks in the area of nuclear (non-)proliferation policy.[9] Under the 1996 Comprehensive Test Ban Treaty, any nuclear explosion used for any purpose whatsoever becomes prohibited. While the treaty is yet to enter into force, the subsequent NPT Review Conferences repeatedly reaffirmed the normative unacceptability of any new PNE conduct.

The technological requirements for NEO deflection by nuclear means often presume another step that collides with the norms of global nuclear order: the need to design and produce new types of NEDs suitable for this specific mission (Birch 2013a). As noted by Oleg Shugin, a Russian scientist and at the time a deputy director of Rosatom, '[i]t will take a nuclear device much bigger than one megaton to intercept an asteroid of more than one kilometer in diameter' (Interfax 2013). The halt of new warhead production and irreversible reductions of respective stockpiles in NWS have been, however, continuously reestablished as one of the key organizing principles of the nuclear disarmament norm, vaguely defined in the NPT Article VI. as a pledge 'to pursue negotiations in good faith on effective measures relating to cessation of the nuclear arms race at an early date and to nuclear disarmament.'[10] After significant reductions in the US nuclear stockpile executed by all post-Cold War US presidents, the Obama administration explicitly pledged in the 2010 Nuclear Posture Review that 'the United States will not develop new

[9] Indian behavior was widely interpreted as a breach of the non-proliferation norm among the NPT members, particularly on the side of the United States and Canada, who supplied India with heavy water and a research reactor, respectively, and saw the test as 'cheating on the basic bargain of the nuclear regime' (Epstein 1975, 1976; Nye 1981). On the consequences of India's 1974 nuclear explosion, see Wan (2014).

[10] The meaning of this article has been subject to numerous interpretations over the course of the last few decades. There were also many attempts to make the normative prescriptions and proscriptions more specific, most prominently in the 'Principles and Objectives for Nuclear Non-Proliferation and Disarmament' adopted at the 1995 NPT Review and Extension Conference, the 'Thirteen Steps' in the Final Document of the 2000 NPT Review Conference, and the 2010 NPT Review Conference 'Action Plan.' Nuclear disarmament, as the fundamental norm of the global nuclear order, was also formally validated in the 1996 ruling of the International Court of Justice and in numerous resolutions of the UNGA and other international bodies. On July 7th, 2017, the UNGA adopted a Treaty on the Prohibition of Nuclear Weapons, as the first international agreement banning nuclear weapons and providing framework for their elimination.

nuclear warheads, [...] will use only nuclear components based on previously tested designs, and will not support new military missions or provide for new military capabilities (DoD 2010, p. 39).'[11]

New types of NEDs tailored for use against NEOs will likely require a renewal of explosive nuclear testing. While explosive testing of nuclear weapons had been a standard practice among NWS during Cold War era, there have been only three countries since the signing of the 1996 CTBT that have engaged in explosive nuclear testing: India and Pakistan in 1998, both using the tests as a one-off status-defining act and subsequently conforming with the non-testing norm through the declaration of testing moratoria (Kienzle 2014; Lantis 2018); and North Korea, a state that withdrew from the NPT in 2003 and has been repeatedly sanctioned for its nuclear and missile tests in breach of the relevant United Nations Security Council resolutions (Grzelczyk 2012; Ogilvie-White 2010; Pollack 2011). In the current state of nuclear affairs, resumption of explosive nuclear testing by any one NWS would be met with strong disapproval from the international community and be widely perceived as a deviant practice in breach of the non-testing norm.

15.5 Transformative Nature of the NPT Regime

The NPT-based non-proliferation regime represents the key component of the global nuclear order and arguably the main source of normative dynamics within this order (Walker 2007a). The regime has evolved quite dynamically since the early 1970s, implementing new rules, initiatives, and procedures to adapt to new requirements of states as well as to changes in the external environment (Müller et al. 2013). Perhaps the most critical moment in the regime's evolution was the 1995 agreement of the state parties to extend the (originally 25-year) duration of the NPT indefinitely, as a part of the package deal that included strengthening the review process and introducing new principles and objectives to assess the implementation of the NPT provisions (Dhanapala and Rydell 2005; Rauf and Johnson 1995; Welsh 1995).

However, this evolutionary interpretation of the regime development should not obfuscate the fact that the normative underpinnings of the NPT are, since the regime's inception, based on a profound transformative logic with a relatively clear-cut ending.[12] The agreement—which eventually reached a level of membership second only to the United Nations Charter—was not meant to simply conserve the nuclear status quo of the time.[13] In spite of the lack of a specific timeframe, the

[11] Admittedly, this US practice of restraint may now be reevaluated in the light of the President Trump's forthcoming 2018 Nuclear Posture Review.

[12] On the 1960s negotiations leading up to the signature of the NPT, see for example Burns (Burns 1969), Epstein (1976), Unger (1976), Shaker (1980), Nye (1981), Müller et al. (1994), Bourantonis (1997), Paul (2003), Krause (2007), Bunn and Rhinelander (2008), Walker (2012), and Popp et al. (2016).

[13] On 'indefinite nuclearism' and its incompatibility with the culture of human rights, see Booth (1999a, 1999b); on the perils of nuclear status quo, see also Hanson (2002) and Thakur (2000).

nuclear status quo was constructed as a form of time-bounded stabilization that should eventually result in a major shift to the 'zero state:' a global nuclear disarmament. In other words:

> The NPT regime has been constructed as a transformative regime, the justice foundations of which were based on 'microjustice' principles [...], with different obligations and rights related to the different, but temporary, status of the two groups of states. As such, the justice within the regime was supposed to be maintained by upholding the principle of nuclear equity for an unspecified, but principally limited, amount of time, with the goal to move beyond nuclear equity and to reach nuclear equality as the final, 'macrojustice' goal of the regime. (Smetana 2015, pp. 51–52)

The inclusion of NEDs as a fully legitimate tool for planetary defense, a principally open-ended task, could serve as a justification for an indefinite retention of (at least a portion of) nuclear arsenals and the related strategic infrastructure. Such proposals, which essentially set aside the idea of a transformational shift to 'global zero,' can only exacerbate the already serious normative conflict in the NPT community over the right approach to nuclear disarmament and bring about a new obstacle to any attempted bridge-building between the deeply divided camps. This arduous conflict recently culminated in the adoption of the aforementioned controversial Nuclear Ban Treaty without the involvement of NWS and its allies (Onderco 2017; Potter 2017; Sagan and Valentino 2017a, 2017b) and can be found behind the persistent blockage of new (arguably pertinent) global non-proliferation and nuclear security initiatives (Müller 2010; Tannenwald 2013). Further polarization would only contribute to the gradual disintegration of the global nuclear order that has (numerous setbacks notwithstanding) served humanity well over the last few decades.

Moreover, the supposed transformation of nuclear order towards global abolition is conceptually based on the diminishing salience of nuclear weapons in world politics and their continued de-legitimization as both military and political tools (Tannenwald 2015). In this context, the advocacy for new missions for existing nuclear arsenals, such as nuclear deflection of 'rogue asteroids,' runs counter to the critically important normative dynamics within the global nuclear order and thereby contributes to the gradual erosion of its normative base. An implementation of practices that would legitimize new uses for nuclear explosions necessarily erodes the normative pillar of non-proliferation strategies by making the acquisition of (structurally ambivalent and inherently dual-use) NEDs justifiable on both moral and national security grounds (Sagan 1996; Tannenwald 2015). This development would contribute to a (partial) de-stigmatization of NEDs and nuclear explosions in world politics, with potentially disastrous consequences for the global non-proliferation efforts. In fact, the non-use of nuclear weapons in military conflict since 1945 has largely rested on the gradual stigmatization of nuclear weapons in world politics (Tannenwald 2005, 2007)[14]—a trend that can be potentially reversed through a societal change of normative perceptions and beliefs towards nuclear-related issues.

[14] For an alternative interpretation of the logic underpinning the nuclear non-use norm, see Paul (2009, 2010), Press et al. (2013), or Sagan & Valentino (2017a, 2017b).

15.6 Nuclear Deflection as a 'Rational' Scientific Argumentation

In debates with planetary defense scientists, the employment of nuclear options against NEOs is frequently portrayed as simply a rational calculation: facing a potential threat, we should pursue solutions that are objectively assessed to be the most effective; if, for a certain subcategory of threats, nuclear weapons are possibly the only available solution at the moment, we should develop this option further; the 'nuclear problem' is a problem of (irrational) politics and not (rational) science. After all, what can be more rational than building on our existing capabilities in order to protect humanity from extinction?

There are certain problems with this line of argumentation, primarily because the practice of science does not take place in a political vacuum. Even in cases of security threats that 'objectively exist' in the natural world, such as global climate change, scientific practice is never completely divorced from the complex net of power relationships, parochial interests, culture, and norms that operate both within and outside of the scientific community (Demeritt 2001). Moreover, even if we subscribe to the objective rationality of the scientific method itself (Latour 1987), science is simply not able to provide a rigorous cost/benefit analysis of all the potential tradeoffs and consequences of subsequent actions that take place in the social world.

Regarding the development of nuclear deflection options, there are different (and necessarily competing) scientific 'rationalities' whose resolutions are also a matter of politics, rather than just an objective scientific assessment. For example, since the most Earth-threatening NEOs can be dealt with by non-nuclear means, the future need for a nuclear option remains a very high-risk and a very low-probability scenario. While the high-risk part of the equation would certainly provide legitimate arguments for the further development of nuclear deflection, its procurement simultaneously raises the probability of another very high-risk event: nuclear war. In this respect, the recent Humanitarian Initiative to abolish nuclear weapons largely draws on the new scientific evidence about the catastrophic consequences of nuclear use and the non-zero probability risk that nuclear war could occur in the future as a result of accident or miscalculation (Borrie 2014; Johnson 2014; Smetana 2016). Despite the scientific evidence and argumentation behind both trajectories, the logic of a global nuclear ban and the development and maintenance of NEDs for NEO deflection necessarily collide.

Moreover, even within the planetary defense community, there is no widespread agreement regarding whether nuclear explosions are indeed necessary for the NEO deflection, and to what extent this capability is potentially replaceable by non-nuclear means in future. In this context, scientists compete over limited resources, and the procurement of one option prevents others from being funded and further developed. To this end, renowned geophysicist and expert on impact cratering H. Jay Melosh suggests that 'the need for deflecting very large objects that might require nuclear detonations is waning and that a reevaluation of realistic needs is very much in order' (cited in Birch 2013b). Even some advocates of nuclear deflection, such as

NASA astrophysicist David Morrison, assert that 'we will soon reach the state of knowledge of asteroids where we are only concerned about smaller impacts, for which [nuclear weapons] are not necessary.'[15]

As noted above, these scientific debates cannot be completely divorced from politics and parochial interests behind individual proposals. After the end of the Cold War and the US imposition of a nuclear testing moratorium, the narratives of planetary defense provide the weapon laboratories—now largely limited in the scope of their nuclear-related activities—with a seemingly legitimate argument for the further research and development of NEDs, newly framed as 'weapons of mass protection' rather than 'weapons of mass destruction.' This is not to say that the rational argumentation of advocates for the development of the nuclear deflection option is merely a veil masking the parochial interests of individual scientists and laboratories; rather, the rational framing provides legitimization for these interests in the broader national politics of resource allocation, a dynamic that needs to be taken into account when assessing the 'objective rationality' of nuclear proposals.

15.7 A Reconciliation?

The aim of this chapter was to discuss the scientific proposals for the development of NEDs against NEOs in the broader context of the normative structure of world politics. Specifically, it focused on the norms and rules of global nuclear order, a complex techno-political space that regulates the conduct of international nuclear affairs. Despite the noble goal of saving the planet from a potential catastrophe, the pursuit of nuclear explosive options represents a profound normative challenge, with numerous risks for international security.

In addition, there are numerous problems connected with the use of NEDs for the purpose of planetary defense that extend beyond the argumentation advanced in this chapter. Primarily, in an anarchic environment such as world politics, the development of inherently dual-use technologies is frequently interpreted as an attempt to militarily prevail under the guise of civilian purpose. The impossibility of definitely proving the peaceful intent of the state developing this technology provides an impetus for arms races and potential military conflict that perhaps would not take place otherwise.

This problem, as well as other concerns raised in this chapter, can likely be solved by employing the logic of collective action—an establishment of an international institution responsible for the development and eventual employment of NEDs against NEOs, under strict verification, monitoring, and procedural rules. In the field of nuclear arms control, this is not a new idea; as early as in 1946, the United States put forward a proposal for the establishment of international control over nuclear arms, widely known as the Baruch plan (Gerber 1982). At the time,

[15] E-mail conversation with David Morrison and Nikola Schmidt, December 2017.

both the US proposal and the Soviet counteroffer eventually fell prey to Cold War competition and mistrust between the two ideological blocs. If these renewed proposals were connected with concrete steps towards complete and irreversible nuclear disarmament of individual nuclear-armed nation states, they would possibly gain broad international support and reinvigorate the multilateral cooperation on non-proliferation and other nuclear-related issues.

Even in the absence of such an ambitious plan, however, the research and development of nuclear options against planetary threats should still seek to minimize the negative impact on existing international norms and rules. This can be done through the increased transparency of individual states' endeavors, trust-building through international collaboration, reliance on existing technology rather than development of new 'toys,' and strict preference for non-nuclear options whenever possible. There needs to be a clear normative commitment to seek alternative solutions and abandon the nuclear pillar of planetary defense once the future state of technology allows it. The imagined Armageddon after the impact of a rogue asteroid should not blind us to the fact that mankind already possesses the means to destroy our civilization in a matter of minutes; the goal of global nuclear order and its norms, fragile and constantly under stress, is exactly to prevent this scenario from taking place.

Acknowledgement I would like to thank Nikola Schmidt, David Morrison, and Daniel Porras for useful comments and ideas. I acknowledge funding by the Charles University Research Centre program UNCE/HUM/028 (Peace Research Center Prague/Faculty of Social Sciences).

References

Abraham, I. (1998). The Making of the Indian Atomic Bomb : Science, Secrecy and the Postcolonial State. London: Zed Books.

Abraham, I. (2006). The Ambivalence of Nuclear Histories. Osiris, 21(1), 49–65. doi:https://doi.org/10.1086/507135

Abraham, I. (2010). 'Who's Next?' Nuclear Ambivalence and the Contradictions of Non-Proliferation Policy. Economic and Political Weekly, 45(43), 18–20.

Acton, J. M. (2009). The Problem with Nuclear Mind Reading. Survival, 51(1), 119–142.

Artemieva, N., & Morgan, J. (2017). Quantifying the Release of Climate-Active Gases by Large Meteorite Impacts With a Case Study of Chicxulub. Geophysical Research Letters, 44(20), 10,180-10,188. doi:https://doi.org/10.1002/2017GL074879

Barbee, B. W., Syal, M. B., Dearborn, D., Gisler, G., Greenaugh, K., Howley, K. M., et al. (2018). Options and uncertainties in planetary defense: Mission planning and vehicle design for flexible response. Acta Astronautica, 143(August 2017), 37–61. doi:https://doi.org/10.1016/j.actaastro.2017.10.021

Birch, D. (2013a). Hunting Rogue Asteroids Could Be A New Use For Nuclear Weapons| HuffPost.

Birch, D. (2013b). The Plans to Use Nuclear Weapons to Blow Up Incoming Asteroids, 1–16.

Biswas, S. (2014). Nuclear desire : power and the postcolonial nuclear order. University of Minnesota Press.

Bolton, D. (2016). Russian scientists discover best way to destroy asteroids with nuclear weapons|The Independent.

Booth, K. (1999a). Nuclearism, human rights and constructions of security (part 1). The International Journal of Human Rights, 3(2), 1–24.

Booth, K. (1999b). Nuclearism, human rights and constructions of security (part 2). The International Journal of Human Rights, 3(3), 44–61.
Borrie, J. (2014). Humanitarian reframing of nuclear weapons and the logic of a ban. International Affairs, 90(3), 625–646.
Bourantonis, D. (1997). The Negotiation of the Non-Proliferation Treaty, 1965-1968: A Note. The International History Review, 19(2), 347–357. doi:https://doi.org/10.1080/07075332.1997.964078
Bucknam, M., & Gold, R. (2008). Asteroid Threat? The Problem of Planetary Defence. Survival, 50(5), 141–156. doi:https://doi.org/10.1080/00396330802456502
Bunn, G., & Rhinelander, J. B. (2008). Looking Back: The Nuclear Nonproliferation Treaty Then and Now. Arms Control Today, (July/August 2008), 1–7.
Burns, E. L. M. (1969). The Nonproliferation Treaty: Its Negotiation and Prospects. International Organization, 23(4), 788–807.
Dearborn, D. S., Patenaude, S., & Managan, R. A. (2007). The Use of Nuclear Explosives To Disrupt or Divert Asteroids. In Planetary Defense Conference. Washington, D.C.
Demeritt, D. (2001). The construction of global warming and the politics of science. Annals of the Association of American Geographers, 91(2), 307–337. doi:https://doi.org/10.1111/0004-5608.00245
Department of Energy. (2013). United States, Russia Sign Agreement to Further Research and Development Collaboration in Nuclear Energy and Security|Department of Energy.
Dhanapala, J., & Rydell, R. (2005). Multilateral Diplomacy and the NPT: An insider's Account. Geneva: United Nations Institute for Disarmament Research.
DoD. (2010). Nuclear Posture Review Report (2010).
Epstein, W. (1975). Nuclear proliferation: The failure of the review conference. Survival, 17(6), 262–269. doi:https://doi.org/10.1080/00396337508441575
Epstein, W. (1976). The last chance : nuclear proliferation and arms control. Free Press.
Fields, J., & Enia, J. S. (2009). The Health of the Nuclear Nonproliferation Regime. The Nonproliferation Review, 16(2), 173–196.
Gerber, L. G. (1982). The Baruch Plan and the Origins of the Cold War. Diplomatic History, 6(1), 69–96. doi:https://doi.org/10.1111/j.1467-7709.1982.tb00364.x
Grzelczyk, V. (2012). Failure to relaunch? the United States, nuclear North Korea, and the future of the six-party talks. North Korean Review, 8(1), 8–21. doi:https://doi.org/10.3172/NKR.8.1.8
Hanson, M. (2002). Nuclear Weapons as Obstacles to International Security. International Relations, 16(3), 361–379.
Harris, A. W. (2015). A Global Approach to Near-Earth Object Impact Threat Mitigation – Final Report.
Huntley, W. L. (2006). Rebels without a cause: North Korea, Iran and the NPT. International Affairs, 82(4), 723–742.
Interfax. (2013, March). Megaton bomb may destroy large asteroid - Rosatom.
Jaipal, R. (1977). The Indian Nuclear Explosion. International Security, 1(4), 44–51.
Jasper, U. (2016). Dysfunctional, but stable – a Bourdieuian reading of the global nuclear order. Critical Studies on Security, 4(1), 42–56. doi:https://doi.org/10.1080/21624887.2015.1106426
Johnson, R. (2014). The Humanitarian Impacts of Nuclear Weapons: an Imperative for Achieving Disarmament. Irish Studies in International Affairs, 25(1), 59–72.
Kaplinger, B., Wie, B., & Dearborn, D. (2012). Earth-Impact Modeling and Analysis of a Near-Earth Object Fragmented and Dispersed by Nuclear Subsurface Explosions. The Journal of the Astronautical Sciences, 59(1–2), 101–119. doi:https://doi.org/10.1007/s40295-013-0008-3
Kienzle, B. (2014). The Exception to the Rule? The EU and India's Challenge to the Non-proliferation Norm. European Security, 24(1), 36–55. doi:https://doi.org/10.1080/09662839.2014.948863
Krause, J. (2007). Enlightenment and nuclear order. International Affairs, 83(3), 483–499.
Lantis, J. S. (2018). Nuclear Cooperation with Non-NPT Member States? An Elite-driven Model of Norm Contestation. Contemporary Security Policy advance online publication 30 January, doi: https://doi.org/10.1080/13523260.2017.1398367. doi:https://doi.org/10.1080/13523260.2017.1398367

Latour, B. (1987). Science in Action: How to Follow Scientists and Engineers Through Society. Harvard University Press.
McDonough, D. S. (2006). Nuclear Superiority: The „New Triad" and the Evolution of Nuclear Strategy. London: International Institute for Strategic Studies.
Mellor, F. (2007). Colliding Worlds. Social Studies of Science (Vol. 37). doi:https://doi.org/10.1177/0306312706075336
Mellor, F. (2010). Negotiating uncertainty: asteroids, risk and the media. Public Understanding of Science, 19(1), 16–33.
Morgan, J., Artemieva, N., & Goldin, T. (2013). Revisiting wildfires at the K-Pg boundary. Journal of Geophysical Research: Biogeosciences, 118(4), 1508–1520. doi:https://doi.org/10.1002/2013JG002428
Morrison, D., & Teller, E. (1994). The Impact Hazard: Issues for the Future. In T. Gehrels (Ed.), Hazards due to Comets and Asteroids (pp. 1135–1143). Tucson: University of Arizona Press.
Mueller, J. E. (2010). Atomic obsession: nuclear alarmism from Hiroshima to al-Qaeda. Oxford; New York: Oxford University Press.
Müller, H. (2010). Between Power and Justice: Current Problems and Perspectives of the NPT Regime. Strategic Analysis, 34(2), 189–201. doi:https://doi.org/10.1080/09700160903542740
Müller, H., Becker-Jakob, U., & Seidler-Diekmann, T. (2013). Regime Conflicts and Norm Dynamics: Nuclear, Biological, and Chemical Weapons. In H. Muller & C. Wunderlich (Eds.), Norm Dynamics in Multilateral Arms Control: Interests, Conflicts, and Justice. Athens: University of Georgia Press.
Müller, H., Fischer, D., & Kötter, W. (1994). Nuclear non-proliferation and global order. Oxford: Oxford University Press.
NASA. (2007). Near-Earth Object Survey and Deflection Analysis of Alternatives. Report to congress. Washington D.C. http://www.nasa.gov/pdf/171331main_NEO_report_march07.pdf
National Research Council. (2010). Defending Planet Earth : Near-Earth Object Surveys and Hazard Mitigation Strategies. National Academies Press.
Nordyke, M. D. (1998). The soviet program for peaceful uses of nuclear explosions. Science & Global Security, 7(1), 1–117. doi:https://doi.org/10.1080/08929889808426448
Nye, J. S. (1981). Maintaining a nonproliferation regime. International Organization, 35(01), 15–38. doi:https://doi.org/10.1017/S0020818300004069
Ogilvie-White, T. (2010). The Defiant States: the Nuclear Diplomacy of North Korea and Iran. The Nonproliferation Review, 17(1), 115–138.
Onderco, M. (2017). Why nuclear weapon ban treaty is unlikely to fulfil its promise. Global Affairs, 0(0), 1–14. doi:https://doi.org/10.1080/23340460.2017.1409082
Paul, T. V. (2003). Systemic Conditions and Security Cooperation: Explaining the Persistence of the Nuclear Non-proliferation Regime. Cambridge Review of International Affairs, 16(1), 135. doi:https://doi.org/10.1080/0955757032000075753
Paul, T. V. (2009). The Tradition of Non-Use of Nuclear Weapons. Stanford: Stanford University Press.
Paul, T. V. (2010). Taboo or tradition? The non-use of nuclear weapons in world politics. Review of International Studies, 36(04), 853–863. doi:https://doi.org/10.1017/S0260210510001336
Pollack, J. D. (2011). No exit : North Korea, nuclear weapons, and international security. London: International Institute for Strategic Studies.
Popp, R., Horovitz, L., & Wenger, A. (2016). Negotiating the Nuclear Non-Proliferation Treaty : Origins of the Nuclear Order. Oxon: Routledge.
Potter, W. C. (2017). Disarmament Diplomacy and the Nuclear Ban Treaty. Survival, 59(4), 75–108.
Press, D. G., Sagan, S. D., & Valentino, B. A. (2013). Atomic Aversion: Experimental Evidence on Taboos, Traditions, and the Non-Use of Nuclear Weapons. American Political Science Review, 107(1), 188–206.
Rauf, T., & Johnson, R. (1995). After the NPT's indefinite extension: The Future of the global nonproliferation regime. The Nonproliferation Review, 3, 28–42. doi:https://doi.org/10.1080/10736709508436604
Rislove, D. C. (2006). Global Warming v. Non-Proliferation: The Time Has Come for Nations to Reassert Their Right to Peaceful Use of Nuclear Energy. Wisconsin International Law Journal, 24, 1069–1098.

Ritchie, N. (2009). US Nuclear Weapons Policy after the Cold War: Russians,'Rogues' and Domestic Division. Oxon: Routledge.
Sagan, S. D. (1996). Why Do States Build Nuclear Weapons? Three Models in Search of a Bomb. International Security, 21(3), 54–86. doi:https://doi.org/10.2307/2539273
Sagan, S. D., & Valentino, B. A. (2017a). Revisiting Hiroshima in Iran: What Americans Really Think about Using Nuclear Weapons and Killing Noncombatants. International Security, 42(1), 41–79.
Sagan, S., & Valentino, B. A. (2017b). The nuclear weapons ban treaty: Opportunities lost. Bulletin of the Atomic Scientists.
Schelling, T. C. (2005). An Astonishing Sixty Years: The Legacy of Hiroshima (Nobel Prize Lecture). The Sveriges Riksbank Prize in Economic Sciences in Memory of Alfred Nobel, 365–375.
Shaker, M. I. (1980). The nuclear non-proliferation treaty : origin and implementation 1959-1979. Oceana.
Smetana, M. (2015). Nuclear infrastructure, strategic hedging, and the implications for disarmament. In N. Hynek & M. Smetana (Eds.), Global Nuclear Disarmament: Strategic, Political, and Regional Perspectives (pp. 44–60). Oxon: Routledge. doi:https://doi.org/10.4324/9781315735856
Smetana, M. (2016). Stuck on disarmament: The European Union and the 2015 NPT Review Conference. International Affairs, 92(1), 137–152.
Stone, R. (2016, October). Russia suspends nuclear R&D pact with United States. Science.
Tannenwald, N. (2005). Stigmatizing the Bomb: Origins of the Nuclear Taboo. International Security, 29(4), 5–49.
Tannenwald, N. (2007). The Nuclear Taboo: The United States and the Non-Use of Nuclear Weapons Since 1945. Cambridge: Cambridge University Press.
Tannenwald, N. (2013). Justice and Fairness in the Nuclear Nonproliferation Regime. Ethics & International Affairs, 27(3), 299–317.
Tannenwald, N. (2015). Delegitimizing the Bomb: The Normative Strategy for Disarmament. In N. Hynek & M. Smetana (Eds.), Nuclear Disarmament: Strategic, Political and Regional Perspectives. Oxon: Routledge.
Tannenwald, N. (2018). The Great Unraveling: The Future of the Nuclear Normative Order. Daedalus, (forthcoming).
Teller, E. (1995). The Need for Experiments on Comets and Asteroids. In Proceedings of the Planetary Defense Workshop (pp. 5–6). Livermore: Lawrence Livermore National Laboratory.
Thakur, R. (2000). Envisioning Nuclear Futures. Security Dialogue, 31(1), 25–40.
UNGA. (2017). Treaty on the Prohibition of Nuclear Weapons. New York.
Unger, B. (1976). The Nuclear Non-Proliferation Treaty Review Conference: An Unsuccessful Attempt to Stem the Tide. World Affairs, 139(2), 87–111.
Walker, W. (2000). Nuclear Order and Disorder. International Affairs, 76(4), 703–724.
Walker, W. (2004). Weapons of Mass Destruction and International Order. The Adelphi Papers, 44(370), 9–19.
Walker, W. (2007a). Nuclear enlightenment and counter-enlightenment. International Affairs, 83(3), 431–453.
Walker, W. (2007b). International nuclear order: A rejoinder. International Affairs, 83(4), 747–756. doi:https://doi.org/10.1111/j.1468-2346.2007.00650.x
Walker, W. (2012). A Perpetual Menace: Nuclear Weapons and International Order. Oxon: Routledge.
Wan, W. (2014). Firewalling nuclear diffusion. International Studies Review, 16(2), 217–228. doi:https://doi.org/10.1111/misr.12133
Welsh, S. B. (1995). Delegate perspectives on the 1995 NPT review and extension conference. The Nonproliferation Review, 2(3), 1–24. doi:https://doi.org/10.1080/10736709508436589
Wie, B. (2013). Hypervelocity nuclear interceptors for asteroid disruption. Acta Astronautica, 90(1), 146–155. doi:https://doi.org/10.1016/j.actaastro.2012.04.028
Wie, B., Barbee, B., Pitz, A., Kaplinger, B., Hawkins, M., Winkler, T., et al. (2014). Final Technical Report of a NIAC Phase 2 Study. Graduate Research Assistants: Alan Pitz (M.S.
Wunderlich, C. (2017). Delegitimisation à la Carte: The 'Rogue State' Label as a Means of Stabilising Order in the Nuclear Non-Proliferation Regime Carmen. In S. Gertheiss, S. Herr, K. D. Wolf, & C. Wunderlich (Eds.), Resistance and Change in World Politics: International Dissidence (pp. 143–186). Springer.

Chapter 16
Dilemmas for Planetary Defense Posed by the Current International Law Framework

Nikola Schmidt and Martin Švec

16.1 Introduction

We have not dedicated substantial space to international law in this book, however, we could not omit the topic entirely while writing about planetary defense in a multidisciplinary manner. The planetary defense endeavor is clearly dependent on more than just one aspect of knowledge. Astronomers can tell us where asteroids are, engineers can design machines to deflect them and political scientists can imagine the best governance structures for achieving public support, but it is international law that tells us where we stand now, and what is or is not possible to do if a threat eventuates tomorrow. The point of international law, for example in the case of nuclear mitigation methods, is to tell us whether it is even worth studying them since, according to international treaties, we will never get the chance to test them in space.

Can we use the traditional collective security measures in the case of an asteroid threat? Is the Security Council the right body to decide? Does an asteroid threat constitute a challenge to the concept of global 'peace and security'? Should states have the right to use the provision of self-defense to legitimize unilateral action against an asteroid? Is a multilateral action necessarily legitimate if some states do not agree? These and many other dilemmas are discussed below.

N. Schmidt (✉)
Department of Political Science, Institute of Political Studies, Faculty of Social Sciences, Charles University, Prague, Czech Republic
e-mail: nikola.schmidt@fsv.cuni.cz

M. Švec
Masaryk University, Brno, Czech Republic

N. Schmidt (ed.), *Planetary Defense*, Space and Society,
https://doi.org/10.1007/978-3-030-01000-3_16

16.2 Five UN Treaties Regulating Behavior in Space

The discussion about an international legal framework for outer space began with the first days of space exploration. Efforts to regulate behavior in space were primarily driven by security concerns, particularly the need to prevent orbits from crossing into the territories of nation-states. The launch of Sputnik, which crossed into the territory of the US, initiated a debate by the US administration on sovereignty and the limit of passage through outer space. The so-called 'Sputnik precedent' led to the establishment of the UN Committee on the Peaceful of Uses of Outer Space (UNCOPOUS).[1] Applicability of international law, including the UN Charter in outer space, has been acknowledged since 1961, when the General Assembly Resolution 1721 A and B (XVI) of 20 December 1961 was adopted. Regardless of its non-binding character, the GA Resolution 1721 was of significant importance for the development of international space law, paving the way for the adoption of five United Nations treaties[2] and sets of principles.[3]

The aim of this chapter is not to provide the reader with a precise legal analysis of the current space law concerning planetary defense in significant detail, but rather to discuss various ideas behind planetary defense efforts in the light of the existing legal framework. The key dilemmas in planetary defense identified by the authors are as follows: Who has the right/obligation to defend the Earth from a K/T class asteroid? Do the different magnitudes of impact varying from local, to regional, to

[1] USA perceived the Sputnik event as a positive precedent giving them opportunity to openly spy on Russia: *"There was no doubt, Quarles confessed, that the Redstone, had it been used, could have orbited a satellite a year or more ago. Ike said that when this information reached Congress they would surely ask why such action was not taken. The President recalled, however, that timing was never given too much importance in our own program, which was tied to The International Geophysical Year, and confirmed that, in order for all the scientists to be able to look at the instrument, it had to be kept away from military secrets. Quarles then accentuated the positive saying the Russians have in fact done us a good turn, unintentionally, in establishing the concept of freedom of international space. The President then looked ahead five years, and asked about a reconnaissance vehicle"* (McDougall 1985, p. 134). However, both sides knew that mutual spying from space would only contribute to better stability of global politics.

[2] Treaty on Principles Governing the Activities of States in the Exploration and Use of Outer Space, including the Moon and Other Celestial Bodies, Agreement on the Rescue of Astronauts, the Return of Astronauts and the Return of Objects Launched into Outer Space, Convention on International Liability for Damage Caused by Space Objects, Convention on Registration of Objects Launched into Outer Space, Agreement Governing the Activities of States on the Moon and Other Celestial Bodies.

[3] Declaration of Legal Principles Governing the Activities of States in the Exploration and Uses of Outer Space, the Principles Governing the Use by States of Artificial Earth Satellites for International Direct Television Broadcasting, the Principles Relating to Remote Sensing of the Earth from Outer Space, the 'Nuclear Power Sources' Principles, the Declaration on International Cooperation in the Exploration and Use of Outer Space for the Benefit and in the Interest of All States, Taking into Particular Account the Needs of Developing Countries.

global influence this right? What technologies are considered legal in the chosen deflection method? If a powerful space faring nation decides to act, does it need to consult the considered actions with other states?

16.2.1 The Outer Space Treaty (Entered into Force in 1967)

The Outer Space Treaty is traditionally recognized as the general legal basis for the peaceful uses of outer space, providing a framework for the developing law of outer space (UN 2002).[4] It is worth mentioning that it was adopted in 1967, just before Americans raised the US flag on the Moon. We can remember the words spoken by Neil Armstrong after that first step: 'That's one small step for a man, one giant leap for mankind'. The word 'mankind' plays a significant role here (note that the audio transmission did not transmit the word 'a' before man, which led to decades-long discussions regarding what Armstrong really meant).

The Outer Space Treaty, in line with the principles acknowledged by the UN, portrays space as *a province of all mankind*, which is crucial for this discussion. Since the treaty dictated that the exploration and use of outer space should be carried out for the benefit and in the interests of all countries,[5] national appropriation of outer space, including the Moon and other celestial bodies, is banned, with no nation permitted to claim sovereignty over these objects.[6] Thus, any mission to defend the Earth will need to be carried out in the interests of all countries.

In addition, exploration and use of outer space, including the Moon and other celestial bodies, shall be guided by the *principle of cooperation and mutual assistance*.[7] All UN treaties on outer space reaffirm the importance of international cooperation in the exploration and peaceful uses of outer space. In the same line, Article IX of the Outer Space Treaty states that if a nation state has reason to believe that an activity in outer space would cause potentially harmful interference with activities of other states in the peaceful exploration and use of outer space, it shall undertake appropriate international consultations before proceeding with any such activity or experiment.[8] Although the wording of the provision ('if a State Party has reason to believe') implies states may exercise their own discretion regarding launch consultations, in light of the other principles of international space law, especially outer space as a province of all mankind, consultations should be understood as a *condition sine qua non* to ensure that the space mission is carried out in the interests of all countries. In this context, there are good reasons to believe that appropriate

[4] United Nations Treaties and Principles on Outer Space, UN 2002. Available at: http://www.unoosa.org/pdf/publications/STSPACE11E.pdf.

[5] See Article 1 of the Outer Space Treaty.

[6] See Article 2 of the Outer Space Treaty.

[7] See Article 9 of the Outer Space Treaty.

[8] *Ibid.*

consultations should precede any mission to defend the Earth from asteroids and comets. These consultations can take a multilateral form or can be dedicated to an international body that sets threshold of risks to national security.

A space mission to defend the Earth from asteroids and comets may be significantly affected by Article IV of the Outer Space Treaty, which says that space should be used exclusively for peaceful purposes and thus conducting military maneuvers shall be forbidden. The Moon and other celestial bodies shall be used by all states parties to the treaty exclusively for peaceful purposes. The establishment of military bases, installations and fortifications, the testing of any type of weapons and the conduct of military maneuvers on celestial bodies shall be forbidden. However, the use of military personnel for scientific research or for any other peaceful purposes shall not be prohibited.[9] The use of technologies for deflection methods to protect the Earth from asteroids and comets are likely to be considered to have a dual-use character (technology that can be used for both peaceful and military aims). Space tools designed to defend the Earth from asteroids and comets may be perceived as space weapons. In this context, diplomatic efforts and international consultations should be considered a precondition for any space mission to defend the Earth.

16.2.2 The Rescue Agreement (Entered into Force in 1968)

The Rescue Agreement addresses assistance to astronauts in the event of accident, distress or emergency landing, the prompt and safe return of astronauts and the return of objects launched into outer space. It builds on the principle of international cooperation in the peaceful exploration and use of outer space. Ratification of the Rescue Agreement has significantly contributed to the establishment of the international legal framework needed for peaceful use of outer space, in line with the principles established by the UN.

16.2.3 The Liability Convention (Entered into Force in 1972)

The Liability Convention elaborates on international rules and procedures concerning liability for damage caused by space objects. According to the Convention, a launching state shall be absolutely liable to pay compensation for damages caused by its space objects on the surface of the Earth. However, in the event of damage caused elsewhere than on the surface of the Earth, a state shall be liable only if the damage is caused by a technical fault, or through human errors made by responsible parties.[10] These principles may have significant consequences for possible space missions to defend the Earth.

[9] See Article 4 of the Outer Space Treaty.

[10] See Article 2 and Article 3 of the Liability Convention.

In the planetary defense effort, a well-discussed scenario is what would happen if a powerful spacefaring nation decided to save the Earth from an asteroid impact, but failed, causing the asteroid to fall elsewhere. Deflection of an asteroid on a terminal trajectory means that the risk corridor of possible impact sites encircling the Earth is moved elsewhere or fully out of range of the Earth, due to low effect of a kinetic impact (see Chap. 7). The scenario is a good argument for deflection methods such as lasers or a gravity tractor, methods capable of pushing (or pulling, respectively) the asteroid for as long as is necessary. Other methods, particularly the kinetic impactor or nuclear deflection methods, are single-use methods without adjustable effectiveness.

In the context of a space mission to defend the Earth, using deflection methods may be held responsible for damage caused on the surface of the Earth (since no such mission has ever been carried out, the risk of causing damage is quite high). Absolute liability, as established by the Liability Convention, necessarily has to be considered by any state seeking to deflect asteroids and comets. Liability implications may discourage nation-states from investing in space defense technologies, as well as launching such space missions.

16.2.4 The Registration Convention (Entered into Force in 1976)

The Registration Convention established a central register of objects launched into outer space to assist in their identification. As regards a space mission to defend the Earth, the Registration Convention does not pose any dilemmas.

16.2.5 The Moon Agreement (Entered into Force in 1984)

The Moon Agreement has an important role to play in the exploration of outer space.[11] However, since the Agreement has not been ratified and is opposed by most states (the US has been a persistent objector), its legal effects are very limited. Nonetheless, political scientists have acknowledged a significant normative power (in political science terms) of the Agreement (Bull 2002).

Given the fact, that the Moon has been the only celestial body beyond the Earth ever visited by human beings, the Moon Agreement addresses security issues in greater details than any other previous UN treaty on outer space. These provisions are particularly relevant for future space missions to defend the Earth from asteroids and comets. According to the Moon Agreement, any threat or use of force or any other hostile act or threat of hostile act on the Moon is prohibited.[12]

[11] See the Preamble of the Moon Agreement.

[12] See Article 3 of the Moon Agreement.

In addition, Article 3 states that the establishment of military bases, installations and fortifications, the testing of any type of weapons and the conduct of military maneuvers on the Moon shall be forbidden.[13] Again, dual-use technologies deployed on the Moon may raise questions as to what extent these activities constitute a 'threat of hostile act'.

In light of current debates on the establishment of a Moon Village, provisions of the Moon Agreement reiterating that 'the exploration and the use of the Moon shall be carried out for the benefit and in the interests of all countries' are becoming increasingly relevant.[14] A particular concern is how these principles should be translated into concrete space missions.

16.3 Planetary Defense Mission Carried Out in the 'Province of All Mankind' and for 'the Benefit and in the Interests of All Countries'

Since the international community's first attempt to address the nonexistent international legal framework in outer space (in form of the UN General Assembly Resolution Declaration of Legal Principles Governing the Activities of States in the Exploration and Use of Outer Space), outer space has been treated as a 'province of all mankind'.[15] In addition, those engaging in activities in outer space need to prove they are carrying them out for the benefit and in the interests of all countries.

Outer space is not the only space without a national sovereignty. An analogy may be drawn between outer space and the international seabed area, regulated by the UNCLOS, and the Antarctic, regulated by the Antarctic Treaty.[16] Both the UNCLOS and the Antarctic Treaty build on the principle of the common heritage of mankind, which deals with international management of resources within a territory, rather than the territory itself (Buxton 2004; Joyner 1986).

A planetary defense mission brings into question the abovementioned principles. Regarding the concrete conditions that must be met in order to carry out a space mission for the benefit and in the interests of all countries, international space law is silent. Since international law, including the UN Charter, is applicable in outer space,[17]

---

[13] *Ibid.*

[14] See Article 4 of the Moon Agreement.

[15] See Article 1 of the Outer Space Treaty; 1996 Declaration on International Cooperation in the Exploration and Use of Outer Space for the Benefit and in the Interest of All States, Taking into Particular Account the Needs of Developing Countries.

[16] See the Antarctic Treaty and the United Nations Convention on the Law of the Sea.

[17] See the UN General Assembly Resolution 1721 A and B (XVI) of 20 December 1961: International cooperation in the peaceful uses of outer space; Resolution adopted by the UN General Assembly 1962 (XVIII): Declaration of Legal Principles Governing the Activities of States in the Exploration and Use of Outer Space.

one would expect that such a benefit should be determined by the United Nations, recognized as 'a center for harmonizing the actions of nations in the attainment of common ends'.[18] The UN outer space treaties themselves do suggest that the UN should play an essential role. For instance, Article 9 of the Moon Agreement states:

> A State Party establishing a station shall use only that area which is required for the needs of the station and shall immediately inform the Secretary-General of the United Nations of the location and purposes of that station. Subsequently, at annual intervals that State shall likewise inform the Secretary-General whether the station continues in use and whether its purposes have changed.[19]

Article 5 of the Moon Agreement obligates state parties to inform the Secretary-General of the United Nations, to the greatest extent feasible and practicable, of their activities concerning the exploration and use of the Moon.[20]

One may claim that the interests and benefits of all countries should be determined by the UN COPUOS, a body that helps to steer interests of countries to follow the provisions of Outer Space Treaty. On the other hand, the UN treaties on outer space do not require the authorization of all countries or international bodies to carry out a space mission. Overall, it is not clear which UN body should determine interests and benefits of all countries, yet there remains wide consensus that every space mission should pay necessary attention to the principles of international space law and be carried out for the benefit and in the interest of all countries, though these have not yet been explicitly proclaimed.

16.4 Dilemmas Posed by Certain Deflection Methods

16.4.1 An Attack on Mankind

Given that international law, including the UN Charter, is applicable in outer space,[21] it faces severe and unprecedented challenges, e.g. planetary defense from asteroids and comets. As has been noted, on many pertinent issues, the existing international space law is often uncertain, if not totally silent (Johnson 2018). Other sources of public international law do not provide the necessary guidance either. Public international law and bodies of legal rules, norms and standards that apply between sovereign states and other subjects of international law have currently not been expected to be applicable in outer space, and they have not anticipated all of the

[18] See Article 1(4) of the UN Charter.

[19] Article 9 of the Moon Agreement.

[20] See Article 5 of the Moon Agreement.

[21] See the UN General Assembly Resolution 1721 A and B (XVI) of 20 December 1961: International cooperation in the peaceful uses of outer space; Resolution adopted by the UN General Assembly 1962 (XVIII): Declaration of Legal Principles Governing the Activities of States in the Exploration and Use of Outer Space.

advancements and technological innovations involved. Thus, an essential question posed on planetary defense is how the international community should respond to a threat of asteroid collision while complying with international law.

Should a threat of asteroid collision with the Earth be considered a threat to international peace and security according to the UN Charter? Affirmative answers would suggest that asteroid collision with the Earth should be addressed by the Security Council, having primary responsibility for the maintenance of international peace and security.[22] Originally, the UN collective security mechanism was designed to address traditional military threats and to maintain international peace and security (Nasu 2011). However, against the background of responding to a diverse range of security challenges (e.g. non-state actors, extensive violation of human rights, interstate conflict, weapons of mass destruction) the traditional concept of peace and security has expanded in order to address dynamic security challenges (Nasu 2011). In line with this dynamic interpretation of peace and security, there is an ongoing debate as to whether the impact of climate change constitutes a threat to international peace and security.[23] As regards an asteroid collision with the Earth, one could claim that due to its potentially devastating consequences on the Earth's surface, it should indeed constitute a threat to international peace and security.

An asteroid collision with the Earth raises the question of whether a state may exercise self-defense under Article 51 of the UN Charter, or under international customary law. However, self-defense is traditionally understood as an exception to the general prohibition of the use of force enshrined in Art. 2 (4) of the UN Charter. In addition, in light of the restrictive interpretation of Article 51 developed by the International Court of Justice, the lawful exercise of the right of self-defense must be in response to an armed attack (I.C.J. 1986, p. 14). Nonetheless, states may invoke circumstances excusing the wrongfulness of conduct that would otherwise violate international obligations of the State concerned, such as self-defense under international customary law, distress or necessity (discussed below).[24]

Against the background of the principles of international space law, it should be emphasized that a coordinated international response should always be prioritized over unilateral action. Uncoordinated responses might lead to anarchic situations, in which all states along the risk corridor (the corridor drawn on the globe delineating countries at risk of a possible impact) act on their own, putting others at risk.

[22] See Article 24 of the UN Charter.

[23] Ban Ki-moon, Remarks to the Security Council on the Impact of Climate Change on International Peace and Security. United Nations Secretary-General. 2011; Security Council, in Statement, Says 'Contextual Information' on Possible Security Implications of Climate Change Important When Climate Impacts Drive Conflict, UNSC/10332 20 July 2011; United Nations Security Council (UNSC) has formally recognized climate change as a destabilizing factor in Somalia, See UN Security Council makes 'historic' warning on climate threat to Somalia, Climate Home News. Available at: http://www.climatechangenews.com/2018/03/28/un-security-council-warns-climate-threat-somalia-peacekeeping/.

[24] See Draft Articles on Responsibility of States for Internationally Wrongful Acts.

16.4.2 Nuclear Deflection Methods: International Law Perspective

Asteroid deflection methods are traditionally classified as methods causing momentum change of an asteroid via mass explosion (including nuclear blast) and methods causing change of an asteroid via mass acquisition or mix expulsion/acquisition (e.g. kinetic impactor, ion beam shepherd) (see Chap. 7). Nuclear methods can be designed to cause a variety of effects on asteroids.

However, deflection methods face various challenges. The nuclear deflection methods may face significant obstacles imposed by international law. Under the 1963 Treaty Banning Nuclear Weapon Tests in the Atmosphere, in Outer Space and Under Water (hereafter referred to as 'PTBT'), any nuclear weapon test explosion, or any other nuclear explosion, at any place under its jurisdiction or control—including outer space—is prohibited.[25] Each state party undertakes not to carry out any nuclear weapon test explosion or any other nuclear explosion, and to prohibit and prevent any such nuclear explosion at any place under its jurisdiction or control.[26] It is worth mentioning that the PTBT does not distinguish between peaceful nuclear explosions and nuclear weapons explosions. Despite the potential benefits of peaceful nuclear explosions, these are banned. Thus, the PTBT may discourage states from investing in research involving nuclear deflection methods.

In addition, the potential classification of a nuclear deflection method as a nuclear weapon would make its application even more difficult. First, the use of outer space for peaceful purposes is a key principle of international space law. Second, Article IV of the Outer Space Treaty places a ban on the placement of nuclear weapons in the Earth's orbit. One could claim that the Outer Space Treaty does not prohibit the passing of nuclear weapons through space and their deployment beyond the Earth's orbit (Hobe et al. 2017, p. 79), in other words, the Outer Space Treaty does not prohibit the use of a nuclear explosive device per se, provided it is not placed in orbit. However, such an interpretation is uncommon among the planetary defense community, primarily because nuclear explosions in outer space are banned by the PTBT.

We may observe a tendency to avoid discussions that categorize space technologies intended to deflect asteroids and comets as weapons of mass destruction. Their stigmatization in being used for military purposes neglects their principal interest of using nuclear energy for the sake of human survival. Moreover, engineered nuclear energy applications for asteroid deflection are significantly different to nuclear bombs. Thus, the space community labels these technologies as Nuclear Explosive Devices (NEDs). NASA identified at least four different NED methods (NASA 2007):

(a) Nuclear Explosive (standoff)—detonate on flyby via proximity fuse
(b) Nuclear Explosive (surface)—impact, detonate via contact fuse
(c) Nuclear Explosive (delayed)—land on surface, detonate at optimal time
(d) Nuclear Explosive (subsurface)—drive explosive device into PHO, detonate.

Nonetheless, as stated in the previous paragraphs, even peaceful nuclear explosions in outer space would constitute a violation of the PTBT.

[25] See Article 1 of the PTBT.

[26] *Ibid.*

16.4.3 The Alternative to the Nuclear Deflection Method: International Law Perspective

Nuclear deflection methods are not the only methods capable of causing momentum changes in asteroids. For instance, direct energy systems that can become comparably effective in the future will be capable of diverting asteroids in the process of their long-distance mining. However, as discussed in Chap. 22, despite these optimistic visions, we are still far away from this future.

Nonetheless, there are additional problems that may arise from the deployment of technologies involved in different deflection methods. It is likely that these technologies could be used for both peaceful and military aims. The dual-use nature of these technologies may be perceived as a threat of force, prohibited by the UN Charter[27] and found in violation of the key principles of international space law, particularly the use of outer space for peaceful purposes and the principle of international cooperation. However, again, public international law provokes more questions rather than answers.

The final perception of the technology may not depend on engineering logic but rather psychological logic. In other words, it is *intent* that determines whether a warhead is a peaceful device or a weapon. For example, the term 'warhead' implies the intention—the head of a bomb to be used in war. The same logic is currently applied to many dual-use assets related to space activities, particularly those involved in cleaning up space debris. Who can be assured about the intention of the operator while the technology has suspicious capabilities such as maneuverability? ESA's activities cleaning up space debris by focusing active debris removal programs on the biggest objects, such as the 8-ton Envisat (Werner 2018), provide a good example of how demonstrated good intentions are not perceived as a threat. Compare this to the maneuver of an unregistered object launched from Russia to the distance of 100 m from Intelsat satellite at the geostationary orbit, which was perceived as a threat (Gruss 2015). Perhaps the fact that Russia is a member of the ESA lends the agency more credibility in such endeavors, and so nobody questions the ESA's orbital active debris removal program. In this context, it could be very difficult to define a device versus a weapon without knowing and verifying the intended purpose.

Nevertheless, the move to refer to tools of asteroid deflection as Nuclear Explosive Devices might have helped pave the road for other perspectives on the same technology with different applications. Precedents can be found in history, especially in space technology development, as the perspective on RTGs—Radioisotope Thermonuclear Generators—has changed dramatically over time.

It is likely that technologies using alternative deflection methods may be understood as a placement of weapons of mass destruction (henceforth 'WMD') in the Earth's orbit, prohibited by Article IV of the Outer Space Treaty. Although there is no definition of WMDs under international law, WMDs are traditionally referred to as a synonym for nuclear, biological and chemical weapons (Carus 2006).

[27] See Article 2(4) of the UN Charter.

According to this definition, conventional dual-use technologies using alternative deflection methods could hardly be considered WMDs. However, application of other definitions, such as WMDs as 'weapons that cause massive destruction or kill large numbers of people, and do not necessarily include or exclude CBRN weapons', may raise questions about potential violation of Article IV of the Outer Space Treaty. That said, technical details of those technologies (e.g. the direction in which a laser is pointed) will determine their future categorization. Interestingly, if we put aside the problem that given the rotation of the Moon a laser on the opposite side of the Moon would not be as useful as a laser on a Lagrange point, we still have a situation that enhances transparency, since such a laser would not threaten the Earth. Furthermore, destruction of deep space scientific probes probably does not factor into the strategic interests of states, and thus would not ruffle policy-makers concerned with national space security.

16.4.4 Circumstances Precluding Wrongfulness

As described in this chapter, there are various international legal obligations preventing states from deploying the technologies necessary to effectively deflect asteroids and comets. The PTBT prohibits any nuclear explosions in outer space, the Outer Space Treaty prohibits deployment of WMDs and the UN Charter prohibits deployment of dual-use technologies that may be controversial and perceived as a threat of force.

Necessity operates to excuse conduct that 'is the only means for the state to safeguard its essential interest against a great and imminent peril' (Evans 2014, p. 466). However, state practice and judicial decisions support the view that necessity may constitute a circumstance precluding wrongfulness only under certain, very limited conditions.[28] The course of action taken must be the 'only way' available to safeguard that interest. In other words, the plea is excluded if there are other means available, even if they may be costlier or less convenient (United Nations n.d., p. 83).[29]

We may draw an analogy between using a nuclear device to deflect an asteroid as part of a planetary defense mission, and the Liberian oil tanker Torrey Canyon accident. In March 1967, the oil tanker Torrey Canyon went aground on submerged rocks outside British territorial waters, spilling large amounts of oil. Since all measures taken by the British Government failed to protect the English coastline, the British Government bombed the ship to burn the remaining oil. Although the British Government did not invoke necessity as a legal justification, it stressed the existence of an extremely danger situation. Since no international protests resulted, the Draft Articles on Responsibility of States for Internationally Wrongful Acts with

[28] The Draft Articles on Responsibility of States for Internationally Wrongful Acts with Commentaries, page 83. Available at: http://legal.un.org/ilc/texts/instruments/english/commentaries/9_6_2001.pdf.

[29] *Ibid.*

commentaries mention the Torrey Canyon accident as an example of a circumstance precluding wrongfulness (United Nations n.d., p. 83).[30] Thus, using a nuclear device to deflect an asteroid as part of a planetary defense mission could be seen as falling within this category.

There is no doubt that invocation of circumstances precluding wrongfulness would require an actual threat. It raises a question whether, for instance, an identification of an asteroid on a collision course to impact the Earth in 10 years could fulfil the requirements set by public international law, since within a decade it might be possible to deal with the threat by other non-military means. In addition, it is very difficult to sufficiently predict the impact of such a collision to objectively determine the risk at stake. The threshold when nuclear detonation is the only solution would be hard to define not only given the limited information from the distant Earth, but also given the consent procedures required. We can deepen our knowledge about the asteroid mainly through rendezvous spacecraft, a journey that would take time, including the time needed to build the spacecraft if an appropriate one does not already exist. The time to gain consent for the action can also be crucial, as the level of threat may change drastically in a short time. For example, we often detect comets when they are already in the inner circle of Jupiter's orbit. If one of these comets was on a collision course, the impact would occur in less than a year. It is worth taking into consideration that comets, in contrast to asteroids, are travelling faster and in more direct angles, making them much more threatening objects. While detection a decade ago would give us an opportunity to deal with the problem using a gravity tractor (in the case of an asteroid), for a closer detection, a single or swarm of kinetic impactors would be the best solution. However, if the object is only detected days or weeks in advance, NEDs might be our only option to deal with the colliding comet.

16.4.5 Deepening Knowledge and the Proper Deflection Method Selection

We are currently operating in an atmosphere of huge uncertainty. The consequences posed by unanticipated asteroids, such as the 2015 TB145[31] that missed us by about a lunar distance with a diameter of up to 600 m and at speed of 35 km/s, are beyond imagining. Collision with any asteroid of this magnitude would cause a regional cataclysm, with hundreds of millions of casualties and implications for the Earth's atmosphere for years ahead. Our lack of knowledge is directly fueling the debate for threat securitization (see Chap. 14), leading to a plea for exceptional politics. In our case, the lack of knowledge understandably leads to a state of exception as the norm (as

[30] The Draft Articles on Responsibility of States for Internationally Wrongful Acts with Commentaries, page 83. Available at: http://legal.un.org/ilc/texts/instruments/english/commentaries/9_6_2001.pdf.

[31] See details about the asteroid here: https://ssd.jpl.nasa.gov/sbdb.cgi?sstr=2015%20TB145.

discussed in Chap. 21) which would eventually cause highly undemocratic decisions that undermine the legitimacy of any governance model. We can avoid such an undesirable development by appropriate investment in observation techniques.

In an email conversation during the development of this book, David Morrison (see Chap. 2) mentioned that deepening our knowledge of the Solar System's asteroid population will reduce the justification for the nuclear deflection method. Here the concerns over political controversies regarding nuclear weapons as an obstacle to scientific progress are not present, because it is the observation that has the potential to enrich scientific knowledge and directly diminish the argument supporting the state of exception. In the light of new findings about near-Earth objects, the global actors should remain bound to their earlier pledges to pursue nuclear disarmament, as that pledge loses ground under the weight of newly acquired knowledge of our Solar System—especially in times when we might get close to a perfect awareness of the Solar System's asteroid population, or at least when astronomers have an idea of how to achieve this (see Chap. 6).

Nuclear deflection offers a substantially more effective option (up to 100 times) than any other type of deflection strategy. Perhaps the first question that should be asked is whether there are currently, or will be in the near future, scenarios that would require them as a response to an imminent threat. As discussed in Chap. 11, the nuclear deflection method is useful against short-time warning asteroids that could only be deflected by kinetic impactors if found earlier. Even then, other options should probably be explored first, not least of all because it might be easier to agree on non-nuclear options for cooperation than nuclear ones.

The concept of HAMMER (Barbee et al. 2018) demonstrates that this forward thinking is present in the engineering community. Hammer can be used for both kinetic impactors and nuclear explosive devices. It can even be scaled according to the particular need while preserving the core engineering concept. This engineering approach provides an option to act ad-hoc. However, if there are no other viable options, then it is still unlikely people will be willing to set aside their skepticism without a tangible threat first, and thus hammer can stay as a delivery vehicle without prepared NEDs. If a hard-to-detect asteroid, comet or other moving body were to enter the atmosphere and cause some damage, then we would argue that world leaders would be more likely to act even unilaterally or by giving approval to the spacefaring nation that may have kept NEDs in secret. However, without some real threat hanging over everyone's heads, it will be difficult to mobilize support behind NEDs in the hands of a select few. One would ask what role fully unpredictable and extremely hard-to-detect comets can play here? Well, this is a delicate situation because, as is well known, comets are faster, bigger, and can approach from any angle from the Oort Cloud surrounding the whole Solar System, up to the distance of 1.5 light years. Meanwhile asteroids lay on a Solar System ecliptic plateau, they are usually detectable after missing Jupiter, thus with short warning time before impact, and are able to be deflected by a swarm of kinetic impactors or even a nuclear subsurface explosion when nuclear ablation might not be enough. Comets, thus, have the potential to keep pouring fuel continuously on the argument for a state of exception, despite their probability of impacting us being only 2% of the likelihood of an asteroid collision.

16.5 From Cooperation to Regime Building

Should the abovementioned uncertainties lead to the establishment of a dedicated planetary defense regime? Since any unilateral action by states with highly developed space programs tends to produce distrust and thus potentially conflict, we believe so. How such a regime would look currently remains unknown, but the proposal to deepen institutional powers (Chap. 21) requires a concerted effort; it will have to be achieved through consensus of all states at COPUOS, and thus would not have the democratic legitimacy Pavel Dufek calls for in this volume (see Chap. 24). As such, this new institution might be a target for many.

Other authors in this book (see Chap. 26) propose a Moon base as a first-line biosphere defense. The concept is described as a scalable system, possibly expandable later to a massive laser capable of diverting distant asteroids. The scalability of the system, and its first deployment as a defense from micrometeorites of the Moon Village for any incoming actors, either public or private, has the potential to create a precedent of transparency through experience of its operation.

NASA called for consent in its 2007 report on NEO threats by stating that 'international coordination' will be necessary before taking into consideration any nuclear weapons applications in space (NASA 2007). The concept of consent when applied to one of the most sensitive topics in international relations probably means a legal consent adopted by an international body such as a Security Council, or by a body that is established for the purposes of the verification intent (see below). This would help to reach an agreement between involved actors. However, according to this perspective, powerful nations might not be willing to let less powerful states dictate what they should and shouldn't do with their nuclear technologies.

We can have reasonable doubts over how likely such consent is in the current international setting. The motives to participate in planetary defense on the national level can be driven by national prestige, the commercial boost multiplier, the advancement of scientific research or simply the demand for a more responsible global governance (Schmidt 2017), and each driver can be triggered by different national political policy priorities. States supporting neoliberal environments will find a way to support asteroid mining and to back companies to reach them, while the same infrastructure can be used for planetary defense. However, more authoritarian states will tend to perceive planetary defense as an opportunity to stimulate their prestige internationally, and such states will be prone to unilateral actions precluding possible cooperative actions based on international consent.

Under the current circumstances, any program to develop a nuclear deflection strategy for planetary protections will be seen with skepticism. What will be important in developing this strategy will therefore be transparency. If everyone knows what is being developed, there will be less fear. To this end, a cooperative structure for developing this plan would be ideal. By inviting many governments to participate, one could mitigate the skepticism behind the employment of nuclear weapons. Nevertheless, it will be difficult to overlook the sensitive nature of any nuclear technology.

Nikola Schmidt recently argued that cooperation would lead to technological interdependence, keeping governments locked in cooperation despite national elections elevating radical politicians (Schmidt 2018). Although we might not be there yet, the current efforts behind the cooperative missions between NASA and ESA, including DART and HERA, show that the will to cooperate on kinetic impactor demonstration missions is strong. Even the combined capacity of two different space agencies has the potential to lay down the basis for a more institutionalized future form of cooperation—in Krasner's words, institutionalizing cooperation into a working coordination (Al-Ekabi et al. 2016, Chap. 3).

16.6 Conclusion

This chapter discussed various international law perspectives related to planetary defense. While it is not clear how to apply the current provision on the case of planetary defense, various views discussed here could open up new discussions. Several provisions in current international law could consider both asteroid collisions and their deflection as threats to global peace and security. Space is a province of mankind, and thus all activities in space should be carried out for the benefit of mankind. While this term is not precisely conceptualized, common sense applies to its interpretation. Adopting measures that would threaten peace and security and would not serve the benefit of mankind should be questioned. For example, if an asteroid threatens the Earth but later the impact risk corridor is narrowed across the territory of a certain country, it would be questionable whether invoking the right of self-defense applies, or whether the state is obliged to discuss the topic multilaterally, since states are obliged under international law to seek harmonization, common ends and benefit to all countries.

We do not need to analyze international law in detail to take a common-sense approach and ask ourselves what would happen if any single state along the risk corridor decided to act on its own: anarchy. Coordination thus seems necessary; however, according to Krasner, coordination requires states to reach harmony first, achieve cooperation second and promote imagination third (see Chap. 20) in order to avoid possible anarchy.

Nuclear deflection methods are in this perspective the fastest way to anarchy, since their usage poses too many problems: even testing them in space is illegal, and their application or even research would de-stigmatize them (see Chap. 15). Their use could also lead to other undesirable situations, such as nuclear fallout. Necessity might legalize their usage, but it is worth pursuing other means that do not pose such challenges but rather open up new horizons for humanity.

Acknowledgments This study was supported by the grant awarded by the Technological Agency of the Czech Republic, project TL01000181: *"A multidisciplinary analysis of planetary defense from asteroids as the key national policy ensuring further flourishing and prosperity of humankind both on Earth and in Space,"* and co-funded by the Institute of Political Studies, Faculty of Social Sciences, Charles University in Prague.

References

Al-Ekabi, C., Baranes, B., Hulsroj, P., & Lahcen, A. (2016). *Yearbook on Space Policy 2014*. https://doi.org/10.1007/978-3-7091-1899-3

Barbee, B. W., Syal, M. B., Dearborn, D., Gisler, G., Greenaugh, K., Howley, K. M., et al. (2018). Options and uncertainties in planetary defense: Mission planning and vehicle design for flexible response. *Acta Astronautica, 143*(August 2017), 37–61. https://doi.org/10.1016/j.actaastro.2017.10.021

Bull, H. (2002). *The Anarchical Society*. Columbia University Press; 3rd edition.

Buxton, C. R. (2004). Property in Outer Space: The Common Heritage of Mankind Principle vs. the First in Time, First in Right, Rule of Property. *J. Air L. & Com., 69*, 689.

Carus, S. (2006). Defining "Weapons of Mass Destruction," 64. http://www.dtic.mil/dtic/tr/fulltext/u2/a446692.pdf

Evans, M. D. (2014). *International Law*. Oxford University Press.

Gruss, M. (2015, October 9). Russian Satellite Maneuvers, Silence Worry Intelsat. *Spacenews.com*. https://spacenews.com/russian-satellite-maneuvers-silence-worry-intelsat/. Accessed 12 July 2018

Hobe, S., Schmidt-Tedd, B., & Schrogl, K. U. (2017). *Cologne Commentary on Space Law: Outer Space Treaty*. BWV Berliner-Wissenschaft.

I.C.J. (1986). *Military and Paramilitary Activities in and against Nicaragua (Nicaragua v. United States of America), Merits*.

Johnson, C. D. (2018). What is allowed in outer space. *Oxford University Press Blog*. https://blog.oup.com/2018/03/outer-space-law/. Accessed 12 July 2018

Joyner, C. C. (1986). Legal Implications of the Concept of the Common Heritage of mankind. *International and Comparative Law Quarterly, 35*(01), 190–199. https://doi.org/10.1093/iclqaj/35.1.190

McDougall, W. A. (1985). *The Heavens and the Earth: A Political History of the Space Age*. Basic Books.

NASA. (2007). *Near-Earth Object Survey and Deflection Analysis of Alternatives. Report to congress*. Washington D.C. http://www.nasa.gov/pdf/171331main_NEO_report_march07.pdf

Nasu, H. (2011). The Expanded Conception of Security and International Law: Challenges to the UN Collective Security System. *Amsterdam Law Forum, 3*(3), 15–31. https://doi.org/10.1525/sp.2007.54.1.23.

Schmidt, N. (2017). Planetary Defense as a Gateway to Space for Commercial and Deep Space Exploration. *New Space, 5*(4), 219–229. https://doi.org/10.1089/space.2017.0005

Schmidt, N. (2018). The political desirability, feasibility, and sustainability of planetary defense governance. *Acta Astronautica*. https://doi.org/10.1016/j.actaastro.2018.06.037

UN. (2002). United Nations Treaties and Principles On Outer Space, related General Assembly resolutions and other documents. New York: United Nations, Office for Outer Space Affairs. http://www.unoosa.org/pdf/publications/STSPACE11E.pdf

United Nations. (n.d.). Draft articles on Responsibility of States for Internationally Wrongful Acts, with commentaries. http://legal.un.org/ilc/texts/instruments/english/commentaries/9_6_2001.pdf

Werner, D. (2018, March 21). ESA to investigate links between debris removal and satellite servicing. *Spacenews.com*. https://spacenews.com/esa-to-investigate-links-between-debris-removal-and-satellite-servicing/. Accessed 12 July 2018

Chapter 17
'No Conscience of Its Own': The Need for Global Space Ethics Review

Evie Kendal

> Molecules are stupid. Industrial poisons, greenhouse gases, and substances that attack the protective ozone layer, because of their abysmal ignorance, do not respect borders. They are oblivious of the notion of national sovereignty. And so, due to the almost mythic powers of our technology (and the prevalence of short-term thinking), we are beginning—on continental and on planetary scales—to pose a danger to ourselves.
>
> –Carl Sagan.

17.1 Introduction

Human interactions with space and celestial objects have previously been limited by what we *can* do, rather than what we *should* do. So-called 'space races' have often shown little consideration of the long-term ethical impacts of space exploration, or the potential dual-use dilemmas that the associated advancements in science and technology represent for global security. It has become apparent in recent times that increased human activity in space poses various legal challenges, leading to a drive to progress the field of space law. In this chapter, I argue for a similar need to codify global space ethics review processes and suggest some possible methods for monitoring and evaluation in this area.

This fragment is taken from John F. Kennedy's public speech at Rice University in Houston regarding the American commitment to the Moon landing: 'We set sail on this new sea because there is new knowledge to be gained, and new rights to be won, and they must be won and used for the progress of all people. For space science, like nuclear science and all technology, has no conscience of its own. Whether it will become a force for good or ill depends on man, and only if the United States occupies a position of preeminence can we help decide whether this new ocean will be a sea of peace or a new terrifying theater of war' (September 12, 1962).

E. Kendal (✉)
Bioethics and Health Humanities, Deakin University School of Medicine,
Waurn Ponds, VIC, Australia
e-mail: evie.kendal@deakin.edu.au

N. Schmidt (ed.), *Planetary Defense*, Space and Society,
https://doi.org/10.1007/978-3-030-01000-3_17

17.2 What is Space Ethics?

As with all ethical theories, space ethics is concerned with what we ought to do. According to Schwartz and Milligan (2016) the emerging field of space ethics 'epitomizes interdisciplinarity,' drawing its proponents from astrobiology, geology, and planetary science on the one hand, and philosophy, law, political science, and science fiction literature on the other. In so doing, the field is well placed to discuss complex questions that require an understanding of both the practical and moral constraints that should guide human space-related endeavors. These include judgments regarding the intrinsic value, if any, of celestial objects and space environments, and whether humans have any ethical obligations to preserve these environments or objects. Comparisons with other forms of human exploration and colonization are easy to find, as are concerns about 'playing god' or interfering with nature.

When it comes to global security concerns, much of the literature that can be classified as space ethics has heretofore been concerned with issues of militarization and the commercialization of space and its objects (Arnould 2011). It is this focus on legal issues, such as property law, that has led Milligan (2014) to claim that development in the realm of space ethics has been 'dramatically outstripped' by that of space law and policy. Nevertheless, progress in the space ethics field has been made through the recent publication of various academic texts and, more notably, UNESCO's 'The Ethics of Space Policy' document (2000). The latter is predominantly concerned with methods of conducting space research, maintaining peace, and avoiding exploitation of natural resources in space. Written by Alain Pompidou on behalf of the Working Group on the 'Ethics of Outer Space' for UNESCO's World Commission on the Ethics of Scientific Knowledge and Technology (COMEST) (Pompidou 2000),[1] the document considers how space ethics transcends national boundaries and interests. It prompts readers to reflect on mankind's potential role in the universe, and asks how we can engage in priority-setting regarding scientific exploration of space while keeping risks to an acceptable level. Of particular interest is the call to share data collected from space-related activities, especially with low and middle-income countries. The need for global cooperation to maximize social benefits and protect space from exploitation is a recurring theme throughout the document and the various conference proceedings contributing to it. While ethical considerations are outlined throughout, many of the key recommendations end up being focused on designing and enforcing legal instruments (Pompidou 2000).

There is also increasing concern within the literature regarding the potential ethical considerations of space colonization, terraforming, and the risk of harm to extraterrestrial environments caused by human interventions, including those intended to protect the Earth from asteroid collisions. This is because any human action necessarily changes the natural course of events—for example, if we deflect an asteroid *away* from Earth, we at the same time deflect it *toward* something or

---

[1] A collaboration between UNESCO, the European Space Agency (ESA), various other UN groups and national space agency representatives, and private industry.

someplace else. Similarly, if we terraform a planet to accommodate human life, we may leave it unable to sustain the evolution of extraterrestrial life. Those who have publicly supported the idea of establishing human colonies on other planets—including Elon Musk, Carl Sagan, and Stephen Hawking, to name a few—have also co-opted the language of normative ethics to promote their agendas. In describing space exploration as a 'moral imperative' aimed at promoting species survival and continued development, Ian Stoner claims these influential figures neglect or obscure the higher ethical obligation to invest in 'planetary defense and civilization protection' to avoid the extinction they fear (Stoner 2017). There is also an inherently anthropocentric aspect to these arguments that warrants ethical evaluation.

Despite continued interest in space ethics among academics, Jacques Arnould (2011) claims there is still a lack of recognition among those working in astronautics of the need to establish ethical review processes to govern human actions in space. He attributes this to the belief that space programs do not align with the categories often associated with ethics review: 'medical research, financial practices, academic councils, etc.' In his book, *Icarus' Second Chance* (2011), Arnould dates the advent of space ethics as a field to the beginnings of the 'space venture itself,' claiming that John F. Kennedy's statement in 1962 that space science had 'no conscience of its own' implied there was an obligation to fill this void as part of the (American) space exploration effort. Arnould's text discusses some potential risks of space exploration, including that nation states might use space for military purposes, or threaten the sovereignty of other states through remote sensing and image gathering technologies deployed in space over certain territories. He makes reference to the so-called 'open skies principle' (derived from the United Nation's 1967 Outer Space Treaty permitting free use of space) as the common defense against suggestions the potential for international spying justifies limiting national space programs. A large portion of the book is devoted to issues surrounding the commercialization of space, including through space tourism, advertising, and private surveillance. Importantly for this book, Arnould's analysis also includes a section on the responsibility of members of the space profession to investigate Near Earth Objects (NEOs) and Potentially Hazardous Asteroids (PHA), to learn how best to prevent catastrophic incidents.

Icarus' Second Chance has been described by Milligan (2014) as a key text delineating the possible scope for appropriate ethical review in this area of scientific study, including in risk assessment, definitions of space debris pollution, and establishing ethical duties and obligations for launch-states. Both Arnould's text and UNESCO's document repeatedly use comparisons between the emerging field of space ethics and existing models of ethics review. Noting that some countries have created space ethics review protocols, while others have noticeably failed to do so, Arnould suggests that the absence of independent ethics review panels associated with international agencies involved in the regulation of space activities can be explained by the fact space ethics rarely involves specific cases that can be subjected to ethical evaluation. He claims that this distinguishes space ethics from areas like medical research ethics and goes on to suggest that when such cases do arise, they can simply be submitted to one of the agency's partner organizations, most of which will already have established ethics review committees (Arnould 2011).

However, as will be demonstrated throughout the remainder of this chapter, existing ethics review systems are ill-equipped to handle some of the unique challenges facing space ethics generally, and planetary defense in particular. Furthermore, if international human research ethics has taught us anything, it is that a lack of consistency and capacity in global ethics review standards often results in exploitation of vulnerable groups. Another element to consider, and one that Pompidou also discusses in the UNESCO document, is that fields like human research ethics arose from historical abuses, such as the Nazi experimentation on Jewish prisoners in concentration camps during World War II (Pompidou 2000).

We currently have an opportunity to establish methods for global space ethics review that are proactive, rather than reactive, taking into consideration the ethical implications of emerging technologies before they are widely implemented, including those involved in planetary defense. Thus, instead of merely responding to abuses, global space ethics review standards could work to prevent these abuses from ever occurring.

17.3 Necessary Conditions for a Space Ethics of Planetary Defense

When considering an ethics of space that would promote the deflection of asteroids and comets to save humanity, a number of normative assumptions must be made. The first is that the preservation of human life is an ethical pursuit—if not morally obligatory, it must at least be morally permissible. This assumption has been challenged by scholars like Marks (2016), who argues that mankind's destructive nature calls into question the morality of defending the species against extinction-level 'asteroidal or cometary impact.' Noting humanity's penchant for violence against the environment, other humans, and non-human animals, Kolbert (2014) predicts that as a species we are already doomed to cause our own extinction. What is at stake in these debates is whether or not humans are worth saving, given this destructive nature. It is also worth reflecting on the fact that the human race may not have risen to its current state if not for the K-T extinction event that wiped out the dinosaurs, which is commonly believed to have involved exactly the kind of large-scale asteroid or meteor impact that planetary defense systems would strive to avoid. While I cannot hope to ethically defend humanity's treatment of the Earth's resources or its continued exploitation of other living beings, any consideration of 'natural law' must necessarily defend the species' right to pursue self-preservation. Moreover, if the practice of ethics carries with it the obligation to seek self-improvement and promote human flourishing, as is often considered the case, the related obligation to preserve moral agency and autonomy could similarly justify the use of planetary defense to secure our survival.

The second assumption needed to support planetary defense is that there are methods of asteroid deflection that are themselves ethically defensible. Such a judgment would typically consider both the method in isolation and its application under a standard cost/benefit analysis. In this way, highly risky interventions, or those that might bring about particularly undesirable side effects, might be deemed morally impermissible, while those considered to represent an acceptable level of risk would be deemed morally permissible. Throughout this book, for example, it has been argued that NEDs fail this test, whereas super powerful lasers do not. Those like Felicity Mellor (2007) who are concerned that 'impact-threat science' is trying to justify all manner of militaristic solutions to a potential global threat, may seek some assurance that limits will be set on the means of intervention. When it comes to any unavoidable negative side effects of a planetary defense system, the doctrine of double effect (DDE) provides one possible method of ethical evaluation that might satisfy this demand. According to Gillon (1986), the DDE was first developed by Thomas Aquinas in an attempt to 'delineate the conditions in which it is morally legitimate to cause or permit evil in the pursuit of good.' There are four criteria in the DDE that an action must satisfy to be considered acceptable (Kendall 2000). These are listed below, alongside a possible application to planetary defense:

- The act when considered independent of its consequences must be of the sort that is morally permissible (e.g. asteroid deflection must itself be morally permissible)
- The intended outcome must be good, while the negative outcome can be foreseen but never intended (e.g. saving human lives must be the intention, any unfortunate loss of actual or potential life must be unintentional)
- The bad outcome must not be the method of achieving the good outcome (e.g. if to save a densely populated area an asteroid must be deflected to strike a less populated area, any loss of life can be tolerated only if it does not constitute an essential methodological step in the process of saving the other lives)
- The benefit of the good outcome must reasonably outweigh the harm of the bad outcome (e.g. saving humanity must justify the ethical cost of any negative side effects of planetary defense)

It should be immediately apparent that the higher the risks involved, the less likely a planetary defense method will be able to satisfy all of these criteria. However, as demonstrated throughout this book, while nuclear options might be disqualified for legal and safety reasons, there are various methods that could be ethically supported when considered through the lens of the DDE. For example, the kinetic impactor method, in which a spacecraft is sent to intercept an NEO, could be permitted if the likelihood of success was high and the risk of deflecting the object to a worse trajectory was minimized. NASA's upcoming Double Asteroid Redirection Test (DART) mission aims to demonstrate the efficacy of this system for planetary defense and alleviate concerns regarding possible misdirection. Similarly, an enhanced gravity tractor—a spacecraft that exerts a gravitational field strong enough to alter the trajectory of a celestial object (Mazanek et al. 2015)—might be supported by the DDE under the same conditions.

The remainder of this chapter will assume the norms discussed above have been satisfied and that there is no compelling reason to ethically condemn planetary defense systems per se. What remains is a consideration of how best to ethically regulate such systems and other human interventions in space.

17.4 Threats to Global Safety and Biosecurity Associated with Unregulated Space Exploration and Use of Celestial Objects

17.4.1 Historical Analogies from Geoengineering

Planetary defenders have cause for alarm when considering the potential threat that unregulated space exploration and use of celestial objects pose for global safety and biosecurity. Due to the scale of its impact, geoengineering represents one of the closest analogies for what might happen if unregulated space travel and research become attainable. Defined by Keith (2000) as the 'intentional large-scale manipulation of the environment,' geoengineering efforts are particularly focused on redressing anthropogenic climate change. In his book *A Dangerous Master* (2015), Wendell Wallach relates how the lack of regulatory and legal instruments has permitted risky geoengineering projects to be completed by wealthy, private citizens with little to no oversight. One example is Russ George's famous iron fertilization experiment in 2012, in which the businessman dumped 100 tons of iron sulfate into the Pacific Ocean with the intention of increasing algae and thus carbon capture capacity. Described by Kristina Gjerde, high-seas adviser for the International Union for Conservation of Nature, as an experiment that '[did] not appear to even have the guise of legitimate scientific research,' the George case clearly demonstrates the need for better regulation when dealing with shared natural resources (Anonymous for Planet Experts 2014).

Cloud seeding represents a weather modification system that functions by introducing silver iodide into clouds to promote the formation of crystals, which eventually has the effect of increasing overall precipitation in a given region. This technique is already being used in several countries, but there are no international regulations monitoring its use or any potential impacts on other regions. Another geoengineering method that could be attempted either by well-resourced individuals or state parties is stratospheric aerosol injection, a technique modelled after natural phenomena in which volcanic eruptions sometimes reduce global temperatures by releasing particles into the stratosphere that reflect sunlight. This is currently considered one of the more credible solar radiation management techniques intended to mitigate the effects of climate change, winning out against alternatives such as space mirrors (Wallach 2015). 'Assisted evolution' experiments, such as the recent genetic modification of coral in Australia's Great Barrier Reef using CRISPR-Cas9 gene editing techniques, are also considered by some to be a form of geoengineering (Cleves et al. 2018). While the expense of engaging in geoengineering makes it unattainable for most citizens, various private corporations already offer cloud seeding services and sufficiently wealthy individuals may soon be in a position to 'purchase' the climate of their choosing. In a similar vein, recent announcements by Russian billionaire Yuri Milner, regarding the Breakthrough Initiatives' mission to reach Saturn's moon Enceladus, highlight that private venture capitalists are already able to mount space missions. If successful, this mission will reach the celestial body before NASA (Bennett 2017). Meanwhile, the affiliated Breakthrough Starshot program has its sights set on Alpha Centauri and the Jovian satellite Europa.

From an ethical perspective, it is noteworthy that most geoengineering processes carry a risk of making the climate less hospitable in some regions, and that scaled testing would involve such a large portion of the globe that it would carry significant risks for human populations. In a similar way, testing that an asteroid defense system worked would eventually require a full-scale deployment of the system, an experiment with potentially catastrophic consequences if executed incorrectly or through dangerous means. As scientific validity, risk assessment, and cost/benefit analyses all form part of a standard ethics review, geoengineering experiments and any rollout of new planetary defense systems could both benefit from such a process. Importantly, concerns about the lack of regulation, particularly for private enterprises, should not be interpreted as a belief such projects would inevitably be unsafe or unethical. The core issue is not that governments and private corporations *will* behave inappropriately, merely that they *could*.

Another risk associated with opening the frontier of space to private interests is a loss of momentum in tackling climate change. Particularly if some citizens are granted access to a 'Planet B' after Earth is no longer habitable, the motivation to reduce carbon emissions and protect the environment may diminish. The new Netflix series *Altered Carbon* (2018) provides a disturbing image of how such a shift may occur along class lines. In this fictional future, each human consciousness is stored in a metal implant, known as a 'cortical stack,' and the extremely wealthy elite are able to transfer their 'stacks' into any number of bodies, or 'sleeves,' in order to achieve a sort of immortality. With death no longer holding meaning for these people, respect for human life and the suffering of others' bodies deteriorates to the point where wealthy citizens start engaging in ritualistic torture of poorer citizens for entertainment. While this is an extreme example, it is not difficult to imagine a reality in which those who build their fortune by exploiting the Earth's natural resources may show less restraint when the impacts of ecological apocalypse won't apply to them. This represents a justice concern that I argue can only be addressed by prohibiting private space exploration without independent ethics review. As with other forms of mandatory ethics review, the purpose of such an activity should be protecting stakeholder interests, promoting shared benefits, and improving transparency and public accountability. Far from hindering beneficial research or necessary defensive actions, global space ethics review would also provide an opportunity to improve methodological rigor through expert peer review, for both privately and publicly funded projects.

17.4.2 Future Threats

In addition to the concerns listed above, there are a number of unique threats space exploration might pose to humanity. The most obvious is the potential for states to weaponize celestial objects, for example, diverting an asteroid or comet to strike an enemy state. Although highly unlikely to occur in practice due in part to the considerable practical difficulties of achieving such a feat, this potentiality must be considered in any ethical evaluation of human space-related activity. There is also the risk that in attempting to prevent an asteroid striking one part of the planet, a state may unilaterally choose to sacrifice another region. Alternatively, a state may launch a nuclear attack on an incoming asteroid that will have flow-on effects for the global population, despite having no capacity to seek consent from affected individuals or those who represent them. For Williamson (2003), any conception of space ethics must take into consideration the 'impact of our actions in space on each other, on each other's property, on the Earth…and on the space environment itself.' In this way, any attempt to deal with a planetary threat must be done in a way that maximizes global benefit rather than merely promoting national objectives.

In terms of biosecurity, there is always the risk that a disease or invasive organism will be introduced to Earth from a space mission, especially if private research facilities are housing specimens for analysis under varying levels of security. As highlighted in the opening quote to this chapter, any breach of biosecurity by one company or country could pose a global health threat. Again, details of the proposed collection, storage, and destruction of samples all feature in ethics review applications, focused on avoiding environmental contamination or threats to human health.

17.5 The Insufficiency of Current Ethics Review Systems to Accommodate Space Ethics

17.5.1 New Players, Unique Vulnerabilities

Having considered some of the ways standard ethics review could prevent negative consequences for human interactions with space and celestial objects, we will now consider the ways in which existing ethics review systems are insufficient for the task of fully promoting space ethics. Hyder et al. (2009) note that ethics review systems consist of legal authorities, institutions, researchers, ethics committees, and participants. In the case of space ethics review, the legal authorities consulted would need to be international and the 'participants' may eventually include extraterrestrial life, even if this only turns out to be bacteria or similar. This immediately leads to the first problem with the dominant principles of biomedical ethics as they apply to the space context.

The first principle to consider, and the one that typically gets the most attention in the bioethics classroom and ethics committee meetings alike, is the principle of autonomy, or 'respect for persons.' Humans have historically been very bad at assigning moral status appropriately to other humans and animals, especially in the research setting, and there is considerable disagreement among ethicists as to what constitutes a 'person.' With most defending some concept of self-awareness as a prerequisite for the label, this criterion is hard enough to judge across terrestrial species, let alone ones that may be completely unfamiliar if ever encountered. Add to this the unique needs of the space environment and it becomes clear that ethics review for space-related endeavors will require people with expertise in astrobiology and related fields. Given the fact any precursors for life found in space could have great evolutionary significance, there would need to be strong advocacy on behalf of what can essentially be conceived of as a 'vulnerable population.' Drawing an analogy to human research ethics, the special protection afforded to fetuses in research is justified on the grounds that they are especially vulnerable to serious injury, and that although they are not 'persons,' they have the potential to become so. A similar level of protection could be designated for fragile potential lifeforms in space, but the dominant conceptions of personhood currently in circulation are not fashioned to include extraterrestrial life.

As the geoengineering example above has demonstrated, people living in low and middle-income countries can also be considered vulnerable, as decisions made about space-related issues may have global implications, but these populations are the least likely to be able to progress their own interests in the matter. This is especially true if they lack the technological development to join the new space race. For planetary defense, in the absence of a global space ethics review system, these populations would be relying on other states to protect them, or at the very least not to sacrifice them to save themselves.

The online poll of the Space Settlement Institute (2018) shows that respondents consider ensuring humanity's survival to be the most important justification for colonizing space, with 1235 votes for this option at the time of writing from 5025 participants (Makukov and ShCherbak 2014). This is a very anthropocentric viewpoint that is unlikely to entertain the possibility that a planet might have a right not to be colonized, and that we could have a duty to allow it to continue its development unhindered. Similarly, the issue of whether an asteroid can be considered to have any rights, particularly if there are possible biosignatures detected on it, might be difficult to explore. Again, the moral intuition that humans have a right to defend themselves against extinction would, at a minimum, justify the destruction of an asteroid threatening to eliminate us, with the DDE permitting any loss of extraterrestrial life as an unfortunate side effect of enforcing this right. Nevertheless, there are justice concerns that would seem to defend the perspective that space and its objects ought not to be exploited solely for human gain, in much the same way that the Earth's resources should be respected for reasons beyond their instrumental value.

17.5.2 New Risks, Unprecedented Levels of Potential Harm, and Invisible Benefits

Beyond autonomy and justice, the principles of beneficence and non-maleficence also carry unique concerns in the space context. Due to the real possibility of humans disrupting the evolutionary and environmental development of other planets and causing ecological damage through mining, there is substantial harm at stake in space exploration from the perspective of the planets and objects visited. However, there is also unprecedented harm possible to Earth should planetary defense systems fail to be properly tested and deployed as needed. This necessitates experimental research trials, including on actual asteroids and comets, in order to test that protection can be achieved. In the case of an immediate threat of collision, the need to coordinate global actions and avoid competing interventions undermining each other reiterates the need for governance systems to be international rather than national.

There are also unique situations in space ethics where it might not be clear whether something should be considered a pro or a con. For example, considering the current levels of pollution on Earth, the potential for space 'landfills' to reduce terrestrial damage could be viewed both positively and negatively. Shifting polluting activities to space might also be difficult to classify according to the principles of beneficence and non-maleficence. If colonies have been established in space, these issues become more complex still. For example, labor rights already differ wildly depending on the country you live in; it can only be expected that the same would apply in space in the absence of consistent regulation. The same arguments regarding opportunity versus exploitation would be applied to extraterrestrial labor as they are for terrestrial labor. As in historical colonization, the use of off-world personnel and resources to benefit Earth's population might become a source of political and social unrest, even if there are benefits conferred to both parties. A related issue is the potential instrumentalization of humans living in space or on off-world colonies to produce medical and engineering knowledge primarily intended for use on Earth.

In terms of cost/benefit analyses, planetary defense systems face the same dilemmas seen in preventive medicine—namely, that if they succeed in their purpose, nothing happens. Therefore, it can be difficult to express to the public and funding bodies the social benefit of designing and implementing these systems. The 'paradox of prevention' in public health provides a good analogy; interventions that have broad overall benefits to whole populations can have minimal observable benefits to individuals (World Health Organization 2002). Fineberg (2013) notes the invisibility of success in preventive medicine works against its interests, as does its lack of drama, long timescales for interventions and incremental statistical results. In the absence of an immediate threat, planetary defense systems face the same obstacles for public support. Furthermore, if planetary security is defined in negative terms, the invisibility issue will also apply. As is discussed elsewhere in this book, Floyd's (2007) concept of securitization states 'security is neither always positive nor negative, but rather is issue dependent.' Such an assessment also applies to the categorization of planetary security measures. According to Charrett (2009) and Roe (2012), to achieve a positive, constructive approach to security requires a reconceptualization of the whole notion, an activity best undertaken outside the confines of an existing normative framework. The need for this to occur within an international, rather than national, discursive space is highlighted by Booth (2007), who claims sovereign states are a source of global insecurity, yet their continuation is assumed in most security discussions. It is the breaking down of such domestic assumptions that a global approach to space ethics review must ultimately achieve.

17.6 Global Space Ethics Review and Governance

17.6.1 Development and Enabling Conditions

The framework for developing human research ethics review systems in low and middle-income countries proposed by Hyder et al. (2009) is easily adapted to the development of a global space ethics review system. This framework outlines the necessary conditions for robust ethics review, including the need for political, economic, and social freedoms among citizens of the relevant jurisdictions, and good training, protection, and transparency for review members and their procedures. Public accountability and a strong sense of community are also listed as enabling conditions, alongside good legal and regulatory oversight, institutional, and personal commitments to the value of ethics review among researchers, and a willingness to follow rules and set a good example among those in leadership positions (Hyder et al. 2009). For the review committee itself, the need to be independent and wield real authority over submission approval is vital for ensuring the review process doesn't dissolve into a rubber-stamping activity.

17.6.2 Membership

Due to the special concerns listed in the previous section, a global space ethics review system would need to have some members not typically found in an ethics committee. This group would be responsible for reviewing all proposed space exploration projects and uses of celestial objects for planetary defense, research, mining, travel, etc. Given the current role of the United Nations in drawing up legal instruments for dealing with matters of space policy, it is recommended the proposed global space ethics review system retain representation from the UN. Other proposed roles include:

- Chairperson
- Lay members of the public to review protocol for public acceptability
- Person in a role of public trust e.g. spiritual leader, carer
- Research experts to review protocol methodology for scientific validity
- Terrestrial lawyer to consider liability issues for and on Earth
- Extraterrestrial lawyer to consider liability issues for and on alternative worlds
- Terrestrial environmental and animal ethicist/advocate
- Extraterrestrial environmental and animal ethicist/advocate[2]

[2]These are roughly adapted from the Australian National Health and Medical Research Council's requirements for human research ethic committee membership but are similar to many other jurisdictions.

17.6.3 Process

The process of review would involve the independent committee considering applications for space exploration and making decisions regarding the safety and potential social benefits of each project. Detailed risk assessments would be made, including consideration of any dual-use dilemmas for technology and equipment designed and used in the project. The potential for profit, including through downstream commercialization, would be considered in light of the 1967 Outer Space Treaty concerning the non-appropriable nature of celestial objects (Sachdeva 2010). Laws permitting the mining of asteroids are often defended on the grounds that they do not conflict with the idea that the asteroid itself cannot be 'owned' (Tronchetti 2014). Establishing the threshold between what constitutes the asteroid and what is merely a product of the asteroid would fall under the purview of the ethics review committee members. A relevant analogy is when human research ethics committees have to arrive at a judgment regarding whether a research participant is being reimbursed for a procedure, e.g. a blood draw, rather than being paid to trade a human tissue, e.g. the blood sample. For jurisdictions in which all trade in human tissues is criminalized, such discussions are more than just philosophical thought experiments. Commercialization considerations will be particularly important, as it is predicted that space will become a booming economic hub in the future, and there are many who are concerned that private corporations can currently make up the rules as they go along (Foster 2016).

The process of global space ethics review would also need to demonstrate responsiveness to emergency situations and have a clear disaster preparedness plan in case of catastrophic incident. Relevant responsibilities would include providing leadership and communicating clearly with relevant stakeholders in the case of a negative incident e.g. an asteroid hit (Walker 2012). A blueprint for such a plan is provided elsewhere in this book.

17.6.4 Monitoring and Evaluation

The final element to be discussed in this chapter is how the proposed global space ethics review system could be monitored and evaluated. For this, I advocate for the use of indicators to measure success. According to the World Health Organization (2011), an indicator is 'a variable that can be measured repeatedly (directly or indirectly) over time to reveal change in a system.' The following list of indicators has been drawn from two similar lists published by OHCHR (2012) and Mainz (2003) and adapted for use to evaluate capacity building for a global space ethics review system:

- Quantitative indicators e.g. number of reviews conducted, percentage of projects satisfying requirements, or length of time members receive training for the committee
- Qualitative indicators e.g. narrative testimonials on the ethics review process and its effectiveness

- Performance and compliance indicators considering the efficacy of changes to the system and the resultant impact on meeting established standards and obligations
- Structural and process indicators demonstrating commitment to enforcing legal instruments, policies, complaints resolution procedures, and budgetary commitments
- Outcome indicators focused on actual versus predicted benefits to society
- Sentinel indicators that identify negative events to trigger further investigation e.g. deaths, near misses, review system failures

These indicators could be used to ensure the review system is functioning within parameters and achieving its mandate, allowing for corrections as needed. The ultimate goal would be to build capacity for independent review on a global scale that would cover all proposed uses for space and its objects, including planetary defense projects.

17.7 Conclusion

Human interactions with space and celestial objects pose a variety of potential threats to safety and biosecurity, however, through the development and implementation of a robust global space ethics review system, these threats can be mitigated. With continuous quality and safety monitoring, such a system could work to protect both terrestrial and extraterrestrial interests and form part of our planetary defense network. As has been demonstrated in this chapter and several others preceding it, methods of planetary defense need to be subjected to legal, political, scientific, and ethical scrutiny if we are to achieve our goal of protecting humanity's interests.

References

Anonymous for Planet Experts. (2014). Two Years After Russ George Illegally Dumped Iron in the Pacific, Salmon Catches Are Up 400%. *Planet Experts*. http://www.planetexperts.com/two-years-russ-george-illegally-dumped-iron-pacific-salmon-catches-400/. Accessed 14 July 2018

Arnould, J. (2011). *Icarus' Second Chance: The Basis and Perspectives of Space Ethics*. Springer Vienna.

Bennett, J. (2017). Breakthrough Initiatives Wants to Launch a Private Mission to Enceladus. *Popular Mechanics*. https://www.popularmechanics.com/space/solar-system/a13579385/breakthrough-initiatives-launch-private-mission-enceladus/. Accessed 14 July 2018

Booth, K. (2007). *Theory of World Security*. Cambridge University Press.

Charrett, C. (2009). A Critical Application of Securitization Theory: Overcoming the Normative Dilemma of Writing Security. *SSRN Electronic Journal*. https://doi.org/10.2139/ssrn.1884149

Cleves, P. A., Strader, M. E., Bay, L. K., Pringle, J. R., & Matz, M. V. (2018). CRISPR/Cas9-mediated genome editing in a reef-building coral. *Proceedings of the National Academy of Sciences*, *115*(20), 5235–5240.

Fineberg, H. V. (2013). The Paradox of Disease Prevention. *JAMA*, *310*(1), 85. https://doi.org/10.1001/jama.2013.7518

Floyd, R. (2007). Towards a consequentialist evaluation of security: bringing together the Copenhagen and the Welsh Schools of security studies. *Review of International Studies*, *33*(2), 327. https://doi.org/10.1017/S026021050700753X

Foster, C. (2016). Excuse Me, You're Mining My Asteroid: Space Property Rights and the US Space Resource Exploration and Utilization Act of 2015. *Journal of Law, Technology & Policy*, 407.
Gillon, R. (1986). The principle of double effect and medical ethics. *British medical journal (Clinical research ed.)*, *292*(6514), 193–4. http://www.ncbi.nlm.nih.gov/pubmed/3080130
Hyder, A. A., Dawson, L., Bachani, A. M., & Lavery, J. V. (2009). Moving from research ethics review to research ethics systems in low-income and middle-income countries. *The Lancet*, *373*(9666), 862–865. https://doi.org/10.1016/S0140-6736(09)60488-8
Keith, D. W. (2000). Geoengineering the climate: History and prospect. *Annual review of energy and the environment*, *25*(1), 245–284.
Kendall, C. E. (2000). A double dose of double effect. *Journal of Medical Ethics*, *26*(3), 204–205. https://doi.org/10.1136/jme.26.3.204
Kolbert, E. (2014). *The Sixth Extinction: An Unnatural History*. Henry Holt and Company.
Mainz, J. (2003). Defining and classifying clinical indicators for quality improvement. *International Journal for Quality in Health Care*, *15*(6), 523–530. https://doi.org/10.1093/intqhc/mzg081
Makukov, M. A., & ShCherbak, V. I. (2014). Space ethics to test directed panspermia. *Life Sciences in Space Research*, *3*, 10–17. https://doi.org/10.1016/j.lssr.2014.07.003
Marks, J. (2016). *Hard Atheism and the Ethics of Desire*. Cham: Springer International Publishing. https://doi.org/10.1007/978-3-319-43799-6
Mazanek, D. D., Reeves, D. M., Hopkins, J. B., Wade, D. W., Tantardini, M., & Shen, H. (2015). Enhanced Gravity Tractor Technique for Planetary Defense. In *4th IAA Planetary Defense Conference*. Fresquati, Roma.
Mellor, F. (2007). *Colliding Worlds*. *Social Studies of Science* (Vol. 37). https://doi.org/10.1177/0306312706075336
Milligan, T. (2014). Space ethics in context. *Space Policy*, *30*(4), 183–184. https://doi.org/10.1016/j.spacepol.2014.11.003
OHCHR. *Human rights indicators: A guide to measurement and implementation* (2012). World Health Organization: Geneva, Switzerland. Available at: https://www.ohchr.org/Documents/Publications/Human_rights_indicators_en.pdf
Pompidou, A. (2000). *Ethics of Space Policy*. UNESCO World Commission on the Ethics of Scientific Knowledge and Technology (COMEST).
Roe, P. (2012). Is securitization a 'negative' concept? Revisiting the normative debate over normal versus extraordinary politics. *Security Dialogue*, *43*(3), 249–266. https://doi.org/10.1177/0967010612443723
Sachdeva, G. S. (2010). Viewpoint: New Ethics for Space Commerce. *Astropolitics*, *8*(1), 49–61. https://doi.org/10.1080/14777622.2010.494518
Schwartz, J.S.J. & Milligan, T (2016). *The Ethics of Space Exploration*. Switzerland: Springer International Publishing.
Stoner, I. (2017). Humans Should Not Colonize Mars. *Journal of the American Philosophical Association*, *3*(03), 334–353. https://doi.org/10.1017/apa.2017.26
Tronchetti, F. (2014). Private property rights on asteroid resources: Assessing the legality of the ASTEROIDS Act. *Space Policy*, *30*(4), 193–196. https://doi.org/10.1016/j.spacepol.2014.07.005
Walker, D. C. (2012). *Mass Notification and Crisis Communications: Planning, Preparedness, and Systems*. CRC Press.
Wallach, W. (2015). *A Dangerous Master: How to Keep Technology from Slipping Beyond Our Control*. Basic Books.
Williamson, M. (2003). Space ethics and protection of the space environment. *Space Policy*, *19*(1), 47–52. https://doi.org/10.1016/S0265-9646(02)00064-4
World Health Organization. (2002). *Choosing Priority Strategies for Risk Prevention in The World Health Report*. Geneva, Switzerland: World Health Organization.
World Health Organization. (2011). *Checklist and Indicators for Monitoring Progress in the Development of International Health Regulations Core Capacities in States Parties (Monitoring Framework)*. Geneva, Switzerland: World Health Organization.

Chapter 18
The Conundrum of Commercial Space

Carlos Manuel Entrena Utrilla

18.1 Introduction

Commercialization is probably one of the most commonly heard words in the space industry. The topic is cause for one of the largest divisions in the industry, between the proposers of commerce, who often pose it as the only realistic approach, and the defenders of the traditional governmental approach, who see commerce as a utopian model that will never become a reality in space. Several commercialization attempts in the past failed to yield a fully commercial space sector (i.e. fully funded by private individuals). However, the many flaws of the various government space programs worldwide make a good case for the development of a commercially sustained space sector, and especially for the development of large projects such as the MMB. Recent approaches to commercialization through private-public partnerships have shown promising results and could be imitated for the MMB. In order to justify the need for commercial space, we have to take a look at the history of the space industry from its very beginnings, the issues affecting the progress of all of the space agencies, and how the involvement of private actors in public space projects can solve these issues and provide further benefits.

18.2 The Eternal Promise of Commercialization

The continuous issue of commercialization is testimony to the anomalous development of the space industry. Starting after World War II, the first successful space programs were based on the German V2 technology, which was arguably the first human object to reach space. The American and Soviet armies seized the V2 rocket

C. M. E. Utrilla (✉)
Independent Researcher, Madrid, Spain
e-mail: centrenautrilla@gmail.com

N. Schmidt (ed.), *Planetary Defense*, Space and Society,
https://doi.org/10.1007/978-3-030-01000-3_18

technology, along with most of the project engineers, as they marched into Nazi Germany. They immediately started developing their own programs, forever tying space technology to governmental use and, more specifically, to military use. Soon after, the USSR launched Sputnik, and the Space Race as we know it began. Although finished just 12 years later with the landing of Apollo 11 in Mare Tranquillitatis, this race defined how we think of space to this very day. Fueled by national pride, the Space Race rushed the development of space technologies with a constant input of seemingly endless government funds and the creation of the large bureaucracies that are today's space agencies. This model provided a fast pace of previously unbelievable achievements: the first artificial satellite (1957), first human spaceflight (1961), first communications satellite (1962), first human spacewalk (1965), first human landing on another celestial body (1969), first space station (1971), and the first missions to the Moon (1959), Venus (1962), and Mars (1971). However, as soon as the incentive of national pride disappeared, the model also found its first and main drawback: the extreme costs.

During the Space Race, cost was essentially no concern. On the American side, the intensive timeline imposed by President Kennedy to land on the Moon 'before this decade is out' meant that the engineering decisions were driven by a reduction of complexity and risk, and not by cost. During this time, NASA's budget grew every year and reached a very comfortable 4.4% of GDP by 1966, a level many times larger than its current share of 0.1%. On the Soviet side, the budget allowed the development of seven Salyut stations, leading up to Mir. With money essentially no object, engineers were free to develop their systems with a pure *engineerial* approach—based only on engineering requirements. US and Soviet scientists developed big (and independent) plans for the expansion into space, with Apollo spacecraft being used to establish semi-permanent Moon bases (Portree 2012, 2013), and private contractors settled on highly expensive projects overseen by the government. The space agencies, and the space industry in general, grew in this environment of monetary abundance for its first decade of existence, during which its mindset and procedures were set.

As soon as the outcome of the space race was clear, when money started to dwindle, both sides of the race looked at new sources of income to sustain themselves and their grandiose plans. The first commercial communications satellite, the American Intelsat I, had already reached geostationary orbit in 1965, and so the concept of space commercialization was already in some people's minds. It seemed like the obvious solution for the agencies' cash needs, and so commercialization was included in governmental plans. The first targets of commercialization were space stations. The US Skylab station, announced just two days after Apollo 11 landed on the Moon, completed several experiments aimed at verifying the feasibility of some on-orbit manufacturing schemes for potential commercial applications, which, though promising, haven't seen commercial use since. On the Soviet side, the Mir station was planned to manufacture products for other Soviet industries in the Kristall module, thus paying for itself with benefits for other programs. Again, this never materialized. The trend has been carried on until the present: the International Space Station (ISS) had among its initial goals to 'open new commercial frontiers' (Zimmerman 2003). This resulted in the denomination of the ISS as a US National Laboratory, managed by the Center for the Advancement of Science in Space (CASIS), and again saw almost no use for most of the station's history.

The launcher segment also saw attempts at commercialization. The Space Shuttle was designed with commercial flights in mind, which in theory would provide it with enough launch demand so that the high flight rate made launches inexpensive. The Shuttle did indeed fly 11 commercial missions in its early life (*Space Shuttle Missions Summary* 2011), until the Challenger accident resulted in a halt to most commercial uses (de Lama 1986). But even before the change in policy, commercial flights never managed to make the Shuttle economically self-sustained. On this aspect, the Soviets were more successful, managing to sell Soyuz launchers for commercial satellites on a consistent basis, along with astronaut seats in Soyuz to other countries and, more recently, to tourists. However, the Soyuz has never ceased to be a fully governmental program, needing public money to stay afloat.

These early failures reduced the interest in commercialization to small pockets of convinced supporters within the agencies, while most people considered space commerce a fool's dream. This idea has been constantly reinforced by the actual situation of the commercial space sector during the decades since the space race. The only feasible business case in space has consistently been communications broadcasts. Any other business has always been too expensive (private space stations), too risky (human spaceflight), or not fully allowed by government regulations (satellite servicing). This has been further complicated by the dual-use nature (potentially commercial and military) of space technology, which has pushed governments to tightly control private initiatives. Most potential markets and applications resulted in government-sponsored monopolies and strong regulatory barriers for any incoming business. The most noticeable example of this trend is the launcher business: the large majority of launchers are government-sponsored, require government funding to exist, and are subject to strict export controls. SpaceX and Rocket Lab have recently started to change that trend, but most launches worldwide are still paid for by governments.

The anomalous history of the space industry's development and its technical nature as both a powerful tool and a sought-after weapon explain the industry's current situation. The long list of previous failures, both in the political scene to secure funds and in the commercial scene to reduce costs, has forced space agencies to limit their activities to the occasional scientific mission and human spaceflight to the International Space Station. This aversion to programmatic risk is compounded by the aversion to technical risk caused by technical failures in human spaceflight, such as the Challenger and Columbia accidents. This pushes agencies to constrain themselves to 'flight-proven' technologies and extensive testing, limiting technical advance, space capabilities, and raising costs overall. In turn, any project involving more than a single probe is deemed too expensive or risky. Larger projects (such as a Moon outpost) require too much time to implement at the current levels of risk-tolerance and are likely to lose political support. The high costs of development impose extremely low acceptable levels of risk. The result is a feedback loop in which high-risk aversion prevents development, and lack of development enforces high-risk aversion. The final straw is provided by the lack of politically interesting achievements from a system that is both extremely risk averse and technically conservative. The agencies' budgets thus come under threat of severe cuts if they fail to deliver, further increasing risk aversion and the overall effect of the feedback loop. This draconian situation is the reason why humans last went to the Moon in 1972 and have not gone back since.

The solution to this problem is, contrary to popular belief, commercialization. International collaboration, usually presented as the only viable option, could also be an option. It aims to share costs between different nations, thus reducing programmatic and political risks and potentially improving the technical solution through better investment in development. However, this is only a way to externalize costs to other agencies, and thus the underlying problem remains. All space agencies (with the potential exception of the Chinese and their current lunar ambitions (*China's Space Activities in 2016* 2016)) are stuck in the risk-aversion loop, and their programs will suffer the same risk and technical constraints as everyone else's. The main (and maybe only) advantage of international collaboration is that it can put governments in a compromise in the political scene, providing an incentive for them to keep programs going in spite of delays, failures, or cost overruns. In the end, though, international collaboration might be able to provide larger projects than a single nation could, but the improvement would just be of a quantitative nature. The development of a project such as the Multipurpose Moon Base will require a qualitative change in our approach to space activities, and this can be enabled by commercialization.

18.3 The Different Models of Private-Public Partnerships

Reasoning from theoretical principles, two different models for commercialization through PPPs are possible, depending on the role of the public partner in the commercial relationship. Both models have been used by space agencies, especially by NASA in recent years, and so we can find examples of previous successes on which to build the development of the MMB. It is necessary to classify the possibilities in order to optimize their use during the development of the MMB. The models are based on whether the public partner provides the demand or the supply of services.

18.3.1 Public-Driven Demand

In a public-driven demand scheme, agencies provide mission objectives and private companies develop capabilities. Thus, the agency is creating a public-funded demand for a service, which companies aim to fulfill. The agency would then support the development of the private capabilities with public funding, knowledge sharing, and other in-kind contributions, and finally provide a long-term commitment to procure services from the private partner once the capabilities are operational. The private partners retain all intellectual property, are free to operate the product or service as they wish, and are encouraged to close the business case outside the public partnership. The best example of this model is the COTS program, where NASA had a demand for orbital transportation services and funded companies to develop them. Public-driven demand is best suited for the initial development stages of any space capabilities due to its advantages:

- It reduces the risk of the private investment by providing certainty in the market
- It reduces the cost of the private development by providing public funding and knowledge
- It provides early monetization opportunities for private investors while the rest of the market develops
- It enables a supply-driven approach for the commercial side once the capability is operational

The benefits of a public-driven demand scheme are ingrained in the COTS program by design or as lessons learned for future programs, as presented in the Lunar-COTS program proposal (Zuniga et al. 2015). The success of the COTS program prompted NASA to proceed with the CCP program, and its lessons can easily be applied to both the Moon and asteroid mining (Utrilla 2017; Zuniga et al. 2015). This speaks to its utility as a tool for the development of capabilities that are required for future expansion, but which are either too expensive for private partners alone or lack a market for private activity to serve. In some cases, both situations are combined (a capability is too expensive for the reduced market it might have) and therefore a public agency can complement the limited market by providing extra demand. The final goal is to create a commercial environment where supply leads to new demand, and where the government helps develop the supply and maintains it for its own use, but the supply provider is always encouraged to find business outside of the government. To this end, a key aspect of the public-driven demand scheme is to nurture a commercially competitive environment from the beginning by backing at least two different private partners to develop the same capability. This not only ensures a competitive environment for later procurement once the program ends, but also provides a level of redundancy for the public partner, which would have at least a second option in case one of the private partners is not able to complete the intended missions, for whatever reason.

18.3.2 Public-Driven Supply

In a public-driven supply scheme, agencies provide some capability for commercial use, and private companies develop mission objectives that can make use of it. The agency is offering its extra supply of a certain capability and companies find ways of making use of it. Additionally, the agency would then support the partner in developing its applications with knowledge sharing, technology transfers, or other in-kind support. The best example of this model is the NanoRacks Space Act Agreement, where NanoRacks has access to an experiment rack in ISS for commercial use ("Nonreimbursable Space Act Agreement between NanoRacks LLC and NASA for Operation of the NanoRacks System aboard the International Space Station National Laboratory" 2009). This model is best used to increase the utilization of already deployed infrastructure and has some advantages:

- It increases the return on investment of public-funded infrastructure by offering unused capacity for commercial use
- It enables potentially new commercial applications that would not be possible without the provided infrastructure
- It provides increased demand for other services along the value chain by enabling new commercial applications (e.g. NanoRacks activity at ISS increases demand for ISS resupply capacity)
- It allows international commercial partners to circumvent legislation aimed at public agencies (e.g. the first Chinese experiment on ISS arrived through a commercial partner, in spite of the ban for collaboration between NASA and the China National Space Administration (Allen-Ebrahimian 2017)).

A public-driven supply scheme is most beneficial when the public partner has developed excessive infrastructure or is no longer able or in need to fully utilize its capabilities. By providing a commercial partner with the permission and the ability to use the deployed infrastructure, it can return to full capacity and potentially help pay for part of its maintenance. The commercial partnership can also use this opportunity to develop its own capabilities to operate in space, with the goal of expanding its offer in the future. For example, NanoRacks has plans now to develop and install its own airlock on ISS to deploy vacuum experiments ("NanoRacks, Boeing to Build First Commercial ISS Airlock Module" 2017), Made in Space is using its experience on microgravity 3D-printing to manufacture fiber optics for use on Earth (Costello 2017), and Bigelow Aerospace has first demonstrated human habitation of its inflatable modules on ISS and will deploy its larger modules to potentially expand ISS in the future (Shear 2016). Public supply of services can then act as a test bed for the development of future commercial applications while maximizing the return on investment of the public funds.

Outside of these two commercialization models, there are other ways for governments to incentivize commercial development, mainly through research grants, tax incentives, or a variation of those. However, we established before that the model most suitable to break the risk-aversion loop of human spaceflight is through PPPs. Therefore, incentive schemes in which the government only provides some kind of monetary support should be discarded. The key to the program is that the public partner shares the economic risk by providing market certainty, while the private partner reduces programmatic and technical risk by acting through commercial (and not political) interests and taking the risk to flight-prove new technologies. Thus, PPPs stand over other incentive schemes in their benefits to space developments. This in turn requires both the public and private partners to have 'skin in the game,' imposing that they provide either the supply or the demand of a service, leaving us with just two options.

The definition of the two schemes are not meant to limit the potential implementations of a PPP, leaving great flexibility in terms of the role of each partner. For example, public partners can vary the level of risk they externalize by the level of monetary commitment they require: by providing a larger share of the funding, and thus requiring more public oversight, the public partner can indirectly incentivize the private partner to assume less technical risks and go for proven solutions, since they are more likely to work within the oversight of an agency. The same goes the other way: by providing a smaller share of the funding, the public partner is putting the private partner under a higher economic risk, which would push it to improve technologies and procedures to reduce cost, but might also force it to leave the program, since it would require too much investment for its part. These trade-offs are nothing more than the usual time-cost-risk trade, and can be found in all the PPPs developed up to date. Higher cost reductions can be obtained by incentivizing risk-taking, while a zero-risk policy will result in cost and schedule overruns. Programs where the agency demanded too much oversight and provided all of the funds, such as the Evolved Expendable Launch Vehicle (EELV) program, resulted in little to no cost reductions compared to a fully public development, and eventually caused the providers to merge into a single provider, United Launch Alliance, to keep the program affordable (Pike n.d.). On the other hand, the COTS program seemed to hit the sweet spot, with large cost reductions and few schedule slips. However, its continuation, the CCP program, kept increasing oversight and requirements imposed on the private partners while providing unnecessary funds, which resulted in large schedule slips of several years (Berger 2017).

The two models also allow mixed schemes. In the case of Made in Space, NASA supplied the infrastructure on ISS, while also providing a demand for their products that would not exist otherwise. This has allowed Made in Space to provide a commercial service that would not exist without the government infrastructure (i.e. providing 3D-printed tools and parts to astronauts on ISS), and to develop new capabilities for commercial use (i.e. microgravity 3D-printing) (Costello 2017). Considering all the potential trade-cost-risk trade-offs and mixed schemes, PPPs can and should be adapted to any particular case to optimize the returns of each particular infrastructure element. However, defining the two models allows us to establish standard procedures on how to act in each step of the development of the MMB, allowing a progressive refinement of the procedures, regulation, and relationships between public and private partners.

The final consideration on establishing the most appropriate PPP model for the development of the MMB is that the final goal must always be a self-sustained cislunar economy. As will be presented in the next chapter, the way for the potential benefits of commercialization to come to pass is to eventually phase out direct government contributions and commitments to commercial partners, and create a self-sustained commercial system that serves the MMB. This requires a free market with free competition, with the public partners acting purely as a final demand with their defense and science activities. The two models for PPPs, public-driven supply and public-driven demand, provide situations in which commercial activities can appear and create new commercial demand. Thus, both models can potentially lead to the final desired result of a free market, but they should be played to their strengths in order to maximize the benefits of commercialization across the development of the MMB.

18.4 The Right Path Towards Commercialization

Any space mission of the scale of the Multipurpose Moon Base (MMB) will need a new approach to current space activities that only commercialization can bring. Previous approaches to return to the Moon testify to this need: both the Space Exploration Initiative and the Constellation program envisioned a return to the Moon as a steppingstone for crewed missions to Mars, and both were cancelled due to constant schedule delays and cost overruns. On the European side, the Hermes program for a crewed spacecraft was also cancelled for the same reasons, as were the Buran and Energia launchers in Russia. This has been a recurrent theme in human spaceflight, where the only project that has come to term in recent years has been the International Space Station, in which international obligations were at play. The pattern has been consistent and is a remnant from the Apollo era:

1. The government establishes a goal based purely on political goals
2. The agencies scramble to present a plan that meets the political goals, making overoptimistic assumptions on the way
3. The program is developed, revealing technical issues and risks that had not been considered before
4. As the agencies try to keep within their acceptable risk levels while meeting the new technical challenges, the schedule starts to slip, and the costs start to rise
5. Rising costs are not matched by political will, leading to underfunding and more schedule slips
6. Reality sets in, and the program is cancelled as politicians realize that their political goals will never be met on schedule or budget

This outcome can be seen as the result of a trade-off between time, cost, and risk, with conflicting priorities. Any project, be it space-related or not, will have to face a trade-off between these three aspects. Given a determined time schedule, risk can be reduced to a certain extent by increasing costs. After that, the only way to further reduce risk is by increasing development times. Similarly, a project done in a short time at a low cost will likely face high levels of risk (technical or otherwise). In the case of traditional space projects, the level of technical risk is fixed at very low levels, imposing either huge costs or long development times (usually both). This conflicts with political goals, which require fixed schedules and low costs, and impose large programmatic risks due to changes in governments and agendas, resulting in the eventual cancellation of most human spaceflight programs. This strife has been pointed out by several authors, most notably Rand Simberg in *Safe is Not an Option* (Simberg 2013), and has been present in the space industry since the Challenger accident.

This pattern stems from doing things the Apollo way, without the Apollo money or commitment. With a fixed schedule, as was the case of Apollo, costs and risk will necessarily be high. This was justified in the Apollo era by the Space Race and the political need to beat the Soviet Union to the Moon. Once the drive of national pride was gone, the main drive became cost. The original Shuttle program saw huge cuts to the original plan, which involved a transfer vehicle to the Moon from the ISS and a lunar base (Launius 2012), leaving it with just the orbiter, and taking risks that are unacceptable today, such as having a crew on the first flight. Soon after the Challenger accident, risk reduction became the main drive, and thus schedules and costs started to slip, and programs were cancelled. This *Apolloism*, as some authors name it (Simberg 2013) has the space agencies developing and maintaining all the infrastructure needed for any mission in a completely centralized manner and based on political goals. This combination of centralization and political direction is what drives the time-cost-risk equilibrium off balance.

Commercialization can overcome the barrier of Apolloism by providing commercial goals and interests instead of political ones. This would tip the balance of the time-cost-risk trade-off by highly increasing risk tolerance and reducing costs. Recent developments in the commercial, entrepreneurial space industry (the so-called New Space sector) have shown that private companies can deliver high quality systems for a fraction of the cost, while accepting a much higher level of risk.

In terms of cost, recent commercialization programs have shown that private-public partnerships can achieve huge cost reductions, while delivering the same capabilities. The most notable case is NASA's Commercial Orbital Transportation Services (COTS) program, which co-funded the development of SpaceX's Falcon 9 rocket and Dragon capsule, and Orbital ATK's Antares rocket and Cygnus capsule. NASA officials from the NASA Ames Space Portal Office estimated that the COTS program achieved a 20-fold (2,000%) cost reduction in the development of Falcon 9 and Dragon, compared to the traditional NASA procurement and development approach (Zuniga et al. 2015).

In terms of risk, private partners can take more risks than public agencies, since they operate with private money and are not subject to such high levels of public scrutiny. The proof of this has been the repeated public failures of SpaceX and Blue Origin in their learning process towards reusability: crash after crash, their very noticeable technical progress all but increasing public support instead of reducing it. Further, SpaceX has suffered two rocket explosions since 2015, one of them carrying a Dragon for NASA, and their quick recovery from these accidents have gained them increased credibility in their technical capabilities within the commercial community, where customers are lining up to fly on the cheaper, reused rockets. The public in general praises failure when it leads to learning and when it comes from private funding, while they are less willing to accept mistakes from public entities.

This increased tolerance for technical risk is paired with a large reduction of the long-term programmatic risk that comes from political commitments. With commercialization, the need for political justification essentially disappears. This prevents programs from becoming a long-term burden to public budgets (as the Shuttle eventually was) and allows national politics to focus on the more relevant things, in this case, scientific exploration and, through science, enabling the effective global governance of aspects that by their nature cannot be limited to national interests. This includes planetary defense, since this agenda cannot avoid tensions in international relations if treated at a national level. This combination of cost reductions and risk acceptance and reduction breaks the risk-aversion loop and allows private companies to benefit from new technologies, new technical approaches, and learning from failures. SpaceX developed the Falcon 1 rocket with fully private funding for a cost of just US $90 million, in only six years (Spudis 2012). This makes it arguably the best example of the potential of private development in space when new technologies and commercial goals are involved.

In spite of its great potential, private initiative on its own still has a hard time operating in space. Falcon 1, impressive as it might have been, was not commercially viable, and SpaceX needed money from NASA's COTS program to keep operating after the first successful launch of the Falcon. This hints at the fact that private companies can, and indeed often do, accept more risk, which is proven by the large number of private space initiatives that have folded before ever reaching space in recent years. Technical challenges in space and the lack of a widespread commercial activity have proven too difficult for most private companies, which operate in what is still a government-dominated field. This is the reason why many people in the space industry, especially in the agencies, still think of space commercialization as a fool's dream, essentially impossible to implement.

Commerce is not hopeless though. Private-public partnerships (PPPs) have proven to be a good approach, providing programs with the benefits of both commercial and government development. Notable PPPs include the already mentioned COTS program. In COTS-like programs, the government calls for private companies to provide a service, and helps them during development with funding and expertise, and after implementation with contracts for initial services that make the business plan work. The key to the COTS program (and PPPs in general) is a phased development that allows incremental development and demonstration of capabilities while providing off-ramps for the government partners to stop funding in case the private partner does not meet objectives. This approach combines the abundance of funds from governments with the risk tolerance of private partners. This has the potential to deliver results on time, provided that both partners fulfill their funding commitments to the program.

While PPPs do not fully eliminate the programmatic risk derived from public oversight, the combination of increased funds (compared to a purely private program) and increased risk (compared to a purely public program) has produced remarkable results. NASA's COTS program produced a launcher with a reusable upper stage and fairing (Falcon 9), a reusable capsule (Dragon), and a backup commercial resupply system for the ISS (Cygnus) for a fraction of the cost, as mentioned before. The follow-up program, the Commercial Crew Program (CCP), will soon restore the US infrastructure of human spaceflight and again provide NASA with large cost savings when launching crew to ISS. In the case of SpaceX, the benefits also include a launcher that is not reliant on NASA for its existence and is therefore financially self-sustained (NASA only accounts for 40% of SpaceX revenues, and most of that is for the Commercial Crew Program (Winkler and Pasztor 2017)). This effectively relieves NASA of the need to support the resupply infrastructure, allowing it to focus more funds on operating ISS and other scientific missions.

NASA officials are already aiming to repeat the success of COTS and CCP for the Moon with the Lunar Commercial Orbital Transfer Services program (Zuniga et al. 2015), and its principles can also be applied to asteroid mining (Utrilla 2017). PPPs are not fully limited to COTS-like programs, as the successes of NanoRacks, Made in Space, or Bigelow's BEAM are other examples of a public agency supporting private development, in this case with contributions of knowledge or infrastructure. PPPs have the potential to unlock the development of the space industry further away from LEO by providing private companies with the incentive to develop a self-sustained space economy, and by providing public agencies with the infrastructure and services they require for advanced exploration efforts.

18.5 Conclusions

Commercialization of space is in essence a requirement for further advancement of humanity into space, especially with larger projects such as the MMB. The current political situation of public space programs cannot tackle the development of projects that require long development times or impose a high level of risk, which creates the need for space commercialization. Even though there is a history of bad approaches to creating commercial activities in space, public-private partnerships have shown promise in recent years, and we can identify a number of ways in which this approach optimizes the risk and cost distribution between the partners to deliver new capabilities in reasonable timeframes. All the PPP programs can be divided in two categories, with specific advantages and applications for each. These two models can help us define a commercialization strategy for the development of the MMB, which will be presented in the next chapter.

References

Allen-Ebrahimian, B. How SpaceX Launched a Chinese Experiment Into Space, Despite U.S. Ban., Foreign Policy (2017).

Berger, E. (2017, January). Technical troubles likely to delay commercial crew flights until 2019. *Ars Technica*.

China's Space Activities in 2016. (2016). Beijing.

Costello, K. (2017, January). Optical Fiber Production in Microgravity (Made In Space Fiber Optics).

de Lama, G. (1986, August). Commercial Use Of Shuttle Draws Fire. *The Chicago Tribune.*
Launius, R. D. (2012). Planning the post-Apollo space program: Are there lessons for the present? *Space Policy, 28*(1), 38–44. https://doi.org/10.1016/j.spacepol.2011.07.005
NanoRacks, Boeing to Build First Commercial ISS Airlock Module. (2017, February). NanoRacks.com.
Nonreimbursable Space Act Agreement between NanoRacks LLC and NASA for Operation of the NanoRacks System aboard the International Space Station National Laboratory. (2009, September).
Pike, J. (n.d.). EELV Evolved Expendable Launch Vehicle. GlobalSecurity.org.
Portree, D. S. F. (2012, August). Before the Fire: Saturn-Apollo Applications (1966). *WIRED.*
Portree, D. S. F. (2013, November). Moonbase Apollo (1968). *WIRED.*
Shear, E. Bigelow Aerospace B330 space hab to get ride on ULA Atlas V 552 in 2020. , SpaceFlight Insider (2016).
Simberg, R. (2013). *Safe is not an Option.* Jackson, Wyoming: Interglobal Media LLC.
Space Shuttle Missions Summary (No. NASA/TM–2011–216142). (2011).
Spudis, P. (2012, July). The Tale of Falcon 1. *Air and Space Magazine.*
Utrilla, C. M.E. (2017). Asteroid-COTS: Developing the cislunar economy with private-public partnerships. *Space Policy, 39–40,* 14–19. https://doi.org/10.1016/j.spacepol.2017.03.001
Winkler, R., & Pasztor, A. (2017, January). Exclusive Peek at SpaceX Data Shows Loss in 2015, Heavy Expectations for Nascent Internet Service. *Wall Street Journal.*
Zimmerman, R. (2003). *Leaving earth: space stations, rival superpowers, and the quest for interplanetary travel.* Washington, D.C: J. Henry Press.
Zuniga, A. F., Rasky, D., & Pittman, R. B. (2015). Lunar COTS: An Economical and Sustainable Approach to Reaching Mars. *AIAA Space 2015 Conference and Exposition,* 1–24. https://doi.org/10.2514/6.2015-4408

Part III
The Architecture of Global Collaboration on Planetary Defense

Chapter 19
The Overview Effect and Planetary Defense

Frank White

19.1 Introduction

The following chapter is a mix of personal experience and scientific research. Its validity has hardly been challenged as astronauts are going to space more and more often and experiencing the same *Overview Effect*. Living in our house for a whole life would leave us in our own shell during the time without significant interest of the outside world. People who travel to meet other cultures find their home in their culture of origin. However, living for months in orbit shows you a different picture of your home: it is a "Spaceship Earth," with an extremely thin atmosphere that protects us from all the threats the universe would otherwise deliver to the surface. From up there, you can see a blue sphere with clouds scudding past, the eyes of hurricanes, night cities everywhere, and even big "fireworks" visible from space as lightning storms. These moments are awe-inspiring and deliver an experience that humans on the ground will most likely not live through. The Overview Effect leaves astronauts with feelings about equality among people, the fragility of their home in the void of the universe and our role in it. If they leave the space profession, they continue to inspire others to go into space while explaining the reasons for their mind shift. What causes this effect? I would argue that it is a powerful experience, and that if more people could replicate it, political consensus to cooperatively keep our world safe from asteroids would be a different challenge. Regular travels to space will change us from *Homo sapiens* to *Homo cosmicus*.

F. White (✉)
Independent Writer, Nyack, NY, USA
e-mail: whitelight2@gmail.com

N. Schmidt (ed.), *Planetary Defense*, Space and Society,
https://doi.org/10.1007/978-3-030-01000-3_19

19.2 Homo Cosmicus

Space travelers who go into Low Earth Orbit or to the Moon have experienced a shift in worldview or identity that I have dubbed "the Overview Effect" (White 1987, 1998, 2014). They come to understand through direct experience, rather than through intellectual knowledge, that we live on a planet that is hurtling through the universe at a high rate of speed. They return from their missions knowing that all of us are, in a very real sense, astronauts on a natural "Spaceship Earth."

Here are a few examples of how the astronauts describe this shift in awareness:

> Don Lind: "You looked down and you could see how incredibly thin the Earth's atmosphere is and realize that if we pollute it, we all breathe it together, and if we are so dumb as to start a thermonuclear war, we all go together; there is no lifeboat, and everybody is in it together."
>
> Sandy Magnus: "It is all connected, it is all interdependent. You look out the window, and in my case, I saw the thinness of the atmosphere and it really hit home, and I thought, 'Wow, this is a fragile ball of life that we are living on.'"
>
> Ron Garan: "I felt for the first time on that mission that I wasn't just looking down at the Earth, I was looking at a planet suspended in space" (White 2014).

Astronauts bring many different insights back from their flights. However, one stands out with powerful relevance to planetary defense: *the Earth is fragile*. As such, it must be protected and defended, because it is the only place that is friendly to human life as far as the eye can see. Of course, the fact is that our planet has existed for 4.6 billion years and the geological/fossil record indicates it has survived cataclysmic meteor strikes, earthquakes, volcanic eruptions, and other events that are simultaneously awe-inspiring and frightening. Still, we have no reason to be complacent. The Overview Effect stands in direct contrast to the experience of human beings on the surface. Like our ancestors thousands of years ago, our senses insist that the planet is a stable platform with the heavens rotating overhead. Of course, we see the occasional meteor or "shooting star" flash through the night sky and burn up in the atmosphere, but there can be no doubt that we seem to be living in a safe place.

The astronauts view a dramatically different scene. They see the Earth against the backdrop of the infinite universe, a sea of stars lighting up the blackness of space. They circle the Earth every 90 min when in orbit, experiencing a sunrise and sunset every 45 min. As astronaut Jeffrey Hoffman puts it, "You see the Sun as a star" (Planetary Collective 2012). Those who traveled to the Moon had the most dramatic picture of the Earth. Although LEO is a very different viewing platform from that of the surface, it is still only a few hundred miles out, and the planet fills the visual field. From the Moon, or even on the way there, one sees the whole Earth and the universe beyond it. Decades ago, Carl Sagan was sitting in the control room and asked to turn around the Voyager to make one last photo of Earth. This picture from the distant reaches of the Solar System gave birth to the famous term "Pale Blue Dot," our fragile home.

19.3 Intellectual vs. Experiential Knowledge

The astronauts truly see the Sun as a star and the Earth as a planet. While we know that this is the reality of the situation intellectually, those of us who have lived our lives on the surface do not know it experientially. Moreover, the space fliers have an immediate recognition that the Earth is a fragile entity. The atmosphere, which seems so thick from the surface, appears impossibly thin from a distance. And yet, that thin layer is all that protects humanity and other life forms from an environment that is essentially hostile to organic entities. As former astronaut Ron Garan puts it, the Earth is a "fragile oasis." There are no guarantees that the planet might not be shaken to its core by internal developments like rapid climate change or external events like a meteor strike (Garan and Yunus 2015).

As Don Lind said, we are like a small lifeboat adrift in a vast and often hostile ocean (White 2014). The metaphor suggests that the lifeboat might sink at any moment, and that is certainly the case. It is also known that there have been "extinction events" in the past. For example, scientists are relatively certain that a meteor crashing into the Earth created such havoc that the dinosaurs lost their grip of dominance among living creatures and slipped quickly into extinction (or evolved into birds).

Rapid climate change has taken place in the past and will recur in the future, regardless of human activities. The ice core research of Mayewski et al. demonstrates that the Earth possesses the capacity to move from an Ice Age to a temperate climate (or back) in the space of just a few years (Mayewski and White 2012). More surprising is the fact that shifts of such magnitude took place in the past, before human beings had the ability to contribute to them. Today, the issue revolves, not around whether rapid climate change events occur, but rather what the human-caused component might be (Mayewski and White 2012). Changing the perspective through a cognitive shift, other authors of this book use the term "paradigm shift" when it comes to motivations behind political decisions. We have become powerful enough to influence the direction of these processes, and power comes with responsibility. If politicians had experiences similar to those of the astronauts, we might have very different situations regarding the policy of planetary defense.

19.4 Security Cosmopolitanism

In his seminal paper, Burke develops a theory of "Security Cosmopolitanism," which covers all of the threats to life on Earth, including those that might arise internally and those that might intrude externally. He convincingly argues that such a theory is necessary to confront the many insecurities now plaguing individual human beings and societies, as well as non-human lifeforms (Burke 2013). Burke offers a convincing argument for his theory, and to his credit, he also reviews some of the criticisms that have been leveled against it. Ultimately, though, the dialogue highlights the resistance of cosmopolitanism's opposite, which is nationalism. Burke strongly suggests that security on a planetary scale will only come when the cosmopolitan mentality becomes more widespread than the nationalist mindset. This is laudable, but perhaps the main flaw in the theory is that there is no clear plan for having this shift take place. That is where the Overview Effect comes in. The Overview Effect is an experience and a concept, and as such it has the potential to fulfill the need highlighted by Burke's theory. Burke argues that security cosmopolitanism will not happen, that it has to be imagined and implemented. The Overview Effect is a reproducible experience that has the power to change people's perspective, change their mindset, and cause the needed paradigm shift.

As Beaver points out, its greatest power lies in its experiential nature (Beaver 2016). Once one experiences the Overview Effect, it restructures one's worldview, and nothing is quite the same again. This is not to imply that every astronaut has an epiphany and devotes his or her life to working for world peace on their return to Earth. As I wrote in *The Overview Effect,* the shift is more subtle and nuanced than that. Subtle though it may be, can we doubt that the shift is away from nationalism and toward cosmopolitanism? We can imagine how cosmopolitanism might come to dominate if more people had the Overview experience. But how will that happen?

The options are clear: as commercial spaceflight comes online, we will be able to send so-called "ordinary citizens" into orbit and beyond. As virtual reality gains a foothold in the marketplace and costs come down, even more people will have access to a simulated Overview Effect experience. The astronaut experience has now been well documented over the 60-plus years of the "First Space Age." However, there is much work to do in order to "bring the Overview Effect down to Earth" and apply this insight to issues like planetary defense. The challenge is that the astronauts only represent a small percentage of the global population, and it is going to be challenging to translate their experiences into everyday reality. What, then, is the key to sharing this extraordinary realization with "ordinary people?"

19.5 The Overview Effect and Cosmopolitanism

One might suggest that the astronauts have directly experienced something that has become the basis for an important theory about the future of nation-states, national security, and the planet itself, i.e. cosmopolitanism. This is a multifaceted and multileveled concept, but its core belief is that all human beings on Earth are members of a common community with a common destiny (Burke 2013).

Theoreticians living on the surface of the Earth have articulated various theories based on cosmopolitanism, but astronauts have experienced the reality of it by seeing the planet from a distance, which shifts their identity. People tend to believe that they belong to a particular social group. However, when they experience anything outside of their last comfortable group, their identity begins to shift. The Overview Effect can bring the shift to the cosmopolitan level. Among the most frequent statements made by space travelers is the fact that the borders and boundaries we draw on maps are invisible from space. In a way, astronauts are the ultimate "Cosmopolitans," the ultimate pioneers of the worldview shift towards cosmopolitan identity. They can talk about it, teach it, and help others to experience it.

In fact, it seems safe to say that astronauts are "Experiential Cosmopolitans" while the rest of us are "Conceptual Cosmopolitans." The basic tenets of cosmopolitanism—we are all part of an organic unity and we are all in this together—have been stated over and over again by astronauts returning from their missions.

Moreover, this cosmopolitan point of view appears to transcend national differences, as demonstrated in several studies of the subject. A study at the University of British Columbia confirmed the transnational nature of the phenomenon. The researchers looked at a wide array of reports by astronauts from many different countries and found that the response to the spaceflight experience remained remarkably constant (Suedfeld et al. 2010).

19.6 The Need for Planetary Defense

The astronauts returned from orbit or the Moon with a clear understanding of our true situation. Human beings and all sentient life are riding through the cosmos in a natural spaceship that may be dramatically affected by events over which we have no control. Scientists are joining the astronauts' warning voices by noting that large asteroids are frequently passing us by, and we have no current plan of action if one of them were on a collision course with Earth.

There are two basic reactions to the realizations prompted by the Overview Effect, which should be examined much more closely in terms of their impact on the notion of planetary defense.

19.6.1 Reaction #1: The Earth is Vulnerable and We Need an Insurance Policy

This point of view is propounded most famously by Elon Musk, founder of SpaceX. It is the underlying rationale for his vision of settling thousands or even millions, of people on Mars. Musk talks about how it is a much more exciting future to contemplate if human beings become a spacefaring species and are out there exploring among the stars. This is the inspirational side of his message, but he also offers a more primal reason: we need an "insurance policy," a "Plan B" in case of an extinction-level event (Musk 2017). Musk was joined in this argument by renowned physicist Stephen Hawking, who said that humanity is unlikely to last another thousand years if we stay put.

Few would find fault in Musk's more positive vision of humanity roaming the stars as a multi-planet species, though there are still some who think we should stay on our home planet. These critics see space exploration as a diversion from solving critical problems on Earth, and this debate has been going on since the earliest days of the Space Age, when governments dominated the field and tax dollars were the principal means of financing space missions. The discussion is worthwhile and will certainly continue, but with the new impetus of private investment, the outcome is not in doubt: humanity is going to leave the Earth, and for many reasons, but making money will be among the most important. We can easily make a link to the age of the Renaissance here and ask how many people questioned the desire of sailors to circumnavigate the Earth? The Chinese had a great age of exploration much earlier, as did ancient civilizations. Centuries later, people are opposing space exploration for a variety of reasons. We have a massive opportunity we live in an age full of possibilities, greater freedom, powerful technology and the capability of people to create new inventions. Right now, we have the chance to trigger a new renaissance (Goldin and Kutarna 2016).

Though we may need Musk's insurance policy, the argument is subtle and needs attention. It is somewhat analogous to both life insurance and health insurance, and the distinctions are relevant. When it comes to life insurance, all of us know that we are going to die. However, we do not know when; it could be tomorrow or it could be in 30 years. Financially, the difference is significant. If I have a $1 million policy and I die tomorrow, having only paid $5000 into it, my family realizes a huge "profit," though we rarely think of it in that way. However, if I have a $1 million policy and I die in 30 years, having paid $500,000 for it, the profit is much less. Moreover, the scenario may differ in different societies. Some never provide you with the opportunity to obtain insurance, and their loved ones receive no benefits.

In spite of this uncertainty, we buy these policies because we are certain of the outcome and we want to provide for our families if we can. Health insurance is similar, but there are key differences. Similar to life insurance, we know we are likely to get sick and need the services of a physician, but we do not know when it will happen or how severe the illness (or accident) might be. Healthcare is provided in different ways in different countries, but one constant seems to hold true: no individual pays the actual cost of their healthcare.

In the United States, health insurance is either a benefit provided by an employer or an entitlement provided by the government. There may be copays and deductibles of various kinds, but almost no one pays the real cost of a doctor visit, unless it is for elective surgery or something similar. In most European countries, national health insurance covers all imaginable treatments. While the situation is the United States is somewhat anomalous, it points to something worth examining: because we do not pay full freight, it *affects our behavior*, which is one of the variables that drives healthcare costs up. For example, we might rethink a visit to the emergency room if the cost to us were $5000 instead of a $100 copay. How does all of this relate to Elon Musk, his "Martian insurance policy," and planetary defense?

We know there have been extinction events on Earth in the past, and there is likely to be another one in the future. In fact, we know for certain that our Sun will go supernova in several billion years, and everything on the planet, even the Earth itself, will die. Thus, planning for some of us to migrate is a bit like life insurance. We know humanity is going to die if we stay here, so we should have a plan to deal with it. The Mars idea is also like life insurance in the following way: I pay for life insurance, but I will not benefit from it, because there is only a return on the investment after I am gone. On the other hand, my family will benefit from it, because they will have the resources to carry on in my absence. Similarly, not all humans will migrate to Mars, but the species will benefit if even a remnant survives in the aftermath of an extinction event.

However, making these plans might affect our behavior around stewardship of the Earth, because we do have an option. Clearly, preserving our home planet becomes less of an imperative if there are alternatives. To be sure, establishing settlements on Mars or on the Moon would not be enough, in the near future, to save the lives of all seven billion people on Earth today. Even Musk's optimistic estimate of one million "Martians" pales in comparison with our planetary population. At the same time, seen from the perspective of our species, the extinction event now becomes a catastrophe of terrible proportions, but not the end of humanity.

By extension, this philosophy also has implications for the notion of planetary defense. In addition to the uncertainty that an asteroid strike, climate change, or something else equally dramatic will occur within our lifetimes, there is now the possibility of an "out." Establishing a presence on Mars will have a cost, and creating a planetary defense system will also incur costs. Given a choice, where should Earthlings invest? Should we do both?

19.6.2 Reaction #2: The Earth is Vulnerable and We Should Preserve It

I have talked with many astronauts in conducting my research on the Overview Effect and have formally interviewed 31 of them. So far, none of those interviewed ever said, "My experience in space tells me we should abandon the Earth and move rapidly out into the Solar System." In fact, astronauts who were interested in the environment before their missions came back with an even stronger concern for the planet's future. And those who were more technically inclined had much the same reaction. If there is any one response that characterizes the Overview Effect, a heightened environmental concern is the most common.

Additionally, none of the astronauts interviewed brought up planetary defense in terms of stopping meteors from hitting the Earth. This remains an issue that still needs to be brought to the forefront of everyone's consciousness. Once the issue is raised, it seems obvious and logical to imagine how it might be accomplished. What is standing in our way? The answer is also clear: planetary defense is, by definition, a planetary problem, and we are still governed by non-planetary entities called countries or sovereign states. In fact, planetary defense is not unlike many other challenges facing humanity today, from global warming to refugees and economic inequality to terrorism. Very few of these problems have a solution that can be confined to a single nation-state.

19.7 Two Approaches to Cooperation

There are two approaches to the question of how Earth's inhabitants might come together to defend the planet and human civilization.

19.7.1 International Cooperation by Countries

International cooperation represents a well-known approach and does nothing to disturb the status quo. Even the most ardent nationalist will likely concede that there are domains where cooperation with other states is in the national interest. The debate then focuses on the degree of cooperation and whether the benefits are equal or not. Trade represents an excellent example of this type of relationship. Countries know they must trade with one another, even when they are engaged in political conflict and even if they are governed by competing ideologies. Without trade, their economies would stagnate and their citizens would suffer. Countries that have tried to cut themselves off from the rest of the world, like North Korea, have paid the price for their isolation.

Climate change, which could itself be an extinction event, provides a case study in international cooperation that is both promising and concerning. Climate change is a natural process inherent to the planet Earth. Discussions about "fighting" it misses the point, because there is no proven methodology to stop it from happening. However, by coming together, the nations of the world might be able to mitigate the harmful effects of specific forms of climate change, including global warming. In fact, they were able to do so with the historic set of accords reached in Paris in 2016. Unfortunately, the United States later abandoned the accords.

Another example of international cooperation, even more successfully implemented, is the International Space Station (ISS). Dozens of nations have worked together since the mid-1980s to build and occupy this orbital outpost, and humans have been living and working there continuously for the past 15 years. During that time, the United States went through five presidential administrations and the Soviet Union fell apart. One cosmonaut, Sergei Krikalev, launched to the station as a Soviet citizen and returned a Russian. As this is written, we are coming to the end of the life of the ISS. The question hangs in the balance regarding what will replace it and whether international cooperation will continue as humans migrate into the Solar System.

These examples should suffice to demonstrate that international cooperation is a reality on planet Earth, and that it takes place without significant disturbance of the current world order. In theory, it should be possible to work on planetary defense within this structure, and perhaps it will take place. However, there is not very much evidence to date that the nations of the world consider the possibility of an extinction event to be very likely, and little action has been taken.

While some believe strongly that a global government is necessary to take the steps needed to protect the planet (see Chap. 24), examples of international cooperation suggest that there are other possibilities.

19.7.2 Transnational Collaboration in a Global Federation

A transnational collaboration is referring to something that does not yet exist, which would be a supragovernmental global authority. By analogy, it might exist in the form of a federal system, which is well known and used worldwide. In the United States, for example, the 50 states have certain powers and the federal government has certain powers, but the federal entity ultimately holds the ability to compel individual states to do its will.

For the planet as a whole, the countries would have to come together and surrender a certain amount of sovereignty to bring a global government into being. The planetary federal government would naturally have the responsibility of protecting the Earth as a whole, which would include planetary defense, however that might be defined. There might be struggles between the global and sub-global authorities, just as there is in terrestrial federal arrangements. However, those differences could be resolved through an institution like the World Court of Justice. While a federal, global government seems practical in theory, getting there from the current international system does not appear to be an achievable goal in the short term. Rather, it seems more strategic to lay the groundwork for a worldview shift that would make such an entity more acceptable to the people of the world—whether it became necessary to have it or not.

This is where the Overview Effect would exert its greatest power. The message that the astronauts have brought back from orbit and the Moon is that we are one species with one destiny. It is a message of unity transcending national boundaries that cannot even be seen from orbit. If more citizens of planet Earth began to adopt this mindset, international cooperation and transnational collaboration would be easier to both achieve and sustain.

When it comes to planetary defense, the result of the protected Earth should be the primary factor, not the institutional structure by which the result has been achieved. Thus, perhaps a shift in consciousness on the part of Earth's human population is more critical than the institutional forms that contain that consciousness. In other words, a "cosmopolitan mindset" is the *sine qua non* of planetary defense, regardless of the institutional frameworks it might foster.

For this reason, we should urge a commitment by those interested in cosmopolitan security to focus attention on creating this shift in consciousness. If we can make it happen, much will follow from it on more specific fronts like planetary defense.

19.8 Conclusion

This chapter introduced to the reader the idea that the cosmopolitan mindset is not wishful thinking, but a shift in thinking that astronauts have been experiencing for decades. Cosmopolitan thinking thus is not a dream but a practical, achievable possibility. If the theory has practical implications, there must be an option to implement it as a policy. Various chapters in this book speculate about which political institutions are the most desirable or what instruments we need to have to find that one asteroid destined to hit us. This chapter showed that these theories can be anchored within the space community, as those who experience space travel return as *Homo cosmicus*.

References

Beaver, D. (2016, March). The Case for Planetary Awareness. *Space Times*, 4–11. http://dev.astronautical.org/wp-content/uploads/2016/03/spacetimes_55-2.pdf

Burke, A. (2013). Security cosmopolitanism. *Critical Studies on Security*, *1*(1), 13–28. https://doi.org/10.1080/21624887.2013.790194

Garan, A. R., & Yunus, M. (2015). *The Orbital Perspective: Lessons in Seeing the Big Picture from a Journey of 71 Million Miles*. Berrett-Koehler Publishers.

Goldin, I., & Kutarna, C. (2016). *Age of Discovery: Navigating the Risks and Rewards of Our New Renaissance*. St. Martin's Press.

Mayewski, P. A., & White, F. (2012). *The Ice Chronicles: The Quest to Understand Global Climate Change*. University of New Hampshire Press.

Musk, E. (2017). Making Humans a Multi-Planetary Species. *New Space*, *5*(2), 46–61. https://doi.org/10.1089/space.2017.29009.emu

Planetary Collective. (2012). *Overview*. Planetary Collective. https://vimeo.com/ondemand/planetary

Suedfeld, P., Legkaia, K., & Brcic, J. (2010). Changes in the Hierarchy of Value References Associated With Flying in Space. *Journal of Personality*, *78*(5), 1411–1436. https://doi.org/10.1111/j.1467-6494.2010.00656.x

White, F. (1987). *The Overview Effect: Space Exploration and Human Evolution*. Houghton Mifflin.

White, F. (1998). *The Overview Effect: Space Exploration and Human Evolution*. AIAA.

White, F. (2014). *The Overview Effect: Space Exploration and Human Evolution*. AIAA.

Chapter 20
Cosmopolitan Rationale for Planetary Defense

Nikola Schmidt

> The greatest problem for the human species, the solution of which nature compels human to seek, is that of attaining a civil society which can administer justice universally.
> – Immanuel Kant

20.1 Introduction

The aim of the following chapter is to introduce a reader to the origins and the most recent theoretical contributions to cosmopolitan theoretical thought (Immanuel Kant, Ulrich Beck and Anthony Burke). It shows how the thousand-years-long tradition of cosmopolitan thinking evolved through Immanuel Kant's enlightened, practical realization that cosmopolitan security thinking is a *strategic necessity* (Burke). Not thinking in this manner means we are caught in the *prison error of identity* driven by *national outlook* (Beck). Planetary defense is not about how high the probability is, as one missed asteroid or sudden comet can cause havoc beyond imagining. Rather, planetary defense is about a mindset that has the potential to solve many other global-scale challenges. Cosmopolitan thinking is once again coming back to the political fore because the (neo)realist conviction about the natural character of nation-states divided by borders looks bizarre under the light of such worldwide issues. Planetary defense, given its binary threat character (the object is, or is not, on a collision trajectory) is a good exercise for politically organizing humanity according to cosmopolitan norms and principles.

As discussed in the previous chapter, astronauts see our planet without borders and are coming back to Earth with a broader view of our destiny. In scientific terms,

N. Schmidt (✉)
Department of Political Science, Institute of Political Studies, Faculty of Social Sciences, Charles University, Prague, Czech Republic
e-mail: nikola.schmidt@fsv.cuni.cz

N. Schmidt (ed.), *Planetary Defense*, Space and Society,
https://doi.org/10.1007/978-3-030-01000-3_20

planetary defense is not about asteroids only. It is also about a coronal mass ejection that will most likely happen in our lifetime, causing damage to satellites; this likelihood of occurrence also applies to other yet-to-be-unveiled, space-related threats. In philosophical terms, planetary defense is about the perspective we are willing to take, a responsibility we recognize—our ingenuity to propose solutions and realize that cosmopolitanism is the political thinking best suited to mankind's survival and flourishing. It is not naïve; it is necessary.

What I have found fascinating about the planetary defense (PD) efforts has a clearly cosmopolitan aim of the passion of the community in answering the question of what humanity *can* do, as we have reached a certain level of knowledge, and *should* do, once we know how to answer a threat of an asteroid on the way. However, each text I have submitted in the last 3 years on planetary defense was criticized based on two basic arguments. First was that we should use the most effective deflection method of nuclear ablation. Second, that the argument for creating a cosmopolitan authority or cosmopolitan global system to deal with planetary defense will never be accepted because the planetary defense community has already spent three decades struggling to explain its agenda to national governments. How, then, can one person argue that we need to build a cosmopolitan body?

Saving humanity from asteroids is naturally a cosmopolitan practice, and cosmopolitan theory has a lot to say on the matter. We can talk about which version of cosmopolitanism, which institutional configuration or whichever system we mean, but our reasoning must always be driven by cosmopolitan thoughts. The beauty of planetary-defense efforts is in their unquestionable features that render nonsensical the notion of managing PD as a national-security issue.

The question of what political system is required for planetary defense to be effective should be approached from a complex and long-term view that maximizes the chosen system's legitimacy and sustainability. As has been argued, planetary defense is cosmopolitan in its nature, and the PD community is thus already working on a cosmopolitan agenda. We need to explore the type of political environment that planetary defense can help establish, rather than creating the simplest reactionary political system within the business-as-usual format of global politics, which can carry many negative consequences.

Cosmopolitanism is not redemptive; it is theoretical thought, not a political system. The necessity of establishing a political system based on cosmopolitan thought must not be the only consideration, but it should be accompanied by a wider pholosophical debate on its effectivness, potential and impact. Moreover, the debate should not be strictly utilitarian but should consider the legacy of political philosophy before proceeding to the discussion of concrete institutional structures. In a nutshell, we should perceive the process towards the cosmopolitan political system as twofold. First, planetary defense is already doing something; *something has been done* in terms of securitization theory as the size of the budget reflects the impact of the speech acts by scientists, with the US pouring $150M a year into a Planetary Defense Coordination Office, and the global epistemic-scientific community becoming stronger and stronger. As a result, planetary defense will unintentionally create a new political environment through its current scientific efforts, for example, by destigmatizing nuclear technology or by building super-powerful lasers. Cosmopolitanism is the best guide to tell us which direction is the right one in various contexts (see Chap. 14). Second, we should think in advance and steer this process toward cosmopolitan ends in our current realist world in order to see the desirable institutionalization and perceive this process of building a cosmopolitan political system not as a redemptive one but as the best one we can imagine. Only then we can avoid the unintended negative political consequences.

20.2 The Moral and Ethical Basis of Cosmopolitanism

Cosmopolitanism can have many meanings. Cosmopolitanism first emerged in the writings of Diogenes (413–327 BCE), who claimed himself to be a citizen of the world (kosmopolitēs). He meant that the morally good are *all friends* and that there is no moral distinction between people, nations or cities, because we all are part of a fraternity of humanity and belong to a single species. Roman Stoics later linked the identities of the universe (cosmos), reason (logos), law (nomos) and citizenship (cosmopolitein): we are part of the universe and can reasonably establish a cosmopolitan citizenship stipulated by law. Stoics disconnected the fate of humanity from God's will, instead connecting it to reason and to the political and legal system, which itself comes with a certain ethical and normative framework.

This framework was later tied to natural law when it fused with Christian universalism during the Medieval Age. By then, it was understood as the *rational offspring* of *universal morality*. There is moral equality, peaceful co-belonging or just war between people and nations. Based on these universal moral presumptions, cosmopolitanism has several central ideas. First, the primary unit of moral concern is an individual human. Second, all people are equal, meaning that the status of each human is equally applied everywhere, and the third idea results from the previous two: as all people are equal, they are necessarily part of a bigger picture, a global civil society—the citizens of the world (Beardsworth 2011).

Christian von Wolff argued in the eighteenth century that if people understand these normative rules as natural law, we are necessarily heading toward a world-state consisting of all nations (Brown et al. 2002, pp. 372–383). However, a world-state in many cosmopolitan writings does not necessarily replace nation-states. For many thinkers, including Kant, a world-state, or federation of nations in Kant's words, will only create universal law above local communities. This chapter differs from Pavel Dufek's work because it aims to discuss what ideas should be allowed and what political structures are then necessary to develop a reliable planetary defense policy, while Dufek's approach is to argue why the world-state is unavoidable (Dufek 2013).

This difference is visible even between strong and weak cosmopolitanism, however, the argument against the world-state is relevant because the menace of a global totalitarian state does not necessarily need to be far away (Zolo 2000). Despite the war in Syria, unending conflict in Afghanistan or resurging human disaster in the

Congo and elsewhere in Africa, we are experiencing an unprecedented peaceful period in history and in current international relations. The state of global peace, integration, globalization and unification shows indications of an emerging cosmopolitan society in Held's terms (Held 2000). Yet the threat of tyrannical rule has greatly increased in the contemporary global political system due to democratic deficits and legitimacy issues. The democratic dilemma between effective and democratic governance systems has been surfacing more frequently (Held 2000).

The most famous philosopher to have developed cosmopolitan thought for current modern reflection was Immanuel Kant. He detached the cosmos from the rule of law and introduced the human as a cognitive agent. For Kant, the human is free to make decisions and fully understands moral consequences. Our rational will is responsible for our awareness of when we do something wrong or when we admire political virtue. Kant recognized the moral, legal and political conditions of world justice. He based his argument on a holistic presumption that everything is interconnected and fully interdependent. A legal system securing some nations and diminishing the freedom of other nations is not morally bearable and is prone to conflict. In such a system, each involved community would presume its interests as morally right, which is an impetus for war.

The space community would do well to remember this. The reasons for dropping the Code of Conduct in Space probably was based on fear by the US that it was signing a binding, though not necessarily "gentleman," confidence-building agreement (Listner 2015). A clash between more pro-cosmopolitan European and more pro-nationalist US perspectives was the reason behind the collapse of the Code of Conduct, though both sides had the same interest—to secure outer space. This can be achieved by a centralized administration overseeing an adopted regime, not by a clash in which one group seeks to secure its rights against the others because it already did something for the security of outer space while others did not. Following a national policy to dominate in space that does belong to mankind is a path to war, regardless of all the efforts the US has put into space debris mitigation or will put into planetary defense. A nation that wishes to lead should seek respect and "being worth to be followed," as opposed to dominance (Johnson-Freese 2013). We should work toward a cosmopolitan authority that treats all equally and thus administers justice universally, which is not going to be an easy task because the universal justice is the greatest challenge to the human species (Kant 2010).

For Kant, these aforementioned dynamics are visible to us all through our reasoning. It is our reasoning that wants us not to live in conflict but in cooperation. Therefore, our final objective is to establish a global civil society under a universal law, in the kingdom of ends. Moral reasoning knows nothing about the working of the cosmos, but it is the rationale, written in speculative ideas, which imagines practical outcomes (Beardsworth 2011). Kant was confident about the universal moral principles tied to our reasoning. In moral pluralism, we accept the fact that there is more than one universal moral law. This is likely the reason why Kant linked his thinking to the New Testament, because it gave him the basis for his argument, anchored in centuries-long moral principles, that rationality leads to a working order, and thus the claim that a certain moral-behavior framework is reasonable. However, moral pluralism inevitably emerges when universal moral principles include everyone's interests.

Kant argued that all people should have the same rights. A human-rights regime that was developed after World War II and affirmed during the Helsinki Accords in 1975 led to fall of the Communist regime in Easter Europe, mainly in Czechoslovakia in 1989, because the regime lost its legitimacy. People like Václav Havel, the first president of post-Communist Czechoslovakia and later the Czech Republic, had been patiently and inexorably questioning the moral foundations of a regime that signs a treaty and does not follow it. His strenuous questioning of the regime put him in jail, but he never ceased to claim that "truth and love will win" because he believed that no other values are more important than sticking to our words. They were the pioneers of the modern age, ready to be imprisoned to stay morally integral to the global normative order they established. Without this global normative order, we would live in anarchy. It is not the international legal system that binds all actors in an orderly manner; it is the normative order based on moral principles that keeps the world in relative peace and prevents full-scale war. The international legal system is just a building block of that normative regime (Bull 2002). If neorealists argue that we live in anarchy, cosmopolitans would align with scholars such as Alexander Wendt and argue that we live in the world we have built, and we should strive for a morally and normatively integral world—a global community of equal people who respect each other.

Even in moral pluralism, we can recognize the universal good in cases where all are involved. In the planetary defense community, all are for example involved in the survival of humanity; however, in the preparation phase and in the phase of certain technologies' deployment, the situation is not the same. In the observation phase, in which no nation threatens the others but cooperates on Solar System observation, we would hardly recognize questionable behavior. Arguing for the necessity of nuclear warheads, operated by nuclear superpowers in case of an Earth-defense emergency, is a completely different situation; claiming that they are needed to save humanity does not change their (stigmatized) normative role at all.

In the end, the ability to overcome socially constructed differences between us will help us to act practically. Ethical analysis of cosmopolitanism as a way of thinking provides an explanation for why any actions in planetary defense that are solely triggered and carried out by one nation, without discussion or consent given by the others, are morally unbearable; such actions exclude all other communities from the political discussion. "Be prepared for a unilateral action" sends the wrong message.

The planetary defense community's core argument remains truly cosmopolitan in nature and has the potential to change our perspective of our role on the Earth—a change toward becoming *homo cosmicus*, perhaps? Let us think big, leaving the prison of national perspectives—the national outlook—and take the cosmopolitan outlook.

20.3 Beck's Cosmopolitan Outlook

Neorealists think that anarchy between states is natural law. They cannot be more wrong, as states are not the natural effect but rather the political end of human intentions and the most recent achievement in a restricted space in which states can freely exercise their "sovereign" politics. Planetary defense shows that exercising politics must and inherently will be global, with no territorial limits. Sovereignty of a nation-state is not a legitimate foundation of its existence but a relic from a time in which leaders designed the world as a community of nation-states because they could not govern the whole world by practical means. The Peace of Westphalia in 1648 divided Europe into sovereign territories ruled by sovereign leaders and stipulated the norm of non-intervention. Today, we live in a different world, one that calls for cosmopolitan answers.

Ulrich Beck developed his concept of risk society (Beck 1992) in the 1990s and argued that we are not heading toward a postmodern, barely discernible, fluid reality of liquid modernity (Bauman 2000), but into a second modernity in which the field of gravity will focus on global threats. These are completely different perspectives. They clearly differentiate between negligent political thinking that will lead to a massive empowerment of global corporations standing above "sovereign" states, incapable of regulating their activity and eating away at the freedoms of citizens in a dystopian future (postmodernism), and thinking that will lead to morally responsible global governance in which humanity realizes that certain challenges will not be addressed without considering an overall perspective of our shared fate (second modernity).

According to Beck, the first modernity was the establishment of nation-states as political structures following the Peace of Westphalia in 1648 (Beck 2006). This first modernity is signified by *territorial prisons of societies* divided by erected political borders. The emerging socially constructed territories over physical surfaces of Earth are then defended by states that encourage the notion of nationality to gain a socially constructed legitimacy. They do so even across various ethnically mixed societies (Afghanistan is a good example) unwilling to be subordinated to the central government. In other words, it is not the society that creates the state, but rather, the state that creates the notion of nationality. Beck calls this dynamic *methodological nationalism* driven by national outlook. The second modernity differs from the first modernity in that states are able to reflexively react on global threats and understand the need to solve these global threats cooperatively; they will develop a consciousness of global responsibility, a cosmopolitan outlook, because "in a radically insecure world, all are equal and everyone is different" (Beck 2006).

The cosmopolitan outlook, according to Beck, is not only a presupposition to act, but also "the result of a conceptual reconfiguration of our modes of perception" (Beck 2006). The cosmopolitan outlook can be constructed through cosmopolitan reality. This new reality, arising from a new perspective, uses a new syntax to respond to the notion of a national ontology, carved in imaginary stones, that is not able to help us solve the most pressing global problems. Building this new perspective, the cosmopolitan outlook, and deconstructing the inner practicality of the national divisions in territorial prisons produces what Beck calls *cosmopolitanization*.

This new perception, a sense of the world's boundarylessness, is the experience and feeling that astronauts talk about when they watch Earth from orbit. We talk about the same sense that every single astronaut experiences as a result of their perspective from the orbit, discussed in detail in Chap. 19. Astronauts have the chance to inspire not only the younger generations to consider space exploration, but also the broader political community to make bold cosmopolitan decisions. Those who have observed every single launch of Falcon 9 might experience similar moments watching the first Block 5 version of the rocket on its way to geostationary orbit: live coverage shows, in a couple of minutes, how a cape disappears between an ocean and a continent, how the rocket leaves the Earth's atmosphere and then how the Earth becomes round, and finally small, in the dark of the universe. Permanent human presence in space, on the Moon or Mars, has the potential to strengthen this sentiment in broader society.

For Beck, cosmopolitanism is not a political agenda to deconstruct nation-states, but rather the contrary. He argues that "cosmopolitanism without provincialism is empty and provincialism without cosmopolitanism is blind." Beck argues that we have to recognize cosmopolitan differences, which are caused by differences between cultures and identities; unifying a variety of societies into one world society will result in a compulsion to redraw old boundaries. Beck talks about the *mélange principle* in which "local, national, ethnic, religious and cosmopolitan cultures and traditions interpenetrate, interconnect and intermingle" (Beck 2006). Perceiving ourselves divided according to boundaries of nation-states, within a territorial social ontology of national outlook, does not bring us solutions on the pressing global challenges; this national outlook is what Beck calls a prison error of identity (Beck 2006).

Cosmopolitan thinkers can be divided into weak and strong cosmopolitans. While the latter tend to argue for full-scale egalitarian society on a global scale, the former argue for basic rights and duties that are universal because we do have different levels of relationships to other people around the world. Strong cosmopolitans would say that there is no native or foreign person, that we live on a *spaceship Earth*, which is simply the only one in our Solar System. The concept of spaceship Earth, widely used in the astronaut community, includes O'Nell's concept of *lifeboat Earth*. In her article, O'Nell argues that to not avoid death is to kill (O'Neill 1975), which is the general motivation of humanitarian aid. We live on one lifeboat Earth, which comes with a responsibility to others. The motivation to save Earth from asteroids is no different—we possess knowledge of the threat and the means to answer it—and we have a responsibility to act. Spaceship Earth was also a term used in Richard Buckminster Fuller's *Operating Manual for Spaceship Earth*, in which he describes our role from a cosmic perspective. The text lucidly illustrates the claim that we not only have a responsibility to act, but we also have the ability to save Earth from ecological collapse; that oneness is necessary for our ultimate survival (Fuller 1969). The 2015 blockbuster movie *Interstellar* made a similar argument, claiming that we need to think as a species or we will simply perish. All of these arguments are a continuation of the Enlightenment philosophy in pictures, imaginations and policy solutions for current global struggles.

Held (2010) makes several critical observations regarding national democracies, pointing out why these systems are a poor channel for global challenges. One of them is the short-termism of representative democracy. The 4-year electoral cycle of the US, for example, does not let us address bigger issues that require forward-looking, long-term, politically unappealing solutions, delivering benefits outside of their political mandate. Thus, the current democratic system removes any rational motivation to undertake these usually tough and expensive steps. This specifically applies to planetary defense, since focusing on human survival and addressing a future asteroid collision requires prioritizing the future. This is problematic for the regular voter as well as for political representatives working within time-limited mandates.

Nevertheless, while continual economic and political interdependence resulting from deepening globalization is making the world more culturally flat, there is not a continual integration of global governance. In the case of planetary defense, the whole of humanity needs to realize it is equally at risk. The fact that humans can be wiped out makes the cosmopolitan outlook on planetary defense policy much easier to imagine and revive. The cosmopolitan outlook can constructively support the moral necessity of developing a global planetary defense policy that emanates from each nation-state's interest as a product of national policies driven by the second modernity (Beck 2006). However, if policy is implemented on a national level with the intention of threatening others, we will witness the most egregious case of extreme-threat politicization with countereffective impacts on global security cosmopolitanization.

Planetary defense is not just an asteroid-collision-avoidance mission that can be studied by science, solved by engineering and somehow signed into law by enlightened politicians. Planetary defense is the kind of multidisciplinary issue where all disciplines need to understand each other for the betterment of humanity: to keep us secure from newly understood threats; to allow us to become a multiplanetary or even interstellar species; to help us thrive and make space for big ideas; to help us see how cooperation is crucial to living in peace; to encourage us to take the cosmopolitan outlook in order to address newly emerging challenges; to persuade us to leave our prison errors of identity; to enable us to adopt consensual policies and global governance models; and to reflect the fact that what we face here is a cosmopolitan reality.

20.4 Burke's Security Cosmopolitanism

In 2013, Anthony Burke published an appellative text in which he claims that his intentions are to "set out a theoretical, ontological, and ethical architecture for security cosmopolitanism" (Burke 2013). Burke is convinced of the necessity to act on a global scale to deal with emerging global threats: climate change, displaced people and refugees, nuclear proliferation and interstate tensions. Security cosmopolitanism builds unintentionally on Ken Booth's emancipatory security, developed in detail in *Theory of World Security* (Booth 2007). However, while Booth argues that there is no other option than to rely on *emancipatory global governance* when we need to solve global security threats, Burke constructs a *global security imperative* based on three ethical principles of security cosmopolitanism inspired by Kant's categorical imperative.

In Booth's terms, humanity needs to think about threats as threats to their potential to flourish. The absence of a threat by disputable means might help us to survive, but it might result in a world in which we might not want to live. This is exactly the reason why planetary defense has to be approached in a cosmopolitan way: first, to prevent negative political consequences, and second, to allow for a broader emancipatory and inclusive conceptualization of security beyond survival (e.g., surviving an asteroid impact) to other threats like coronal mass ejection or orbital debris. Booth's cosmopolitan security includes the assumption that while security is socially constructed, security is not political in its nature but in its practice. This means pursuing security does not always need to be in conflict with other policies, or as the Copenhagen School of security says, essentially contested, if we look beyond its constituting politics. Security is thus contingently contested (Booth 2007; McSweeney 2013). This means that security policies can be compatible or complementary and not in inherent conflict with the national interests of other states as long as the policies incorporate a cosmopolitan perspective. This is the moral basis of caring about collisions with extinction-level asteroids, city-killers or smaller objects of negligible impact on a global scale as well as issues of orbital debris.

The difference between the impact of various asteroids is linked to the idea of the concentric-circles model of obligations discussed by Linklater (2002), where obligations to fellow citizens have priority over those toward the whole of humanity (Linklater 1998, p. 105)—an argument of weak cosmopolitans against global egalitarianism. Anthony Burke's security cosmopolitanism replaces this concept with the networks perspective, in which we have the opportunity to take preferential care of our fellow citizens; since human existence is one of interlinked and interdependent reality, this fellowship extends to all humanity, as it is hard to say who is a fellow citizen and to what extent we depend on each other. A centuries-old sentiment common to many cultures suggests that we should be good people before we ask others to follow suit, and if we are good, we do not need to ask them or convince them. Such a strategy is fruitful: others will follow in our footsteps, and good brings good. The idea is not one of deconstructing nation-states but of constructing a new global society and promoting the concept of factual interdependencies (Burke 2013) or communities of shared fate (Held 2010). It also goes beyond the discussion of global egalitarianism criticized by Nagel (2005) and opens up the debate about inevitable global challenges to humanity as a whole.

Texts such as Anthony Burke's article Security Cosmopolitanism leads us to post-international politics, which is not centered on (isolated) nation-states (Grande and Pauly 2005) but which tells the nation-states that they have a responsibility beyond their borders and can form global governance structures based on moral commitments—every action matters in this mission. Burke sees the opportunity to reduce insecurity across mankind through security cosmopolitanism, in which three transformations have to happen. First, the transformation of identity-forms and policies of states. Second, the transformation of norms emanating from moral commitments, leading to structures and operations of global governance. Third, the reflexive transformation of an actor's perception of their actions, including states, corporations or non-state actors that could have an impact on the security of people and ecosystems (Burke 2013).

One could argue, especially the realists, that only states have the necessary power to change global policies. Yet Trump's withdrawal from the Paris Agreement shows us that states will not necessarily play a significant political role in solving global challenges if they refuse to act. The billionaires of American cities have demonstrated their readiness to face climate change regardless of the will of the federal government. Elon Musk founded the company SpaceX with the idea of saving humanity in case of an extinction event, and he repeats this rationale every chance he gets (Musk 2017). Yuri Millner funds Breakthrough Initiatives, throwing resources behind the search for extraterrestrial intelligence and sending nanoprobes to Proxima Centauri using a super-powerful laser. Mark Zuckerberg seems to be interested enough to pay for the laser's construction. These are just a few of the philanthropic projects seeking breakthroughs in mankind's destiny.

Power, as studied thoroughly by Arendt (1972), pp. 36–37, cannot be measured by military capabilities only; power is an instrument of rule. Using Voltaire's terms, power "consists in making others act as I choose;" similarly, Clausewitz reminds us that power figuratively signifies "power of a man over a man" (Strausz-Hupé 1956). Arendt was one of the first to openly say that power does not necessarily imply violence or military force. Regarding Burke's security cosmopolitanism, power in cosmopolitan terms means a readiness for actions that are powerful enough to change the world toward the higher good—a *cosmopolitan consciousness*, an intentional adoption of a cosmopolitan outlook and our cosmopolitan duty.

Here, the argument relates to two ethical obligations regarding positive and negative duties (Shapcott 2010). In the positive duty, actors are obliged to act and answer particular moral calls. The positive duty of a state can be the obligation to save its citizens from terrorist captivity. Based on cosmopolitan values, such a positive duty has developed into the norm of Responsibility to Protect (R2P). Under R2P, states are obliged to act even against other states that fail to provide security to their citizens—in this example, to prevent the captivity of its own citizens. The R2P concept has become a controversial policy because it violates the sovereignty of other states, and the results of military humanitarian interventions have not necessarily led to a clear, successful end-state. If one dies unnecessarily, all broader intentions are canceled out.

The move from taking care of the citizens of one's own state to all citizens—global citizens with no regard to their nationality—is an important milestone in history. What is quite moving in Burke's article is the clear break from classical cosmopolitan thought, focused on mankind, its wellbeing and global justice, to his call that we should respect the entire biosphere. During the development of this volume, there has been ambivalence about this. We would like to stipulate that our reference object here is the biosphere as a higher entity above the humanity. We sometimes reference the fate of humanity; however, it is the obligation of humanity to reflect on our own behavior and how this behavior resonates in the fate of biosphere that we are inherently a part of.

Burke does not share the idea of Beck's enlightening function of cosmopolitanization and risk-cosmopolitanism that "generates an unavoidable pressure to cooperate"(Beck 2006). On the contrary, Burke argues that "security cosmopolitanism is not going to arrive [but] it must be imagined and created with a combination of creativity, agency and moral and strategic caution" (Burke 2013). This is not necessarily in conflict. In the case of climate change, the politics to keep temperatures below a certain threshold need a cosmopolitan outlook and realization that current human activity will burn us alive if we do not act to slow or reverse climate change; the science needs to confirm the process. From this perspective, the cosmopolitan outlook is an intentional imagining of strategic caution based on accumulating proofs. The asteroid impact works quite a different way—a clear imagining that, given the number and character of asteroids around us, one single event can occur. The possibility of such an extinction event needs to be imagined without a single confirmed detection of an asteroid on an impact trajectory; the cosmopolitan outlook is what the community has taken for granted and calls for a global action before we miss this one rock (most likely) on the way to hitting us. An asteroid collision has already been socially constructed through the cosmopolitan outlook as a global-scale threat that should unite people around one table. In the development of the moral imperative to act, statistical-risk analysis is not relevant here in the development of the *moral imperative* because one asteroid can cause a game over scenario, in the same way a lottery jackpot is won from time to time. The moral conviction to act, to use the available and emerging scientific knowledge to find the solution, is what should drive the community. Starting with a proven asteroid on the way does not work (the future needs to be imagined, as Burke argues); in the case of climate change, the scientifically proven climate dynamic is a required condition in order to take action (it generates unavoidable pressure to cooperate, as Beck argues).

Burke precisely combines technological, political and strategic choices. In planetary defense, this translates to questions such as, what technology would we like to have in space in order to deal with the asteroid threat? What policy and strategic alliances do we need to proceed without causing political distrust of these actions but rather, to deepen wider political confidence? As we mentioned at the beginning, Burke has drafted three important ethical principles that should underline global ethics of security cosmopolitanism (Burke 2013):

1. "The responsibility of all states and security actors is to create deep and enduring security for all human beings in a form that harmonizes human social, economic, cultural and political activity with the integrity of global ecosystems."
2. "All states and security actors have fundamental responsibilities to future generations and the long-term survival of global ecosystems: to consider the impact of their decisions, choices and commitments through time."

thus illegitimate. National democratic systems already do not ensure lawful and wise application of such power, and the current sovereignty-based system does not ensure defense against hegemony (Archibugi 2004). There is no reason why we should assume that the modus operandi would change in planetary defense. Further, if planetary defense truly considers human survival as its goal, it needs to adopt a broad, inclusive understanding of security that is only possible through the cosmopolitan approach and realization that mere survival is not the goal (see Chap. 14).

The pro-cosmopolitan effect of planetary defense efforts is important for two connecting reasons. First, to deflect an asteroid will require a sound, multilateral global consensus regardless of whether a nuclear method is used to avoid negative political consequences. Second, a cosmopolitan approach to security is required if planetary defense is to be effective and reflexive of its epistemic qualities. However, contrary to the argument of Dufek's Chap. 24 calling for a world government, Petr Bohacek and I argue in Chap. 21 that a global government is not necessary in this scenario, but rather, that a broadly inclusive Planetary Council should be established.

20.6 Conclusion

This chapter introduced various cosmopolitan thinkers and their ideas to explain why planetary defense is clearly and inherently a cosmopolitan issue. The biggest mistake would be to treat planetary defense as another scientific challenge to be solved by the most effective method. Planetary defense has the potential to become a substance for a great global political change mainly because the problem is binary: the asteroid is or is not on a collisions course. This advantage leaves behind any discussions of whether the problem is here or not. The point is that we do not need to detect an asteroid but instead need to realize that these events occur from time to time. We need to understand that it may happen and that we have the capability to prepare and take action.

If we imagine what can be achieved (not necessarily only mitigated), we are imagining a future in Burke's terms, where we accept the fact that we as a species can act to avoid cataclysmic event. It is not a challenge—it is more of an opportunity to take Beck's cosmopolitan outlook and act as one. We must forget the socially constructed nation-state borders and rationally follow Kant's moral imperative. This is not only a scientific mission or an opportunity to change the global political landscape; this is moral imperative, a strategic necessity, Burke's *global categorical imperative*.

Acknowledgment This study was supported by the grant awarded by the Technological Agency of the Czech Republic, project TL01000181: *"A multidisciplinary analysis of planetary defense from asteroids as the key national policy ensuring further flourishing and prosperity of humankind both on Earth and in Space,"* and co-funded by the Institute of Political Studies, Faculty of Social Sciences, Charles University in Prague.

References

Archibugi, D. (2004). Cosmopolitan Democracy and Its Critics: A Review. *European Journal of International Relations*, *10*(3), 437–473. https://doi.org/10.1177/1354066104045543

Arendt, H. (1972). On Violence. In *Crises of the Republic* (pp. 105–198). San Diego, New York, London: Harcourt Brace Jovanovich.

Bauman, Z. (2000). *Liquid Modernity. Contemporary Sociology* (Vol. 30). https://doi.org/10.2307/3089803

Beardsworth, R. (2011). *Cosmopolitanism and International Relations Theory*. Polity.

Beck, U. (1992). Risk Society: Towards a New Modernity. London: SAGE.

Beck, U. (2006). *The cosmopolitan vision*. Polity. https://doi.org/10.1007/s13398-014-0173-7.2

Booth, K. (2007). *Theory of World Security*. Cambridge University Press.

Brown, C., Nardin, T., & Rengger, N. (2002). *International Relations in Political Thought: Texts from the Ancient Greeks to the First World War*.

Bull, H. (2002). *The Anarchical Society*. Columbia University Press; 3rd edition.

Burke, A. (2013). Security cosmopolitanism. *Critical Studies on Security*, *1*(1), 13–28. https://doi.org/10.1080/21624887.2013.790194

Dufek, P. (2013). Why strong moral cosmopolitanism requires a world-state. *International Theory*, *5*(02), 177–212. https://doi.org/10.1017/S1752971913000171

Fuller, R. B. (1969). *Operating Manual for Spaceship Earth*. Pocket books. http://designsciencelab.com/resources/OperatingManual_BF.pdf

Grande, E., & Pauly, L. W. (Eds.). (2005). *Complex Sovereignty*. Toronto: University of Toronto Press. https://doi.org/10.3138/9781442684201

Held, D. (2000). The changing contours of political community: rethinking democracy in the context of globalization. In B. Holden (Ed.), *Global Democracy, Key debate* (pp. 7–31). London and New York: Routledge.

Held, D. (2010). *Cosmopolitanism: Ideals and Realities*. Polity.

Johnson-Freese, J. (2013). *Space as a Strategic Asset*. New York: Columbia University Press.

Kant, I. (2010). Idea for a Universal History. In D. Held & G. W. Brown (Eds.), *The Cosmopolitan Reader* (pp. 17–26). Polity.

Kant, I., & Wood, A. W. (2002). *Groundwork for the Metaphysics of Morals*. Yale University Press.

Linklater, A. (1998). *The Transformation of Political Community: Ethical Foundations of the Post-Westphalian Era*. University of South Carolina Press.

Linklater, A. (2002). Cosmopolitan Political Communities in International Relations. *International Relations*, *16*(1), 135–150. https://doi.org/10.1177/0047117802016001009

Listner, M. (2015, October 16). The International Code of Conduct: Comments on changes in the latest draft and post-mortem thoughts. *The Space Review*. http://www.thespacereview.com/article/2851/1. Accessed 10 January 2016

McSweeney. (2013). Security, Identity and Interests. *Journal of Chemical Information and Modeling*, *53*(9), 1689–1699. https://doi.org/10.1017/CBO9781107415324.004

Musk, E. (2017). Making Humans a Multi-Planetary Species. *New Space*, *5*(2), 46–61. https://doi.org/10.1089/space.2017.29009.emu

Nagel, T. (2005). The Problem of Global Justice. *Philosophy & Public Affairs*, *33*(2), 113–147. http://www.jstor.org/stable/3558011.

O'Neill, O. (1975). Lifeboat Earth. *Philosophy & Public Affairs*, *4*(3), 273–292.

Shapcott, R. (2010). *International Ethics: A Critical Introduction*. Polity. http://books.google.com/books?id=GhBEsRunj54C

Strausz-Hupé, R. (1956). *Power and Community*. Praeger. https://books.google.fr/books?id=Tp4dAAAAMAAJ

Zolo, D. (2000). The lords of peace: from the Holy Alliance to the new international criminal tribunals. In B. Holden (Ed.), *Global Democracy, Key Debate* (pp. 73–86). London and New York: Routledge.

Chapter 21
Dawn of Cosmopolitan Order? The New Norm of Responsibility to Defend Earth and the Planetary Council

Nikola Schmidt and Petr Boháček

> Cosmopolitanism without provincialism is empty and provincialism without cosmopolitanism is blind.
>
> –Ulrich Beck

21.1 Introduction

Generally, the planetary defense community and national delegates that meet every year at the UN COPUOS are skeptical towards the establishment of any cosmopolitan-like authority anytime soon. This is understandable, as they are part of the current governance mode consisting of a system of nation-states rather than its opposition. In contrast, every student group at any space conference these two authors have attended in the last few years that was given the task of solving space-oriented global problems such as orbital debris, coronal mass ejections, or asteroid threat, tended to come up with the 'obvious solution' of establishing a global authority. The proposal is usually driven both by a rational engineering logic of effectiveness and a sense of cooperation for the sake of human knowledge about deep space and our role in it. The magical essence of space community and its passion to explore, understand and go beyond frontiers is hardly comparable to any other epistemic community in the world, which is giving it specific power for a fundamental change.

N. Schmidt (✉)
Department of Political Science, Institute of Political Studies, Faculty of Social Sciences, Charles University, Prague, Czech Republic
e-mail: nikola.schmidt@fsv.cuni.cz

P. Boháček
Institute of International Relations, Prague, Czech Republic

N. Schmidt (ed.), *Planetary Defense*, Space and Society,
https://doi.org/10.1007/978-3-030-01000-3_21

On the other side of the academic society, cosmopolitan thinkers generally agree that without collective consciousness, we will not survive the twenty-first century (Held 2010a). We can even find a similar argument outside of influential social scientists. Stephen Hawking, for example, claimed that from the orbital perspective, we will perish in 250 years without a significant change in policy towards more collective decision-making. Davis Held's argument focuses on the moral consciousness of the international community (Held 2010b) and the costs of global inaction (Held 2010a); Ulrich Beck adopts the cosmopolitan outlook (Beck 2006); Thomas Pogge uses the theory of social justice (Pogge 2008); and Kenneth Booth proposes security as the basis for human flourishing. Planetary defenders expose the asteroid threat by saying that they see asteroids coming and are developing a database to deliver the most precise scientific knowledge to decision-makers. Every single day we read articles about the Earth reaching the highest temperatures in the history of mankind (Samenow 2018), but we are still in a gridlock at a time when we need to cooperate the most (Hale et al. 2013). So why do we hesitate redistributing power to institutions capable of dealing with global challenges?

Cosmopolitan scholars discuss the moral foundations of political power and appeal to a broader consensus between political actors. They portray the decay of the nation-state, pointing to historical heritage from the peace of Westphalia as the only reason why nation-states perceive themselves as sovereign over their territory. But in the current spatially interconnected world, their legitimacy has vanished due to the disconnect between space and political authority. Geographically delimited nation-states can only deliver on issues and represent citizens that lay within their borders. It is merely imagination that leads us to believe nation-states are in power in the face of global challenges. Similarly, the legitimacy of international organizations is also questioned by their democratic deficit and lack of coercive power and authority, resulting in the ineffective delivery of global solutions.

Studies on International Organizations (IOs) show us that states claim to have built IOs to project their interests globally, while in fact they have given them power to deal with certain problems they cannot solve on their own. The UNHCR secures the rights of refugees in front of (all) states' administrations; the FAO distributes seeds and supports agriculture development through hard-negotiated government-sanctioned access to people whose nation-states fail to alleviate their hunger; the ICJ points at dictators and delegitimizes their fraudulent, power-oriented justifications for violence. These moves are not coming from nation-states themselves, they are perceived as an intervention into state sovereignty, an excuse used for frequent noncompliance with international obligations. State sovereignty is not a carte blanche. In light of global challenges, ongoing globalization, and emerging collective problems, it has become short-sighted to trust that states with physically restricted legitimacy would be capable of solving these issues on their own.

The inherent tensions between two founding principles of the UN—human rights and sovereignty—have been reconfigured significantly after the genocide in Rwanda in 1994. The historical tragedy produced momentum toward adopting the norm of a Responsibility to Protect in 2005. And while the norm has been severely damaged by failures, e.g. in Libya, we should ask ourselves whether we need to experience a

Rwanda-like disaster on the planetary scale (an impact of a large asteroid) before adopting a similar norm for Earth: the Responsibility to Defend Earth "to Defend" here does not mean defense from a physical threat, or defense from a sudden use of force. What we mean by is defense of Earth's interests as a whole, defense from asteroids, defense from power exploitation—defense of all-inclusive cosmopolitan values of humanity in the universe.

In light of cosmic threats, delegates representing states at international fora would never admit the fact that they are representing obsolete political actors within a body that has already surpassed their own purpose. This is because they would first, often absurdly, secure their national interests, before seriously discussing a policy of global consequences where borders do not exist. The status quo lays in performative dances of state representatives meeting at evening banquets, an enclosed family of bureaucrats playing according to the rulebook and treading carefully so as not to upset nation-states, but that does not deliver the capability to act when it is imminently necessary. How could the current international system act in case of an asteroid on a collision course?

Calls for global decision-making above the state-centric inter-governmentalism have moral, epistemic, legal, economic and security arguments that can be found amid the increasing de-legitimization of territorially bound state sovereignty caused by the incongruence of space and politically legitimate power. Changing our collective behavior as a species is necessary now more than ever. This chapter will explore the structural problems of global governance and provide some cosmopolitan alternatives before moving on to propose the three-layer Planetary Council[1] as a starting point for the debate on planetary defense governance.

21.2 Theoretical and Empirical Foundations of the Proposal

Concepts of cosmopolitan ethics, security and democracy are not only best suited to address major governance issues of planetary defense but can also unlock new forms of organizing human activity in Earth and space. Yesterday's system of governance, based on territoriality instead of functionality and reality, is unsuitable for the world of today and space efforts of tomorrow. One of the main issues is the need to balance the legitimacy of political systems between effectivity and democracy, a problem that is enhanced by the current lack of effective global decision-taking and global democratic decision-making processes. This issue seems even more challenging in the face of democratic deficits and the ineffectiveness of contemporary global governance. This chapter discusses several theoretical concepts that can address the aforementioned governance challenges, including security cosmopolitanism, vertical and horizontal dispersion of authority, sectoral constitutionalization, and embracing

[1] We would like to thank Pete Worden here, as it was him who proposed the term Planetary Council in Prague in June 2018 just before the publication of this volume. We spent several days with long-lasting inexhaustible and extremely inspirational discussions.

positive security, including what Nyman and Burke (2016) terms 'security-plus.' All of these concepts are backed by the creation of a norm to Defend Planet Earth as the guiding principle for planetary defense.

21.2.1 Democratic Quality and Security in Global Governance

Global governance faces many issues regarding democratic quality. The demand for quick decision-making in the face of a threat can lead to exceptional politics. Putting too much authority into the hands of one entity generates fears of tyranny, while establishing democracy on the global scale increases the democratic dilemma between achieving policy effectivity and participation, and between the need to act and democratic quality.[2] Burke's proposed security cosmopolitanism (Burke 2013a, 2015) and adequate security conceptualization (Nyman and Burke 2016) developed with Jonna Nyman and building on the Welsh School perception of security (Booth 2007), can address these concerns as well as the democratic deficit[3] of global governance.

The democratic deficit of the contemporary intergovernmental decision-making on global affairs is clear in international organizations. The concept of double representation, where all citizens are represented by government-appointed envoy outside of their direct influence, does not give an appropriate and democratic representation to global citizenry. Not every vote is weighted equally on the global scale, as tiny states have the same voting strength as the biggest countries. National representatives decide according to their national mandates dictated by the national majority, and a majority of nation-states rarely means a majority of the global population (Marchetti 2007). This diminishes the accountability of international policy-making, which is not subject to the influence of global population, but rather spatially divided diverse groups. It is critical to ensure accountability of planetary defense governance that will warrant that all issues of planetary defense are treated in relation to humanity, and not just to the benefit of the most powerful, technologically advanced, or capable nation-states.

The contemporary state-centric system leads to these pathologies in many spheres of global governance, whether it is security, environmental protection, the economy, and so on. Expectations that planetary defense systems built on the current political power setting would somehow function differently are baseless. However, the cosmopolitan nature of planetary defense requires security cosmopolitanism, which is based on inclusive all-human cosmopolitan principles addressing the contemporary democratic deficit in global affairs. While we do not aim to sketch out what this structure can be, we hope to provide some ideas for developing cosmopolitan norms ensuring accountability and thus basic democratic quality.

---

[2] The requirement to act quickly does not necessarily go along with democratic quality. However, we do not plan to elaborate on democratic quality in detail. In the literature, the concept generally measures freedom, the rule of law, vertical accountability, responsiveness, equality, participation, competition, and horizontal accountability (Diamond and Morlino 2004).

[3] Deficit of democratic quality.

One immediate issue is that there is no system where people can globally voice their concerns, due to the abovementioned double representation and the fact that state democratic regimes do not have the capacity or even legitimacy to deal with issues outside of their territory. Were the population allowed to voice its concerns in a democratic way over global challenges, the coercive and legitimate power of these challenges would be self-created through democratic means. We do not think this can happen naturally, but has to be created through realizing Anthony Burke's vision of security cosmopolitanism (see Chap. 14). Burke's argument is not about the securitization of an asteroid impact or any similar threat, but about establishing a mindset that would stimulate the cosmopolitan perspective (see Chap. 18). We aim to elucidate the image of a truly cosmopolitan body that would deal with such issue in an effective, rational, and especially democratic and accountable manner. A way that would prevent democratic deficits and any chance of misuse of power on a global scale.

A frequently mentioned issue linked to global governance is the fear of a global tyranny. Pavel Dufek makes a sequence of remarkable arguments in this book in favor of the inevitability of the world state governance model to address global issues (Chap. 24). Despite the well-formulated theoretical argument, the probability that a world state will emerge soon is in our opinion close to zero. A world government would mean the focus of power in the hands of one entity. Many cosmopolitan thinkers are reluctant to call for global cosmopolitan coercive authority, which carries highly negative connotations and represents the main area of criticism of cosmopolitanism by the realist school of thought. The threat of power centralization brings the threat of authoritarian rule (Zolo 2000).

Cosmopolitan scholars are far from achieving consensus regarding the world government model. Some authors assume limits to democracy, justice, and even coercion on the way to a global political system (Nagel 2005; Rodrik 2007; Wendt 2003). David Held points out that under the pressure and attack of tyrannical dictatorships, the cosmopolitan system would require some sort of coercive power, but stresses deliberative democracy as the most inclusive, dispersed model (Held 1995). Yet, this question represents the known democratic paradox that requires a seemingly hypocritical suspension of some rights, free speech and basic democratic principles in regard to fascist ideologies to preserve the system's values. Moreover, planetary defense does raise the question of quick and effective decision-making in the face of a fast approaching threat, along the lines of the democratic dilemma between effectivity and democratic participation. Planetary defense has the potential to bring either the 'effective' fascist system or an inclusive cosmopolitan democracy. Accepting this possibility only increases the need for political science debates over the issue, as the epistemic community of planetary defenders is particularly well placed to raise awareness of asteroid threats[4] and secure significant funds for their objectives.[5]

[4] Adopted as a global Asteroid Day by UN in 2016.

[5] US NASA PDCO—Planetary Defense Coordination Office will probably (proposed in time of writing) receive $150M in 2019 while it was working with a budget of $60M in 2017, $50M in 2016 and $4M in 2010.

Held (2010b) views deliberative democracy as a key component of cosmopolitan structures. The deliberative principle allows for better participation and gives voice to all interested actors, but also requires that decisions must be rationally justifiable based on shared principles and common values. These values are usually associated with the *Demos,* or national identity, that creates a sense of trust and solidarity and is required for compliance with final decisions—hardly achievable on the global level due to different historical perspectives, cultural diversity and linguistics. Michael Zürn (2000) argues that while national identity was a social prerequisite for the historical forming of nation-state democratic institutions, the institutions themselves can now create the sense of a shared identity. The necessary point is not the national identity itself, but the shared willingness to suppress individual preferences for the benefit of the other. We argue that human survival and flourishing developed through the cosmopolitan norm of Responsibility to Defend Earth can set the basic shared principle required for deliberative democracy. Given that social construction and cultural conditionality of risk perceptions are based on the values of a society, developing a common shared cosmopolitan perception is possible. Planetary defense is here to shape the perception, which is finally the base of its awareness program. However, awareness without normative reflexivity is baseless: it can raise private funds after successful PR campaigns, be abused by delusional people striving for power, or, as we purport, provide the foundation to bring cosmopolitan democracy.

The call for exceptional politics addressing exceptional situations such as an asteroid threat in an appropriately fast manner under a centralized global authority is an extremely sensitive question. This is the issue of size and scale pointed out by Robert Dahl in his critique of international democracy, which stresses the increasing difficulty of acquiring informed consent, satisfying participation of stakeholders, and overall democratic quality with the growing body of voters (Dahl 2001). This issue constitutes another problem for global democracy. Cosmopolitanism, however, combines cosmopolitan ethics with security-plus, which conceptualizes security as seeking all-human flourishing and not a mere absence of threat, pursued under negative security. Negative security may legitimize an authoritarian rule under conditions of radical uncertainty of possible exceptional security situations. This can include declarations of state of emergency or crisis that suspend some regular democratic processes. An asteroid threat should never be used to create such an exceptional political structure or to decrease democratic plurality or liberal values. Yet, it is impractical to demand global democratic consent regarding the method of mitigating the effects of a fast-approaching asteroid. This is a delicate dilemma, but there are solutions. The best approach is to work on a multipurpose long-lasting infrastructure with civilian missions, which may be used for defensive measures as well (see Chap. 26).

Direct democratic participation and processes are usually circumvented in the face of threats, which are used as the reason for extreme politicization of security—the securitization process—which justifies exceptional procedures. Such politicization of security becomes negative, as it promotes undemocratic and fast-tracked decision-making with little deliberation (Roe 2012), and limits pluralism and political participation. In his chapter, Pavel Dufek suggests that planetary defense requires macrosecuritization that would limit pluralism and self-governance since it pits them against each other: first, the needed securitization lowers self-governing and, second, protecting pluralism leads to an inadequate politicization of the threat and mobilization to deal with it. We propose that cosmopolitan security can bridge this gap. Its effectivity originates from inclusivity, accountability, and sustainability, rather than the pursuit of quickly achievable goals in a manner that is hardly sustainable or accountable.

As mentioned, negative security is linked to the mere absence of threat. We suggested above, and throughout this book, that planetary defense requires a broader understanding of security—security-plus or positive security (Chap. 14)—to be legitimate and effective. Positive security emphasizes the inclusive nature of cosmopolitan ethics and the need to protect the biosphere against all asteroids and space threats, but mainly it produces an enabling state for further human/biosphere flourishing. The ability to manipulate asteroids will also carry with it the capacity to bring them closer to Earth for mining (Schmidt 2017). The security-plus concept has legal connections, as it fits the Outer Space Treaty's emphasis on ensuring the benefits of space are accrued for all humanity. This aspect requires better conceptualization in the light of current efforts to mine asteroids. However, securitization does not need to be understood as a negative phenomenon of fast-tracked and undemocratic processes, as has been suggested by the Copenhagen School. As Rita Floyd describes (2007), it can also be more positive than politicization if we adopt a consequentialist approach to its evaluation, based on its capability to address threats.[6]

This would mean judging the value of securitization based on its intended and unintended consequences (the difference between utilitarianism that prioritizes the intended ones). Its evaluation wouldn't be limited to its outcomes (defense of the planet through whatever undemocratic practices), but would be enriched with its context and processes to decide whether it is positive or not. Planetary defense policy cannot be focused on the mere absence of threat, due the radical uncertainty of the nature of the threats. It has the enabling power to internalize security-plus and security cosmopolitanism as the norm on which global efforts for space colonization, sending probes to near stars, or mining of asteroids can, should and must be developed. This approach will make the decision-making processes broadly acceptable, explained as a care for all cities and individuals to promote their free and equal wellbeing. Asteroid mining can plug directly into efforts for survival and flourishing of cities, which may be a gamechanger in global politics, as cities will soon host the majority of the global population and garner increased political power.

Cosmopolitan security must be plural and inclusive to be effective, accountable, and acceptable. This means that planetary defense should not be negatively securitized to create a state of exception as the norm. Neither should securitization come at the expense of self-government, or non-securitization at the expense of human survival. Negative and positive security can be complementary if we judge them

[6] See Chaps. 11 and 14 for details and theory explanation related to this argument.

based on the context: cities' survival, expansion into the Solar System through in situ asteroid mining, exploration of nearby stars, or even thinking about building visionary energy sources close to the Sun. All of these ideas are achievable if we begin thinking about security as a species, not as states.

Several cosmopolitan ideas can thus address the democratic deficit of global governance. Any adopted planetary defense policy must be justifiable even in the absence of threat and can only be effective and implementable with the security-plus principle in mind, thereby requiring day-to-day governance structures. Moreover, possessing power for action against a threat that never eventuates given the fast-tracked undemocratic measures deployed against it, could become less defensible with time. But positive securitization is always reasonable, as it continually promotes human wellbeing and extends it into other areas such as space migration, colonization, and mining, with *human flourishing* principle in mind. The security-plus approach avoids exceptional political procedures and requires an inclusive, democratic, and positive decision-making procedure, which would serve as one of the checks on the global conglomeration of power. Moreover, its alignment with the Outer Space Treaty's focus on benefits to all mankind gives it a legal basis. These concepts can provide the basis to develop the Responsibility to Defend Earth as a new fundamental norm guiding our global political efforts. Notwithstanding, according to the democratic dilemma, increasing democratic quality necessarily entails a lowering of effectivity and authority, issues we will address in following paragraphs.

21.2.2 Global Governance and Sources of Authority

Besides the democratic deficit, global governance and international institutions suffer from a lack of authority. This authority is either political (assigned from states), or epistemic (originating from scientific expertise) (Zürn 2018). But both these sources of authority are being challenged. The effectivity of IOs originates from constituting state-centric path dependency, reliance on power awarded by nation-states, and duplicity or overlapping agendas. Meanwhile, scientific authority is context-dependent, politicized, and frequently negatively securitized (Rychnovská et al. 2017). Cosmopolitan distribution of sovereignty to other relevant and powerful actors beyond nation-states as well as utilization of strengths of International Organizations can improve these issues.

Despite the growing number of IOs, NGOs, and activist groups dealing with the climate change, this structural growth has not succeeded in creating effective policies or inclusive governance (Held 2010a, b). IOs currently only possess reflexive authority based on demands, not orders, and thus depend on national compliance dictated by national interests. International organizations are caught in the circle of institutional authority emanating from the nation-state system, since their power depends on how IOs are perceived by the strongest nations, who grant them more rights only if they benefit from IOs reciprocally (Zürn 2018). It is therefore hard to imagine that international organizations would limit the autonomy of nation-states, since they depend on them for their own authority (Archibugi 2004). However, addressing global issues in an adequate way requires reducing the position of state sovereignty, since states would otherwise act against a proposed policy that, for example, threatened their key industry sector.

Besides the abovementioned concerns, governance of planetary defense through international organizations brings other problems that are well discussed in the political science literature. IOs tend to fall into having cultural preferences, normalize deviance from norms by certain actors, and develop their own internal worlds due to bureaucratic insulation (MN Barnett and Finnemore 1999). They have a tendency to become self-producing entities with self-licking issues to justify their existence (Kessler and Guillaume 2012), just like most other functioning bureaucracies. And as international organizations with adequate relevance and power are dependent on the state-centric system, they are victims of path dependency that ensures they honor and protect their original framework and fail to be reflexive (Dryzek 2016). It is difficult to think of an effective planetary defense without normative reflexivity or the ability to adapt to new realities shaped by science. This reflexivity should not be bound by politically assigned authority but rather the epistemic one, by the social scientists with natural scientists. This firm foundation in science provides an opportunity for planetary defenders to shape the mindset of people in bureaucracies, international organizations, and future global cosmopolitan governance structures.

International organizations do have the advantage of being able to organize epistemic communities of particular scientific discipline (Adler and Haas 1992) and develop critical wells of knowledge. Scientific-based entities, such as the likes of SMPAG or IAWN, may lack the true supranational, state-independent nature and democratic quality to limit exclusive expertise and possible hegemonic hijacking, but they have the potential to organize knowledge independent of nation-states. As mentioned above, security cosmopolitanism based on inclusive conceptualization is not compatible with a coercive political authority, but it still requires a global epistemic community to be energized and empowered by a particular imaginable case. The binary character of the asteroid threat makes it easier to imagine for everyone.

A scientific argument on empirically observable threats can thus be considered as an objective threat (Floyd 2011) that can avoid politicization by adopting a cosmopolitan inclusive format. However, when an epistemic scientific community is involved in decision making, epistemic authority should possess accountability. A well-studied dilemma is when science transforms into politics without proper reflexivity (Jasanoff 2004). However, it can be a positive connection between power and knowledge, in contrast to ongoing trends to rationalize political stances as scientific, and can serve as a departing point for a cosmopolitan body for planetary defense with imbedded inclusive cosmopolitan principles to address the contemporary democratic deficit in global affairs. As Nikola Schmidt has argued elsewhere (2018), planetary defense does not have to deal with equally complex knowledge as in the case of climate change. Planetary defense is a simple binary problem—is an asteroid on a collision trajectory or not? Any governance efforts are thus much more related to the willingness of involved actors, because the scientific background of the problem developed for policy-making is significantly less questionable and easily comprehensible. Planetary defense can thus serve as an exercise bench for more complex global issues. It is also a very important advantage that planetary defenders should be in the unique position of being able to demonstrate how to shape our cosmopolitan future.

As Schmidt discussed earlier (Schmidt 2016), the production of a *good knowledge*, a knowledge that is useful as it reflects on the problem and answer in ways that make the problem and its solution comprehensible, leads to a situation in which the particular institution can manipulate the produced knowledge, disbalance the weight of various necessities, and produce biased outcomes for the policy makers. This dynamic is not necessarily limited to institutions, as it can be observed in any organized entities, including epistemic communities. However, when states do not deliver practical knowledge, the source of epistemic authority can move to private enterprises where good oversight is less likely and all discussion of democratic qualities are left behind. Backed by enormous amounts of personal wealth, SpaceX is changing space flight and the prospect of Mars settlements; Blue Origin aims to build a Moon colony; and Breakthrough Initiatives, has a goal of discovering extraterrestrial life, sending messages to space, or reaching the closest stars with nanoprobes called 'space chips' within our lifetime. These kinds of companies are radically shifting the source of knowledge. In this way, private enterprises become consumers of the knowledge produced by epistemic authorities but exist beyond the reach of democratic principles.

Although we should not create too many obstacles for people willing to make breakthrough changes in mankind history, accountability processes should still apply to them. Cosmopolitan governance ideas include philanthropes as serious actors capable of driving game changing projects. As knowledge releases objectivity and its value-free characters from their dependence on political discourse, practice, and context, more democratic deliberation, involving private enterprises, is the key. Even epistemic authority has to be inclusive and cosmopolitan in order to be effective and constitutive. It also cannot be limited to nation-states, who naturally draw borders around a governed agenda on a global scale to avoid intrusions into their geographically limited national interests.

In Chap. 22, Joe Pelton talks about a 'global program of action' supported by 'substantial financial resources committed to identifying cosmic hazards and undertaking planetary defense.' Despite the wide array of issues, IOs are good at setting the agenda for such a program, a type of a non-coercive action. Besides agenda-setting, IOs impact the decisions of nation-states through other means, including social shaming, norm socialization, information, and economic coercion, and also through discourse, policy proposals, and technical or scientific support (Dellmuth et al. 2018). Through non-coercive efforts, IOs do make authoritative decisions regarding every aspect of global affairs through rule-making bureaucracy (Michael Barnett and Finnemore 2004). Peter Gourevitch's (1978) second-image reversed theory, applied to climate change by Sjur Kasa (2013), shows that pressures in the form of international organizations concerning global challenges, specifically the United Nations Framework Convention on Climate Change, can positively impact domestic politics and progress towards a global norm. International organizations have been erected to support the initial norms. For example, the EU is actively and legally enforcing these norms against the will of nation-states. Developing the right norms and setting the agenda are key moves through which IOs can advance an adequate approach to planetary defense – the norm of Responsibility to Defend Earth.

Contemporary IOs, dependent on power from states, fail to produce strong enough pressure to comply with measures that can be undesirable from a national point of view. To be effective, a depoliticized epistemic authority should thus be accompanied by a clear normative argument—the Responsibility to Defend Earth. This norm can lead to a change of state behavior to a more responsible approach that takes the cosmic point of view (Burke 2013b). Further, to disperse vertically the state-centric authority of IOs limited to power granted by states, a new cosmopolitan body should also include private entities, local or regional bodies, and non-governmental actors. Yet even a scientific authority will be vulnerable to power disputes and horizontal dispersion of authority, an issue addressed in the next section.

21.2.3 The First Steps Toward the International Organization

When planetary defense is governed by an intergovernmental scientific elite—*the planetary defense epistemic-scientific community*—without any kind of democratic oversight or legal constitutional checks, there is a low probability of any significant progress. Nationally assigned delegates encircled by territorially limited democratic legitimacy will be vulnerable to politicizing science and opposing any rational argument for the sake of preserving the power in the hands of their geographically bound governments and constituents. Combining epistemic authority with adequate checks can be achieved not only by including non-state actors and adopting cosmopolitan norms, but also through *sectoral constitutionalization*. This would spread the authority horizontally according to different issue-specific areas. Importantly, a sectoral constitutionalization ensures that power is never vested in one entity through bold, vigorous, and oppressive hierarchy or chaotic and disorganized anarchy (Deudney 2007).

Sectoral constitutionalization builds on the example of the International Telecommunication Union, which shows us that technology is able to create conditions of imperative cooperation between states, but only in a limited and deeply technical area. While states are not willing to give up sovereignty directly, they can agree to respect certain norms in specific issue-related areas where cooperation is unquestionably deemed essential, like in the case of ITU's radio frequency rules and technical standards of communication. A functioning communication system is thus predicated on global agreement. This need for compliance gives the technical body political power from nation-states, but also a chance to generate the revenue necessary to manage such technical issues. The ITU's introduction of fees for satellite orbit registrations addressed the issue of skyrocketing applications for satellites, thus improving management and generating revenue for maintenance. With the critical overcrowding of the orbits above Earth and the growing indispensability of satellite communication to everyday life, the introduction of fees for orbital licenses, or even launch licenses, can similarly be justified by the need for sectoral management. This includes management of orbital observation, space safety and awareness (SSA), satellite collision prevention, and most importantly, debris removal. With such tools and norms, extending these functions to asteroid mining seems the next logical step. Since without a global adaptation of ITU's technical norms no one would be able to communicate, and without global orbital management and debris removal no satellite would be safe, we can say that without a global asteroid observation and threat mitigation, no one could be living in the future. Fees collected for debris removal from launch licenses should be used for supporting the same institutional structures dealing with an asteroid threat, which is the basis of a social contract between a state and its citizen.

Marchetti (2007) problematizes the creation of issue-specific entities, as it excludes those not directly involved in the matter. The same principle can be applied to the creation of the specific scientific authority we discuss. The sector-specific inclusive and deliberative authority is also an argument against expert elitism in decision-making—the warning against the emergence of epistemic authorities without democratic oversight. The *sectoral issue-based approach* still aims to avoid elitism and exclusion by giving voice only to those with technical knowledge or contribution means in a specific area (observation or deflection methods) by setting the cosmopolitan ethics as the basic norm. Even on the national level, there are areas that are left out of political participation for various reasons. Professional administrations, judicial enforcement, or police forces are sectors that require impartial decision-making, but that are guided by liberal democratic values. Finally, democracy is not limited to direct elections, but to the values the society pursues. This normative layer of modern democracy should ensure that scientific and expert opinions prevail over political bias, while also protecting minority groups from tyranny or power grabs by interest groups, as Moravcsik (2002) has argued. Some institutions are not fit for direct democratic oversight, specifically to preserve democratic legitimacy in a wider sense; representation does not have to – in some areas – mean more democracy. The normative guidance of cosmopolitan ethics stipulated in the norm of Responsibility to Defend Earth aims to secure the sincerest intentions.

Current efforts to establish the Space Mission Planning Advisory Group (SMPAG) as a body capable of managing asteroid deflection do not envisage it as a decision-making body surpassing the UN Security Council in case of an imminent threat. Rather, it would help states to steer their interests and understand their intentions, deepen transparency, and shape mutual predictability of each actor. Therefore, the body will not significantly decrease the undesirable possibility of a unilateral action by a strong space power willing to act if necessity calls upon them. Such a unilateral scenario involving nuclear weapons without a proper, transparent, inclusive, and democratic network is highly undesirable. De-stigmatization of the nuclear taboo and upsetting geopolitical balance in favor of the willing state could have further severe political consequences. However, as previously mentioned, necessity might call upon us, and the only mechanism that can avoid global political turbulence caused by the use of nuclear weapons, such as NEDs, is confidence and transparency. Even this it is not a reliable instrument. Joe Pelton argues for deepening the powers of SMPAG and IAWN, understood as a preliminary step toward the establishment of an international body in charge of the mitigation of NEOs (Jakhu and Pelton 2017). However, he does not propose any ideas regarding how we can create an independent governance body on planetary defense without the exclusive and direct influence of territorially bounded states, which are shaped by the games and short-termism of domestic politics, state-centric security interests, democratic deficits, and accountability. Especially in the case of NEDs, it is unimaginable that any global authority will be trusted with their operation anytime soon.

On the other hand, the more we focus on the agenda of planetary defense as a mere scientific endeavor, the more it will be perceived as the cosmopolitan elitism of a very select group of scientists, and consequently lacking democratic legitimacy and accountability. And while asteroid detections through the lenses of astronomers do not pose any tensions between the key actors or states, carrying out planetary defense measures through means of unprecedentedly powerful technologies would require a supranational authority. Bringing together nation-states, commercial private entities, and other non-governmental bodies is the method we propose. Nation-states must play an active role, with each nation contributing their specific expertise. This demonstrates how the epistemic community can be practically dispersed through different actors. If conglomerated into an international body and then specifically divided through sectoral constitutionalization, the expertise can possess adequate checks and legitimacy.

Technology has politics, because the choice of technology can directly cause inequality or different undesirable political implications (Winner 2003). Planetary defense policy architecture needs to be based on sound cosmopolitan norms with the epistemic authority dispersed horizontally and vertically to avoid politicization and dominance of national interests. We should recognize that proper selection of technology has the potential to stimulate particular politics: inclusive cooperation with accountability in mind and long-term sustainability. From the beginning, such technology must be designed according to security-plus principles and thus for multipurpose civilian applications aimed at human flourishing: planetary defense, asteroid mining, sending probes to stars, Solar System exploration, energy transfer, or other space efforts. Then we can begin thinking about Dyson spheres, because the idea of military application of technology between nations (not humans) will dissipate. Given the violent history of international relations dominated by territorially bound national interests, there is no reason to believe that this system can fare better in space. The described principles represent a clear starting point, not only for planetary defense but also for a new mode of global governance.

21.3 Semi-Cosmopolitan Governance System for Planetary Defense and Other Space Activities

This next section explores the application of cosmopolitan theories to planetary defense and extends them to other areas of global governance. There is much more to unpack concerning how to ensure accountability through cosmopolitan democracy in planetary defense. The traditional features of national democracy seem to be inapplicable on the global level, as stressed by influential political scientist Dahl (2001). But the limits of human imagination should not be taken as a given in exploring better political alternatives. Democracy is not a fixed system or institution, but rather is a learning process, notes Held (2010b). Thus, the cosmopolitan effort needs to trickle into several spheres of what we describe as a semi-cosmopolitan system—a system that takes only selected issues to be governed through cosmopolitan means while recognizing the persistent power of nation-states.

The topic of planetary defense is not only a question of ensuring the survival of humanity, but also promoting its flourishing and wellbeing. This project interprets asteroid impact mitigation as following the discussed security-plus and security cosmopolitanism concepts. As security is contingently contested (see Chap. 11) and dependent on its constituting politics, it can be mutually compatible on the global level if its political processes are global as well. These practices must reflect on the current pathologies of global governance, as explained above, while also utilizing its strengths. One such strength is the ability of IOs to set agendas and establish norms. Meanwhile, private commercial entities boast innovative, effective approaches, and in combination with philanthropes, direct financing. The issues span from the state-centric governance that lacks effectivity and (global) democratic legitimacy, as well as possessing national outlooks that constitute a negative security conceptualization that is ill-suited to dealing with global threats. The prevention of nuclear de-stigmatization, negative political consequences, including a hegemonic rise, and building of planetary defense in an effective, inclusive, democratic, and cosmopolitan approach, should be kept in mind.

Our proposal aims to bridge the gap between space and politics, as political authority and democracy remain territorially constrained, as well as between knowledge and power, as scientific arguments are plagued by politicization from nationally bound constituting political practices. One of the key concepts to achieve this is the horizontal and vertical dispersion of sovereignty to other actors and issue-specific bodies. It is precisely planetary defense that has the potential not only to fulfil biological preconditions for sustaining life, but also to enable social and ethical conditions for preserving human life. Nothing less than human flourishing should lay at the beginning of every discussion about governance. We therefore hope to initiate an innovative and open discussion on how we ought to organize human activity. Building on these features, we propose the creation of the three-layer Planetary Council consisting of epistemic authority, technical commission, and cosmopolitan assembly (Fig. 21.1).

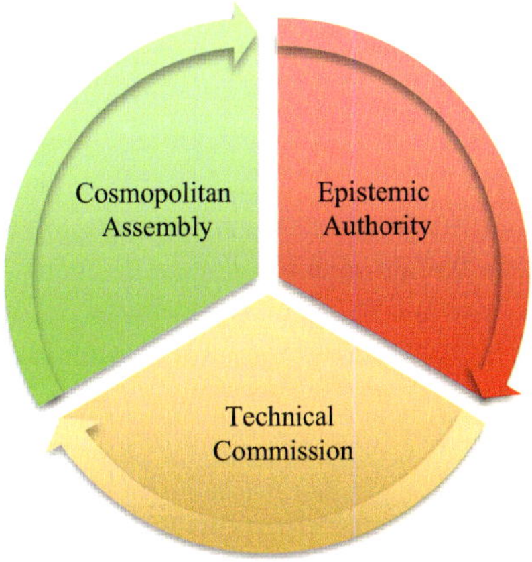

Fig. 21.1 Structure of the Planetary Council

21.3.1 Epistemic Authority

The epistemic authority, the first out of the three bodies of the proposed Planetary Council, would provide the positive connection between power and knowledge. Scientific deliberations would be open to different actors with adequate expertise to be involved in the process, ensuring inclusiveness based on knowledge not territorial allegiances. This would serve to avoid politicization of scientific arguments while also providing democratic oversight (from other bodies, see below). Moreover, it would contribute to the development of the cosmopolitan norm, built on the normatively and legally defensible security-plus concept, and the empirically observable cosmopolitan nature of cosmic issues. This norm, the Responsibility to Defend Earth, would thus consist of more than just the need to defend us from asteroids, but also from anything threatening overall human flourishing. It would therefore need to uphold humanity's ingenuity, curiosity, and willingness for societal progress. It would be more than just a forum of national delegates, including also private entities, national academies of science, other governance units, experts, and members of the public involved in the issue.

Over time and its own institutionalization, the authority can adopt a role similar to bureaucracies, congressional research offices, and apolitical administrations tasked with monitoring, evaluating, and examining observational data, developing risk assessments, resilience concepts, impact scenarios, and most importantly, mitigation efforts. The epistemic authority emerges from the inclusive epistemic community when this community produces knowledge addressing a problem or threat that has been generally accepted by policymaking bodies as legitimate (Kastenhofer 2011). When the community becomes a source of advice to policymakers, it gains authority. With knowledge unavoidably comes power. And while states might resist being involved in a structure where they are not the unique decision-makers, the empowerment of the epistemic authority through private entities, especially in space industry, will become inescapable.

In fact, such an expertise-driven epistemic community is already emerging and is discussed bi-annually at Planetary Defense Conferences; however, it is currently lacking the social aspect. Nevertheless, the fact that knowledge has become part of the National Near-Earth Object Preparedness Strategy and Action Plan in USA (The White House 2018) makes this point very clearly. The epistemic authority will use scientific arguments and deliberation to reach a consensus on scientifically observable threats and mitigation methods. As mentioned above, topic-specific bodies should not be exclusive to those possessing expert knowledge or vice versa, however, apolitical expert decision-making is required in some areas to paradoxically deliver more democracy, as is the case for judiciary or apolitical domestic bodies. Science is bound by the will of scientists to uncover new realities but still remains dependent on government financing formulated by the national political agenda. Bringing these efforts to an apolitical global body could unlock the potential of global scientific efforts.

This horizontal dispersion of authority in the form of described *sectoral constitutionalization* will also aim to generate an integrated governance approach between sectors, similar to the SMPAG and IAWN, without complex political unease and under conditions where cosmopolitan concepts, norms, and practices can develop easier. The division of the body called *the epistemic authority* into issue areas led by various epistemic communities would limit any power usurpation by one body or politicization of scientific arguments for a proposed policy. We do not deal with the problem of representation behind the sectoral constitutionalization in detail, reserving it for future analysis regarding the need for role-dependent representation balance. Nevertheless, the risk of political bias emerging from epistemic authorities can be balanced by having more than one working group assigned to each specific issue, fostering deliberative discussions between experts in the epistemic communities.

In decision-making, we follow the approach taken by the European Union, which divides its decisions between regulations and directives. The former is obligatory to member states immediately, while the latter proposes a policy direction for a law to be adopted on the national level.[7] While the decisions taken need to reflect the principle of subsidiarity on a global level, there is no reason to apply directives on a global level that can be better dealt with on the national level. The Planetary Council should probably have both instruments. For example, the warming of the ocean is not in question, whereas the discussion over its mitigation and appropriate climate change policy is complex enough to keep it as a direction, and not a regulation. But finding the right balance between the two measures will be key and dependent on deliberations with other bodies of the Planetary Council. The epistemic authority could issue regulations and directives to be carried out by another body, the *technical commission* that reserves the power to directly question the epistemic authority.

[7] The European Union also uses decisions that apply to selected actors, recommendations that are not binding, and options for possible statements made by anybody. For details see https://europa.eu/european-union/eu-law/legal-acts_en.

Disputes should be settled between the three bodies; we do not elaborate on this in detail here, reserving the question of dispute resolution for further analysis.

21.3.2 Technical Commission

The second out of the three bodies of the proposed Planetary Council would be composed of nation-states as the most capable governing units, with a horizontal dispersion of state sovereignty between issue-specific areas based on sectoral constitutionalization. Therefore, power struggles could be narrowed into disputes between emerging institutions (similar to Directorate-Generals in EU) over their competencies, similar to domestic institutional power tussles between governmental ministries. This can limit the competition between national interests that lack a legal or legitimate solution or global perspective on governed problems. But the key role of the technical commission, as the body representing the capabilities or the 'hardware,' will be to balance out national, commercial, and global interests by bringing these actors together and capitalizing on their strengths. When it comes to expertise, businesses can afford only the kind of expertise that is directly needed to make profit, while international organizations tend to deepen the expertise in general, including elements that are not necessarily needed at the time. States perceive their participation in sharing expert knowledge as a strategic necessity and achieve this by building scientific centers. Moreover, states might come under pressure from scientific communities to develop a policy environment that allows national scientific centers to help them participate on a global project.

For nation-states, the concepts of burden-sharing or pooling and sharing used for conventional collective defense models in NATO can serve as an inspiration. Further, NATO divides expertise through its Centers of Excellence spread across member states—a model currently pursued by the authors of this chapter in the Czech Republic to motivate the involvement of the ELI Beam and HiLase world-class laser centers in planetary defense and space exploration efforts. If different nation-states (or groups of nation-states that are willing to share their expertise and participate on the particular issue area) host a different technical issue area—for example, laser research centers for super powerful laser—they are becoming inherently interdependent and thus less prone to possible conflict when their governments change. The technical commission would be the organizing platform that would through this division of labor bridge the disorganized national expertise in the space sector, which is dominated by secrecy and national security prism and global overlapping and duplicity in relation to global space objectives. Smaller states understandably develop specific expertise that is customized for their economy in relation to local and global market demands.

The proposed *technical authority* in the Planetary Council can motivate states to be more involved in global collaborative efforts. Large Technical Complexes described by Mayer and Acuto (LTC) (2015) prove that states are interested in participating in efforts such as the CERN or ITER, usually under pressure from the lobby of the scientific community. Unsurprisingly, such interest does not originate from general national policy, which is vulnerable to fluctuating plans and visions, as we know very well from the constantly changing policies of NASA, caused by decisions made in Congress or its own appointed directors. However, for states these efforts can be attractive from a political and strategic view, and not only because of sharing scientific knowledge. The growing link between scientific and technical development and global security issues is constituting this strategic necessity, with territorially limited interests fearing exclusion. While Burke (2013c) describes the strategic necessity of dealing with global challenges through imagined normative security cosmopolitanism, we see this strategic necessity to participate in LTCs dealing with global challenges (normative necessity) as justifiable on the national interests basis (rational necessity) directly contributing to the creation of Beck's cosmopolitan outlook.

The examples of the ITU's success in governing radio frequencies and orbital licenses points us to the example of technological imperatives that mandate states reach a consensus with the coercive power. This can be developed into a norm in planetary defense, as the cosmopolitan cooperation also has economic justification. Globally adopted technical norms enable global communication, and global orbital management and debris removal enable global satellite protection. A global system of planetary defense would enable nothing less than human survival and human flourishing. Moreover, administrative fees for upholding the management of radio frequencies or orbital management are easily justifiable with the shared common goal—fees for planetary defense systems would have the same rationalization in the Planetary Council. However, if the fees can grow with growing businesses in asteroid mining, our species will be seriously heading towards the Dyson spheres. Nation state systems will never bring us there.

When it comes to legitimacy, states may lean towards democratic legitimacy. But democracies usually do not have the effectivity to make radical changes, and states are not interested in handing over their sovereignty. Momentum, quick changes, and innovations are more the advantage of businesses because of their short chain of command, long-term planning, and management oriented purely on outcomes for the advancement of the business. From the commercial perspective, Breakthrough Starshot is a good example of how a philanthrope can fund development of a super-powerful laser without demonstrated intentions to use it for planetary defense, even though it can be. And while the last US NEO strategy (The White House 2018) even omits the laser deflection as an option, Breakthrough now has room to proceed with a civilian project and is building up its authority by developing a laser of multipurpose usability (see Chap. 26).

We expect philanthropes and commercial entities to play a significant role no matter if they build rockets, plan for space settlements, or send probes to stars, as they are willing to launch their space ambitions with or without states (*sic erat scriptum!*). Yet, they need a stable and working regime, without regulatory barriers but with sufficient protection, where their endeavors can flourish. The discrepancy between international and national space law creates gaps that lead to forum shopping, where commercial entities search for a stable space business environment. Asteroid mining is a good example, as interpretation of the Outer Space Treaty by the United States or Luxembourg create incentives for commercial actors amid the absence of sound legal definitions on what constitutes 'appropriation' by other means, or how to provide the benefits of space to all mankind while it is its province. Global cooperation between private and public bodies in a cosmopolitan structure based on the security-plus approach stressing inclusive all-human flourishing seems like an ideal place to address this and other issues of international space law. A good example is the emergence of the Hague Working group, which already provokes national ministries who defend their status quo. Once standards and governance modes are developed and carried out by those who possess asteroid mining capabilities, nation-states will protest, resign, but finally obey new rules or participate in their making. Actors act not only according to their intentions, but also to the possibilities they have. The shift in space capabilities from only nation-states in the past to private entities needs to be reflected in the modes of governance.

To summarize, the Technical Commission would deal with the hardware—the technical capabilities for planetary defense and their development. The body would be horizontally dispersed through various issue-specific areas and vertically dispersed through different stakeholders. Nation-states should realize that, thanks to the globalization efforts driven by liberal economic principles after World War II, they are in a world in which they are not capable of defending us from asteroids but are well prepared to develop business environments supporting our vision of horizontally dispersed, issue-specific areas contributing to Planetary Council. All this would bring together not only the diverse capacities of nation-states into a joint, organized module, but also technically strong and capable private entities of the likes of SpaceX, Blue Origin, and Breakthrough. As history shows us, game-changing events happen when select, powerful persons decide for a change. Working according to scientific knowledge with normative security-plus grounding of the epistemic authority, the technical commission would propose concrete development plans for planetary defense efforts. And while the Epistemic Authority would ensure its focus on cosmopolitan human flourishing, the next body, the Cosmopolitan Assembly, would warrant as a safeguard for distributing the benefits of such efforts to all mankind and approve measures taken by the Technical Commission.

21.3.3 Cosmopolitan Assembly

The third out of the three bodies of the proposed Planetary Council is designed to enhance democratic quality, requiring that all stakeholders affected by a decision can in some way influence it and participate in cosmopolitan securitization and inclusiveness. While recognizing the current legitimacy and representative value of nation-states, they do not provide a full political participation, accountability, and rights of all affected by global issues; the inclusivity of the process needs to be extended, but also remain rational and realistic. Following Beck's quotation from the beginning of this chapter that 'Cosmopolitanism without provincialism is empty and provincialism without cosmopolitanism is blind,' we believe that we do not need to deconstruct nation-states, but deconstruct their ultimate sovereignty to issues they are the best to deal with, with global issues definitely not falling into this category. Given the threat of the so-called city killers, nation-states should not be the only ones involved as representatives in the process, but also representatives of megalopolis areas, megacities, and large urban areas, given that 60% of people globally will live in cities by 2030 and one in every three will inhabit a city of above 500,000 (United Nations 2016). This can provide accountability and inclusivity of inhabitants, extend the reference object of security threats from nation-states, and vertically spread sovereignty from its lone holders—the nation-states—to other relevant actors.

As with other complex issues, we decided to leave the principles of how the cosmopolitan assembly is created—whether through direct elections, selection from among already elected representative bodies in megapolis and nation states, or a combination of methods—for further analysis. However, making sure that those affected by a policy have a say in its formulation is critical. Despite stressing cosmopolitan inclusivity through security cosmopolitanism, involvement of various non-state actors and adoption of the global norm of Responsibility to Defend Earth, the Planetary Council needs some more direct accountability, not simply by a direct vote but also through certain empowered voices, as it is in the European Union. The voting procedures are left for a further debate, but the Cosmopolitan Assembly should serve as a check on the other two bodies and approve their efforts in order to ensure the distribution of goods across the global population.

The spread of scientific information and critical knowledge originating from the Epistemic Authority, and the opportunity to collaborate and carry out ambitious, commercially sound, and economically beneficial projects through participation in the Technical Commission, will serve as an incentive to keep actors involved in the Planetary Council. The three bodies of the Council would progress towards cosmopolitan security and democracy, while avoiding the negative political consequences and pathologies of global governance described above. The horizontal and vertical dispersion of sovereignty will extend the reference objects to other involved stakeholders and ensure more democratic quality beyond nation-states as the only political actors of the global community through technically imperative cooperation and include affected stake-holders beyond states. This will also contribute to the nurturing of the norm of Responsibility to Defend Earth and an inclusive cosmopolitan conceptualization and addressing of security. Combining a shared rational and scientifically observable base, technically imperative cooperation on a semi-cosmopolitan basis while preserving more inclusive participation in cosmopolitan assembly, will create a multilateral and multi-stakeholder structure that would ultimately lower the negative political risks and de-stigmatization of nuclear weapons use for planetary defense. More importantly, it will also set a base for further spread of cosmopolitan principles into areas beyond space activities.

We assume that the epistemic community would be firstly formed in the likes of the Hague Working group before being institutionalized through practice, produced knowledge, and acceptance of states in the form of an international organization (Table 21.1). This would set the basis for a full development of the three-layer Planetary Council (Figs. 21.1 and 21.2).

Table 21.1 Advantages and disadvantages of current global actors for their participation in cosmopolitan governance

	States	International organizations	Businesses
Key advantages	Territorial accountability	Institutional stability	Effective, innovative, ambitious
	Main power and democratic legitimacy holders	Cosmopolitan outlook	Free of national lens
	Norm initiators	Norm enforcers	Norm renderers
	Scientific centers	Coordinating global expertise	Issue-specific expertise
Key disadvantages	Fluctuating politics	Legitimacy through states	No democratic oversight
	Disorganized expertise	Low accountability	Focused on its own interests
	National outlook, territorially limited authority	Prone to institutional pathologies	Dominated by commerce

Fig. 21.2 A roadmap to the Planetary Council

21.4 Conclusion

This chapter did not aim to propose a detailed cosmopolitan political structure for planetary defense, as such efforts would be beyond its scope. Its central purpose was to sketch out the main argument regarding why planetary defense needs to be approached from a cosmopolitan outlook, and how different cosmopolitan concepts can be used to address inadequacies in global governance of space issues. The main goal was to introduce some innovative ideas to start the debate about a cosmopolitan planetary defense through the creation of a Planetary Council. The idea that space governance, including planetary defense, would be governed adequately according to the contemporary state-centric world order with its violent, divisive, and clashing history is naïve. In other areas of global governance, the system does not work. Utilizing the true nature of planetary defense to nurture shared values and norms under the Responsibility to Defend Earth is a starting point for global cosmopolitan democracy, effective addressing of new challenges, and inclusive and appropriate decision-making. The proposed creation of a three-branched Planetary Council aims to stimulate the discussion.

Any other answer than a cosmopolitan one risks a unilateral response to planetary defense. That would destigmatize nuclear taboos and carry unintended political consequences that can unlock ways for hegemonial tyranny under the pretenses of human survival, especially should it follow the contemporary international climate and political system to planetary defense. Since multilateralism and intergovernmentalism will not cut it, as the dominance of national interests does not allow for adequate compromises and keeps the state-centric system prevailing, a new cosmopolitan approach is critical. To ensure legitimacy and benefit to all humanity, accountability between decision-takers and decision-makers on the global cosmopolitan scale is the only way to secure inclusion of all individuals, their survival, and their flourishing. Without democratic quality in the planetary defense system, we cannot have legitimacy nor effectivity. To truly protect the planet from the variety of known threats (an extinction-size asteroid, city killers, coronal mass ejections, or to set space regimes for asteroid mining or sending probes to stars) or the known ones (requiring the system or institution to be reflexive) we have to adopt the cosmopolitan conceptualization of security that shifts the focus from nation-states to individuals, along with the planet Earth and entire biosphere. We must integrate experts from various disciplines not as mere advisors but as makers of the policy itself through their own dedicated bodies.

We see the constitutive power of planetary defense toward cosmopolitan global governance because it provides the needed legitimacy before the decision to act required by a detected asteroid on a collision course. The way we organize human activity and our collective behavior should not be guided by territoriality, but functionality based on our shared destiny as a species, as can be achieve through a body such as the Planetary Council. Our effectiveness should be measured by nothing less than the survival and flourishing of human life. There is no better place to start this change than planetary defense.

Acknowledgments This study was supported by the grant awarded by the Technological Agency of the Czech Republic, project TL01000181: *"A multidisciplinary analysis of planetary defense from asteroids as the key national policy ensuring further flourishing and prosperity of humankind both on Earth and in Space,"* and co-funded by the Institute of Political Studies, Faculty of Social Sciences, Charles University in Prague.

References

Adler, E., & Haas, P. M. (1992). Epistemic Communities, World Order, and the Creation of a Reflective Research Program. *International organization*, *46*(01), 367. https://doi.org/10.1017/S0020818300001533

Archibugi, D. (2004). Cosmopolitan Democracy and Its Critics: A Review. *European Journal of International Relations*, *10*(3), 437–473. https://doi.org/10.1177/1354066104045543

Barnett, M., & Finnemore, M. (1999). The politics, power, and pathologies of international organizations. *International organization, 53*(4), 699–732.
Barnett, M., & Finnemore, M. (2004). *Rules of the World: International Organizations in Global Politics*. Ithaca and London: Cornell University Press.
Beck, U. (2006). *The cosmopolitan vision*. Polity. https://doi.org/10.1007/s13398-014-0173-7.2
Booth, K. (2007). *Theory of World Security*. Cambridge University Press.
Burke, A. (2013a). Security cosmopolitanism. *Critical Studies on Security, 1*(1), 13–28. https://doi.org/10.1080/21624887.2013.790194
Burke, A. (2013b). The good state, from a cosmic point of view. *International Politics, 50*(1), 57–76. https://doi.org/10.1057/ip.2012.28
Burke, A. (2013c). Security cosmopolitanism. *Critical Studies on Security, 1*(1), 13–28. https://doi.org/10.1080/21624887.2013.790194
Burke, A. (2015). Security cosmopolitanism: the next phase. *Critical Studies on Security, 3*(2), 190–212. https://doi.org/10.1080/21624887.2015.1065109
Dahl, R. A. (2001). Is Post-national Democracy Possible? In S. Fabbrini (Ed.), *Nation, Federalism and Democracy*. Bologna: Trento: Editrice Compositori.
Dellmuth, L. M., Gustafsson, M. T., Bremberg, N., & Mobjörk, M. (2018). Intergovernmental organizations and climate security: advancing the research agenda. *Wiley Interdisciplinary Reviews: Climate Change, 9*(1), 1–13. https://doi.org/10.1002/wcc.496
Deudney, D. (2007). *Bounding power: Republican security theory from the polis to the global village*. Princeton University Press.
Diamond, L. J., & Morlino, L. (2004). An Overview. *Journal of Democracy, 15*(4), 20–31. https://doi.org/10.1353/jod.2004.0060
Dryzek, J. S. (2016). Institutions for the Anthropocene: Governance in a Changing Earth System. *British Journal of Political Science, 46*(4), 937–956. https://doi.org/10.1017/S0007123414000453
Floyd, R. (2007). Towards a consequentialist evaluation of security: bringing together the Copenhagen and the Welsh Schools of security studies. *Review of International Studies, 33*(2), 327. https://doi.org/10.1017/S026021050700753X
Floyd, R. (2011). Can Securitization Theory be Used in Normative Analysis? Towards a Just Securitization Theory. *Security Dialogue, 42*(4–5), 427–439. https://doi.org/10.1177/0967010611418712
Gourevitch, P. (1978). The second image reversed: the international sources of domestic politics. *International Organization, 32*(4), 881–912. http://journals.cambridge.org/production/action/cjoGetFulltext?fulltextid=4305592. Accessed 5 March 2014
Hale, T., Held, D., & Young, K. (2013). *GRIDLOCK: Why Global Cooperation is Failing when we Need it Most*. Cambridge: Polity.
Held, D. (1995). *Democracy and the Global Order: From the Modern State to Cosmopolitan Governance*. Stanford University Press.
Held, D. (2010a). Reframing Global Governance: Apocalypse Soon or Reform! In G. W. Brown & D. Held (Eds.), *The Cosmopolitanism Reader* (pp. 293–211). Polity.
Held, D. (2010b). *Cosmopolitanism: Ideals and Realities*. Polity.
Jakhu, R. S., & Pelton, J. N. (2017). *Global Space Governance: An International Study*. (R. S. Jakhu & J. N. Pelton, Eds.). Cham: Springer International Publishing. https://doi.org/10.1007/978-3-319-54364-2
Jasanoff, S. (2004). *States of Knowledge: The co-production of science and social order*. Routledge.
Kasa, S. (2013). The Second-Image Reversed and Climate Policy: How International Influences Helped Changing Brazil's Positions on Climate Change, 1049–1066. https://doi.org/10.3390/su5031049
Kastenhofer, K. (2011). Risk Assessment of Emerging Technologies and Post-Normal Science. *Science, Technology & Human Values, 36*(3), 307–333. https://doi.org/10.1177/0162243910385787
Kessler, O., & Guillaume, X. (2012). Everyday practices of international relations: People in organizations. *Journal of International Relations and Development, 15*(1), 110–120. https://doi.org/10.1057/jird.2011.29

Marchetti, R. (2007). Global governance or world federalism ? A cosmopolitan dispute on institutional models Cosmopolitan Dispute on Institutional Models Ã, (December 2014), 37–41. https://doi.org/10.1080/13600820600816282

Mayer, M., & Acuto, M. (2015). The Global Governance of Large Technical Systems. *Millennium - Journal of International Studies*, *43*(2), 660–683. https://doi.org/10.1177/0305829814561540

Moravcsik, A. (2002). In Defence of the "Democratic Deficit": Reassessing Legitimacy in the European Union. *Journal of Common Market Studies*, *40*(4), 603–624. https://doi.org/10.1111/1468-5965.00390

Nagel, T. (2005). The Problem of Global Justice. *Philosophy & Public Affairs*, *33*(2), 113–147. http://www.jstor.org/stable/3558011.

Nyman, J., & Burke, A. (2016). *Ethical Security Studies: A New Research Agenda*. (J. Nyman & A. Burke, Eds.). Routledge.

Pogge, T. (2008). *World Poverty and Human Rights* (2nd ed.). Cambridge: Polity.

Rodrik, D. (2007). The inescapable trilemma of the world economy. *Dani Rodrik's Weblog*.

Roe, P. (2012). Is securitization a 'negative' concept? Revisiting the normative debate over normal versus extraordinary politics. *Security Dialogue*, *43*(3), 249–266. https://doi.org/10.1177/0967010612443723

Rychnovská, D., Pasgaard, M., & Berling, T. V. (2017). Science and security expertise: Authority, knowledge, subjectivity. *Geoforum*, *84*(June), 327–331. https://doi.org/10.1016/j.geoforum.2017.06.010

Samenow, J. (2018). Red-hot planet: All-time heat records have been set all over the world during the past week - The Washington Post. *The Washington Post*. https://www.washingtonpost.com/news/capital-weather-gang/wp/2018/07/03/hot-planet-all-time-heat-records-have-been-set-all-over-the-world-in-last-week/?noredirect=on&utm_term=.57d5a43cb75f. Accessed 6 July 2018

Schmidt, N. (2016). *The Birth of Cyber as a National Security Agenda (PhD Thesis)*. Charles University.

Schmidt, N. (2017). Planetary Defense as a Gateway to Space for Commercial and Deep Space Exploration. *New Space*, *5*(4), 219–229. https://doi.org/10.1089/space.2017.0005

Schmidt, N. (2018). The political desirability, feasibility, and sustainability of planetary defense governance. *Acta Astronautica*. https://doi.org/10.1016/j.actaastro.2018.06.037

The White House. (2018). *National Near-Earth Object Preparedness Strategy and Action Plan*. http://www.whitehouse.gov/ostp. Accessed 6 July 2018

United Nations. (2016). The World's Cities in 2016: Data Booklet. *Economic and social affair*, 29. https://doi.org/10.18356/8519891f-en

Wendt, A. (2003). Why a World State is Inevitable. *European Journal of International Relations*, *9*(4), 491–542. https://doi.org/10.1177/135406610394001

Winner, L. (2003). Do artifacts have politics? *Technology and the Future*, *109*(1), 148–164. https://doi.org/10.2307/20024652

Zolo, D. (2000). The lords of peace: from the Holy Alliance to the new international criminal tribunals. In B. Holden (Ed.), *Global Democracy, Key Debate* (pp. 73–86). London and New York: Routledge.

Zürn, M. (2000). Democratic Governance Beyond the Nation-State: The EU and Other International Institutons. *European Journal of International Relations*, *6*(2), 183–221. https://doi.org/10.1177/1354066100006002002

Zürn, M. (2018). A Theory of Contested Global Governance, *9*(1), 138–145. https://doi.org/10.1111/1758-5899.12521

Chapter 22
Global Space Governance and Planetary Defense Mechanisms

Joseph N. Pelton

22.1 Introduction

For centuries now, humans have falteringly unraveled the nature of the cosmos and threats from outer space. Eventually, they learned that the Earth revolves around the Sun. They learned that there have been five mass extinction events over the past one and a half billion years, of which four involve climate change and temperature increases. The last and most deadly one involved a giant asteroid some 6 km in size that wiped out the dinosaurs and over 70% of all species on the planet, i.e. the K-T mass extinction event (Encyclopedia Britannica 2017).

Space scientists have learned that the Earth is protected by its natural magnetic field that wards off massive solar storms called coronal mass ejections, and this protective shielding altered with the Earth's magnetic poles shifts on sequences measured in tens of thousands of years. Satellite systems designed to detect nuclear explosions have now taught us that near-Earth objects (i.e. asteroids) hit the world at least four times more frequently than earlier thought. On the plus side, we have also learned that the passive gravitational fields of the Sun, Jupiter and Saturn provide a natural protective capability to capture dangerous asteroids and comets and swallow them up before they become dangerous to Earth.

This hard-won knowledge of the world's cosmic neighborhood and the many dangers that it holds for mankind has taken centuries to acquire. Most of what we know with authority regarding cosmic hazards has only come with the advent of the Space Age and infrared space telescopes, sophisticated satellite experimentation and scientific instrumentation observation. It is only within the last 60 years that we have learned most of what we know with regard to mass exterminations, changes to the Van Allen Belts and changes to the geomagnetosphere, the physics of the Sun and solar storms, asteroid and comet behavior, air, water and land impact processes for

J. N. Pelton (✉)
International Association for the Advancement of Space Safety (IAASS), Arlington, VA, USA
e-mail: joepelton@verizon.net

N. Schmidt (ed.), *Planetary Defense*, Space and Society,
https://doi.org/10.1007/978-3-030-01000-3_22

asteroids, natural protective shielding of the Earth and orbital space debris. There is a reason for new urgency in learning more about space hazards and planetary defense.

The recent scientific knowledge and space experimentation that has fueled our information about the nature of space hazards does not come at a time of stability in the development of human civilization. Indeed, the other half of the story of human civilization and vulnerability to cosmic dangers comes from the growth rate and nature of Earth-based society. In 1800, there were only 800 million humans residing on spaceship Earth. By 1900 there were 1.8 billion people, and today there are amazingly 7.5 billion souls on this small six sextillion metric ton planet. The exponential growth of the human population that might reach nine billion by 2050 and 10–12 billion by the end of the century, coupled with ever greater dependence on modern infrastructure, urbanization and the constraints imposed by climate change, adds up to equation for ever larger exposure to massive natural disasters and especially to cosmic hazards. This could well be a major asteroid or comet strike, major coronal mass ejection disasters amplified by shifts in the Earth's magnetosphere, or even the persistent threats posed by climate change, which is an unrecognized form of cosmic danger.

The bottom line is that unfettered population growth, climate change, ever greater dependence on modern infrastructure, urbanization and population concentration, when linked to the increased knowledge we have of various types of cosmic hazards, all add up to greater vulnerability for all humans. This vulnerability has increased and will increase ever more rapidly in coming decades regardless of one's country of origin or residence, and indeed wherever a person might live on spaceship Earth. Scientists, engineers and physical researchers have led us on a voyage of discovery about cosmic hazards (Pelton and Allahdadi 2015). Now, political scientists, political figures and various elements of the United Nations Organizations such as the UN Committee on the Peaceful Uses of Outer Space and efforts related the Sustainability Goals for 2030 are increasingly becoming aware of various types of problems posed by cosmic hazards. This has, among other things, resulted in discussions or at least the recognition that new or improved forms of global space governance (GSG) might be needed to take on what might be called the Cosmic Hazards agenda and to seek consensus around the world about what could be done to ameliorate, mitigate or even eliminate these type threats.

22.2 Purpose of this Chapter

The purpose of this particular chapter is to provide the latest information about what can be done around the world, especially within appropriate international fora, to address the various types of cosmic hazards that threaten our world that is becoming ever more vulnerable to cosmic threats. The secondary purpose of this chapter is to help people understand what is changing to dramatically increase the various types of so-called cosmic threats that will increasingly threat human society on Earth.

The UN Organization through the General Assembly and other forums has recognized the seriousness of cosmic threats. Further UN bodies have begun to take proactive steps to respond. It has endorsed the creation of a new International Asteroid Warning Network (IAWN) as well as a Space Mission Planning Advisory Group (SMPAG) starting with initial action in December 2013. It has taken a number of steps forward either to better understand and identify cosmic threats or to mitigate these dangers. Sometimes the effort have been indirect, such as through the adoption of Sustainability Goals for 2030, or more directly targeted, such as efforts taken by the UN Committee on the Peaceful Uses of Outer Space (COPUOS) to establish a Working Group on the Long Term Sustainability of Outer Space Activities (LTSOSA) and to begin work to consider or adopt the recommendations of this working group.

In addition to efforts within the UN, individual nations, and most often national space agencies acting under explicit guidance by national legislatures, have undertaken to identify more clearly cosmic hazards and useful response programs. These space agency efforts have primarily focused on threats from near-Earth orbits and potentially hazardous objects, such as from asteroids and comets, orbital space debris and solar storms. Most of these efforts are to detect the nature and scope of these threats, but in some cases this also involves undertaking research regarding how these risks could be mitigated (Fig. 22.1).

There are a number of other actors from around the world that are engaged in addressing space hazards, carrying out research and discovering improved ways to protect against cosmic dangers. These include international scientific and space-

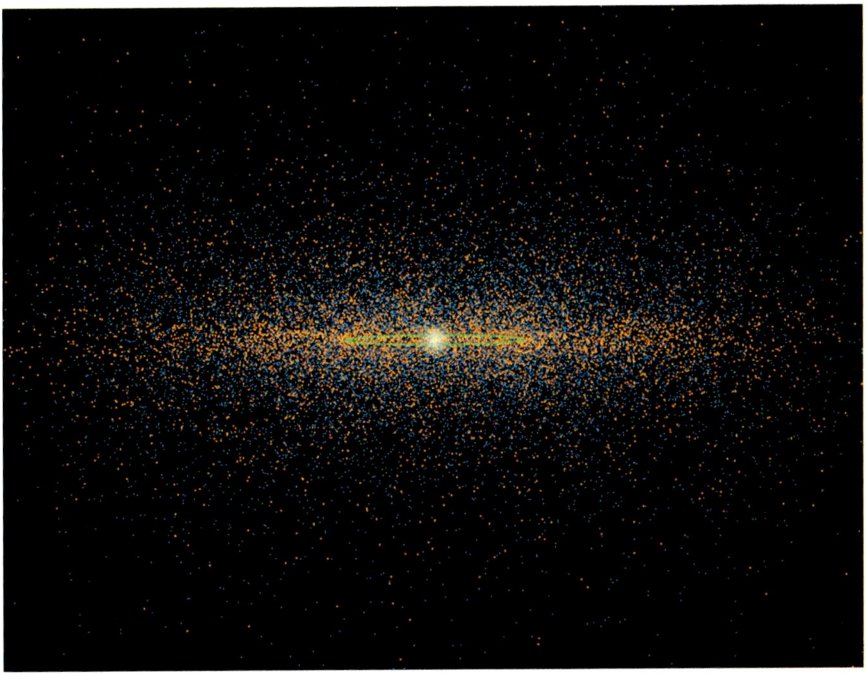

Fig. 22.1 A view of asteroids seen from the Solar System Plane (this includes some 500,000 asteroids shown in blue that are sufficiently large to be city killers) (Graphic courtesy of NASA)

related bodies such as the Committee on Space Research (COSPAR), the International Council of Scientific Unions (ICSU), the International Association for the Advancement of Space Safety (IAASS), the International Academy of Astronautics (IAA) and the International Astronautic Congress of the International Astronautical Federation.

There are also specialized agencies of the United Nations that have a clear-cut interest in space-related research and studies, such as the International Telecommunication Union (ITU) and the World Meteorological Organization. Further, mechanisms have been set up by national or regional space agencies to coordinate their work with regard to cosmic hazards, such as the Inter-Agency space Debris Committee (IADC). In addition to all of these official, quasi-official and non-governmental organizations (NGOs), private organizations also make important contributions to the identification of cosmic hazards and improved warning and response capabilities. These include the Safeguard Foundation of Italy, the Minor Planet Center of the Harvard University-Smithsonian Observatory, the B612 Foundation, university research centers and academic and private observatories.

22.3 The Need for Improved Global Space Governance Related to Cosmic Hazards and Planetary Defense

Other parts of this book explore the increased knowledge we have acquired with regard to cosmic hazards and particularly asteroid threats, as well as new means and methods that could be employed for effective planetary defense against these hazards. The focus of this chapter is on the processes, standards and governance systems that are in place or under development with regard to such protective systems at the national, international and private institutional level of activity. Many of the first efforts to address cosmic hazards and planetary defense started at the national level and within national governmental hearings that pertained to national space agencies. There were also international scientific conferences that took up these issues and concerns. Finally, these issues made their way to international institutions and the United Nations consideration of these important issues. Thus, this discussion will be presented in historical order, showing the evolution of concern and protective actions and improved global space governance innovations that might be considered and implemented.

22.3.1 Legislative Hearings and Actions Concerning Space Hazards in the United States

The first formal US Congressional hearings with NASA officials that resulted in legislative action to initiate an organized search for potentially hazardous asteroids occurred in 1994, when the US Congress directed NASA "to develop a plan to discover, characterize and catalog potentially hazardous NEOs larger than 1 km in size." Efforts began in 1995. In 1998, NASA formally established a Near-Earth Object (NEO) program. As has been mentioned in previous chapters, the specific US Congressional directive was to discover at least 90% of 1-km-sized NEOs within 10 years. NASA managed to fulfill this mandate by 2010 (Loff 2014).

Subsequently, hearings indicated that an inventory limited to 1-km-sized asteroids potentially harmful to Earth was not considered sufficiently comprehensive and that a much larger database was required.

Former Congressman George Brown, who headed the Congressional Science and Technology Committee for many years, was memorialized by US legislation passed in 2005 to assign a more comprehensive task to NASA and Earth observatories to chart 90% of all near-Earth objects larger than 140 m by 2020 (National Aeronautics and Space Administration Authorization Act 2005). This Congressional legislation, although based in part on scientific input from NASA, had no specific rationale other than the fact that clearly an inventory of 140 m potentially hazardous asteroids represents a much more protective search directive than the previously adopted 1-km guideline. In essence, this new limit for a comprehensive search for PHAs was one that NASA could reasonably apply to their search capabilities, which included resources such as the repurposing of the so-called Wide-field Infrared Surveyor Explorer (WISE) space telescope to carry out this asteroid survey (Greicius 2017).

This program has continued to grow and expand. In fact, NASA has created at the Jet Propulsion Labs a Center for Near-Earth Objects (CNEOS), which collects information on potentially hazardous asteroids and displays the latest information on its web site. The latest report from this Center on Near Earth Asteroids as of mid-year 2017 is provided in Fig. 22.2 (Chodas 2018).

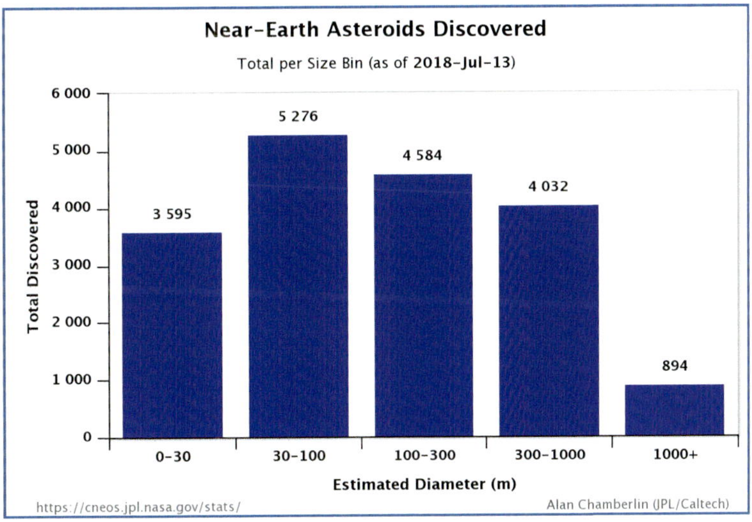

Fig. 22.2 Report on near-earth asteroid discovered as of July 2018 (graphic courtesy of NASA)

This program is often referred to in the press and in popular accounts by its informal name "Spaceguard." This name is actually derived from the Arthur C. Clarke science fiction novel *Rendezvous with Rama* (1972). In this story, set well into the twenty-first century, a new global initiative is begun to mount a planetary defense program and to discover potentially hazardous asteroids on a systematic global basis. This new "Spaceguard" program is finally organized just after a catastrophic asteroid impact occurs in the novel (see Chap. 2).

NASA, together with ground observatories and other space agencies, particularly the European Space Agency, has now been working on the identification of potentially hazardous asteroids for well over a decade. The good news is that a great deal of data has been acquired and stored at the Minor Planet Center in Cambridge Massachusetts (Harvard-Smithsonian Observatory), the CNEOS at JPL and the Spaceguard Foundation in Italy. This effort is closely linked to the newly created and UN-sanctioned International Astronaut Warning Network (IAWN), which will be discussed shortly.

The European Space Agency has created its own office for the coordination of NEO detection and analysis. This Coordination Office and how it relates to other scientific units, space situational awareness activities, the press, political entities and the rest of the world is shown in Fig. 22.3.

The bad news is that since 2005, a great deal more information has been attained about asteroid threats, and this information indicates a much larger threat. We know

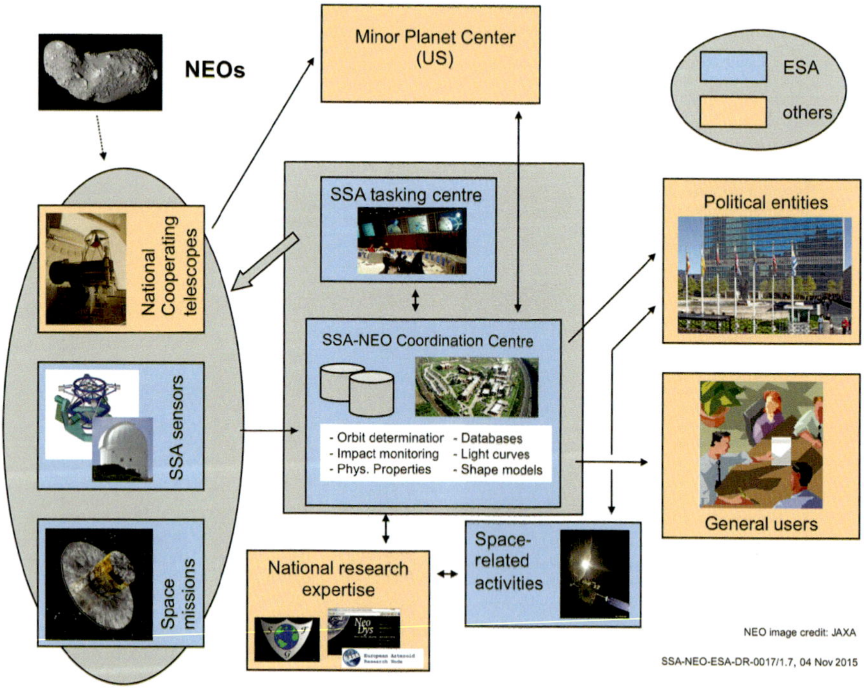

Fig. 22.3 The European Space Agency NEO coordination flow diagram (graphic courtesy of ESA)

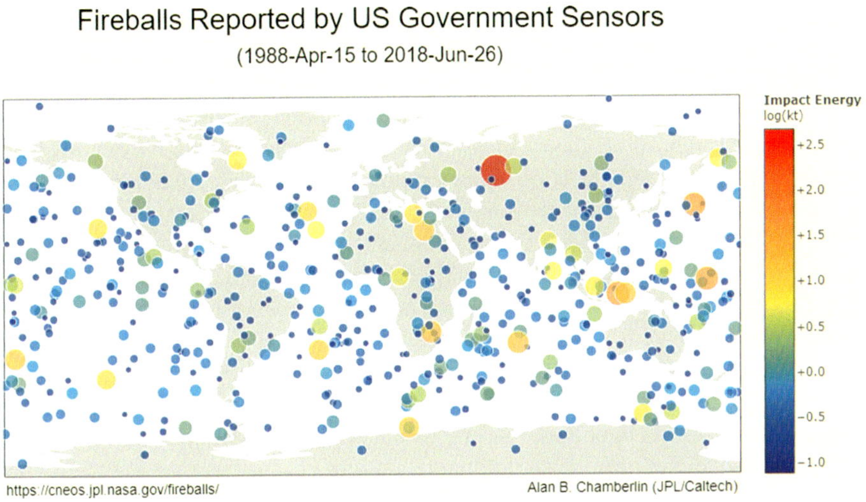

Fig. 22.4 Asteroid impact on earth over a two decade period (data source CNEOS JPL NASA)

that the universe of potentially hazardous asteroids is much larger than had been previously thought. We know that the Tunguska event early in the twenty-first century that devastated an area equivalent in size to the San Francisco Bay area was about 40 m in diameter and that depending on their relative velocity to Earth, asteroids 25–35 m in diameter are truly city killers. As we add more and more megacities (i.e. cities more than 10 million) each year to our planet, our vulnerability to cosmic hazards continues to grow.

The sensors mounted on GPS Navstar satellites that are designed to detect nuclear explosions have fortunately sensed few nuclear tests. The data from these sensors, however, has indicated that asteroid, bolide and meteor impacts on Earth are four to even ten times more common than previously thought. Indeed, every day the Earth is impacted by about 100 tons of stardust and cosmic particles.

The following chart as provided in Fig. 22.4 shows the distribution of disintegrated asteroids that impacted the Earth, the atmosphere or oceans over a two-decade period. The graphic illustrates the size, energy level and time of day occurrence for collisions by large-scale bolides and asteroids (NASA JPL 2014).

22.3.2 *Public Awareness of Cosmic Hazards and the Need for Planetary Defense*

Charts such as that of Fig. 22.4 need to be widely distributed across the planet to show the public the extent to which there is a need for global agreement for an active planetary defense system. Only with widespread public appreciation of the size and immanence of the danger can further progress be made to agree on the need, urgency and required public investment in planetary defense programs. Without such widespread recognition of these facts, we will not see such programs implemented on a worldwide basis. Thus this general public awareness of cosmic hazard is critical.

Threat awareness is a vital step in being able to move ahead to achieve widespread global recognition of the need for more concerted efforts that are more than lip service. In light of all the competing needs related to food and water, shelter, disease and healthcare, education, energy, transportation, and communications and national defense, the current tendency is to take actions that seem responsive but cost very little. Actions by the UN to endorse the International Asteroid Warning Network (IAWN) and to create the Space Mission Planning Advisory Group (SMPAG) seem useful and responsive but involve virtually no expenditure of significant public funds. It is essential to have broad and well-focused public support that is truly urgent and focused in order to obtain support within the United Nations—and especially the 83 current national members of the UN Committee on the Peaceful Uses of Outer Space—for a global program of action that involves the potential expenditure of billions of dollars.

One of the important steps to help the public understand cosmic threats, especially from potentially hazardous asteroids, has been to create what is known as the "Torino Impact Hazard Scale," as is depicted in Table 22.1 (discussed in detail in Chap. 12). This was originally adopted in 1999 and has subsequently been updated based on the latest information that has now been acquired. Scientists actually employ the more precise and thorough Palermo Technical Impact Hazard Scale, which indicates the likelihood of a near-Earth object hitting earth using a logarithmic and other key data such as their energy level in their scientific catalogues. Only asteroids with a positive number above zero are consider a potential threat on the Palermo Scale.

The Torino Scale, however, is akin to the Richter Scale in that it shows the level of threat on a range of zero to ten and allows laymen to understand clearly and easily the level of threat and the likelihood of these events occurring. If every school across the world taught students about the Torino Scale, they would be implementing a key first step towards educating the world about cosmic threats and the need for planetary defense.

22.4 Assessing Asteroid and Comet Impact Hazard Predictions in the Twenty-First Century

Table 22.1 Torino impact scale for potentially hazardous asteroids and comets

No hazard (white zone)	0	The likelihood of a collision is zero, or is so low as to be effectively zero. Also applies to small objects such as meteors and bodies that burn up in the atmosphere as well as infrequent meteorite falls that rarely cause damage
Normal (green zone)	1	A routine discovery in which a pass near the Earth is predicted that poses no unusual level of danger. Current calculations show the chance of collision is extremely unlikely with no cause for public attention or public concern. New telescopic observations very likely will lead to re-assignment to Level 0

(continued)

Table 22.1 (continued)

Meriting attention by astronomers (yellow zone)	2	A discovery, which may become routine with expanded searches, of an object making a somewhat close but not highly unusual pass near the Earth. While meriting attention by astronomers, there is no cause for public attention or public concern as an actual collision is very unlikely. New telescopic observations very likely will lead to re-assignment to Level 0
	3	A close encounter, meriting attention by astronomers. Current calculations give a 1% or greater chance of collision capable of localized destruction. Most likely, new telescopic observations will lead to re-assignment to Level 0. Attention by public and by public officials is merited if the encounter is less than a decade away
	4	A close encounter, meriting attention by astronomers. Current calculations give a 1% or greater chance of collision capable of regional devastation. Most likely, new telescopic observations will lead to re-assignment to Level 0. Attention by public and by public officials is merited if the encounter is less than a decade away
Threatening (orange zone)	5	A close encounter posing a serious, but still uncertain threat of regional devastation. Critical attention by astronomers is needed to determine conclusively whether or not a collision will occur. If the encounter is less than a decade away, governmental contingency planning may be warranted
	6	A close encounter by a large object posing a serious but still uncertain threat of a global catastrophe. Critical attention by astronomers is needed to determine conclusively whether or not a collision will occur. If the encounter is less than three decades away, governmental contingency planning may be warranted
	7	A very close encounter by a large object, which if occurring this century, poses an unprecedented but still uncertain threat of a global catastrophe. For such a threat in this century, international contingency planning is warranted, especially to determine urgently and conclusively whether or not a collision will occur
Certain collisions (red zone)	8	A collision is certain, capable of causing localized destruction for an impact over land or possibly a tsunami if close offshore. Such events occur on average between once per 50 years and once per several 1000 years
	9	A collision is certain, capable of causing unprecedented regional devastation for a land impact or the threat of a major tsunami for an ocean impact. Such events occur on average between once per 10,000 years and once per 100,000 years
	10	A collision is certain, capable of causing global climatic catastrophe that may threaten the future of civilization as we know it, whether impacting land or ocean. Such events occur on average once per 100,000 years, or less often

22.5 International Efforts to Identify Cosmic Hazards and Undertake Planetary Defense

The international astronomical community has, of course, been well aware of the cosmic hazards and the need for planetary defense for many years. The International Council of Scientific Unions (ICSU), the Committee on Space Research (COSPAR), the International Academic of Astronautics, the Association of Space Explorers, the Secure World Foundation, The Space Generation Advisory Council and other similar bodies have perceived the threat and the need for increased vigilance. The advent of the Space Age and increased awareness of such aspects as the Van Allen Belts (Explorer 1 in 1958), the recognition that there are more potentially hazardous asteroids than previously thought and new space-based capabilities have prompted a great deal of new information about space hazards. The added capability proved by the repurposed NEOWISE spacecraft (the Near Earth Orbit Wide-Range Infrared Surveyor Explorer) has dramatically increased our ability to detect additional asteroids, as depicted in Fig. 22.5.

In addition, new initiatives such as the International Association for the Advancement of Space Safety (IAASS), new international conferences on cosmic hazards and planetary defense and global events such as Asteroid Day have helped to move ahead efforts to cope with planetary defense.

At COPUOS in 2001, following the UNISPACE III Conference, the so-called Action 14 was formed to address the cosmic hazard of potentially hazardous asteroids and comets. It was officially created with the following membership: Australia, Brazil, China, Czech Republic, Finland, Germany, Iran, Japan, Kazakhstan, Lebanon, Malaysia, Nigeria, Pakistan, Poland, Russian Federation, Saudi Arabia, Syrian Arab Republic, he United Kingdom and the United States of America. In addition, the following organizations were also a part of Action Team 14: the

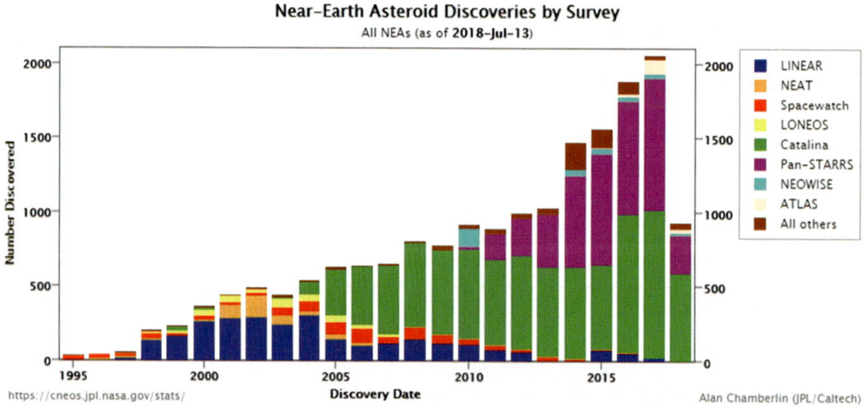

Fig. 22.5 Detection of asteroids by land observatories and NASA NEOWISE Program (graphic courtesy of CNEOS JPL NASA)

Association of Space Explorers (ASE), European Space Agency (ESA), Committee of Space Research (COSPAR), International Astronomical Union (IAU), National Space Society, Space Generation Advisory Council, European Space Science Committee, European Science Foundation and the Safe Guard Foundation.

Since 2007, The NEO issue has been officially dealt with by the Science and Technical SubCommittee of COPUOS through a Working Group on NEOs, with its overall work being endorsed and guided by COPUOS.

The Action Team 14 looks at potentially hazardous asteroids in a comprehensive way and has divided its work into the following six parts, which began to consider mitigation and policy that went beyond simply identification of NEO threats:

- NEO Detection and Remote Characterization
- Orbit Determination and Cataloguing
- Consequence Determination
- In-situ Characterization
- Mitigation
- Policy

As a result of the work of Action Team 14, a resolution was adopted by the UN General Assembly in December 2013 (A/RES/68/075). This resolution on near-Earth object impact threat had been endorsed by the Scientific and Technical Subcommittee at its 50th session and endorsed by the Committee at its 56th session. This resolution supported the creation of the International Asteroid Warning Network (IAWN) as well as the Space Mission Planning Advisory Group (SMPAG), which was seen as a significant step forward by the United Nations in accepting the gravity of cosmic hazards and the need to plan for preventive action.

The Association of Space Explorers (ASE), which includes most of the astronauts and cosmonauts that have flown into space, was particularly active in lobbying for action on this topic by the UN COPUOS and the UN General Assembly. Just subsequent to this adoption of the UN Resolution, the ASE issued a press release urging significant follow-up actions. This release included the following significant statement:

> Because near-Earth asteroid searches have focused almost exclusively on large objects with global destructive potential, 99% of the objects big enough to level a major metropolitan area remain undiscovered. As technology improves and hundreds of thousands of new asteroids are found, the global community will likely be confronted by one posing a worryingly high probability of striking Earth (Turnage and Jones 2014)

If one reviews the data provided in Fig. 22.2, one might conclude that there are a smaller number of potentially hazardous asteroids that are below 30 m in diameter (i.e. 2949 now identified) than asteroids that are say 30–100 m in diameter (i.e. 4475 now identified). This is dramatically not the case. In fact, there is a strong inverse ratio. The smaller the diameter of asteroids, the more there are of them. As the size decreases by a factor of ten, the population of these smaller asteroids goes up by about a factor of ten. The reason that the recorded number of smaller asteroids is low is simply because they are harder to identify with today's technology. NASA is currently developing concepts to define a new infrared telescope, named NEOCAM (see Chap. 4), that is geared to find asteroids that are 140 m in diameter, in accord with instructions from the US Congress. In contrast, the B612 Foundation has undertaken to define and construct an infrared telescope that would be able to locate and determine the orbit of near earth asteroids down to 35 m in diameter. These smaller asteroids might possess a hundred times less kinetic energy than 140-m asteroids. Nevertheless, these 35-m potentially hazardous asteroids have sufficient energy to still be major city killers that could wipe out a megacity with 10 million inhabitants. Thus, the number one priority is to be able to identify, track and calculate the threat level of much smaller asteroids than is currently defined by space agencies such as NASA.

The object of such a "Wide Survey" under the European Space Agency Space Situational Awareness (SSA) program would seek to identify smaller asteroids under the size of 50 m. The logic behind such an effort is that it could guarantee the European competitiveness in discovering a peculiar class of objects, i.e. those very small in size (<50 m) closely approaching our planet. These asteroids are likely to become the source of the most frequent alerts, as demonstrated by the 2008 TC3 case, which was the first meteorite fall observed when the relative meteoroid was still in space, as well as a similar event involving asteroid 2014 AA, whose associated meteorites could not be recovered as they presumably splashed down in the Atlantic Ocean (Pippo and Perozzi 2015).

22.6 Charting the Path Forward: Global Space Governance and Planetary Defense

The UN General Assembly Resolution to endorse the establishment of the International Asteroid Warning Network (IAWN) and the Space Mission Planning Advisory Group (SMPAG) are clearly positive steps forward. Likewise, progress by the UN COPUOS Working Group on the Long Term Sustainability of Outer Space Activities and adoption of a number of their recommendations is another positive sign. Nevertheless, serious problems remain to be addressed, and until a global program of action is agreed and substantial financial resources committed to identifying cosmic hazards and undertaking planetary defense, there remains a long way still to go.

The book *Global Space Governance: an International Study* devoted an entire chapter to examining this topic. At the end of this book's chapter, the international team of analysts developed two charts. The first chart outlined current concerns and problems related to cosmic hazards and efforts at planetary defense. The second chart sought to develop suggestions and recommendations of what might be done to improve efforts in this key field that might be essential to the long-term survival of the human race and life as we know it. These key charts that outline key problems and actions for the future are provided as Tables 22.2 and 22.3 (Jakhu and Pelton 2017).

By the end of the twenty-first century, the world will be much more heavily dependent on a wide range of modern infrastructure, will host some 10–12 billion people that will be 75% urban and will include perhaps a 100 megacities each with populations of over 10 million inhabitants. Spaceship Earth will by 2100 be much more vulnerable that the world that existed in 1900. In that world, the global population was less than 25% urban. There were only 1.8 billion people, and their jobs, livelihoods and access to food were far less dependent on the electrical grid, broadband telecommunications and information networks.

In June 2018 the UNISPACE +50 event was held in Vienna, Austria. This important global meeting of space leaders from around the world sought to establish the agenda for action by the UN COPUOS through 2030. These discussions and deliberations in Vienna strove to coordinate the longer-term agenda to be carried out by COPUOS in support of the UN 17 Sustainable Development Goals for Planet Earth. This overall action agenda seeks to address among other items cosmic hazards and planetary defense. A copy of the book *Global Space Governance: An International Study* was distributed to the 83 national delegations represented on the UN COPUOS. It is hope that the action agenda outlined in Tables 22.2 and 22.3 will be considered for possible implementation as part of this ongoing process to meet the overall UN goals set to be accomplished by 2030.

22.7 Conclusions

The Space Age has enabled many new satellite applications that are now vital to global communications, information networking, remote sensing, meteorological forecasting, and positioning, navigation and precise timing. These new useful satellites enable global commerce, national defense, rescue operations, avoidance of loss of life due to significant storms and much, much more. Scientific satellites have also provided vital new knowledge about Earth, the Solar System and the great cosmos beyond. This new information has allowed us to understand the nature and great hazard presented by cosmic threats to our six sextillion-ton spacecraft that orbits the Sun each year in a perilous path. Many thousands of potentially hazardous asteroids, solar storms and more present us with cosmic hazards against which we currently have no particular planetary defense.

As we better understand the nature of asteroid and comet threats in terms of potential Earth impact, and as we better understand the risks presented by solar storms, shifts in the Earth's magnetosphere that helps to defend us against cosmic threats, we are better prepared to undertake what can truly be labeled planetary defense.

The nature of these threats is quite different, and this means that both the detection of cosmic threats and the creation of protective systems that might constitute a future planetary defense are also quite different. The first step is to create public awareness that real threats of potentially devastating effect do exist and to create improved means to detect this risks on a more systematic basis. The next step is to create

Table 22.2 Key problems and challenges presented by cosmic hazards

Key elements of cosmic hazards			
Issues and challenges	Technical challenges (natural and manmade)	Economic, business, demographic, health and other challenges	Risks/threat mitigation/equity issues
Potentially hazardous asteroids and comets	Weakening of Earth's magneto-sphere	High cost of tracking and detection systems (space and ground based)	(1) Extreme potential destruction of urban infrastructure and massive loss of life. (2) Asteroid defense systems and technologies can be seen as space weapons. (3) Unsuccessful defensive actions could expose a country to liabilities. (4) Current international agreements to defend against cosmic hazards are generally inadequate. (5) There are no current agreements that allow the use of UN peacekeeping forces or international resources to recover from a major asteroid strike. (6) There are no massive insurance funds (i.e. trillions of dollars) that could be deployed to respond to a major asteroid strike
Solar flares, solar energetic particles, and coronal mass ejections from the Sun	Weakening effectiveness of Earth's magnetosphere, seasonal ozone holes in polar regions	Increased incidence of skin cancer; genetic mutations High cost of space systems to study solar phenomena; Much higher cost to develop protective systems against extreme solar weather	(1) No known means to forestall or even forecast extreme solar weather events. (2) Techniques to forestall extreme solar events and climate change would be expensive and there are no globally agreed systems to finance nor implement. (3) Planetary defense may require time critical response. (4) There are no current agreements that allow the use of UN peacekeeping forces or international resources to recover from a major solar flare event. (5) There are no massive insurance funds (i.e. trillions of dollars) that could be deployed to respond to major asteroid strike

(continued)

Table 22.2 (continued)

Key elements of cosmic hazards			
Changes in Earth's Vulnerability to these Cosmic Hazards	Earth magnetosphere has decreased 15% in its deflective strength.	Spread of urban development; Population growth and exposure of vital infrastructure (i.e. satellites, electrical grid, and electronics and computer chips)	(1) If Earth's magnetic poles shift, its deflective strength might be reduced 20 times more and for hundreds of years during shift. (2) Modern economies vulnerable to natural EMP (particularly satellites and electric grid) increasing every year

Table 22.3 Proposed initiatives to cope with cosmic hazards and strengthen planetary defense

Proposed international actions related to cosmic hazards and planetary defense	
Proposed action	Entities to take proposed action
Space agencies should all explicitly adopt planetary defense as a prime strategic goal	All space agencies and national legislatures
Create Inter-Agency Committee to address all forms of cosmic hazards	Inter-Agency space Debris Coordinating (IADC) Committee should move to create Inter-Agency Committee on all space hazards and planetary defense
UN COPUOS Action Team 14 should be given a new and stronger terms of reference that takes into account the most recent actions by the UN General Assembly as of December 2013	UN COPUOS and its Committees and Working Groups. Discussion within COPUOS and UN Office of Disarmament of new elements of an International Code of Conduct or amended language to a PPWT
The two improved asteroid detection infrared telescopes (Sentinel and NEOCAM) should be built and launched as soon as possible	B612 foundation and U.S. Government/NASA
Coordinated National and International Research Program on Cosmic Hazards and Planetary Defense (involving potentially hazardous asteroid & comet detection, extreme solar weather, study of changes to Earth's protective systems, and new natural and human-devised protective strategies and systems against cosmic hazards)	All national governments through legislative and budgetary action. All relevant research institutes and laboratories around the world. The United Nations General Assembly and COPUOS and Action Team 14. The International Astronomical Union (IAU) and Committee on Space Research (COSPAR). Professional bodies such as the IAASS, IAA, ISU
After various studies and assessments as to threats from cosmic hazards have been undertaken and coordinated. And after further capabilities have been developed to cope with both Near Earth Objects and extreme solar storms, a 'UNISPACE + 50' Conference should consider, among other items how to mitigate the risks associated with cosmic hazards—NEOs, Orbital Space Debris and especially solar storms and the weakening Earth's magnetosphere	A 'UNISPACE + 50' Conference might consider new arrangements, goals and objects, and risk minimization and mitigation against solar hazards that might follow the precedent of the Action Team 14 model that addressed NEO hazards. At its 2016 General Session COPUOS recommended "International framework for space weather services " for adoption by the UN General Assembly as one of seven thematic priorities to be pursued through the UNISPACE+50 process

improved means to detect such cosmic hazards more broadly and more accurately as much in advance as possible. The final and most difficult step is to begin to systematically implement planetary defense systems that can protect us from the most damaging cosmic threats. This requires international cooperation, improved global space governance, new and better space technology and the appropriate mechanisms and systems to allow planetary defense systems to be created and deployed.

It is true that planetary defense systems could we be expensive, i.e. billions of dollars to design, deploy and operate, but it is also true that planetary defense systems can not only help to increase our knowledge of the universe, but could also present new economic and resource opportunities. The tools and technology needed to defend against potentially hazardous asteroids and comets might also be deployed to enable cost-effective and viable space mining activities. Efforts to understand solar storms, coronal mass ejections and radiation patterns from the Sun might not only help to protect vital electrical, information and telecommunications infrastructure but also might be used to mitigate global climate change and even geo-engineer a breathable atmosphere on Mars.

In short, it might be wise and foresighted to see cosmic hazards and planetary defense not as a huge expense to be undertaken to defend Earth against an extreme assault, but rather as an investment opportunity to create a new trillion-dollar space economy that returns tens of dollars in dividends for every dollar invested. What is the value of saving the inhabitants and infrastructure of an entire city such as Tokyo, London, Paris, Beijing, Mexico City, Rio, Moscow, Cairo, Lagos or Cape Town? What is the value of being able to recover vast resources from the skies? What is payoff if we could make the entire planet of Mars inhabitable for humans without wearing space suits?

Today, concrete answers about the payouts from such investments are as imponderable as Thomas Jefferson's decision to buy the Louisiana Purchase or Columbus's voyage to the New World. Some decisions to move ahead to develop new technology and system capabilities seem like an inherent part of human destiny. The concerted global effort to pursue a better knowledge of cosmic hazards and to undertake a systematic planetary defense is a key step in the human experience and future evolution of Homo sapiens.

References

Encyclopedia Britannica. (2017). K-T extinction. In *Encyclopedia Britannica*. https://www.britannica.com/science/K-T-extinction

Greicius, T. (2017). Wide-field Infrared Survey Explorer (WISE). *NASA*. https://www.nasa.gov/mission_pages/WISE/main/index.html. Accessed 16 May 2018

Jakhu, R. S., & Pelton, J. N. (2017). *Global Space Governance: An International Study*. (R. S. Jakhu & J. N. Pelton, Eds.). Cham: Springer International Publishing. https://doi.org/10.1007/978-3-319-54364-2

Loff, S. (2014). NASA's Search for Asteroids to Help Protect Earth and Understand Our History. *NASA*. https://www.nasa.gov/content/nasas-search-for-asteroids-to-help-protect-earth-and-understand-our-history/. Accessed 16 May 2018

NASA JPL. (2014). New Map Shows Frequency of Small Asteroid Impacts, Provides Clues on Larger Asteroid Population. *NASA*. https://www.jpl.nasa.gov/news/news.php?feature=4380. Accessed 16 May 2018

National Aeronautics and Space Administration Authorization Act (2005). https://www.congress.gov/bill/109th-congress/senate-bill/1281

Paul Chodas. (2018). Center for NEO Studies. *NASA*. https://cneos.jpl.nasa.gov/. Accessed 16 May 2018

Pelton, J. N., & Allahdadi, F. (2015). *Handbook of Cosmic Hazards and Planetary Defense*. Springer International Publishing.

Pippo, S. di, & Perozzi, E. (2015). The European Operational Initiative on NEO Hazard Monitoring. In J. N. Pelton & F. Allahdadi (Eds.), *Handbook of Cosmic Hazards and Planetary Defense* (pp. 615–636).

Turnage, A., & Jones, T. D. (2014). *Vital Next Steps in Global Response to the Asteroid Threat*. Associatino of Space Explorers.

Chapter 23
New Enemy, Old Pact? Past Defense Pacts and Future Planetary Defense Treaty

Tomáš Bruner

23.1 Introduction

To create a suitable framework for international cooperation that would safeguard planetary defense, one must address several interconnected questions: first, when is it the right time to establish adequate legal frameworks for international cooperation; second, how can we ensure that all the participating nations will contribute the same efforts and resources to the common defense project; third, how can we regulate decision making, and fourth, how can we safeguard that the outcomes of the project, e.g. in the form of advanced military technology, shall not be misused for the benefit of several leading nations, or even one single nation, to the detriment of others.

This chapter draws a parallel between past defense pacts and a possible future planetary defense treaty, i.e. a treaty that would establish a legal basis for how and when to act in case of a NEO threat. To draft hypothetical answers to the questions above, the chapter quantitatively investigates selected defense pacts drawn between nations from 1815 to 2003, and qualitatively examines five defense pacts that are currently in force. Comparing when and how those pacts were achieved offers valuable insights about the possible creation of a future planetary defense treaty.

This chapter intends to contribute to this debate. Its first section argues that a cosmopolitan regime would need a specific written agreement, which we may provisionally call the 'Earth Defense Treaty.' The second section creates a historical analogy, comparing such a treaty to a defense pact. It explains what findings could be brought through this parallel to justify its relevance and contribution to the debate. The third section quantitatively examines 74 multilateral past defense pacts

T. Bruner (✉)
Charles University, Prague, Czech Republic
e-mail: tomas.bruner@fsv.cuni.cz

N. Schmidt (ed.), *Planetary Defense*, Space and Society,
https://doi.org/10.1007/978-3-030-01000-3_23

concluded between 1815 and 2003 and assesses the correlation between threat presence and a completed defense pact. The cosmopolitan view expects that countries would adopt a treaty establishing a NEO defense regime preventively. The section therefore asks whether there are any precedents for such preventive action, or whether the past defense pacts were always concluded as an immediate, realistic reaction to threats. It also quantitatively analyzes the number of multilateral defense pacts that included clauses regulating common contributions and cooperation. The fourth section looks at chosen defense pacts qualitatively and examines how the pacts dealt with decision making, prevention of freeriding, and prevention of abuse. These topics would have to be dealt with in the Earth Defense Treaty as well.

23.2 Why 'Earth Defense Treaty'?

Burke defined cosmopolitanism as a prescriptive approach demanding political and normative change. This change should be brought by 'concerted collective effort of scholarship, analysis, and social response,' which is the most suitable way to address global insecurities (Burke 2015, p. 193). A common treaty regulating Earth Defense would establish an adequate legal basis for such a collective social response. Falling into the category of a progressive development of public international law,[1] the treaty would introduce new norms to the Outer Space regime and reflect cosmopolitan values in two ways: by enabling collective military action in Outer Space, and by coordinating efforts.

Firstly, it would enable collective Outer Space military action that is ambiguously regulated or even prohibited by present treaty-based and customary international law. The current legal setting, based mainly on Cold War Treaties, hinders the development of planetary defense systems (PDS) and prohibits the deployment of weapons[2] in space (Remo 2015, p. 38). Waiting for a change in customary law is logically not an option here; therefore the treaty would progressively change international law with much desired precision (Knox 1998; Kunich 1997). Zaitsev expressly recommends that the conclusion of an international treaty '[o]n the principles ensuring the Earth's defense against asteroid and comet impact hazards' would equip the participating countries with legal instruments to develop the PDS. The instruments should be grouped around particular areas: scientific, technical, organizational, and financial. The treaty should also safeguard against a single

[1] Progressive development of international law delimits the opposite of a slow, gradual development of international customs. It refers to when a treaty is adopted that suddenly changes the current practice of states.

[2] The qualification of a weapon may cause problems; certain systems such as lasers may serve peaceful as well as military purposes. Moreover, necessity as a general condition precluding wrongfulness might represent an exemption from this prohibitory regime in case that the collision with NEO would seem inevitable. For further discussion on weapons in space see cf. Tannenwald (2004).

military alliance or country from among its members solely developing, deploying, and controlling the PDS[3] (Zaitsev 2011, pp. 437–438).

Moreover, thanks to the Earth Defense Treaty, the actions of the international community would be coordinated (Smith 2004) with prescribed equal sharing of financial burdens (Zaitsev 2011, p. 438). Already in 2001, the action team on near-Earth objects (NEOs)[4] recommended to coordinate mitigation actions among the countries that could be affected and those capable of participating in active deflection (Remo and Haubold 2014, p. 85). The treaty would be an ideal tool for this purpose. Moreover, it could serve as an assurance that the activities are for the common good and will not be used for individual benefit to the detriment of others. Without this treaty-based assurance, singular defense activities may become a destabilizing factor, trigger arms races, and destroy the balance between leading powers, not only in space but also on Earth. Already Rosas (1983, pp. 362–363) has warned against placing highly destabilizing potential weapons in orbit. Tannenwald (2004) also emphasized that the placement of weapons in Outer Space could have far reaching security implications. Similarly, the US argument that ballistic missile defense is aimed against ‘rogue states’ did not convince Russia (Hynek and Stritecky 2010; Thangavelu et al. 2015).

The feasibility of achieving such a treaty is yet to be examined. Billings (2006) asserts that intense international cooperation in space, with just a few strains, has been an attribute of Outer Space exploration. She also refers to Article IX of the Outer Space Treaty as a relatively successful example of an International Community agreement on a planetary defense norm. This article states that the member states shall refrain from contaminating the terrestrial environment by extraterrestrial materials (physical, chemical, or biological, should there be some) in order to protect the biosphere against adverse outside materials (Billings 2006, pp. 432–433). Nevertheless, a weaponized deflection system is a more sensitive issue than peacetime research cooperation and material containment. States will react more cautiously and might tend to pursue their own interests more vigorously. Therefore, it is necessary to carefully consider which obligations could form a plausible part of the eventual Earth Defense Treaty for future member countries, and which could rather distract them—all of this prior to the stipulation of the actual wording of the treaty.

By considering these issues sufficiently in advance, we could ensure smoother adoption and avoid the scenario when ‘legal detailia are debated to the detriment of life on Earth’ (Knox 1998, p. 43). Comparison of the eventual Earth Defense Treaty to past defense pacts may significantly contribute to this task. The following section explains how we can learn from past defense pacts when drafting a possible future Earth Defense Treaty.

[3] Thus, the treaty would firstly impose an obligation upon its members. Later, this obligation could transform into a custom with wider applicability.

[4] An expert body established after the Third UN Conference on the Exploration and Peaceful Uses of Outer Space.

23.3 Why is an 'Earth Defense Treaty' Comparable to a Defense Pact?

An Earth defense treaty could be defined as a legal agreement, governed by the Vienna Convention on the Law of Treaties of 1969 or respective customary law, intended to regulate and prescribe certain kinetic military activities of the signatory states in order to establish and enhance defense against common threats coming from outer space, mostly in the form of NEOs. Offense pacts are defined as 'alliances that promise assistance in conflict initiated by a member state' (Johnson and Leeds 2011, p. 47), with the respective definition of a defense pact thus becoming 'alliances that promise assistance in conflict initiated by a third state/party against a member state.'

The definition of a defense pact displays several similarities to the eventual Earth Defense Treaty.[5] Firstly, a defense pact is concluded in the form of an international legal agreement. Secondly, it regulates and prescribes kinetic and military activities for states if certain conditions are met. Thirdly, its basic aim is to counter a danger, either a situation of an armed attack against the sovereignty and territorial integrity of a member state ('old enemy') or a situation of NEO impact ('new enemy'). Both parties in a defense pact, and in the eventual Earth Defense Treaty, pledge to support a prescribed response to this specific impetus. As a result, past defense pacts may serve as suitable precedents for knowledge gathering and hypothesis generation about the future Earth Defense Treaty.

An additional argument in favor of this analogical enquiry is offered by Peterson (1997, p. 247): 'Reasoning by analogy, which permits the transfer of assumptions and postulates from a well-known field to an unfamiliar one, provides the necessary cognitive resources for developing a working conception of the new issue or problem.' Peterson claimed that in the 1950s, the states had two possibilities to conceptualize Outer Space and its legal regime according to what they had already known: national airspace, and high seas legal regime. While the Soviets proposed that Outer Space be governed by a similar regime as national airspace, the USA countered that Outer Space should be treated as commons, similarly to the high seas or common heritage of mankind. The US conception prevailed.[6]

The political actors build their cognitive frames upon the existing regimes and transferred the knowledge of the source area to the target area (Peterson 1997, pp. 246–248). A similar process is likely to occur to states when negotiating the Earth Defense Treaty. We may therefore learn from past defense pacts by asking the

---

[5] The sole words 'pact' and 'treaty' may be to a large extent treated synonymously. Cf. article 2, para 1, letter A/ of the Vienna Convention on the Law of Treaties: '"*Treaty*" *means an international agreement concluded between States in written form and governed by international law, whether embodied in a single instrument or in two or more related instruments and whatever its particular designation.*' Therefore both 'treaty' and 'pact' refer to agreement between states governed by international law.

[6] Eisenhower succeeded with the argument that under the Soviet perception, flight of Sputnik over US territory would represent a violation of international law (further reading Divine 1993).

questions: When were the defense pacts concluded? What obligation and language were the states most willing or reluctant to accept? The answers shall help us hypothesize about possibilities for the future Earth Defense Treaty.

Lachs (2010, pp. 20–21) quotes Lord Mansfield ('There is nothing in law so misleading as a metaphor or an analogy') and warns that analogy may be a source of great fallacy if applied imprudently: '"mechanical transfer" of institutions from one environment to another is of little avail: it may lead to distortion and even seriously stunt the development of the new branch of law.' We agree with Lachs that analogy may force a researcher to overlook, ignore, or even deny important qualities of the new domain, and in so doing may restrict possible solutions and lead to self-fulfilling prophecies. On the other hand, Lachs (ibid.) provides certain criteria that may safeguard the use of analogy in regards to Outer Space. Firstly, it must be applied progressively and with creativity, not remaining bound to old frameworks. Secondly, it must distinguish different qualities of Outer Space and respect the mutatis mutandis (apply only when reasonable) doctrine. This chapter intends to respect those roles. Moreover, it claims that the states will most likely resort to analogical perceptions of Outer Space, which is a possibility to be counted on, but not because analogy is necessarily the best approach. Both in this chapter and in this book, there are suggestions about how to overcome the analogical approach and lean towards cosmopolitan development.

23.4 Conditions of the Conclusion: How Severe a Threat is Necessary to Bring About a Successful Pact?

This section examines what role threats played in the adoption of defense pacts. According to our results, we may investigate how tangible the NEO danger must be to force states to adopt the Earth Defense Treaty.

Generally speaking, there are different causes of pact conclusion, such as (I) ideological, economic, and worldview proximity of parties, (II) bargaining over material resources or immaterial gains, (III) possible balancing and or bandwagoning, and (IV) presence of actual threat or actual attack (see Table 23.1).

Each of the causes is favored by different theories of international relations. The security communities approach (I) prefers the first cause and expects that countries are willing to enter into a common defense system, once they share simi-

Table 23.1 Different causes of defense pacts conclusion

	IR theory	Explanation of defense pact emergence
(I)	Constructivism, liberalism	Shared identity, ideology and economy
(II)	Marxism, geopolitics	Bargaining over material and immaterial resources
(III)	Structural Realism	Balancing and bandwagoning
(IV)	Defensive realism	Presence of a threat of certain severity

lar conceptual views and identity. This constructed perception of homogeneity makes the countries willing to defend what they perceive as their shared home (Adler 1997, pp. 249–251). The cosmopolitan security view would fit into this category. Identity and determination to defend planet Earth as our home could play a significant role, however, the common identity of all the 'earthlings' is currently not sufficiently established to facilitate the treaty adoption as a sole catalyzer. This common identity could develop in the future. In the cosmopolitan view, debates on planetary defense could be encompassed by the public sphere as defined by Habermas (2015) and lead to the gradual establishment of a common cooperation and treaty-based Earth Defense Project. Other parts of this book elaborate on the plausibility of such a development. This chapter recognizes that common identity might play an important role, but focuses mainly on the level of threat necessary to cause a comparable action in the past.

The second cause, advocated by Marxist or geopolitical theories (II), presupposes that the countries will create alliances to gain smoother access to resources or pledge their military support to gain other advantages. Pure Earth Defense could hardly bring any material resources. Therefore, this theoretical underpinning offers only limited insights into the dynamics of Earth Defense Treaty conclusion. Of course, technology developed for NEO deflection could be later used for extraction of resources from NEOs. Participation on the development of such technology could represent a bargain for states. The insights from this theoretical grounding remain valid, but as already explained, the primary focus of this chapter is the role of threat in a treaty conclusion. And for this purpose, this theoretical underpinning offers only limited insights.[7]

The third cause of defense pact conclusion is theoretically built on the concepts popularized by Kenneth Waltz (1979, p. 26) and his structural realism (III). It assumes that states may either oppose their stronger rivals by forging an alliance to counter the threat (balancing), or align with the rival, realizing that the costs of opposing them would be greater anyway (bandwagoning). The contribution of this systemic theory, no matter how persuasive in cases of interstate relations, is limited in the case of external, non-systemic threats, such as NEOs. The theory would predict that all the countries will rally against this threat, balance against it, or possibly struggle for hegemony in this alliance against the threat. The threat itself would be decisive for the pact conclusion, which is covered by the fourth cause.

This section emphasizes the fourth cause (IV), expecting a causal link between threat severity and the conclusion of a defense pact. This hypothesis draws mainly on the ideas of defensive realism. It presupposes that any threat may create fear (on fear in realism cf. Pashakhanlou (2017)) that forces states to an action—the adoption of a defense pact. This section identifies the correlation between pact conclusion

[7]As explained below, this chapter primarily inquires how much threat is necessary to cause a pact (and analogically Earth Defense Treaty) conclusion. Those pact concluded preventively might have been built on common identity or common struggle for resources. Those factors could also be emphasized when promoting Earth Defense Treaty adoption. However, the level of threat is more comparable and as empirical records show, significant.

and threat presence and severity. By respective findings we may infer how severe the NEO danger must be to persuade countries that the treaty must be concluded.

If the correlation between a threat and pact conclusion is nearly absolute (pacts are concluded only if severe threats or direct attacks are in place), we may infer that the Earth Defense Treaty conclusion is less likely, unless the NEO threat begins to materialize. The weaker the correlation is, and the more preventive defense pacts we find, the more likely it is that the Earth Defense Treaty might be concluded in the near future even without direct threat by NEOs.[8]

The research reported in this chapter quantitatively examined the defense pacts concluded between 1815 and 2003. It was sorted from The Alliance Treaty Obligations and Provisions Projects (ATOP Project) (Leeds et al. 2002) according to the following criteria:

- pact must be in form of a written treaty, verbal pacts are excluded because of their high volatility and weak record of real conclusion and content;
- pact must be defense pact, or mixed (defense/offense), offence pacts might have been concluded according to different dynamics, especially regarding the motivation of states to gain the spoils of war, which would be impossible in the case of planetary defense;[9]
- pact must be multilateral; bilateral pacts were excluded because the Earth Defense Treaty would be by its very nature multilateral; different dynamics stand behind the conclusion of a bilateral pact (the relationship is easier to manage, special circumstances between the members could represent a spurious variable).

By this selection, 74 multilateral defense pacts were sorted out for analysis. Table 23.2 displays the findings. Similarly to the ATOP project, the data are divided into periods of 1815–1913, 1914–1944, 1945–1989, and 1990–2003. Each pact was assigned one of the following values according to the situation that the parties were in at the time of pact adoption:

- 1—the pact was concluded purely preventively, no direct or recurrent threat was present;
- 2—the pact was concluded directly after a war, as a reaction to that previous war, to fix status quo;
- 3—the threat to signatories was exact and present, yet not immediate (rhetoric threats, unfriendly acts, challenges to political order);

[8] Other causes (I) to (III) are disregarded for the purpose of this research, as they are less relevant for outer space-based threat. Ideological proximity (cosmopolitan motivation) and motivation by resources (Marxist motivation) can be important factors supporting the Earth Defense Treaty conclusion, will be just briefly discussed in the conclusion, they are discussed elsewhere in this book and this chapter focuses on the link between threat and treaty conclusion.

[9] I.e. Marxist motivation to gain the resources might have been the decisive factor in case of offense pacts, which makes them incomparable to Earth Defense Treaty, which would be more treaty-based. Access to resources would thus play a less significant role here.

Table 23.2 Correlation of concluded pacts with threat presence

	1815–1913	1914–1944	1945–1989	1990–2003	Total
I—the pact was concluded preventively	3	2	8	5	18
II—the pact was concluded as a result of previous war	5	0	0	0	5
III—the threat to signatories was exact and present, yet not immediate	4	1	15	0	20
IV—the threat to signatories was immediate and severe	6	1	7	0	14
V—the pacts were concluded when at least one signatory was in inter-state war that related to pact conclusion	7	8	0	2	17

- 4—the threat to signatories was present and severe (e.g. arms races, possibly proliferating conflict, internal conflict);
- 5—the pact was concluded when at least one signatory was in inter-state war that was related to pact conclusion.

In the majority of cases, a certain level of threat had been present at the time of pact conclusion. Eighteen out of the 74 multilateral defense pacts were concluded preventively, which represents approximately ¼ of the all the pacts (see Fig. 23.1).

The pacts concluded in individual period mirror the situation of that particular era (see Fig. 23.2). Between 1815 and 1913, we may observe evenly differentiated situations. Some pacts such as the Holly Alliance were concluded post-bellum; as a reaction to previous conflict, they aimed at fixing the status quo. Others were adopted to drag other members to ongoing war. There were three preventive defense pacts, such as repeated reinstatement of a defense pact between German states[10] in 1820 and 1850, or a defense pact of 1883 between Austria-Hungary, Romania, Germany, and Italy (entered in 1888) safeguarding the security of newly born Romania.

Between 1914 and 1944, most of the pacts were concluded in relation to existing conflict (fully materialized threat), as represented by two world wars. Only two defense pacts were concluded preventively, among them the 1925 defense pact between Germany, Belgium, France, UK, and Italy, where Germany, France, and Belgium promised nonaggression and all the parties pledged to defend the others against those who violate this obligation.

[10]Austria, Baden, Bavaria, Prussia, Hesse Electoral, Hesse Darmstadt, Saxony, Wurttemberg, Hanover, Mecklenburg-Schwerin. Although concluded preventively, the defense pact conclusion might have been facilitated by desire of small and relatively weak German states to strengthen their security.

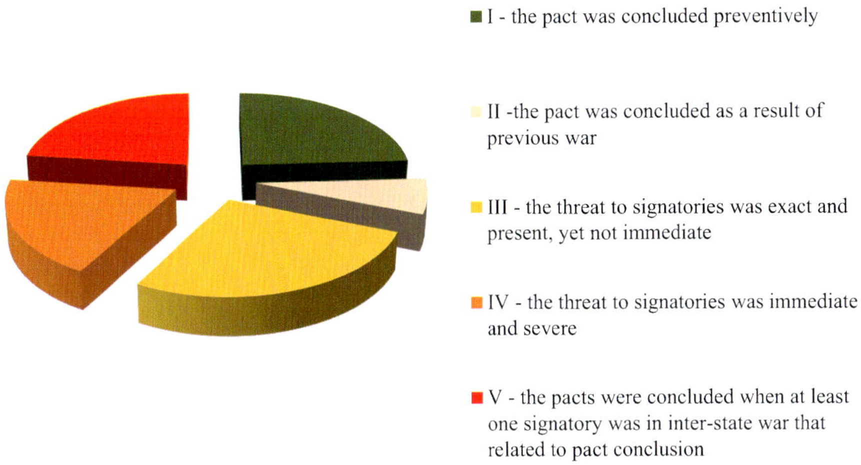

Fig. 23.1 Overall correlation of concluded defense pacts with threat presence

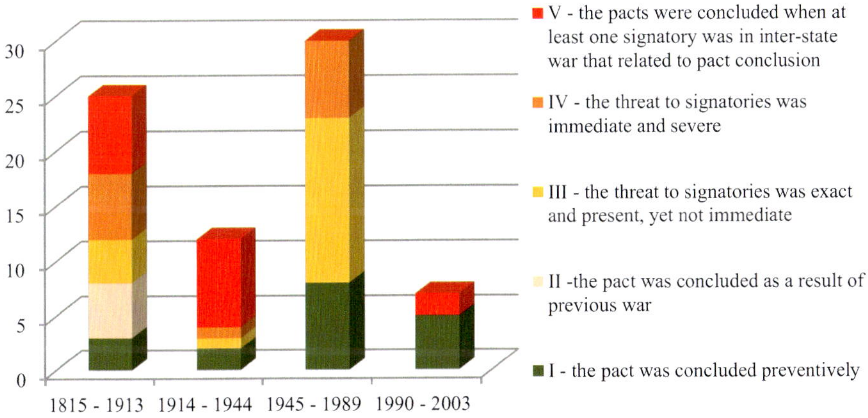

Fig. 23.2 Defense pacts and threats in different periods

The Cold War era (1945–1989) brought symptomatic balancing of power between two superpower blocs, and 15 defense pacts balancing specific power threats stemming from the opposite ideological camp. However, eight preventive defense pacts were concluded primarily by nonaligned states that did not face any clearly delimited or imminent threat.

Finally, after the fall of the Eastern Bloc, five preventive pacts were concluded, mainly to substitute the Soviet Union as the previous joint security provider.

The ratio of preventively concluded defense pacts grew and peaked in the last examined period, when the countries sought collective security after the decline of their previous security provider—the USSR (see Fig. 23.3). From the data analysis,

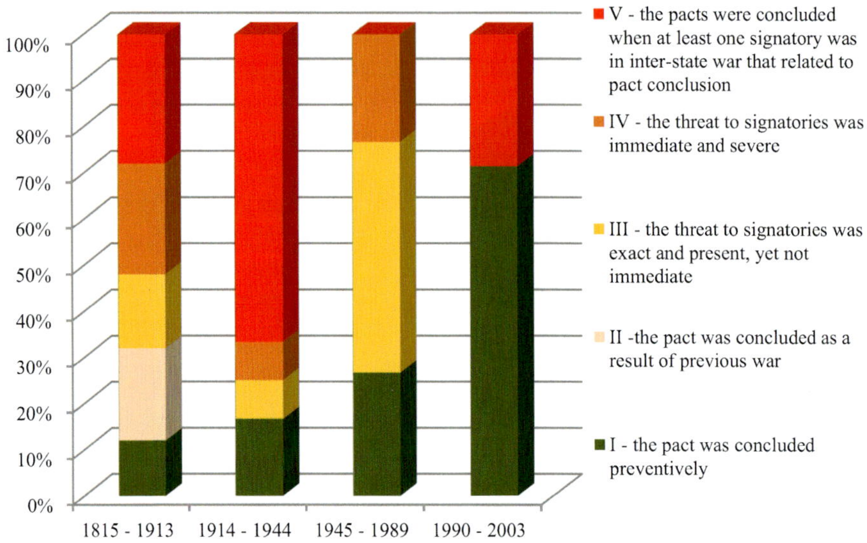

Fig. 23.3 Ratio of preventively concluded defense pacts

we may infer the following assumptions about the Earth Defense Treaty. Purely preventive conclusion of this treaty is less probable, but not impossible. There is certainly an existing tradition of preventive multilateral defense pacts. Although the countries in 18 cases might have been in weaker structural positions or suddenly found themselves without previous security provider, they concluded the defense pacts without any specific threat aimed against them. As will be discussed below, this means that there is a certain probability that a defense treaty such as the Earth Defense Treaty could be concluded preventively, without interference of a direct NEO threat. It establishes a certain chance that that cosmopolitan approach prescribing collective action without imminent threat might just succeed, and that the Earth Defense Treaty could be adopted preventively in advance, without specific emergence of NEO danger.

However, it is also necessary to deal with the fact that the conclusion of the majority of defense pacts is correlated with direct and imminent threat or with already existing emergency. This suggests that many countries may refuse to accept any obligation unless the threat of collision with a NEO is clear and tangible. Therefore, the Earth Defense Treaty could be prepared in advance in two or three versions with different degrees of cooperation and imposed obligations. Thus, it could remain in standby mode, waiting to see if a NEO danger materializes, while enhancing monitoring and rapid-reaction capabilities. Moreover, it could be discussed with the expert bodies, such as the UN Conference on the Exploration and Peaceful Uses of Outer Space and the UN International Law Commission. Having two or three drafts prepared would significantly speed up the process of treaty adoption in case a NEO was detected. Similarly, breaking the treaty into several parts and

Table 23.3 Past defense pacts with/without the obligation to contribute to the joint project

Does a pact specify contribution levels?	1815–1913	1914–1944	1945–89	1990–2003	Total
I—no	12	10	19	4	45
II—only by vague language	7	0	1	0	8
III—yes, by specific obligations	6	2	10	2	20

concluding each of them separately might prove useful.[11] The core principles of the Earth Defense Treaty could also be pronounced as a soft law declaration in the meantime, or wait in the form of a draft similar to other important documents.[12] This would minimize unintended consequences of treaty adoption by safeguarding that the wording would not be prepared entirely under time pressure. Moreover, this approach might help to avoid an unpleasant scenario when member states would get stuck on details, hegemons would be playing a game of chicken over leadership in NEO deflection, and/or smaller states would struggle to find qualified experts to professionally comment on proposals, while an asteroid is approaching.

The quantitative research also took the opportunity to examine whether the defense pacts included an obligation to provide specific contributions to a joint defense project (see Table 23.3). This inquiry provides a valuable hint for how willing the states were to accept a specific and clearly defined duty. 20 out of the 73 pacts[13] included a precise obligation to contribute certain financial, personal, or other resources to the common defense projects. By contrast, eight other pacts included only vague language pledging material support to the project, without any specification (e.g. 'countries will pay all the material and financial costs necessary to achieve given goal'). Embedding joint financing in the treaty is therefore feasible, although the cases where countries left the financing unregulated and at their convenience prevail.

Last but not least, the quantitative research scrutinized whether past defense pacts included any particular clauses establishing closer cooperation that would be necessary for joint Earth Defense, namely: transfer of technology, integrated command of military forces, subordination of national forces to forces from another state, coordination of armaments, regulations of weapons and means of warfare, establishment of joint military bases, enabling military presence on the territory of other countries, direct coordination of military efforts, and coordination of disarmament after the conflict. The results are displayed in Table 23.4. Most frequently,

[11] Comparative process-tracing analysis of conclusion of selected pacts could shed further light on the way how the pacts are negotiated and concluded and to what extent those hypotheses are useful. This chapter primarily focused on the correlation between pact conclusion and threat.

[12] E.g. draft articles on international responsibility for internationally wrongful act prepared by International Law Commission and waiting in standby mode for the right time to be proposed and adopted.

[13] One pact was not included into analysis here (overall amount is not 74), because its actual wording is not known, therefore we cannot find out whether the obligation to contribute to joint defense efforts were included.

Table 23.4 Past defense pacts and mutual cooperation of members

Amount of multilateral def. pacts establishing/regulating	1815–1913	1914–1944	1945–1989	1990–2003	Total
Transfer of technology among parties	0	0	0	1	1
Integrated command of military forces	1	0	8	4	13
Subordination of forces to leadership from another state	3	2	2	1	8
Armaments	0	0	2	0	2
Weapons and means of warfare	0	0	2	0	2
Joint military bases	0	0	3	0	3
Presence on other's territory	2	3	6	0	11
Coordination of military efforts	2	0	4	2	8
Disarmament after the conflict	5	1	0	0	6

countries were willing to give a legal grounding to joint command of military forces and agreed to accept the forces of other countries on their territory, coordinate their military effort, and subordinate their forces to leaders from different states.

Nevertheless, countries were reluctant to legally regulate the transfer of technology to their allies, as well as armaments and weapons, and means of warfare. While the latter two areas might be seen as sufficiently regulated by other normative systems (international arms control law, humanitarian law), the minimal number of pacts establishing transfer of technology suggests that countries were not willing to accept an obligation to share their technological advances on a multilateral contractual base. Thus, military technology sharing could play a dual role in the case of an Earth Defense Treaty. On the one hand, it could be hard to persuade the states to accept a contractual obligation to share their technology with others. Although the sharing of peace-serving technology is common, in the case of the military technology necessary for NEO deflection, countries could be reluctant.[14] On the other hand, once shared technology would be contractually guaranteed, it would motivate newcomers to join the treaty system to benefit from the provided technological knowledge. The transfer of technology could also occur on a less formal, non-contractual basis.[15]

[14] An option could be to regulate the technology sharing by subordinated acts that would be less formalized and binding, thus the states would feel higher level of discretion and conformity when accepting a treaty.

[15] Nevertheless this would just strengthen the argument that states prefer informal technology sharing to legal clauses establishing formalized technological cooperation.

23.5 What Wording Shall the Eventual Earth Defense Treaty Include?

The following section qualitatively examines the actual wording of the five defense pacts currently in force and the subsequent executive documents and arrangements. It studies whether or not joint capacities are developed, how decisions are made, and how the pacts prevent freeriding and abuse. The criteria for selection of the examined pact were: (1) still valid pact; (2) multilateral, with as many parties as possible; (3) geographical coverage of all the regions. According to those criteria, the following pacts were selected (for overview see Table 23.5):

- Inter-American Treaty of Reciprocal Assistance (1947)
- North Atlantic Treaty (1949)
- ANZUS (1951)[16]
- Collective Security Treaty (1992)
- African Union Non-Aggression and Common Defense Pact (2005)

The findings displayed in Table 23.5 can be summarized in the following manner. Firstly, military cooperation and joint capacities development is usually described only vaguely in the primary document. In the case of ANZUS it is missing entirely. More exact obligations concerning joint capacities development were usually adopted later in a subordinated agreement. Therefore, the recommendation for the Earth Defense Treaty would be to include a soft obligation formulated in general language to regulate the joint capacities development. Specific details could be debated later; once the countries gain more confidence in the mechanism of cooperation they could adopt a subordinated accord. This is of course only an option in cases where there is enough time left and no direct threat of collision with a NEO exists.

Secondly, common financing systems and freeriding prevention remain unregulated in the case of the Inter-American Treaty and ANZUS. The NATO and AU nonaggression pacts leave detailed regulation of common financing to subordinated agreements. NATO has mechanisms in place to cope with political freeriding prevention, and the AU nonaggression pact expects that debtors have less opportunity to get elected to a decision-making body. The most inspiring way of freeriding prevention is provided by the Collective Security Treaty. The Treaty itself regulates the levels of contributions and stipulates that failure to contribute for 2 years results in withholding of the debtor's right to vote or nominate its citizens to functions within the organization. This sanction might be more effective than threats of exclusion from the joint defense system.

[16] The US and New Zealand no longer maintain security relationship under this treaty, but it is still valid between the US and Australia and Australia and New Zealand.

Table 23.5 Summarization of findings

Pact	Mutual military cooperation and capacities development clause?	Freeriding prevention clause(s)?	Abuse prevention clause?	Decision making organ/mode:
Inter-American treaty of reciprocal assistance	No	No	Collective defense does not cover the situations when the member state would provoke an attack (*art. 9(a)*)	*Organ of Consultation (art. 17)* consists of member states representatives, voting by *2/3 of members*
NATO	*Art. 3* promises general support; 2014 Whales summit declaration in *para 67–70* describes multinational and national capability development and deployment initiatives, earlier subordinated documents include common technology deployment, joint exercises information sharing etc	*Art. 3* promises general support; 2014 Whales summit declaration is *para 14 and 15* prescribes that 2% of members' GDP must be spent on related military expenses. This represents indirect contribution to NATO capabilities. Direct funding entails common, joint and ad hoc funding for specific purposes provided by agreements. If the countries fail to fulfil their funding obligations they have a political responsibility to Defense Ministerial meetings and Head of State and Government Summits	General political responsibility and legal responsibility for *mala fide* application of the treaty	*Consensual decision making* on multiple levels
ANZUS	No shared capacities, no integrated structures, only *ad hoc* exercises of US and Australia	No	General political responsibility and legal responsibility for *mala fide* application of the treaty	Council (*article 7*), consisting of foreign ministers, *only consultative* role

(continued)

Table 23.5 (continued)

Pact	Mutual military cooperation and capacities development clause?	Freeriding prevention clause(s)?	Abuse prevention clause?	Decision making organ/mode:
Collective security treaty (as amended)	*Art. 6* entrusts the coordination of activities to the Council of Collective Security; *art. 7* provides that placement and functioning of objects of common defense will be regulated by special agreement(s); subordinated Charter of the Collective Security Treaty Organization (2002) provides a legal basis for specific cooperation especially *art. 7 and 8*: technology + information sharing, training and joint military exercises; the alliance newly develops rapid response capabilities	*Art. 24* of the Charter of the Collective Security Treaty Organization (2002)—prescribes common financing; *art. 25* of this Charter states that the failure to contribute for two consequent years results in retention of the debtor's right to vote and nominate its citizens to functions within the organization	General political responsibility and legal responsibility for interpretation and application of the treaty *mala fide*	Council of Collective Security (*article 3*) consisting of the heads of member states, decisions of the council and other bodies adopted by *consensus*
African Union non-aggression and common defense pact	*Art. 4(a)* lists African Army and African Standby Force, *articles 10* and *11* prescribe info and tech sharing and other support. *Art. 13* of the earlier Protocol relating to the establishment of the Peace and Security Council of the African Union ("the Protocol") establishes African Standby Force (2002)	*Article 21* of the Protocol regulates contributions to Peace Fund. Additionally Specific Fund may be established for a specific purpose. If the country does not provide specific contributions to the funds, the possibility that it would be elected into Peace and Security Council diminishes (*art. 5/2(f)* of the Protocol)	General political responsibility to the AU and legal responsibility for interpretation and application of the treaties *mala fide*	*Art. 9* provides for Peace and Security Council (further regulated by the Protocol relating to the establishment of the Peace and Security Council of the African Union, *art. 8, para 12*) voting *consensually* and cooperating with the African Union bodies

The fact that countries were often reluctant to accept a detailed obligation regarding cooperation and financing in the primary document could be problematic, as the Earth Defense Treaty would need to be adopted under the time pressure caused by an approaching NEO. This represents another reason why the draft treaty should be prepared in standby mode, to leave states sufficient time to consider their position on cooperation and common financing.

Thirdly, the pact did not include specific clauses describing what an abusive application would be and what sanctions would follow this conduct. The Inter-American Treaty excluded provoked attacks from the benefit of the common defense system, implicitly presupposing that provoking such an attack and relying on joint defense would be an abuse of the treaty. If other pacts were interpreted or applied with bad intention ('mala fide'), it would result in general political and legal responsibility of the acting member state (cf. article 31, para 1 of the Vienna convention on the law of treaties). Even though prevention of abuse might be a pressing issue in the case of an Earth Defense Treaty, the defense pacts suggest that legal prevention might not be the best option for countries. In the five given cases, the countries were reluctant to regulate abuse through specific wording. High technical interdependence and assurances together with mutual monitoring and transparency might represent much better and more attractive tools for abuse prevention than specific treaty-based clauses.

Finally, three out of the five defense pacts concluded with a consensual decision-making process. ANZUS established a council of ministers with only consultative functions, and the Inter-American Treaty embedded a mechanism of qualified voting by two thirds of member states. Consensual decision-making seems most palatable for the states. However, in the case of the Earth Defense Treaty, the decisions of a qualified majority might be better. The qualified majority may avoid a scenario where one or two countries hold the entire planet hostage, demanding certain benefits in exchange for support of the common project. Should the strongest powers object to a qualified majority voting, they may be entitled to unilateral veto to satisfy their demands.

23.6 Conclusion

Although the analogy between past defense pacts and a future Planetary Defense Treaty may have its limits, it is a valuable stepping stone for considerations of how the Treaty could be concluded and what it could look like.

Quantitative analysis revealed that approximately one quarter of the examined multilateral defense pacts have been concluded preventively, without exact threat. Therefore, there is a chance that the Planetary Defense Treaty could be concluded preventively in truly cosmopolitan manner. To deal with the fact that the conclusion of three-fourths of the multilateral defense pacts correlated with a threat, we propose to prepare at least a draft of the Earth Defense Treaty, or several versions of this draft, in advance. Should the countries dedicate their effort to more pressing issues and refuse to deal with this matter without a tangible threat, the wording can remain in standby mode and facilitate the discussions in case of a sudden NEO appearance. This would also help to avoid the scenario when political representatives 'play chicken' over the leading role in the deflection program. Quantitative inquiry also revealed that the clauses prescribing common levels of contribution and certain military cooperation are scarcer, but not isolated.

Quantitative analysis showed that countries prefer to include exact obligations regarding military cooperation, common financing, and freeriding prevention in subordinated agreements that are accepted later, once the cooperation unfolds and parties get accustomed to each other. This represents another reason why discussion of the Earth Defense Treaty should start as early as possible. In case of abuse prevention, countries completely relied on general political and legal responsibility. Therefore, we propose that the Earth Defense Treaty also need not necessarily work to regulate abuses. Other extralegal mechanisms may sufficiently prevent the abuse, e.g. technical interdependence and mutual monitoring. In regards to decision making, countries preferred consensual decisions. Nevertheless, for the Earth Defense Treaty, qualified majority voting could prevent members from bargaining for their consent. To make this option acceptable for the major powers, the treaty could equip them with powers of veto.

Generally speaking, the research indicated that there were several historical examples of rather cosmopolitan behavior. However, threat-related and rather hesitant conduct was more frequent. Hopefully, this could be changed by emphasizing a common identity and Earth as our shared home,[17] which might cause a shift from a culture of countering threats to a cosmopolitan culture of cooperative prevention. The cosmopolitan view of security will not take hold overnight. Therefore, this chapter recommends preparing the treaty with moderate expectations regarding states' willingness to adopt it without tangible threat and to accept specific detailed obligations. To put it in Wendt's words, we should expect (Lockean) anarchy, and allow states sufficient time to make cosmopolitan cooperation out of it.

References

Adler, E. (1997). Imagined (Security) Communities: Cognitive Regions in International Relations. *Millennium: Journal of International Studies*, *26*(2), 249–277. https://doi.org/10.1177/03058298970260021101

Billings, L. (2006). To the Moon, Mars, and Beyond: Culture, Law, and Ethics in Space-Faring Societies. *Bulletin of Science, Technology & Society*, *26*(5), 430–437. :https://doi.org/10.1177/0270467606292504

Burke, A. (2015). Security cosmopolitanism: the next phase. *Critical Studies on Security*, *3*(2), 190–212. https://doi.org/10.1080/21624887.2015.1065109

Divine, R. (1993). *The Sputnik Challenge*. Oxford: Oxford University Press.

[17] Cf. the first cause of pact conclusion listed in Sect. 23.3 and Table 23.1.

Habermas, J. (2015). *The Structural Transformation of the Public Sphere: An Inquiry Into a Category of Bourgeois Society*. Wiley.

Hynek, N., & Stritecky, V. (2010). The rise and fall of the Third Site of Ballistic Missile Defense. *Communist and Post-Communist Studies*, *43*(2), 179–187. https://doi.org/10.1016/j.postcomstud.2010.03.002

Johnson, J. C., & Leeds, B. A. (2011). Defense pacts: A prescription for peace? *Foreign Policy Analysis*, *7*(1), 45–65. https://doi.org/10.1111/j.1743-8594.2010.00122.x

Knox, J. S. (1998). *Planetary Defense: Legacy for a Certain Future*. Maxwell Air Force Base, Alabama. Retrieved from http://www.dtic.mil/get-tr-doc/pdf?AD=ADA397269

Kunich, J. C. J. C. (1997). Planetary Defense: The Legality of Global Survival. *The Air Force Law Review*, *41*(41), 119–162. http://heinonlinebackup.com/hol-cgi-bin/get_pdf.cgi?handle=hein.journals/airfor41&section=6

Lachs, M. (2010). *The Law of Outer Space: An Experience in Contemporary Law-Making*. Leiden, Boston: Martinus Nijhoff Publishers.

Leeds, B., Ritter, J., Mitchell, S., & Long, A. (2002). Alliance treaty obligations and provisions, 1815-1944. *International Interactions*, *28*(3), 237–260.

Pashakhanlou, A. H. (2017). *Realism and Fear in International Relations. Morgenthau, Waltz and Mearsheimer Reconsidered*. London: Palgrave Macmillan.

Peterson, M. J. (1997). The use of analogies in developing outer space law. *International Organization*, *51*(2), 245–274. https://doi.org/10.1162/002081897550357

Remo, J. L. (2015). The dilemma of nuclear energy in space. *Bulletin of the Atomic Scientists*, *71*(3), 38–45. https://doi.org/10.1177/0096340215581359

Remo, J. L., & Haubold, H. J. (2014). Threats from space: 20 Years of progress. *Bulletin of the Atomic Scientists*, *70*(4), 85–93. https://doi.org/10.1177/0096340214539125

Rosas, A. (1983). The Militarization of Space and International Law. *Journal of Peace Research*, *20*(4), 357–364. https://doi.org/10.1177/002234338302000406

Smith, W. (2004). Earth-Moon Defense Management. In *2004 Planetary Defense Conference: Protecting Earth from Asteroids*. Reston, Virigina: American Institute of Aeronautics and Astronautics. https://doi.org/10.2514/6.2004-1422

Tannenwald, N. (2004). Law Versus Power on the High Frontier: The Case for a Rule-Based Regime for Outer Space. *Yale J. Int'l L.*, *29*, 363–422.

Thangavelu, M., Wilson, T., Hussein, A., Aliaj, B., Entrena, C., Lee, C., et al. (2015). *READI - Roadmap of Earth Defense Initiatives*. Athens (OH). https://isulibrary.isunet.edu/opac/doc_num.php?explnum_id=722

Waltz, K. (1979). *Theory of international politics*. McGraw-Hill.

Zaitsev, A. V. (2011). Defense against Asteroid-Comet Nucleus Dangers – A Real Task of Today. In R. Ragaini (Ed.), *International Seminar on Nuclear War and Planetary Emergencies. 43rd Session*. World Scientific Publishing Co. Pte. Ltd.

Chapter 24
Why a World State Is Unavoidable in Planetary Defense: On Loopholes in the Vision of a Cosmopolitan Governance

Pavel Dufek

24.1 Introduction

The idea of global planetary defense (henceforth PD) against threats from outer space as explored in previous chapters requires an effective global decision-making system that would both implement the necessary technology (or oversee its implementation) and deploy it in case of emergency. The proposal (Chap. 21) is based on the assumption that such a political body would be 'semi-cosmopolitan,' i.e. short of a full-fledged global political authority—a 'world-state'—represented by a world government, although it implies 'more centralized [global] power(s).'

The main claim of this chapter is that planetary defense against asteroids, as variously envisioned in the chapters in this book, cannot be implemented under such a model of democratic global governance. All relevant indices point to the necessary establishment of a *global political authority with legitimate coercive powers,* one that retains basic elements of statehood, which is a similar argument to the one I explored in some detail in an earlier paper on global justice (Dufek 2013). It remains to be seen whether such a political system can be in any recognizable sense democratic. Put bluntly, I finds it unconvincing that planetary-wide, physical-threat, all-comprehensive macrosecuritization (Buzan and Waever 2009) ('planet-wide securitizations,' as labelled in Chap. 11), coupled with deep transformations of international law, global centralization of core decision-making powers, de-stigmatization of nuclear weapons, and the like, can proceed, succeed, and be implemented in a non-hierarchical international system where PD constitutes only one regime among many and states basically remain the major actors—sovereign *principals*. This is the

P. Dufek (✉)
Department of Political Science, Faculty of Social Studies, Masaryk University, Brno, Czech Republic
e-mail: dufek@fss.muni.cz

N. Schmidt (ed.), *Planetary Defense*, Space and Society,
https://doi.org/10.1007/978-3-030-01000-3_24

main sense in which proposals in support of a semi-cosmopolitan model of global governance driven by cosmopolitan ethics, as discussed in Chap. 21, are too idealistic.

My field of expertise is political theory/philosophy and political science, and I believe this background allows me to spot important conceptual, normative, and institutional hurdles that await any such extremely ambitious project with clear political ramifications. On a lower level of generality, the present chapter therefore provides an exposition of a number of important issues that also need to be seriously reflected upon by advocates of planetary defense if their case is to be robust, persuasive, and ultimately successful, as regards the implementation of such a politically sensitive planetary defense policy. Although rationally and scientifically robust, the PD project suffers from over-simplification, as well as naivety with respect to how both international and domestic politics works—especially if the resultant model of governance is to be democratic. It is impossible to insulate the scientific enterprise from the broader societal context within which it operates, and which it aims to transform.

It needs to be stressed that this chapter offers neither a critique nor a glorification of the idea of a world state/world government, understood as a model of political decision-making on the global level; in this regard I remain agnostic.[1] Instead, this chapter argues that in the case of PD (and arguably other global-scale issues as well), an effective, stable, and legitimate solution cannot be based on the seemingly attractive, yet vague and ultimately unworkable idea of 'global governance without government'—that is, exercise of functions normally entrusted to state governments, such as dealing with political, economic, environmental etc. problems, without any such formal governmental body in place. 'Government' would be replaced by a multilevel network of cooperating actors, bodies, and agencies.[2] Defenders of such a vision of PD make their own case even more difficult by insisting on a democratic character of the global decision-making framework, which creates further conceptual, normative, and institutional problems.[3]

For the purpose of clarity, this chapter divides the issues into five broad groups: law, politics, identity and society, economics, and meta-theory. As will become clear, the respective points are not strictly separated and often overlap, so that a problematic feature in the realm of law and legislation affects—one might say infects—the circumstances of political action, as do unresolved issues of national, global, or regional identities. Other combinations are of course possible as well. Some arguments that will be developed in detail below include claims from instability; efficiency; power politics; collective action problems; conflicts among political issues; democratic legitimacy; and common identity.

[1] For a recent discussion of the desirability of a world-statist solution to global problems see (Nili 2015).

[2] The classic here is 2. (Rosenau and Czempiel 1992). See also (Rosenau 2006). For a widely cited conceptual clarification see (Rhodes 1996).

[3] For an internal critique of how loosely the concept of global governance has been used in the study of international relations see (Weiss and Wilkinson 2014).

The main message of this chapter could be summed up in the following way: If PD is a good idea and a desirable goal, which it very well may be, then we should be honest with ourselves and avoid trying to have our cake and eat it too, because not all good things go together in planetary defense endeavors.[4]

24.2 Law and Politics

We start with the realm of law, and international law in particular. It sometimes gets lost on us that modern democratic politics, and the modern state as such, are fundamentally intertwined with modern law: In a sense, the modern state is first and foremost a *legal* state, which means that political decisions have to be promulgated and implemented strictly in the form of a legal norm, be it constitutional laws, regular laws, or subordinate legislation. The rationale behind the primacy of law—we usually speak of the *rule of law*, the *Rechtsstaat*, or the *prééminence du droit*—is to prevent unchecked discretion of the rulers, as well as to stabilize legal expectations and ensure legal certainty (Tamanaha 2004; Zolo 2007). From this follows the paramount role of the judiciary, and less optimistically, the always present tension between what the law in force requires versus what the ever-changing democratic will of the people wants. Although it is an open question how to approach law and legislation on the supranational level where there is no single source of legitimate legislation, it is an eminently important one. The absence of a legal framework both enabling and constraining the exercise of democratic politics makes room for the game of power politics, and this eventuality should be taken seriously by the PD community.

24.2.1 Shared Framework or Fragmentation

The guiding idea behind this argument is that PD entails an overarching legal framework shared and respected by all actors, as is the case with other global issues (Held 2010; Kumm 2016). This will require reconstruction or transformation of both domestic and international law in whose language PD-related policies will be formulated. Given that PD aims to transform both policy priorities across existing countries and symbolic and cultural self-understanding of the respective actors (governments, peoples, social groups, interest groups, individuals etc.), while providing for efficient global decision-making machinery, we cannot expect the extant body of international law to be sufficient for such a momentous shift. Empirically speaking, such preconditions are not there.

---

[4] I am paraphrasing here the title of an important article on democracy promotion. See (Grimm and Leininger 2012).

commendable moral ideal, a shared, though presumably thin global identity is a fragile utopian goal, as the recent wave of populism, nationalism, and cultural and economic protectionism amply attests to. Achieving such a shared feeling of global commonality is certainly a tall order (more on this in Sect. 24.4 below), rendered even more difficult by such explicit rejection on a unified legal system.

24.2.4 The Point and Purpose of Law

If 'achieving social order' is one of the basic purposes of law (Fuller 1969, Chap. 3), then it needs to be made public and well known, so that everyone affected can, at least in principle, verify which legal rules are in force. Only legal rules that are accessible by their recipients can be thus binding for them. (To complicate things a bit, it could be argued that only those legal rules that the majority of other actors can reasonably expect to uphold are binding; this point is discussed in Sect. 24.3.4.) If PD is to operate within the bounds of legality, and if the resulting legal rules are to be binding for all relevant actors (sometimes that might directly mean all individual human beings), then PD-related decisions need to be promulgated publicly, continually, globally, authoritatively, and consistently. How this could be done in a decentralized system of governance remains a matter of concern.

An objection might be pressed—and this is indeed a fairly common response in contemporary legal and political theory—that such a view of law and legal obligation remains locked in the old hierarchical understanding of law. My response is that I simply cannot see how long-term global cooperation with respect to PD can proceed effectively in a non-hierarchical system of international law. Insofar as the foundational rationale of judicial decision-making is to aim for 'authoritative resolution of a legal problem' (Ondřejek 2016, p. 125), and as long as conflicts among different norms or legal orders are inevitable in a decentralized system of global governance, then global cooperation in such a huge task of coordination as represented by PD is difficult to envision. Flexibility and open-texturedness of the emerging global system of law might be considered virtues in many areas of human activity (such as, say, commercial law), as they arguably involve engagement with different value systems and their cultural backgrounds. PD is, however, one of those areas where these qualities become vices. Development, deployment, and control of such sensitive technology as laser beams and nuclear weapons are simply not the right objects for legal and political experiments, not least because the common goal is fully known beforehand.

24.2.5 Securitization, Pluralism, and Self-Government

There are, of course, other rationales behind the decentralized approach. One basic point of global governance models concerns the importance of preserving autonomy and self-government on lower levels of political organization, from regional formations all the way down to municipalities, in order to avoid the danger of global authoritarianism and retain both the idea and practice of democracy. The principle of *subsidiarity* has a foremost place in such visions and has been commonly invoked with respect to the European Union (MacCormick 1999).[9] One weighty reason behind this principle is moral, cultural, religious diversity. Liberal conscience counsels toleration or even celebration of diversity (within certain limits specified by liberalism itself). If, however, PD presupposes *securitization*—more precisely, existential-threat macrosecuritization on a planetary level (see Sect. 24.3.3)—which I argue it necessarily does, then fairly limited room remains for pluralism and self-government, simply because of the unconditional and depoliticizing nature of securitization. The threat is labelled as objective, and the discretionary (i.e. insulated from normal democratic control and accountability) decision on how to cope with it is thus rendered immune to protest and disagreement. Of course, asteroid threats are claimed to be genuinely objective, as they can be explained and calculated in precise scientific language, including the possibilities of collision aversion and/or asteroid trajectory alteration. Such securitization might very well be a just one, as Nikola Schmidt argues in Chap. 14;[10] nevertheless, it would still imply unconditional compliance with the selected course of action by all involved actors, from states to international organizations to individual human beings. Unless PD is properly securitized (i.e. if it is 'merely' politicized, introduced as a relevant political topic in public debates), it is unclear why it is to take immediate priority over other policy goals, such as eradication of world poverty (see Sect. 15.2.7 for a discussion of competing securitization initiatives).

In effect, this creates a dilemma between two scenarios: Securitization at the expense of self-government, and no securitization at the expense of survival of humanity (or so the argument goes). The point is not that the former course of action is not preferable, but that we cannot have both subsidiarity/self-government and PD-level securitization at once. Incidentally, all this also seriously weakens the case for a 'judicial dialogue' across the many levels and sectors, which is precisely the way defenders of global legal pluralism/polycentrism envision the possibility of binding common decisions. Plainly, a dialogue is of little use if we know that enforceable decisions are made elsewhere and on highly different grounds.

24.2.6 Global Rule of Law and Constituent Power

The preceding sections employ the notion of a rule of law. This is an inconspicuous concept; however, the fact is that without a rule of law, there can be no democracy, no protection of human rights, and arguably, no reciprocal social cooperation, at least not a democratic one. It is not by chance that the rule of law constitutes a

[9] The principle originated within Christian (mostly Catholic) social teaching/ethics; see (Anzenbacher 1998, p. 210).

[10] See also (Floyd 2011).

conceptual, normative, and institutional pillar of liberal constitutional democracies, and a precondition of consistent enforcement of citizens' fundamental rights.[11] It could even be argued that establishing a functional rule of law is the most basic challenge for any stable transnational or cosmopolitan legal and political system (Palombella 2009). As with previous issues, weighty reasons have been put forward by scholars of legal and political theory against a monistic conception of a global legal system, one imposed from above by the highest-level institutions, such as the current UN Security Council.[12] They claim that such a rule of law would be unilateral, rooted in the allegedly obsolete conception of modern sovereign statehood, and would thus undermine autonomy and self-government based the principle of subsidiarity.

Herein lies the paradox. In order to fulfil its rule of law-specified and law-constrained functions, law needs to be the same for all affected subjects and its force needs to be inescapable, at least in principle. This is highly dubious in a fragmented system of global governance that lacks constitutionalized and legalized enforcement power.[13] The PD regime would be only one among many, without having any clear priority that it implicitly requires. If, alternatively, the recognition of valid norms were to proceed via an emergent consensus on substantive normative commitments (e.g. the content of the *ius cogens* and *erga omnes* obligations (Palombella 2009)); then we would quickly enter the domain of (international) political theory, which has been centrally concerned with ways of accommodating normative conflicts among actors as well as among orders on which various actors operate. By entering this field, questions and problems for PD supporters multiply in both number and nature, as will be clear from Sect. 24.3 below.

To take stock before advancing further, I am of the view that robust global democratic governance cannot do without elements of (post-national) statehood, even if this does not amount to a blind replication of a sovereign state on the global level. Coupled with the securitization-fueled urgency of PD, it seems obvious that a solid global rule of law cannot be squared with a decentralized model of lawmaking: the two rationales simply collide. Consequently, PD calls for a significant transformation of how international and/or global politics is done and understood by both its protagonists and the global audience. Such a shift cannot be rationally designed and swiftly executed, because the evolution of such complex systems as international law and politics is simply beyond the capacities of any existing actor. Some hope has been invested into reconstruction of the United Nations (Held 2010, p. 230), and much hope into the European Union as a laboratory of global governance. Yet the outlook is ambiguous, as Brexit has most recently shown.

The inescapability of political-theoretical considerations of this kind is further confirmed by looking into the process of norm generation in the global realm

[11] See e.g. (Loghlin 2010, Chap. 11).

[12] The infamous *Kadi case* which concerned the hierarchy between international law—specifically UN Security Council resolutions—and EU law (and by implication, of domestic law as well) is probably the clearest recent example. See (Avbelj et al. 2014; De Búrca 2010).

[13] For some further reflections on the rule of law in global context see (Scheuerman 2002).

(Patberg 2016). On the liberal, inter-governmentalist model of global governance, which seems to inform the vision defended in this book, the bulk of public legal and administrative regulations are produced by countries' executives or executive-authorized agencies (where the latter can be national, transnational, or global). One problem here is the tendency of international bodies toward becoming independent on their supposed principals (see further Sect. 24.3.2 below); another problem concerns the legitimacy of legal norms thusly generated. Put schematically, the *chain of legitimization* is too long, too abstract, and too incomplete for such norms to gain or perhaps just retain democratic legitimacy. Technically speaking, a country's executive as one branch of *constituted* power directly or indirectly co-creates on the supranational level new bodies, norms, and corresponding duties, the latter of which are meant to bind the citizens ('the people') who are in fact the *constituent* power—i.e. logically and normatively prior.[14] The problem lies in the fact that the constituent power has little real influence over such global-level decisions, or little opportunity to effectively contest them; national parliaments that normally represent the people are notoriously weak actors in international politics. This may result in various pathologies of power exercised by trans- or supranational bodies, as discussed further in Sect. 24.3.2. After all, even in constitutional democratic states themselves, there is preciously little public deliberation and thus democratic control over the goals and purposes of political action in the international/global realm. Widespread indifference to EU-related issues among citizens of EU member states is an illustration of this problem.

The upshot is that we need global constituent power—a *global demos* no less, whatever kind of actors and procedures this entails—if the global rule of law is to function properly and global law is to be perceived as legitimate.

24.2.7 Conflicts of Regimes, Goals, and Purposes

All this is because, as already hinted, conflicts among regimes, actors, levels of decision-making, as well as goals and purposes are to be expected, if only for reasons of sheer complexity of global governance in an expected multilevel pluralist setting. Scholars have recently begun to speak about nonhierarchical *regime complexes* or *regime complexity*, which depicts the 'presence of nested, partially

[14] This is a common distinction in constitutional law literature. In a constitutional democracy at least, *constituent* power is embodied in the sovereign people who is the foundational source of all power in a state. The people as the *sovereign* "gives" itself the constitution which in turn gives birth and form to the political system, including the highest constitutional institutions (branches of power, constitutional court, central banks etc.) and the division of power among them. All these are created in the act of constitutional self-giving, and as such are *constituted* powers. This is a very interesting if highly abstract and philosophically challenging issue concerning the deepest sources of legitimacy in a democratic society (recall occasional tensions between parliaments and constitutional courts), and here things are somewhat simplified. Cf. (Arato 2016, pp. 1–74; Loghlin 2010).

overlapping, and parallel international regimes that are not hierarchically ordered' (Alter and Meunier 2009). Under such fragmented conditions, clarity of legal obligations erodes. Some authors add that such a labyrinthine system of governance is open to exploitation by great powers (Drezner 2009). More sympathetically minded alternatives such as *global experimentalist governance* aimed at 'participatory and multilevel collective problem solving' (De Búrca et al. 2014, p. 477), assume a highly decentralized system of rule-making and rule-enforcement, which again goes directly against what PD appears to require.

More specifically, once we look into contemporary realities of power and politics in the international realm, as well as into the issues that motivate global political activism, it becomes clear that PD will be merely one among many competing macrosecuritizations. Paradoxically, given the nuclear-destigmatizing ethos of PD carried by the hope in a rationally justifiable mitigation method, the anti-nuclear movement, coupled with fears that destigmatization might undermine the non-proliferation regime, may actually rival PD as regards (macro-)level, degree, and perhaps acceptance as well (Buzan and Waever 2009, p. 259). Moreover, it is conceivable and actually quite likely that competing versions of a desirable PD regime itself will be put forward and steadfastly stuck to by their supporters—including, as it were, various IR research communities with their idiosyncratic assumptions and/or normative preferences. But of course, PD is precisely the kind of issue that requires consensus, cooperation, and orchestrated political and financial effort. Will there be an actor capable of and legitimized to decide which course of action is to be taken with respect to a macrosecuritized topic such as PD?

If the answer is affirmative, then we are faced with something or someone akin to a global sovereign. Recall that PD-related macrosecuritization is to take absolute priority over competing macrosecuritizations, not to mention regular political goals. Such a state of affairs can hardly be achieved by simply coming up with just another international regime, especially if the new global political authority were to carry democratic legitimacy, not least in order to keep other powerful actors in check.

24.2.8 State of Exception as the Norm

Let us assume that PD-based macrosecuritization has been successful and the global audience consents to granting unquestionable priority to PD-related policies and/or actions.[15] This means, among other things, that the regime has been isolated from regular politics and made impervious to change by normal (constitutional) means. What will be the legal status of the newly securitized regime? Chapter 21 mentions a 'special legal regime,' but also hints at the possibility—forcefully campaigned for by the PD community—that PD measures (especially in its nuclear-friendly version) could rather operate 'beyond the pale of legal regulation, that is, in a *state-of-*

[15] See Chap. 11 for the nuts and bolts of securitization theory.

exception mode. This is presumably for reasons of efficiency for decision-making procedures. Here, there seem to be two areas of concern or at least uncertainty.

First, a successfully securitized extralegal global PD regime would be difficult to desecuritize, and perhaps desecuritization would be undesirable, given the seriousness and omnipresence of the threat. In such a case, we would have a state of exception that became the norm. Intuitively, however, there is something deeply unsettling about the possibility of a perpetual state of exception. Indeed, a standard objection against normative use of securitization theory per se points out that securitization proceeds at the expense of regular democratic politics. The omnipresence of a securitized asteroid threat would seem to preclude any moves towards strengthening democratic legitimacy of decision-making structures and procedures on the global level. Quite the contrary, argue critics of securitization: It would represent a step towards global hegemony by whoever is able to exploit and perhaps control the language of securitization.[16] Nurturing collective fears of immigration as an imminent danger to citizens of European countries is a useful analogy: Different beliefs and policies follow from framing immigrants as (variably) 'people in need,' 'terrorists,' 'victims of Western colonial imperialism,' 'parents of four,' 'a cheap labor force,' 'a threat to employment rates,' 'a threat to social security systems,' or 'an enrichment of diversity.'[17]

Second, paraphrasing the famous proposition by constitutional and legal theorist Carl Schmitt (Schmitt 2011, p. 5), whoever decides on the introduction of a (global) state of exception becomes the sovereign power in the given realm; it is also to be expected that whoever he or she is, this actor will control the means of defense against extraterrestrial threats and will decide on their deployment. Insofar as the comprehensive global nature of the threat means PD ought to assume priority over other concerns and policy goals, those endowed with the power to decide on PD will in effect structure limits and possibilities of political action in other areas as well.

Thus arises a global sovereign power as a result of PD-based macrosecuritization. Although we cannot simply assume it will become authoritarian or despotic, abuse of power is always a distinct possibility, and this is why thinking about the institutional context of PD needs to be more acutely aware of the legal-political context. The role and secure status of political opposition, without which a healthy democracy can hardly be imagined (Shapiro 1999, pp. 39–45), is one such issue for further reflection.

24.3 Politics and Democracy

Democratic cosmopolitan governance has recently become almost a magic formula for political theorists. Although some of the doubts expressed in the previous Sect. 24.2 apply here as well, there are a host of discipline-specific questions to be asked.

[16] For a more discussion of such claims see (Roe 2012).

[17] I use these examples in a slightly different context in (Dufek 2018, p. 62).

Most generally, insofar as policy (including budgetary) priorities need to be set among competing goals and purposes, we are speaking about political decisions, which necessarily precede efforts of bureaucratic-rational policy administration and implementation.

24.3.1 Political Representation and Legitimacy

In Sect. 24.2.6, the notion of a *chain of legitimization* was mentioned, concerning the implicit transfer of legitimacy from the domestic level, where elections ensure reasonable proximity between citizens and institutions of state power, to international organizations and other supranational bodies whose democratic legitimacy is at best indirect, mediated via selected domestic actors (especially the executive in its foreign policy role). Similarly to the domestic case, values such as long-term stability or political loyalty invite a search for ways of linking the decisions made on transnational/global level to interests, preferences, or identities of those in whose name they are made. After all, even if we cannot reproduce on the global level the type of democracy we are intimately familiar with on the nation-state level, presumably greater rather than lesser responsiveness, accountability, and responsibility is called for, so that the lack of democratic proximity is partly offset by other means. This is the domain of (a theory of) political representation that obviously applies to PD as well, if democracy is a political regime (model of political decision-making) worth retaining.

There are essentially four ways of approaching the challenge. One possibility is to reject these desiderata and argue that some kind of paternalism is required—that is, thinking and acting *for* (meaning *instead of*) ordinary people who are incapable of recognizing their true interests. Some hints in this direction can be traced in Schmidt's contributions to the planetary defense issue; cf. the ideas of 'technology interdependence' or the disdain for 'delusional people' (Schmidt 2018). But this will be hard to square with the democratic ethos that otherwise permeates the international community, a great majority of international political theory (including EU studies), and also the general ethical background of the present PD project.

Another possibility is to argue that since the PD regime will be securitized in case of its successful implementation (thus acquiring priority over other regimes and policy goals), regular democratic politics consisting of bargaining, deliberation, bartering, compromises, fulfilled and broken promises, clever subterfuges, and so on loses relevance (as it should be with securitized issues anyway). Yet this implies that a state-of-exception becomes the norm (see Sect. 24.2.8), otherwise PD would have to return to the normal (politicized) stage after some time. Not only would issues of representation immediately reappear, but also the capacity to efficiently deal with PD-related tasks and challenges would be lost. Obversely, however, we have seen that a permanent state of exception is hardly desirable from a democratic point of view, and this is why the problem of democratic representation cannot be simply blocked off in that way.

A third option would be to follow recent attempts to separate representation from the electoralist channel it has become identified with, and argue that various actors, such as scientists or (groups of) activists, can claim to represent the interests of the humanity, without having been provided any authority at all to do so via the electoral mechanism (Saward 2010). This approach to representation has been labelled *constructivist*. The problem with constructivism concerns the absence on the global level of a robust *electoral* mode of representation, which arguably forms the precondition for the alternatives to flourish. While constructivists present their dynamic version of representation as complementary to the standard electoralist model, they stop short of claiming that we can actually do without elections; these are taken as necessary, even though they are not by themselves sufficient for democratic representation. But if this is the case, then supranational constructivist political representation is radically incomplete, inasmuch as electoral representation is basically nonexistent there, save for the notoriously problematic case of the European Union (Mulieri 2013).

The fourth possibility has already been hinted at: the emergence of a global constituent power that would serve as the source of political legitimacy, analogously to the constitutionalist fiction of a people upon which existing constitutional (liberal) democracies rest. Let us assume for the sake of argument that such a scenario is conceivable in principle.[18] The question then naturally arises: who or what will be the *constituted* power? If a global parliament, then a global government comes as a natural corollary; after all, what use is a legislative body without an efficient executive to implement the legislation? If it is to be 'merely' a global government of sorts, then not only does the role of the UN General Assembly become unclear, but the very notion of a legislative body also loses meaning. And, has been discussed above (Sect. 15.2.2), the indispensability of the judiciary. In short, it seems difficult to imagine under this fourth scenario a system of political representation that would be significantly different from the domestic model.

24.3.2 Agent–Principal Relationship and Independence of International Bodies

Some contributors to this volume seem to assume in a broadly inter-governmentalist manner that global bodies authorized to make or carry out PD-related decisions would be at least accountable to the governments of involved countries under the classic principal–agent model of delegation: Principals set goals, while agents are authorized to seek them and are in turn held accountable by their principals. Yet research into institutions, such as the IMF or the UN Secretariat, reveals that they tend to become at least partially autonomous actors independent of the will or interest of their founders (i.e. principals) (Barnett and Finnemore 2004). International

[18] For an interesting discussion see (List and Koenig-Archibugi 2010).

organizations, as agents possessing more or less legitimate authority and capable of defining 'problems' and mobilizing resources for their solution, wield power, or the ability to force other actors to behave in a way they otherwise would not. Creating and enforcing obligatory rules regarding human rights and humanitarian intervention are two prominent examples of this power.

There is no reason to assume that duties or obligations related to PD would be different, and as long as PD can be categorized as a 'huge, aspirational task,' then even more power is probably required (Dufek 2013, p. 197). Yet the problem of accountability looms large. As long as there is no meaningful political layer over the global bureaucratic/administrative/scientific networks (understood as embodiments of rationality and impartiality), then it becomes difficult to assign responsibility if things go wrong in some way, as they very well might. Recall that responsibility is one of the desiderata applicable to democratic representation on the global level, especially in situations that call for deployment of sensitive technology. Barnett and Finnemore elaborate on the inevitable pathological tendencies within the bureaucratic juggernauts that are international organizations (in their case, IMF and two UN agencies), singling out (1) routinization and ritualization of internal practices and (2) compartmentalization resulting in tunnel vision by the agencies as well as their subunits (Barnett and Finnemore 2004, pp. 38–41).[19] Several more particular dysfunctions are the consequence, which should cause some concern for those who view trans- and supranational actors as the transmission belts for lower-level goals and preferences.

24.3.3 *Macrosecuritization = Macrodepoliticization = Dedemocratization*

It is worth repeating that PD presupposes a planetary-wide, physical-threat, all-comprehensive macrosecuritization in order to have a reasonable chance for success. It was argued above (Sect. 24.2.5) that such a move is incompatible with normative and political pluralism, including local or regional self-government based on the principle of subsidiarity, if the result is to be stable (uncontested) in the long term. Here we reemphasize the worry over *macrodepoliticization*, since depoliticization in the given realm is what this securitization entails.[20] Schematically, depoliticization stands simply for 'taking an issue off the political agenda,' or alternatively, 'making it a public non-issue,' thus avoiding democratic political conflict of interests, opinions, preferences, identities etc. Appeals to scientific facts or rational inevitability (they usually go hand-in-hand) are quintessential depoliticizing moves, perhaps most resolutely employed by environmentalist groups. The scientific, rationalistic discourse attached to PD-type macrosecuritization obviously follows this

[19] See also (Barnett and Finnemore 1999).

[20] I leave aside the debate on the "politics of securitization" e.g. (Wæver 2011) as it addresses different types of issues.

depoliticizing logic, not least because it is meant to achieve precisely that.[21] As might be clear from the preceding sections, detachment of decision-making from public involvement comes at a hefty price, even though it might be deemed necessary by involved actors. Ultimately, behind all this lurks *rule by experts*, a kind of scientific-administrative epistocracy which by definition needs no democratic legitimacy. While the case for such epistocracy might very well be rationally convincing, the resulting political regime would be *anything but democratic*—in contrast to the original vision of PD embedded in multilevel global democratic governance.

The PD community could collectively reply that the relevant agenda will be limited, both sectorally and with regards to its scope, not least because the envisioned type of global political authority is semi- and not fully cosmopolitan. This response is highly implausible, first because it is question-begging: Whether the suitable political counterpart is semi-cosmopolitan, multilevel global governance, or full-fledged global political authority, is precisely the point of contention, and cannot be thus used as an argument (as it would be supporting itself). Most of what is said in this chapter is meant to show that the latter option—the 'world-state model'—is much better equipped to make PD policy legitimate, efficient, stable, and ultimately successful. Second, and similar to nuclear weapons macrosecuritization, it makes sense to construe PD as a widely inclusive (i.e. comprehensive) macrosecuritizing move across multiple sectors, as its very rationale is to minimize 'the number of separate concerns, issues and conflicts that achieve strong securitization separately from the macrosecuritization' (Buzan and Waever 2009, p. 258). This means that due to its all-encompassing nature and scientific interdisciplinarity,[22] PD has to be accorded complete or near-complete priority vis-à-vis other candidates for securitization or even macrosecuritization. I truly do not find it convincing that such an ambitious plan as PD can be successfully implemented in a weakly centralized system of governance, and certainly not in a genuinely democratic one.

24.3.4 Public Goods Production and Collective Action Problems. Expectations of Respect for Rules

One central reason for such skepticism has to do with public goods provision. 'Public good' in the sense used here is a technical term in economic science, defined by two basic attributes. PD seems to constitute a (rare) *pure* public good, as it is both *non-excludable* (no one can be intentionally excluded from its enjoyment once supply has started) and characterized by *non-rival consumption* (its enjoyment does not

[21] See also Chap. 15 on weapons of mass protection and the risk of rational science argumentation.

[22] Seen from this angle, the present chapter might perhaps also be read as an appeal to the importance of social sciences and humanities in this interdisciplinary undertaking.

reduce the amount left for others) (Gaus 2008, p. 88).[23] Yet this obviously gives rise to a textbook incentive to freeride, especially in 'Prisoner's Dilemma' types of situations, which are supplied generously by international politics (Gaus 2008, p. 109). The plurality of independent actors in international politics—in the first place sovereign states, but not exclusively—is not kept in check by an efficient superior authority. If the standard response from within normative political theory is correct, then the very existence of government, in our case global government, endowed with sufficient power to deter potential freeriders is not only (morally) permitted but also rationally required. Arguably, the more essential the public good is, – technically speaking, the more we can expect that having the good (either paid or unpaid for) is strictly preferred by relevant actors to not having the good –, the stronger the case for global government (and by implication, global taxation and administration and all the related stuff) becomes. Of course, PD claims that the good it is primarily meant to secure (the very physical existence of mankind) is conceptually and normatively prior to any other conceivable human good.

Two points seem to weaken this line of argument. First, there are seemingly excludable technological and commercial benefits of PD, the enjoyment of which the parties to the larger deal would not want to be excluded (as argued in Chaps. 21 and 26). This by itself should produce incentives not to defect (freeride), even under the current interstate system. At least with respect to technology, this fix would work for only a limited timeframe, for reasons of technology diffusion and theft. As regards extraterrestrial extraction of resources (Moon base etc., see Part III of this book), no one should be surprised if those actors who contribute the most—presumably current great powers—would want to benefit the most, to the effect that access to the benefit would be highly unequal, further weakening the case for wide voluntary cooperation (not to mention competition among great powers).

Second, it might be argued that there are cases where public goods can be provided without contribution from each and every actor (e.g. funds and capacities by 150 out of 200 actors, such as states, are suffice to provide the good). But this is a slippery slope: For a social rule (norm) to be considered authoritative and binding by the actors, they need to expect observation of the rule by a sufficiently large set of other actors (Bicchieri 2006, p. 11; Gaus 2011, p. 165; Hart 1994, p. 56). If non-defecting actors see that one can get away with defecting, then the stability and perhaps very existence of the rule is put in jeopardy. All this seems to point to the indispensability of a legitimate political authority on the global level (in the case of PD). Although the arch-globalist David Held agrees that public goods provision as a positive externality of great powers' national interests can no longer be assumed in an increasingly multipolar and heterogeneous world (Held 2016, pp. 843–845), he nevertheless thinks that especially 'in the face of incontrovertible security threats,' relevant powerful actors can 'gain much more from cooperation than from conflict' (Held 2016, p. 844) (see also Chap. 21 of this volume on the motivation not to be excluded from benefits of cooperation). In other words, it is in the interest of

[23] By implication, there are also *impure* public goods that are either non-rivalrous but excludable (such as copyrighted works), or non-excludable but subject to congestion (such as fish stocks).

great powers to keep producing public goods, which means that we can have our cake (comprehensive global governance) and eat it too (i.e. avoid world statism).

Equipped with this assumption, Held et al. furnish their cosmopolitan visions with a wide array of norms and principles to be observed by all actors. Although to my knowledge, no leading author has directly engaged the issue of PD, the general normative outlook of cosmopolitan democratic work is very much in line with the one elaborated in other chapters of this book. I hope to have shown why a mere existence of a norm (written or customary) cannot by itself assure compliance by a sufficient number of relevant actors. In a case such as PD, where what is at stake is control and deployment of extremely powerful (thermo-) nuclear weapons, it seems to me that the ideal existence of an impartial, external, and supreme authority capable of enforcement constitutes a precondition for willing cooperation.

24.4 Democracy, Identity, and Society

24.4.1 Global 'We' as a Prescriptive Project

Section 24.2.3 discussed the possibility of material points of intersection as a path towards global cooperation, adding that this consideration works towards a 'common global identity.' No such 'global we' is currently available, apart from philosophical conceptions of what has been termed *moral cosmopolitanism*, whose links to political practice are loose at best.[24] Talk of a 'multinational global cosmopolitan village' (see Chap. 26) sounds too journalistic for my liking; mostly members of affluent countries' middle and higher strata see the world this particular way, and the ongoing 'immigration crisis' in Europe (as of 2018) attests that a widespread 'cosmopolitan we-feeling' is mostly a philosopher's pipedream (see also Sect. 24.4.3 below). Put another way, the 'perspective of humanity' is an essentially *prescriptive* notion, packed with not insignificant controversial normative baggage. We can look to difficulties in procuring even the most basic human rights worldwide, or the failure of most developed (well-off) countries to achieve the target of development aid (0.7% of gross national income), to see how different the world is.

Opponents might reply that a threat of 'planet-killer asteroid' magnitude would bring humanity together. Perhaps it would; the catch is that the designing and building of the PD infrastructure, including the overarching political architecture, cannot wait until the threat materializes, and as such all reservations noted in this chapter apply. Forging common identities is not a task that can be accomplished overnight, as will be further pointed out in Sect. 24.4.5. Historic experience of modern state-

[24] Moral cosmopolitanism stands for the belief that all human beings are members of a moral community of mankind, which means that they share certain morally relevant characteristics which should bear upon further moral and political considerations (as regards, say, duties to immigrants, human rights policies, poverty reduction and so on). See (Caney 2005; List and Koenig-Archibugi 2010; Pogge 2008).

hood suggests that the existence of a common political-institutional framework serves as a useful, and perhaps indispensable, instrument for constructing common identities. All the hype about 'common European identity' built around what the political philosopher Jürgen Habermas and others have called *constitutional patriotism* assumes that there is a set of universally valid constitutional values and principles that is embedded in and promulgated by shared political and judicial institutions (Habermas 2011; Müller 2009). For a task of PD proportions, 'mere' constitutionalization of world politics is insufficient and would need to be accompanied by establishing a robust system of global political institutions.[25]

24.4.2 Public Identity and Internalization of Public Rules

The birth of common cosmopolitan identity thus needs to be assisted somehow. In fact, one often overlooked yet fundamental element of liberalism as political theory[26] is complementarity of private and public individual identity, where the latter corresponds with the individual's internalization of shared public rules and norms that set up the basic normative framework of a given polity. This is in fact a precondition of the realization of individual liberty, and it has a lot to do with the importance of the rule of law (see Sect. 24.2.6). To drive the point home, there has to be a legitimate political authority which ensures that these rules and norms are publicly promulgated, respected, protected, and enforced, not least because they may entail both rights and obligations. As pointed out above (Sect. 24.3.4), the mere existence of a norm cannot by itself ensure compliance. If there is no such authority on the global level, and if cosmopolitan liberalism is deemed a desirable normative vehicle for dealing with global challenges and issues, then it follows that some sort of global political body needs to be created (Bláhová and Dufek 2018). Not surprisingly, this perspective overlaps with reflections on the rule of law above (Sect. 24.2.6), as well as with the 'guiding idea' that an overarching legal framework is required (0).

24.4.3 In-Groups and Out-Groups; Global Political Identity

Another set of reasons for the previous conclusion is provided by social psychological research, which plentifully reveals that people are generally preprogrammed to define their social identity (self-understanding) in terms of social groups defined by

---

[25] For a defense of constitutionalization of world politics, see (Habermas 2008).

[26] Liberal political ideas inform the bulk of cosmopolitan political thought, so that cosmopolitanism is usually a shorthand for *cosmopolitan liberalism*. The part of liberalism most relevant here is its emphasis on basic rights and liberties of the individual, as well as insistence of fundamental moral equality of all human beings. These ideas normatively co-ground constitutional democracies.

objective or quasi-objective traits (ethnic, geographical, gender etc.), and in opposition to individuals/groups who do not fit into this self-definition. This is the basis for the *in-group* vs. *out-group* distinction explored by social psychologists and neuroscientists (Gaus 2017; Greene 2013; Haidt 2012; Tajfel et al. 1971). Such a self-identification with in-group traits inevitably carries moral distinctions: 'We' are *better* than 'the others,' because we do things *right* while they do things *wrong*, where 'better' ultimately means following 'our way.' The upshot is that the prescriptive project of a 'global political identity' is highly ambitious, in the sense that it challenges certain deep human inclinations without actually providing the tools for overcoming them (apart from idealistic hopes). This finding should further dampen the global governance-based optimism about PD.

24.4.4 Reasons vs. Motivation for Respecting Norms

Many contributions to this volume assert that participating in PD (i.e. not freeriding) is the rational option, and that any rational actor will be inevitably compelled to conclude that doing his/her part is advantageous. The problem is that *reasons for action* cannot be equated with *motivations to act*. Akrasia (weakness of the will) is one trivial example why this is so; (fear of) relative power gains in international politics— i.e. my gain is another's loss, and vice versa—is another. Only fully idealized actors (see Sect. 15.6.1 below) can be expected not only to calculate flawlessly what is in their interest, but also to consistently act upon it (while we need to assume that their individual interest overlaps with common good, i.e., with the realization of PD, which is an unalterable, necessary goal). Recall that conflict among goals, purposes, regimes, or even macrosecuritizations is to be expected. How are we to ensure that those issues that urgently need our attention, capacities, and resources will be preferentially engaged? One type of answer would point to cosmopolitan civic education in existing societies (Price 2008, p. 206; Scheuerman 2011, pp. 106–113; Ypi 2008, p. 55); however, convergence of school curricula is a touchy issue even in contemporary EU, whose member states are comparably close in cultural beliefs and historical experiences of their respective societies.

24.4.5 The Constructivist Hope and Timeframe of Change of Self-Understanding

The constructivist appeal to malleability of identities and self-definitions seems plausible and even convincing as a normative-theoretical position. After all, I have myself found refuge in the constructivist camp in my texts (Dufek 2013; Dufek and Mochtak 2017). The problem with constructivism in PD is the urgency of the practical task, which does not correspond with the much larger timeframe needed for evolutionary transformation of identities. Even Alexander Wendt's famous constructivist argument that a world state is ultimately inevitable (see Sect. 15.2.1) assumes a fairly long run; Wendt's initial guess was 100–200 years. Therefore, PD-related visions cannot rely on a parallel cultivation of cosmopolitan sensibilities, and for the abovementioned reasons, this should either weaken the ambitions accordingly, or push towards a different, more realistic world-statist perspective.

24.5 Economics and Finance

24.5.1 Funding Before Commercial Use of Extraterrestrial Resources Commences; Global Taxation

Funds required for setting up and maintaining an effective system of PD would probably be astronomical (pun intended). Committed chapters in this volume unfortunately say few specific things on that matter, although they include interesting discussions of commercial exploitation of space outposts (see Chaps. 18 and 25), or fees from launching rockets (see Chap. 21). Although commercial use of new technologies or the Moon hub certainly provide a plausible idea of how to reduce or offset costs, return on investments will not start accumulating until after the whole technology is deployed and tested, assuming that everything goes without a hitch—a bold assumption in matters of extra-terrestrial exploration. Until then, either voluntary contributions by involved actors (countries, firms, individuals) or specifically targeted tax money is needed to keep the whole enterprise afloat. If the argumentation on public goods provision in Sect. 24.3.4 is correct, then relying on voluntary contributions is risky: Imagine a successful populist like Donald Trump arguing space exploration, not to say PD, is a complete waste of resources that should be used elsewhere (like a boost to the military budget). Commercial enterprises, on the other hand, can go under for numerous reasons (such as wrong investment decisions as regards research, development, and production). At any rate, it remains to be seen why large businesses should care about producing public goods at the expense of increasing profit. Funding of PD needs to be continuous and secure, and the firmest source of funding would thus seem to be some system of targeted global taxation, each actor paying their share of money each year.[27]

The question arises: can global tax collection do without global tax collecting corps authorized to issue sanctions? It seems not, and for sanctions to have any purchase there needs to be real possibility of their enforcement, which is doubtful in a weakly centralized system—recent experience with human rights protection/enforcement worldwide again provides a sobering reminder. Similarly, if the tax money is to remain targeted and not dissolve into completely unrelated projects, as

[27] Global taxation belongs among David Held's proposals for democratic transformation of the global order. See Held (2010).

is normal with national budgets, some global body would need to make sure it is consistently channeled to the desired destination. We have to keep in mind that the issue in question is one of a macrosecuritized, planetary-wide, existential-threat, and the means of dealing with that threat once it materializes are highly destructive weapons whose potential abuse, and the ensuing reaction by nuclear powers, invokes our worst nightmares. Under such conditions, it seems unreasonable to leave decisions on the allocation of funds to a plurality of self-interested actors. The European Union might again serve as an analogy: The lesson from the debt crisis, and particularly the Geek bailout quagmire, seems to be that the EU/Eurozone needs a much tighter common fiscal policy, which is just a different term for pressures towards further centralization of crucial fiscal decisions.

24.6 Metatheory

24.6.1 Ideal Theory, Full Compliance, and Limits of Moral Possibility

PD visions proceed on the level of ideal theory, which means in the present context that first, relevant actors are fully rational and voluntarily comply with the demands and obligations placed upon them (see Sect. 24.3.4 on collective action problems); and second, that theoretical visions of PD as such are utopian, in the sense that they first construct a desirable world that is in relevant respects remarkably different from our own, and only then questions whether it is possible to achieve it in reality. In order to answer the latter question affirmatively, some link between idealized actors and their real-world counterparts needs to be provided.

The approach canvassed in this volume underplays the indispensability of *political agency*, that is, the fact that there needs to be a real-world actor or a group of actors capable of not only making the ideal plan a reality but also of overseeing its long-term stability (which includes widespread acceptance/legitimacy). This is a precondition for both effectiveness and a 'motivationally sustainable development' of a particular conception of PD (this in fact applies to any comparable global political project).[28] It is not enough to claim that the desirable course of action is simply 'rational' and 'inevitable' for major actors, if by that we do not imply some metaphysical notion of historical necessity or causality (which would be questionable for separate reasons). Much more needs to be said as to how procurement of an ambitious goal such as PD can proceed in a world where power competition, rather than cooperation, arguably still intensifies.

[28] Although Ypi's subject matter is global justice, I see no reason why the argument cannot be generalised. See Ypi (2012).

24.7 Conclusion

By way of conclusion, I shall briefly sum up the argument and add one generalizing remark. This chapter has provided various legal, political, and sociocultural reasons why planetary defense against asteroids requires a centralized global political authority, as opposed to a multilevel decentralized system of global governance that informs the political-institutional dimension of the PD project. It has also explained that the democratic legitimacy of any system of supranational decision-making represents a tough nut to crack, one that cannot be solved by invoking (controversial) normative visions supplied by cosmopolitan-minded political theorists. Both these claims are further reinforced by the fact that PD relies on existential-threat macrosecuritization.

If the political-institutional background of PD is anything but self-evident, then PD is anything but a done deal, its rational-scientific merits notwithstanding. The message here is that the PD community should be more aware of the social and political context of their own enterprise, a context that implies that not all good things go together in planetary defense. Moreover, to the extent that so-called *large technical systems* (LTS) provide the technological muscle for global governance—and PD infrastructure is a LTS of the highest order—they are enmeshed in the 'politics of science and technology' that inevitably accompanies any such complex projects (Mayer and Acuto 2015).

The generalization is this: In the diachronic perspective, large technical-political formations rarely stick to the initial design, due to both technological advances and sociopolitical changes. While the PD project as outlined in the present volume is strong in incorporating the former dimension, it needs to work out its sociopolitical interface. In the long run, evolution beats rational design, and my Wendtian wager is that unless the asteroid strikes first, which would obviously change the game a bit, PD will end up in a world-state—a possibility the PD community should take more seriously.

Acknowledgments The text is an output of a research project supported by the Czech Science Foundation (code GA16-13980S). I thank Nikola Schmidt for many suggestions regarding the shape of my argument, as well as for heated debates, both past and future, about real and imaginary limits to idealistic visions of global governance.

References

Alter, K. J., & Meunier, S. (2009). The politics of international regime complexity. *Perspectives on Politics*, *7*(1), 13–24. doi:https://doi.org/10.1017/S1537592709090033

Anzenbacher, A. (1998). *Christliche Sozialethik: Einführung und Prinzipien*. Paderborn: Ferdinand Schöning

Arato, A. (2016). *Post Sovereign Constitutional Making: Learning and Legitimacy*. Oxford: Oxford University Press

Avbelj, M., Fontanelli, F., & Martinico, G. (2014). *Kadi on trial: a multifaceted analysis of the Kadi trial*. Abingdon: Routledge

Barnett, M., & Finnemore, M. (1999). The politics, power, and pathologies of international organizations. *International organization*, *53*(4), 699–732

Barnett, M., & Finnemore, M. (2004). *Rules of the World: International Organizations in Global Politics*. Ithaca and London: Cornell University Press

Belling, V. (2014). Exekutivní vládnutí jako nový model politiky v Evropské unii? Krizová politika EU a její dopady na politický systém. *Mezinarodni Vztahy*, *49*(4), 9–27

Bicchieri, C. (2006). *The Grammar of Society: The Nature and Dynamics of Social Norms*. Cambridge: Cambridge University Press

Bláhová, S., & Dufek, P. (2018). Identita v liberální politické teorii a dilema kosmopolitismu. *Filosofický časopis*, *66*(3 or 4 (Forthcoming))

Buzan, B., & Waever, O. (2009). Macrosecuritization and security constellations: reconsidering scale in securitization theory. *Review of International Studies*, *35*(April 2009), 253–276. doi:https://doi.org/10.1017/S0260210509008511

Caney, S. (2005). *Justice Beyond Borders. A Global Political Theory*. Oxford: Oxford University Press

Catherine, L. (2012). World Government. In E. N. Zalta (Ed.), *Stanford Encyclopedia of Philosohy*. https://plato.stanford.edu/entries/world-government/

Craig, P. P., & Búrca, G. De. (2012). *EU Law: Text, Cases, and Material* (6th ed.). Oxford: Oxford University Press

De Búrca, G. (2010). The European court of Justice and the international legal order after Kadi. *Harvard International Law Journal*, *51*(1), 1–49

De Búrca, G., Keohane, R. O., & Sabel, C. (2014). Global experimentalist governance. *British Journal of Political Science*, *44*(3), 477–486. doi:https://doi.org/10.1017/S0007123414000076

Drezner, D. W. (2009). The power and peril of international regime complexity. *Perspectives on Politics*, *7*(1), 65–70. doi:https://doi.org/10.1017/S1537592709090100

Dufek, P. (2013). Why strong moral cosmopolitanism requires a world-state. *International Theory*, *5*(02), 177–212. doi:https://doi.org/10.1017/S1752971913000171

Dufek, P. (2018). Lidská práva, ideologie a veřejné ospravedlnění: co obnáší brát pluralismus vážně. *Právník*, *157*(1), 50–70

Dufek, P., & Mochtak, M. (2017). A case for global democracy? Arms exports and conflicting goals in democracy promotion. *Journal of International Relations and Development*. doi:https://doi.org/10.1057/s41268-017-0114-0

Floyd, R. (2011). Can Securitization Theory be Used in Normative Analysis? Towards a Just Securitization Theory. *Security Dialogue*, *42*(4–5), 427–439. doi:https://doi.org/10.1177/0967010611418712

Fuller, L. L. (1969). *The Morality of Law* (2nd ed.). New Heaven: Yale University Press

Gaus, G. (2008). *On Philosophy, Politics, and Economics*. Belmont: Thomson Wadsworth

Gaus, G. (2011). *The Order of Public Reason: A Theory of Freedom and Morality in a Diverse and Bounded World*. Cambridge: Cambridge University Press

Gaus, G. (2017). The Open Society and Its Friends. *The Critique*, *January 15*. http://www.thecritique.com/articles/open-society-and-its-friends/

Greene, J. (2013). *Moral Tribes: Emotion, Reason, and the Gap Between Them and Us*. New York: Penguin Books

Grimm, S., & Leininger, J. (2012). Not all good things go together: conflicting objectives in democracy promotion. *Democratization*, *19*(3), 391–414. doi:https://doi.org/10.1080/13510347.2012.674355

Habermas, J. (2008). A Political Constitution for a Pluralist World Society? In J. Habermas (Ed.), *Between Naturalism and Religion: Philosophical Essays* (pp. 312–352). Cambridge: Polity Press

Habermas, J. (2011). *Zur Verfassung Europas: Ein Essay*. Berlin: Suhrkamp

Haidt, J. (2012). *The Righteous Mind: Why Good People are Divided by Politics and Religion*. New York: Pantheon Books
Hart, H. L. (1994). *The Concept of Law* (2nd ed.). Oxford: Clarendon Press
Held, D. (2010). *Cosmopolitanism: Ideals and Realities*. Polity
Held, D. (2016). Elements of a theory of global governance. *Philosophy & Social Criticism*, *42*(9), 837–846. doi:https://doi.org/10.1177/0191453716659520
Kumm, M. (2016). Constituent power, cosmopolitan constitutionalism, and post-positivist law. *International Journal of Constitutional Law*, *14*(3), 697–711. doi:https://doi.org/10.1093/icon/mow050
List, C., & Koenig-Archibugi, M. (2010). Can There Be a Global Demos? An Agency-Based Approach. *Philosophy and Public Affairs*, *38*(1), 76–110. doi:https://doi.org/10.1111/j.1088-4963.2009.01174.x
Loghlin, M. (2010). *The Foundations of Public Law*. Oxford: Oxford University Press
MacCormick, N. (1999). *Questioning Sovereignty*. Oxford: Oxford University Press
Mayer, M., & Acuto, M. (2015). The Global Governance of Large Technical Systems. *Millennium - Journal of International Studies*, *43*(2), 660–683. doi:https://doi.org/10.1177/0305829814561540
Mulieri, A. (2013). Beyond Electoral Democracy? A Critical Assessment of Constructivist Representation in the Global Arena. *Representation*, *49*(4), 515–527. doi:https://doi.org/10.1080/00344893.2013.846276
Müller, J. W. (2009). *Constitutional Patriotism*. Princenton: Princeton University Press
Nili, S. (2015). Who's afraid of a world state? A global sovereign and the statist-cosmopolitan debate. *Critical Review of International Social and Political Philosophy*, *18*(3), 241–263. doi:https://doi.org/10.1080/13698230.2013.850833
Ondřejek, P. (2016). Státní moc a mezinárodní právo mezi nadřazeností a dialogem. In J. Kysela & P. Ondřejek (Eds.), *Kolos na hliněných nohou? K proměnám srtátu a jeho rolí* (pp. 101–129). Praha: Leges
Palombella, G. (2009). The rule of law beyond the state: Failures, promises, and theory. *International Journal of Constitutional Law*, *7*(3), 442–467. doi:https://doi.org/10.1093/icon/mop012
Patberg, M. (2016). Against democratic intergovernmentalism: The case for a theory of constituent power in the global realm. *International Journal of Constitutional Law*, *14*(3), 622–638. doi:https://doi.org/10.1093/icon/mow040
Peters, A. (2009). Membership in the Global Constitutional Community. In J. Klabbers, A. Peters, & G. Ulfstein (Eds.), *The constitutionalization of international law* (pp. 153–178). Oxford: Oxford University Press
Peters, A. (2015). Constitutional Fragments: On the Interaction of Constitutionalization and Fragmentation in International Law. *CGC Working Paper*, (No. 2), 1–42
Pogge, T. (2008). *World Poverty and Human Rights* (2nd ed.). Cambridge: Polity
Price, R. (2008). Moral limit and possibility in world politics. *International Organization*, *62*(2), 191–220. doi:https://doi.org/10.1017/S0020818308080132
Rhodes, R. A. W. (1996). The New Governance: Governing without Government. *Political Studies*, *44*(4), 652–667. doi:https://doi.org/10.1111/j.1467-9248.1996.tb01747.x
Roe, P. (2012). Is securitization a 'negative' concept? Revisiting the normative debate over normal versus extraordinary politics. *Security Dialogue*, *43*(3), 249–266. doi:https://doi.org/10.1177/0967010612443723
Rosenau, J. (2006). *Study of World Politics, Volume II: Globalization and Governance*. Abingdon, UK: Routledge
Rosenau, J., & Czempiel, O. (Eds.). (1992). *Governance without Government: Order and Change in World Politics*. Cambridge: Vambridge UP
Rosenfeld, M. (2008). Rethinking constitutional ordering in an era of legal and ideological pluralism. *International Journal of Constitutional Law*, *6*(3–4), 415–455. doi:https://doi.org/10.1093/icon/mon023

Saward, M. (2010). *The Representative Claim*. Oxford University Press
Scheuerman, W. E. (2002). Cosmopolitan Democracy and the Rule of Law. *Ratio Juris*, *15*(4), 439–457. doi:https://doi.org/10.1111/1467-9337.00218
Scheuerman, W. E. (2011). *The Realist Case for Global Reform*. Cambridge, UK: Polity
Schmidt, N. (2018). The political desirability, feasibility, and sustainability of planetary defense governance. *Acta Astronautica*. doi:https://doi.org/10.1016/j.actaastro.2018.06.037
Schmitt, C. (2011). *Political Theology: Four Chapters on the Concept of Sovereignty*. Chicago: University of Chicago Press
Shapiro, I. (1999). *Democratic Justice*. Yale: Yale University Press
Somek, A. (2012). Monism: A Tale of the Undead. In M. Avbelj & J. Komárek (Eds.), *Constitutional Pluralism in the European Union and Beyond* (pp. 343–380). Oxford: Hart Publishing
Tajfel, H., Billig, M. G., Bundy, R. P., & Flament, C. (1971). Social categorization and intergroup behaviour. *European Journal of Social Psychology*, *1*(2), 149–178. doi:https://doi.org/10.1002/ejsp.2420010202
Tamanaha, B. (2004). *On the Rule of Law: History, Politics, Theory*. Cambridge: Cambridge University Press
Wæver, O. (2011). Politics, security, theory. *Security Dialogue*, *42*(4–5), 465–480. doi:https://doi.org/10.1177/0967010611418718
Walker, N. (2008). Beyond boundary disputes and basic grids: Mapping the global disorder of normative orders. *International Journal of Constitutional Law*, *6*(3–4), 373–396. doi:https://doi.org/10.1093/icon/mon016
Weiss, T. G., & Wilkinson, R. (2014). Rethinking Global Governance? Complexity, Authority, Power, Change. *International Studies Quarterly*, *58*(1), 207–215. doi:https://doi.org/10.1111/isqu.12082
Wendt, A. (2003). Why a World State is Inevitable. *European Journal of International Relations*, *9*(4), 491–542. doi:https://doi.org/10.1177/135406610394001
Wendt, A. (2011). Why a World State is Inevitable. In L. Cabrera (Ed.), *Global Governance, Global Government. International Visions for an Evolving World System* (pp. 27–63). Albany: SUNY Press
Ypi, L. (2008). Statist cosmopolitanism. *Journal of Political Philosophy*, *16*(1), 48–71. doi:https://doi.org/10.1111/j.1467-9760.2008.00308.x
Ypi, L. (2012). *Global Justice and Avant-Garde Political Agency*. Oxford: Oxford University Press
Zolo, D. (2007). The Rule of Law: A Critical Reappraisal. In P. Costa & D. Zolo (Eds.), *The Rule of Law: History, Theory and Criticism* (p. 24). Dordrecht: Springer

Chapter 25
The Moon Base as a Commercial Hub

Carlos Manuel Entrena Utrilla

25.1 Introduction

The main limitation of current space settlement plans is, without doubt, the cost. Even though some people might argue that we lack some key technologies for long-term human habitation, such as closed-loop life support or advanced propulsion, the criterion that determines which technologies are necessary is at its core a trade-off between cost and risk. This means that, if a given program needs to be implemented with current technological levels, a desired level of risk can usually be achieved by increasing the cost, which is the current practice among public entities. If these expenses are to come from public sources, the costs need to be further justified in terms of the priority of the project against all the other aspects where a government could invest taxpayer money. Planetary defense can provide a good cause for governments to finance the creation of a lunar outpost, but the high costs of the project might threaten continuous support and strain the budgets of the participant countries at some point in the future, and would certainly spark tough political debate over its need against the asteroid threat. The solution to this is to bring commercial partners into the picture.

Outsourcing several activities of the lunar outpost to private companies, for example cargo delivery or resource production, would reduce costs for the main partners, while at the same time putting in place the necessary infrastructure for the creation of a commercial economy in cislunar space. Private companies such as SpaceX have recently shown that they can develop technologies for a much lower cost, accepting a much higher level of risk than public space agencies. Moreover, several studies such as the Evolvable Lunar Architecture or the International Lunar Research Park have already shown that commercial partnerships for the delivery of cargo to orbit and the lunar surface can tremendously reduce costs when compared

C. M. E. Utrilla (✉)
Independent Researcher, Madrid, Spain
e-mail: centrenautrilla@gmail.com

N. Schmidt (ed.), *Planetary Defense*, Space and Society,
https://doi.org/10.1007/978-3-030-01000-3_25

to an end-to-end governmental architecture that includes developing all the elements, from heavy launchers to lunar landers. Thus, including businesses into the program would provide cost reductions and higher risk tolerance in the initial stages, and eventually produce more reliable technology and economic gain for the government partners to use.

A planetary defense goal could also provide the final element in the value chain for commercial activities in the cislunar region, thus creating an initial demand for the different commercial activities. This chapter presents several proposals for public-private partnerships that could be the basis for the commercialization of the lunar outpost, such as the Moon Village proposal from the current ESA Director General Jan-Dietrich Werner, and the possibilities that they present for the development of the multipurpose Planetary Defense base. The end result of a commercialization policy would be a lunar outpost that is not purely dependent on governmental budget, but is instead economically self-sustained and able to finance itself in the long term, thus guaranteeing long-term habitation and maximizing its output in terms of defense, science, and commerce.

25.2 Commerce as an Enabler for the Lunar Outpost

Commerce, more precisely commercialization through PPPs, will be an essential enabler for the Multipurpose Moon Base. The MMB program would be outright impossible under an *Apolloist* approach, given the high costs and the large number of services and infrastructure to be developed. In terms of infrastructure, the MMB would require at least the following elements:

- Earth orbital launch infrastructure (crew and cargo)
- Orbital gateway stations (at least one in high Earth orbit)
- Lunar landers and landing pads (crew and cargo)
- Lunar habitation modules
- In situ resource utilization facilities
- Scientific laboratories and other facilities
- Planetary defense systems (beginning with soft version against micrometeorites)
- Support systems: communications (including any satellite relays), life support, power production, surface transportation, and surface construction

Developing and maintaining all these infrastructure elements presents a challenge in cost and political commitment. Even if an international agreement for collaboration were reached, developing all the elements would take a relatively long time (at least a decade for initial in situ production, according to the estimates of the Evolvable Lunar Architecture (Miller et al. 2015)), during which the political commitment of the partners to the program would be tested against potential cost and schedule overruns. The costs, even if spread out among several partners, would potentially reach tens of billions of dollars per year for the complete infrastructure.

This would hardly fit within current agency budgets and would need to be politically justified even while the base is still not operational. After deployment, the situation might become worse: once the MMB is fully operational and becomes routine, public interest would quickly disappear, while the main benefits are mostly invisible to the public. There is exceptionally good science to be done on the Moon, as presented in *The Scientific Context for Exploration of the Moon* of the National Research Council (NAP 2007), but politicians might find the political justification of the MMB as a science and defense base harder than its technical implementation, given the low likelihood of an asteroid threat and the indirect nature of the scientific benefits.

This unstable and unsustainable situation is the main reason why the MMB needs to be economically self-sustained, with limited need for government support. A fully government-funded MMB would likely fall into the risk-aversion loop, further worsening the situation. As we have seen, the commercial approach to development, for example based on the PPPs presented in the previous chapter, would solve the risk-aversion loop situation, and make for a more sustainable approach to the MMB.

Infrastructure could be commercially sustained instead of relying on government funding. Governments could focus purely on the defense and science infrastructure, which would be a fraction of the cost of maintaining the complete infrastructure to support them. The wealth provided by commercial entities would justify maintaining a defense and science outpost, and would provide more direct benefits to the economy in the form of taxes and highly qualified jobs, which are easier to justify politically than a purely public space endeavor. Finally, a self-sustained space and lunar commercial infrastructure would enable other applications to appear, such as asteroid mining and on-orbit manufacturing, and could provide support for missions to Mars as plenty of others argue (Utrilla 2017a, b; Miller et al. 2015; Zuniga et al. 2015).

As we have seen, experience in previous programs in human spaceflight and Moon settlement (the risk-aversion loop) shows us that the MMB would require commerce to be politically feasible in the current mindset in Western societies. On the other hand, recent experience presents a potential model to reach economic feasibility through PPPs. This model was elaborated for the case of a lunar base in several proposals (Miller et al. 2015; Pittman et al. 2016), showing its applicability to the lunar case. The conclusion is that commerce is an indispensable element in the development of MMB, acting as an enabler for the other activities: defense and science. Beyond that, commerce would as well as enable a fully self-sustained economy in cislunar space, guaranteeing the long-term sustainability of the MMB. In sum, the effect of commerce would be:

- Reducing costs, freeing public agencies from maintaining basic infrastructure (transportation, communication network, resource production, and other surface activities) and allowing them to focus on science and to start thinking seriously about planetary defense in global governance terms.

- Eliminating programmatic risk, freeing basic infrastructure from the need to fight other national priorities for their pieces of government budgets in a constantly changing political environment.
- Mitigating technical risk, exploiting the risk-prone development of private companies.
- Enabling other advanced applications such as on-orbit manufacturing, asteroid mining, and Mars missions, providing cost reductions and a customer base for private initiatives to become economically feasible.

A potential model for how commerce would be able to support the development of the MMB is presented in Sects. 25.3 and 25.4 of this chapter, including the evolution of the commercialization approach along the implementation of the MMB.

25.3 The Lunar Outpost as the Driver for Commerce

The commercialization model based on PPPs can provide the basis for the development of the Multipurpose Moon Base but cannot be realistically extended indefinitely. PPP programs like COTS rely on the commitment from the public partner to long-term contracts, with the private partner to supply services and encouraged to find business outside the government. In the most successful case, SpaceX, COTS developed a commercially sustainable launcher. But that was not the case with SpaceX's competitor Orbital ATK, whose sole business for Antares relies on NASA missions. If and when NASA stops procuring ISS resupply missions, the future of Antares and Cygnus will be uncertain, mainly due to the lack of demand for resupply missions outside the International Space Station. This shows that even a program as successful as COTS might fail to produce real commercial activity (i.e. with no government involved) in the sector in which it operates.

On the other hand, other examples of PPP, such as those of NanoRacks or Made in Space, rely on the public partner providing readily existing infrastructure for the private partner to develop its activities. This might be a way to improve the usefulness of past public investment and can provide commercial activity (NanoRacks has been commercially providing launch and experiments in ISS since 2009 ("Nonreimbursable Space Act Agreement between NanoRacks LLC and NASA for Operation of the NanoRacks System aboard the International Space Station National Laboratory 2009"), but it is not an option during the development base of the lunar base, when it is more interesting for the public partner to share development costs. Collaborations based on a knowledge exchange, such as the Red Dragon missions (NASA 2016), present a similar situation. This model can reduce the costs and risk of development for the private partner, but relies on previous public investment (to achieve that knowledge) and brings little benefit to the public partner.

Considering the potential aspects of commerce as an enabler (self-sustained and private infrastructure, private development, and enabling new applications), the ultimate goal of any PPP program must be the development of a self-sustained economy in the sector of interest—in this case, cislunar space. A self-sustained cislunar economy, i.e. a set of interrelated activities of production, resource allocation, exchange, and distribution of goods and services in cislunar space (Utrilla 2017a, b), is a requirement for the different private partners to find enough demand to sustain the infrastructure on a purely commercial basis. The benefits of private development will only be incentivized if the private partners are brought in the picture from the beginning, instead of relying on infrastructure deployed by the public partners. Finally, the existence of a self-sustained commercial infrastructure is what allows other applications (on-orbit manufacturing, asteroid mining) to appear, since they require the infrastructure to be cost-effective. Thus, the presence of a self-sustained economy in cislunar space might be the only way to achieve the benefits of commercialization and enable the deployment of the MMB.

We can see a small-scale example of this reasoning again in the COTS program and SpaceX. The firm managed to develop a system that does not depend exclusively on NASA funding to sustain itself by successfully entering the commercial satellite launch market. SpaceX disrupted the existing market by providing much lower costs than previous competitors, and managed to capture a noticeable share of the market away from the previous incumbent, Arianespace, who controlled 60% of the worldwide commercial launch market (Knapp 2014). By providing commercial launch services, NASA does not need to be an anchor customer for SpaceX and is free to not provide funds to sustain that piece of infrastructure if it so wishes. This is the final goal of a sector commercialization, but in the case of cislunar space and the MMB, there is no existing market for a private partner to turn to. The development effort must focus then on creating these markets, so that public funds can progressively be shifted towards the goals of public interest of the MMB: defense and science.

In order to develop a self-sustained economy in cislunar space, the commercialization effort must focus on activities that create value in space for space. This means that the value chains of the commercial activities happening in cislunar space must end in space, as opposed to Earth, where all current commercial activities focus. This argument, developed by the author in his past work at NASA (Utrilla 2017a, b), can be justified on a number of technical aspects that limit the cost effectiveness of any space activities carried out from Earth. These are all related to the unique phase of spaceflight that all Earth hardware must go through: launch.

- *Launch costs*: at a minimum of US$ 1700/kg to LEO (Falcon Heavy (SpaceX 2012)), this is the main constraint to how much mass can be placed in orbit. Prices rise exponentially for the lunar surface (US$ 1.2 million/kg). These prices might be justifiable for highly advanced GEO satellites, probes, or other high-tech components (robotic rovers) but become the largest fraction of the cost for "dumb" mass such as habitat structures, propellant, or human consumables. There are ways to reduce these costs on Earth, such as reusable launchers (Post 2015) and distributed launch schemes (Kutter et al. 2015), but ultimately the best solution is to avoid launching as much dumb mass as possible by producing these elements in situ.

- *Impedance matching*: the strict requirements imposed on spacecraft by the launch environments, for example in terms of resistance to acceleration loads or vibrations, suggests that spacecraft should be built in space for space, limiting launch and reentry (or landing and takeoff on the Moon) to vehicles designed especially for that purpose. This concept, termed *impedance matching* (Bienhoff 2010; Simberg 2010), calls for specialized spacecraft for each use case in general. For example, building in microgravity allows for larger structures that only need to support tensile loads, and limiting reentry to a number of specific spacecraft allows the reuse of most space vehicles in space. The result is nothing else than a reduction of costs by eliminating conflicting requirements in spacecraft design (operation in microgravity vs. operation in Earth's gravity) and by providing more efficient vehicles for each situation.
- *The tyranny of the fairing*: this is the term used to refer to the fact that all hardware launched to space must fit inside the limited size of our launch fairings. The larger ones provide just a 5-m diameter and limited length (16 m) (United Launch Alliance 2013). This causes most rocket launches to be limited by volume before being limited by mass, further increasing launch costs. It also constraints the size of the vehicles and structures we can launch, requiring complex on-orbit assembly operations. This constraint can be eliminated with space-built structures—built in space for space—similarly to the requirements of impedance matching. The result is not so much a direct reduction in cost, but through an improvement in the capabilities of our potential spacecraft and an optimization of the designs.

The role of the limitations of launch costs and the requirements of impedance matching can be found in the (arguably) only financially viable lunar base proposals. Launch costs are the largest bill item in both ILRP and ELA (Miller et al. 2015; Pittman et al. 2016), even considering reusable vehicles and block-buy purchases. Both models also consider dedicated space tugs and lunar landers (impedance matching), whose development takes most of the development time. Eventually, both also consider in situ manufacturing of large structures for the long-term evolution of the station, to avoid the constraints of launch fairings.

These three aspects impose a hard limit on the cost-effectiveness of any space missions launched from Earth, and therefore on the feasibility of any business case. This constrains potential commercial markets to focus on Earth, since it presents the largest market available. The best representation of this is the current New Space scene, where large projects for satellite internet constellations manage to raise large sums of money (US$ 1.2B in the case of OneWeb (Caleb 2016)) while other projects such as lunar missions (Astrobotic, Moon Express), asteroid mining (Deep Space Industries, Planetary Resources) or Mars missions (SpaceX) struggle to find funding due to the limited nature of their markets, and must find different markets to finance their ultimate endeavors (DSI pivoted to manufacturing spacecraft components, Planetary Resources is entering the Earth observation business with Ceres, and SpaceX will finance its Mars mission with an satellite internet constellation). If we are to develop viable businesses to support the MMB, we need to focus on space-for-space activities (those that create value in space for space), in order to avoid the limitations imposed by launch. To this end, any potential business plans require the existence of a final customer in space, in which value chains end.

We find then that commerce can only act as an enabler for the MMB if two requirements are met: there must be a commercial market outside the lunar base, and the commercial market must be based on space-for-space applications. These two requirements will define the development of a financially sustainable MMB, which presents the perfect solution: while commerce enables the development of the MMB, the final use of the MMB as a defense and scientific base drives the sustainability of commerce. Governments can finance the development of the MMB through PPPs, and progressively move their investment down the value chain until they become just the final customers. In the MMB, defense and science would ask as drivers for commerce, while commerce acts as an enabler for science and defense. The interconnection of the three activities solves all the issues that have plagued past lunar mission proposals. Planetary defense provides a reduction of programmatic risk over the long term, while commercial involvement through PPPs mitigates it in the short term. Commerce also reduces costs for both science and defense, while newly developed defense capabilities and scientific discoveries provide a fertile ground for further development of commerce, for example with applications such as asteroid mining. By providing a demand for goods, transportation, and services on the lunar surface, the MMB would drive the establishment and growth of a self-sustained cislunar economy to support it.

25.4 A Roadmap for Commercial Development of the Multipurpose Moon Base

Considering the two models for PPPs discussed in Chap. 18, and the goals and requirements of the Multipurpose Moon Base, a model for a commerce-enabled development of the MMB can be defined. The end result of implementing the model would be a lunar base with a commercially sustained infrastructure supporting the scientific and defense activities, whose costs are shared by the governments but additionally allows commercial scientific activities and commercial uses of the planetary defense systems. The model will consider different schemes of partnerships depending on the application and the phase in the implementation of the MMB.

25.4.1 Commercialization During Development

The development phase of the MMB will likely be characterized by a stepwise approach to the final capabilities of the base as both a planetary defense center and a scientific outpost. The infrastructure cannot be deployed all at once, and earlier elements are required for the deployment of the more advanced ones. For example, a gateway station might be required before the lunar landers can be affordably operated, in the case that these are design to only operate close to the Moon. Remember that the minimal infrastructure elements required for the fully operational MMB are the following:

1. Earth orbital launch infrastructure (crew and cargo)
2. Orbital gateway stations
3. Lunar landers and landing pads (crew and cargo)
4. Lunar habitation modules
5. In situ resource utilization facilities
6. Scientific laboratories and other facilities
7. Planetary defense systems
8. Support systems: communications, life support, power production, surface transportation, and surface construction

Of all these, only the orbital launch and Earth-side communications (i.e. the Earth-based elements) are operational today: we have launch vehicles, and agencies have networks for deep space communications. The rest rely on the existence of the previous ones (i.e. habitation modules rely on lunar landers, which in turn rely on gateway stations, and Earth communications and launchers), imposing a progressive development of the base. Finally, the support systems are required by all the surface modules (and a separate version of them by the orbital gateway), and must be developed in parallel to all the surface elements for these to operate. A quick look at the infrastructure requirements of the MMB tells us the considerations needed to select the ideal PPP model:

- Except for Earth launch and communications, all the elements are nonexistent and we lack the required capability to implement them.
- Except for the first three elements (launch, stations, and landers) and some support systems, the technology needs to be developed from scratch. While we have the required level of knowledge concerning the required technology, we have no practical experience with habitation modules, in situ resource extraction, or off-Earth scientific facilities, not to mention planetary defense systems, and all the systems must be developed from the beginning.
- Except for the launch of cargo, there is no existing market for any of the elements.

The lack of a market, and the fact that the technologies must be developed from zero, impose a high level of technical and development risks, while increasing the costs of development. The only existing demand would then be the public agencies developing the MMB, which calls for a public-driven demand model. As we have seen, the public-driven demand model reduces the costs of development, while reducing risks for the private partner by providing certainty in the market. Combined with a progressive development approach, this would yield a series of partnerships to develop each step of the infrastructure leading to the lunar surface, each relying on the commercial activity generated by the previous one. The only elements of the infrastructure that could potentially not be developed through a COTS-like program would be those that must be centralized or are key to the mission, e.g. the gateway stations, the defense character of deployed systems, and the scientific facilities. The centralized nature of these elements constrains one of the key aspects of the public-driven demand model: the creation of a commercially competitive environment. Thus, for these three elements the best aspect for development might be direct public funding.

As the previous chapter discussed, the COTS program is the best example of a public-driven demand model, yielding cost reductions of around 2000% (20 to 1) and in-schedule development (Zuniga et al. 2015). Its main characteristics are the typical of public-driven demand: shared development costs and risks, private intellectual property, long-term commitments to procure services, and encouragement to close business cases outside the public partner's market. However, these characteristics were achieved through an unconventional approach to PPPs, which included pay-on-performance milestones to reduce risk for the agency, a type of contract designed to offer maximum flexibility for the design solutions. This combination of firm-fixed prices and reduced agency oversight managed to get the best of the public-driven demand approach making a COTS-like program ideal for the development of the MMB.

25.4.2 Commercialization During Operations

Once the initial capabilities of the base are operational, the situation changes. A proper commercialization scheme should have yielded a commercial capability to transport goods to and from the Moon, as well as a number of support systems that sustain initial operations. The development activity is then constrained to the expansion of the base and to the full commercialization of the infrastructure (i.e. elimination of reliance on government demand and transition to free market, the consolidation of infrastructure services as purely commercial services).

This situation of simultaneous expansion and consolidation during the base's operations imply that there is not a single scheme that could fit all the options. Consolidation of the base and commercial operations would imply a scheme of public-driven supply (concept explained in detail in Chap. 18), where the base can increase the demand for transportation and other services (power, life support) by supplying infrastructure for scientific or other applications. Consolidation would also imply that the private partners should gradually eliminate their reliance on the public demand, which also pushes for the public partners to provide demand indirectly by supplying infrastructure, instead of by directly procuring those services for applications other than science and defense.

On the other hand, expansion of the base would call for a public-driven demand scheme (explained in detail in Chap. 18) in which private companies develop capabilities to satisfy the demand of public agencies as they did during the initial deployment. However, this contradicts the goal of reaching a self-sustained commercial infrastructure. The best option might be then to limit public involvement to activities related to science and defense, and let commercial partners handle the rest, supported by the market created by the newly available services of the MMB. The public partners would then assume a purely regulatory and coordination role in which the authority controlling the MMB ensures that any commercial expansion of the base's activities does not interfere with science or defense, does not conflict with technical requirements (e.g. overloads the power supply systems), is in accordance with the regulation applying to the base, and provides all of these services to the market as well as to the public partners. Outside these considerations, the MMB should incentivize private partners to further develop the commercial activities on the Moon and on cislunar space, leading to the fully independent cislunar economy that would let the MMB flourish in the long term.

This approach, in which initial stages of development are lead through a COTS-like program, and more advanced activities are coordinated through a central authority, is akin to the structure proposed by the Evolvable Lunar Architecture in the implementation of their base (Miller et al. 2015), and similar to a potential implementation of ESA's Moon Village, though its use of a centralized authority remains to be seen (discussed in Chap. 26). The main difference here is the consideration that the base should expand beyond its initial purpose on a fully private basis, driven by commercial interests but still supporting by public needs such as planetary defense. This is not considered an option in other proposals, which just consider the lunar base as a government endeavor to support some other application (e.g. propellant production for Mars missions). We think that this is a mistake, given that the long-term sustainability of the lunar base depends on its economic feasibility, and therefore requires the development of a self-sustained cislunar economy to support it at affordable costs. Even though commerce might emerge without a public trigger related to public interests, we understand this codependence as a mutual constitutive precondition for human expansion into space. The MMB aims to reach this long-term sustainability by having defense and science act as drivers for commerce, allowing further commercial development of the lunar base beyond its initial defense and science applications.

25.4.3 Commercial Development of the MMB

A precise development plan including timelines and costs would depend on the exact technical details of the final implementation. We can develop nonetheless a roadmap for commercialization that would determine which capabilities should be implemented and how. This roadmap would provide guidance for public partners and regulators as to which partnerships are needed at any point and how they should be executed (e.g. public-driven demand or public-driven supply). We can define this roadmap by assigning a PPP scheme for the development of each element and then identifying different stages in the deployment of the MMB, defining which elements of the infrastructure are to be developed in each stage.

The assignment of a PPP scheme to each of the development elements can be done according to the role of each element in the general plan for the MMB, along with aspects relating to market, private interest, and technological maturity, as proposed in (Utrilla 2017a, b; Zuniga et al. 2015). However, the final goal of the MMB is to act as a driver for the market, providing a suitable market for each of the commercial activities, and we can assume that the implementation of a project such as the MMB would leverage current private interest (all of the infrastructure elements have at least one company pursuing them) and incentivize a number of companies to join the program. Thus, these criteria would not really make a difference between the different elements. Moreover, the idea of using commerce as an enabler for the MMB calls for a bottom-up implementation that is not based on current interests of the industry but rather on the needs of the MMB. A different criterion is needed.

The final goal of the MMB can be the defining factor in how each element is commercialized. The MMB aims to act as a scientific and defense outpost relying on commercialization for long-term sustainability. This requires that some elements are centralized and under direct control of the MMB authority. The scientific facilities and the defense systems would be developed with public funds, as would the orbital gateway as a node of the transportation infrastructure and the habitation modules that house the personnel operating the facilities, which in turn implies centralized life support, since this system is closely related to the design of the habitation modules and should be deployed with it. If these elements (gateway station, habitation modules, life support equipment, and science and defense facilities) were developed with public funding, the only potential commercialization scheme is a public-driven supply scheme: the MMB authority develops and deploys the infrastructure and then offers part of it for commercial use to increase the return on investment and maximize its occupation.

On the other hand, we argued before that the MMB would need a commercially self-sustained infrastructure to be affordable in the long term. To this end, all other infrastructure elements should be developed from a commercial point of view from the beginning, implying a public-driven demand scheme and a COTS-like model. This would include all the transportation methods (i.e. landers), the systems for in situ resource utilization and its supporting facilities (power, communications) and the rest of the surface activities (construction and transportation). These activities also offer potential for the development of a competitive environment, and their commercialization provides the most potential for future applications outside the MMB to appear (for example, the commercial expansion of the base). Table 25.1 summarizes the PPP scheme that should be used for each of the different elements (crew launch is considered commercially available from the beginning, provided by the CCP participants).

Having settled on the different PPP models, the next step is defining the development stages of the MMB. As discussed previously, the final goals of the MMB as a defense and scientific outpost require a phased development approach. It is unrealistic to attempt to land the large directed energy systems of the defense outpost in one piece. Many of the components would have to be built or assembled in situ, and the power requirements of the systems would demand a large infrastructure for

Table 25.1 Most suitable PPP scheme for each infrastructure element

Infrastructure element		PPP implementation
Earth launchers	Cargo	Existing commercial
	Crew	Existing commercial
Orbital gateway stations		Public supply
Lunar landers	Cargo	COTS-like
	Crew	COTS-like
Lunar habitation modules		Public supply
In situ resource utilization		COTS-like
Scientific facilities		Public supply
Planetary defense systems		Public supply
Support systems	Communications	COTS-like
	Life support	Public supply
	Power production	COTS-like
	Surface construction	COTS-like
	Surface transportation	COTS-like

power generation to be in place. As for science, scientific operations can begin as early as there are crews on the surface, but the full potential of the MMB as scientific outpost would demand science-dedicated crews to be present, instead of them sharing the time with other activities such as surface construction. The final objectives of the MMB then would require some time and would need other infrastructure to already be present.

We can establish a basic roadmap for the MMB by defining which infrastructure should be deployed first. A number of studies for the establishment of off-Earth settlements, either on the Moon or Mars, have shown that in situ propellant production is the activity that brings the largest cost reductions to any extraterrestrial base. Propellant comprises the heaviest component of most spacecraft, and its production in situ greatly reduces the mass that needs to be launched from Earth. These cost reductions have been repeatedly shown to enable the deployment of the rest of the base, and to enable continuous operation over long periods (Drake et al. 2010; Utrilla 2017a, b; Miller et al. 2015; Pittman et al. 2016; Zubrin and Wagner 1996; Zuniga et al. 2015). If the first stage of the MMB is to deploy the initial infrastructure needed for the full deployment of the base, it will then focus on establishing propellant production on the lunar surface, and a transportation network that can deploy components for it, including the necessary site preparations, such as terrain flattening.

Once propellant production is established, the next step would be to complete the requirements for the MMB to start operating as a defense and science outpost. This would include expanding the surface infrastructure with crew habitation, surface construction and transportation, and begin the deployment of the necessary systems for science and defense. This phase would be the last part of the development and would end with the initial operational capability for science and defense. The next phase would then just be full operations including any base expansion that would be to take place. We can thus define three different development phases, presented in Table 25.2.

Table 25.2. Phases in the implementation of the MMB

Initial development	Initial capabilities	Full operations
• Develop technologies and capabilities for space transportation and surface operations • Deploy initial basic surface infrastructure • Goal: propellant production	• Build defense and science capabilities • Expand surface infrastructure and capabilities • Goal: initial operational capability of defense and science outpost	• Full operational capability of defense and science outpost • Commercial expansion of the base

- *Initial development*: this first stage is dedicated to the development of the basic infrastructure that is required to deploy the science and defense capabilities. This includes the elements of the transportation network, initial resource extraction and propellant production, and general site preparation activities (construction of landing pads, terrain flattening, installation of power production, etc.). As mentioned, propellant production is an enabling activity for the affordable deployment of the rest of the base and thus should be deployed first. This stage would also take care of any technology development for activities in the second stage. The result of this development stage should be a functioning lunar base that can support the development of the defense and science installations.
- *Initial capabilities*: the second stage of development is dedicated to defense and science capabilities. Once the basic transportation and surface infrastructure are available, the crew habitats can be deployed, enabling the first science applications and human-assisted construction. This is accompanied by a progressive expansion of the surface infrastructure, which needs to grow to meet the requirements of the large defense facilities. Power production would be progressively increased, as would be propellant production capabilities, cargo processing facilities (e.g. landing pads), and other surface facilities (e.g. habitats and construction systems). This stage ends once the defense system has reached initial operational capability and the outpost supports science-dedicated operations, although these two events may not happen at the same time.
- *Full operations*: the final stage in the development is when the development ends and the MMB reaches full operational capability. At this point, all the basic infrastructure should be operated commercially, and the defense and science functions are the only input of public funds into the base. The MMB should then enter a phase of commercial expansion, where the capabilities already in place drive the supply and demand of commercial activities on the Moon and cislunar space, with the aim of creating a self-sustained cislunar economy. This stage signifies the transition into long-term operations of the lunar base, and the end of the development that we are considering in this analysis.

With this definition for the three development stages, we can quickly see which infrastructure elements need to be developed at each stage. Starting from the commercially available Earth launch capabilities (crew launch is considered commer-

Table 25.3 Status of the infrastructure element during each development stage

Infrastructure element		Initial development	Initial capability	Full operations
Earth launchers	Cargo	Commercial	Commercial	Commercial
	Crew	Commercial	Commercial	Commercial
Orbital gateway stations		Deployment	MMB operations	MMB operations
Lunar landers	Cargo	Deployment	Commercial	Commercial
	Crew	Tech development	Deployment	Commercial
Lunar habitation modules		Tech development	Deployment	MMB operations
In situ resource utilization		Deployment	Commercial	Commercial
Scientific facilities		Tech development	Deployment	MMB operations
Planetary defense systems		Tech development	Deployment	MMB operations
Support systems	Communications	Deployment	Commercial	Commercial
	Life support	Tech development	Deployment	MMB operations
	Power production	Deployment	Commercial	Commercial
	Surface construction	Tech development	Deployment	Commercial
	Surface transportation	Tech development	Deployment	Commercial

cially available from the beginning since the CCP participants should start flying by 2018), the *initial development* stage would see the deployment of the transportation network and the basic surface infrastructure, that is, the gateway stations, the lunar cargo landers, the ISRU equipment, and communications and power for it. Technology development for the elements to be deployed in the next stage should also start. Once the deployment is complete, each element would transition to fully commercial operations during *initial capability*, except the gateway station, which would remain under MMB control. *Initial capability* would also see the deployment of the crew landers, habitation modules, life support, surface construction and transportation, and the defense and science facilities. Finally, in *full operations,* any element developed through a COTS-like program would be operated commercially, and the centralized elements would be dedicated to MMB operations, with commercialization through a public-driven supply scheme. Table 25.3 presents a summary of the status of each element of the infrastructure during the each development stage.

With the information from Tables 25.1 and 25.2, the roadmap can be finally integrated, and the different stages of the development detailed. The implementation would go as follows:

1. Stage 1: Initial development
    a. Implementation of COTS-like programs for:

Lunar cargo landers
ISRU equipment
Surface power production
Surface communications

   b. Implementation with fully public funding of gateway station
   c. Start technology development for Stage 2

2. Stage 2: Initial capability

   a. Transition to full commercial operations of:

Lunar cargo landers
ISRU equipment
Surface power production
Surface communications

   b. Implementation of public-driven supply commercialization of gateway station
   c. Implementation of COTS-like programs for:

Lunar crew landers
Surface construction
Surface transportation

   d. Implementation with fully public funding of:

Lunar habitation modules
Life support infrastructure
Scientific facilities
Planetary defense system

3. Stage 3: Full operations

   a. Transition to commercial operations of:

Lunar crew landers
Surface construction
Surface transportation

   b. Implementation of public-driven supply commercialization of:
   c. Lunar habitation modules

Life support infrastructure
Scientific facilities
Planetary defense system

   d. Commercial expansion of the base

This roadmap ends with a fully operational MMB, acting as a defense and scientific outpost driving other commercial activities, and ready for commercial expansion. Each stage of the implementation of the MMB would combine a number of public-driven demand and public-driven supply programs, along with fully public-funded infrastructure deployment and technology development. A model like this can now be used to guide a technical implementation plan that would take into account current technology levels, a technology development plan considering synergies between different elements, and a schedule for deployment. With this first roadmap, the authorities behind the MMB would be also able to start bringing private partners in, preparing the landscape for the many PPPs that would be needed for the MMB, while agreeing on common interfaces and procedures. Further, this roadmap can inform private investors of the intentions of their public partners, and serve as long-term commitment from the public partners to the development of the different partnerships. This is a key requirement in order to bring private investment in at an early enough stage.

25.5 Conclusions

As we have seen, commercialization of space activities can act as an enabler for the implementation of the Multipurpose Moon Base and should be considered in the base's planning from the very early stages. An optimal commercialization model would be based on public-private partnerships, which would enable the implementation of the defense and science infrastructure while developing a self-sustained cislunar economy that can maintain scientific and defense activities on the Moon for the long term. The activities of the MMB on the lunar surface would in turn provide a market for the emerging cislunar economy, acting as a driver for commerce and providing a system that can flourish over time. Based on this mutual dependency of the three main activities of the MMB, we can create a commercialization roadmap for the basic infrastructure, taking into account the different development stages of the base and the nature of each element of the infrastructure. This roadmap combines different models for PPPs and aims to establish a commercial base for the defense and science activities to be deployed affordably. This base can then expand and take the lead in a flourishing cislunar economy, guaranteeing the long-term sustainability of the defense and science activities.

References

Bienhoff, D. (2010). *LEO Propellant Depot - Space Transportation Impedance Matching*

Caleb, H. (2016). OneWeb gets $1.2 billion in SoftBank-led investment. *Space News*

Drake, B. G., Hoffman, S. J., & Beaty, D. W. (2010). Human exploration of Mars, Design Reference Architecture 5.0. In *2010 IEEE Aerospace Conference* (pp. 1–24). IEEE. doi:https://doi.org/10.1109/AERO.2010.5446736

Knapp, A. (2014). Europe's Arianespace Claims 60% Of The Commercial Launch Market. *Forbes*. http://www.forbes.com/sites/alexknapp/2014/09/08/europes-arianespace-claims-60-of-the-commercial-launch-market/

Kutter, B., Monda, E., Wenner, C., & Rhys, N. (2015). Distributed Launch-Enabling Beyond LEO Missions. In *AIAA SPACE 2015 Conference and Exposition* (p. 4593)

Miller, C., Wilhite, A., Cheuvront, D., Kelso, R., McCurdy, H., & Zapata, E. (2015). *Economic Assessment and Systems Analysis of an Evolvable Lunar Architecture that Leverages Commercial Space Capabilities and Pubic-Private Partnerships*. NexGen Space LLC

NAP. (2007). *The Scientific Context for Exploration of the Moon*. Washington, D.C.: National Academies Press. doi:https://doi.org/10.17226/11954

NASA. (2016). *Nonreimbursable Space Act Agreement between the National Aeronautics and Space Administration and Space Exploration Technologies Corp. for Commercial Space Capabilities Collaboration*. https://www.nasa.gov/centers/johnson/pdf/173097main_tSpace_SAA.pdf

Nonreimbursable Space Act Agreement between NanoRacks LLC and NASA for Operation of the NanoRacks System aboard the International Space Station National Laboratory. (2009)

Pittman, R. B., Harper, L. D., Newfield, M. E., & Rasky, D. J. (2016). Lunar Station: The Next Logical Step in Space Development. *New Space*, *4*(1), 7–14. doi:https://doi.org/10.1089/space.2015.0031

Post, H. (2015). Reusability: The Key to Making Human Life Multi-Planetary. *SpaceX*. Text

Simberg, R. (2010), *Impedance Matching Between LEO and BEO*

SpaceX. (2012). Falcon Heavy. http://www.spacex.com/falcon-heavy

United Launch Alliance. (2013). Atlas V. ulalaunch.com. https://www.ulalaunch.com/rockets/atlas-v. Accessed 14 July 2018

Utrilla, C. M.E (2017a). Establishing a framework for studying the emerging cislunar economy. *Acta Astronautica*, *141*, 209–218. doi:https://doi.org/10.1016/j.actaastro.2017.10.005

Utrilla, C. M. E (2017b). Asteroid-COTS: Developing the cislunar economy with private-public partnerships. *Space Policy*, *39–40*, 14–19. doi:https://doi.org/10.1016/j.spacepol.2017.03.001

Zubrin, R., & Wagner, R. (1996). *The case for Mars*. New York: Touchstone

Zuniga, A. F., Rasky, D., & Pittman, R. B. (2015). Lunar COTS: An Economical and Sustainable Approach to Reaching Mars. *AIAA Space 2015 Conference and Exposition*, 1–24. doi:https://doi.org/10.2514/6.2015-4408

Chapter 26
The Multipurpose Lunar Base as a First-Line Biosphere Defense and as a Gateway to the Universe

Nikola Schmidt, Carlos Manuel Entrena Utrilla, Petr Boháček, Jackelynne Silva-Martinez, and Pete Worden

> You want to wake up in the morning and think the future is going to be great—and that's what being a spacefaring civilization is all about. It's about believing in the future and thinking that the future will be better than the past. And I can't think of anything more exciting than going out there and being among the stars.
>
> —Elon Musk, CEO and Lead Designer, SpaceX

26.1 Introduction

The discussion about the most powerful mitigation efforts for planetary defense has circled around nuclear technology, plagued by many negative political consequences. The lack of imagination should not be taken as a barrier. Is an establishment of a multipurpose lunar base a plausible solution? And how should we approach it? Developing the idea of Moon settlements is nothing new, but despite various concepts being developed since the 1960s and 1970s, no government efforts have been able to achieve much of progress. With private entities and philanthropes

N. Schmidt (✉)
Department of Political Science, Institute of Political Studies, Faculty of Social Sciences, Charles University, Prague, Czech Republic
e-mail: nikola.schmidt@fsv.cuni.cz

C. M. E. Utrilla
Independent Researcher, Madrid, Spain

P. Boháček
Institute of International Relations, Prague, Czech Republic

J. Silva-Martinez
International Space University, Houston, TX, USA

P. Worden
Breakthrough Prize Foundation, Breakthrough StarShot, Moffett Field, CA, USA

N. Schmidt (ed.), *Planetary Defense*, Space and Society,
https://doi.org/10.1007/978-3-030-01000-3_26

Fig. 26.1 Laser firing against an asteroid. Credits: visualization by Martin Vaněk, AD13

turning their eyes on the Moon, what is the best way to organize human activity on the Earth's satellite? This chapter aims to answer these questions by discussing contemporary psychical, physiological, technical, political and operational challenges and opportunities of establishing a multipurpose lunar base with the intent to defend Earth (Fig. 26.1).

26.2 Why a More Expensive Laser on the Moon Instead of Nuclear Explosive Devices on the Earth?

26.2.1 Differences of Perception Regarding Technology From the Cosmopolitan Versus National Perspective

The negative political consequences of deflecting an asteroid using Nuclear Explosive Devices (NEDs) will persist until we have a cosmopolitan authority. Nuclear technology in the hands of nation-states represents Weapons of Mass Destruction (WMD), regardless of the intentions of those possessing them, and thus do not qualify for the term Nuclear Explosive Device (NED). We still live in a world of nation-states—a realist world—in which such technologies are necessarily used in national defense. Achieving transparency regarding nuclear technology is complicated, and it would be unwise to depend on functional transparency measures in planetary defense. The nuclear-ablation method might be the most effective against asteroids, but it will also be controversial, cause political concerns, create negative precedents for future operations in space, destigmatize taboos (Chap. 15) and hinder global space efforts.

Efforts to send probes to the stars for scientific exploration and not as part of a security response can be wholly forgotten, because we will be living in a world focused on who should be responsible for maintaining nuclear arsernals for use on asteroids, instead of what benefits can be gained through studying nearest planetary systems. Laser technology can be used for asteroid ablation and mitigation, and nanoprobes used for propulsion as well. Thus, the technology we choose will influence more than just asteroid threat mitigation. Technology is inherently political, and the choice of technology carries clear political implications (Winner 2003).

Each of these anxieties diminishes if we seek *good* effects through the things we do and decisions we make. Adopting policies that cause international political destabilization cannot be excused by the argument that the focus of the policy is mankind's survival. This is the very definition of negative securitization, security in its bad form (Nyman 2016), because in the end this approach would erode the political system itself.[1] The argument is especially strong considering that the observation of asteroids has skyrocketed in the last few years, awarding us with an unprecedented awareness about the majority of asteroids flying around us. The estimation that an extinction-size asteroid will not impact the Earth is about 95% in the forthcoming millennium. The justification for building a nuclear deflection system is based on its effectivity, giving us the last resort measure, but to create political instability is to weaken political systems possibly to the point of collapse. A responsible approach would be to choose a kind of technology that does not cause predictable political instability. This would mean adopting the security-plus concept that seeks human flourishing through stable political procedures instead of negative security.

We live in radical uncertainty, with asteroids flying around us, yet this does not mean that the global political system of planetary defense has to be that way as well. Defense emanating from scientific and civilian exploration missions would be the better choice (Fig. 26.2).

---

[1] Negative security seeks a mere absence of the threat while resorting to fast-tracked, exceptional and undemocratic decision-making justified as necessary to address it. For the discussion of negative vs. positive security see Chap. 20.

Fig. 26.2 A laser array surrounded by radiators for heat dissipation. Overall view. Credits: engineering design of the laser array: Philip Lubin. Visualization and the radiators by Martin Vaněk, AD13

26.2.2 The Hazard of Unilateral Action Using Nuclear Explosive Device

Beyond the technology itself lies the problem of action by a space super power. The idea that any unilateral action involving a nuclear warhead as a mitigation device that can save humanity will be precise and effective is erroneous. Such an action does not account for the negative political consequences that could lead to the monopolization of power and rise of tyranny through technology, unless it arises from an inclusive, democratic global system with a broader understanding of security. We argue that the last-resort action with NEDs in the possession of nation-states should be fully excluded if we are to develop a working and long-lasting global policy of planetary defense before we face an imminent threat.

The norms written in international treaties that stigmatize nuclear weapons work for the sake of world peace. The case of South Africa shows that the stigmatization does not work only against the technology itself, but understandably also against the governments possessing them. South Africa quickly realized that being in the 'nuclear club' is not necessarily an advantage, but rather is a situation that brings about various responsibilities. In fact, we might be currently observing a comparable process in North Korea. Kim Jong-un might have found that possessing nukes has not brought about the desired national security—quite the contrary. However, he might like to charge something more for denuclearization.

Fig. 26.3 Artist depiction of a Bigelow Aerospace inflatable habitat. The dome is an artist's impression of the future habitat with a service rover. Credits: Bigelow Aerospace (inflatable habitat). Other structures and visualization by Martin Vaněk, AD13

Possessing nuclear technology implies significant security tensions between states, and always will, unless we live in a more united world. Only in a politically inclusive, emancipatory, cosmopolitan system could WMDs potentially become NEDs. But first, we need to develop a functioning, accountable, ethical, legitimate, inclusive, democratic, stable and consensual planetary defense governance system. We can achieve such an objective with technologies that are not stigmatized, with science, with our engineering virtue and through governance that holds cosmopolitan ethics in mind (Fig. 26.3).

26.3 From Security Communities to Communities of Fate

The Cold War bipolarity created conditions under which certain groups of states lowered their Westphalian enmity toward each other, allowing the development of what Karl Deutsch described as *security communities*, largely built upon the example of NATO. As the North Atlantic Alliance continues to function on the basis of collective defense, sharing and pooling of resources and expertise, and burden sharing, it seems a suitable concept for planetary defense. According to Deutsch, their existence is preconditioned by "the compatibility of major values relevant to political decision-making," and certain levels of interactions and transactions to achieve a peace between individual states (Deutsch 1957). The security community concept was reintroduced in 1998 through a social constructivist lens by Adler and Barnett, who emphasized shared identities, values and meanings as the main drivers for the constitution of Security Communities over shared threat perceptions, while pointing to supranational institutions as the shaping force of states' behavior, absorbing some state sovereignty (Barnett 1998). However, Ditrych (2014) in his revisit of the concept identifies three critical challenges for future research agendas within security communities: first, the restoration of rigorous and transdisciplinary research methods, pointing to humanistic methodologies or international political sociology; second, the need to rid the concept of inherent realist perspectives preserved by the adoption of states as the only organizational units, whether represented in the form of loosely cooperating nation-states, or vertically moving state sovereignty to a supranational state body; and third, research should also focus on possible disintegration of security communities to complement the current lone emphasis on their integration (Ditrych 2014). The following paragraphs will build on Ditrych's suggestions about security communities, bridging them with David Held's idea of communities of fate (Held 2010), which are applicable for all-human efforts such as planetary defense.

Firstly, the application of the sociological approach brings up individual perception, practices and Foucault's subjected knowledges, rather than focusing on institutional actors or nation-states. As communities of practice, not necessarily of epistemics, increasingly define what knowledge is, we ought to focus on the available moves of individuals that define their practices, instead of their mere intentions (Kessler and Guillaume 2012). In the sphere of Moon settlement, space mining, or any space activity, nation-states remain constrained by instable budgets, short-termism or lack of political will—an obstacle that private entities and philanthropes do not confront. As a result, these nongovernmental bodies are becoming the shaping actors in the field. It is regulations, business barriers and security of their commercial activity, rather than security of a nation-state, that plays a defining role in constituting the practices and moves of their community. For these private entities free of the national security prism, the territorially unbounded nature of the space domain validates their practices, not nationally oriented domestic politics. In this view, state-centric conceptions of security become inadequate and ultimately burdensome for such a community when we focus on individual actors and their moves. The inescapable, borderless nature of the environment becomes constitutive of knowledge and practices. In this sense, the empirical dimension of sociology's focus on individual behavior can enrich the normative political philosophy of cosmopolitan theory (Inglis 2014).

Secondly, the constructivists' emphasis on shared values and identity symbolizes persistent realism, or more accurately methodological nationalism, in the concept of security communities. It originates from the historical forming of nation-states as the only known globally applied governing units, where shared values, culture, language and collective identity played a critical role in the construction of their institutions and fundamental principles. However, that does not mean this relationship is exclusively causal in this way. The social dimension of the nation can emerge as an effect of the efforts of institutionalization. Thus, institutions and bureaucratic principles can themselves be social prerequisites for a political or Moon community. As Zürn (2018) argues, a global *Demos* of shared historical experience, vernacular and homogeneity, is not an exclusive social prerequisite for the construction of democratic governing institutions. Accepted principles can assume the role of constituting social prerequisites. For a democracy, as Zurn shows, it is the will to suppress individual preferences for the benefit of the other. For security communities, it seemingly does not have to be shared values. We can note deep value-based divergences between the unilateral Trump administration and remaining multilateral-in-core NATO allies, or despotic Turkey; the absence of shared values between the US, Israel and Saudi Arabia in their security cooperation in the Middle East; or political tensions between Eastern European illiberal and Western members states in the EU. In these cases, we can identify converging interests despite divergent politics. Similarly, for planetary defense, cosmic hazards can become a common threat perception forming the basis for international cooperation, without the usual dominating national practices. In this way, the goal of protecting the shared fate of humanity, free of geographical divisions, becomes the social prerequisite for the community.

Lastly, we can find disintegration pressure on communities in resurrecting nationalism and protectionism, arguably fueled by the increasing democratic deficit of growing global decision-making when restricted to nation-states. Unequal distribution of voting power between smaller and bigger nation-states, double representation that strips voters of influence on global issues, and global decision-making inequality enforced by bargaining processes are the results of the contemporary global governance system based on geography instead of function and participation. The current incarnation of globalization stimulates de-democratization, which has been exploited by rising nationalist politicians in recent years. Decreasing democratic quality thus represents one of the main issues a security community ought to address, with cosmopolitan principles based on function, inclusivity and deliberative democracy (as discussed in Chap. 21). This way, the institutions, principles and planetary scale security projects can nurture a shared identity in the future.

The practices, available moves, influences and perceptions of private entities in the space domain extend the number of actors and limit the persistence of the national outlook. Common global threat perception thus becomes a constituting principle, not historically observed shared values that can come later through institutionalization or national security concerns. Yet, the lack of global democratic quality encouraging nationalism and protectionism functions as the main disintegrating force. Looking at security communities through these ideas, ridding the concept of geographical boundaries, territorially limited political power, democratic deficits and state-centrism, they rather resemble what David Held describes as overlapping communities of fate. Such communities build on a shared fate beyond national outlooks and historical realism, while emphasizing democratic quality on a global level—a concept suitable for inclusive and effective Moon endeavors for the benefit of all-humanity. A concept that intentionally securitizes the asteroid threat with *sincere intentions*, with an institutionalization such as the Planetary Council in mind and nourishment of human virtue towards space science and exploration. A concept that does not focus on a mere absence of threat and with a proper awareness strategy for planetary defense. Not the one that shows dinosaurs in cartoon-like images catching asteroids in fishing nets, but one with a serious *necessary strategy* based on a *global categorical imperative* that can make a real political impact for the benefit of mankind.

26.4 Science of Survival on the Moon

While we have some ideas of how humans will survive in the lunar environment, long-term habitation will allow us to refine that understanding and begin developing research programs to quantify the issues that affect humans living and thriving in an environment, such as the Moon.

The primary lunar environmental challenges to human life are: low gravity (one-sixth *g*), various forms of radiation (solar, galactic), carcinogenic effects of the regolith (lunar soil) and temperature extremes. Physiological damage can lead to psychological challenges. Living in close quarters with the same people in isolated environments for long-duration missions can be challenging and may affect human performance. We have some experience with this from habitation in Antarctic missions, the ISS, submarines and even prisons. Human responses to harsh environments vary, and psychological stressors can in turn lead to further physiological damage (Fig. 26.4).

Crews for space missions will be selected according to psychological profiles, disciplinary training, team care and teamwork skills, cultural competency, communication skills, leadership and followership talents, and other relevant experience. While conducting scientific experiments, the crew will be exposed to the remote conditions of mission exploration. The design of systems and structures, as well as mission planning operations, will need to follow a Human Systems Integration (HSI) approach, a process intended to integrate human capabilities and limitations within and across all systems throughout the mission lifecycle. HSI is an enabler to systems engineering, which by definition facilitates the realization of successful systems. Today's technology and mission objectives are no longer dependent on a single system but rather multiple complex systems, or a system of systems that need to deliver unique capabilities. In fact, the paradigm shift is toward capability and how the system of systems can operate when integrated. This may require devoting more time up front to perform modeling and simulation, which would include human in the loop and usability evaluations, and models to close the design space, getting closer to an optimized result. In other words, we are shifting our efforts from just designing for performance to designing for affordability based on the capability of the different systems that may be operating within other settings (Silva-Martinez

Fig. 26.4 Fuel containers, labs, pipelines and landed Big Falcon Rocket in the background. Credits: engineering design: SpaceX (BFR). Visualization and other structures by Martin Vaněk, AD13

2016). We need to realize that as humans we cannot eliminate all possible problems, especially in a partially known environment with many uncertainties. However, we can reduce those uncertainties and lifecycle-cost impacts by using HSI activities during early phase decision-making. Throughout the other phases of the mission's lifecycle, it is also necessary to continuously reevaluate our assumptions and reassess any modifications that may impact other parts within the system of systems, in this case, a multipurpose Moon Base.

Sherwood (2005) provides an excellent discussion of structures and possible issues from the perspective of a space architect. Any lunar structure will be designed for and built with the following prime considerations.

26.4.1 Safety and Reliability

Human safety and the minimization of risk to acceptable levels are always at the top of the list of considerations for any engineering project. The Moon offers new challenges to the engineer-designer. Minimization of risk implies structural redundancy, and when all else fails, easy escape for the inhabitants. The key word regarding risk here is 'acceptable.' It is a subjective consideration, deeply rooted in economics. What is an acceptable level of safety and reliability for a lunar site, one that must by its very nature be considered highly hazardous? Such questions go beyond engineering considerations and must include policy considerations: Can we afford to fail?

Even the meaning of failure is relative. Human error is one of the weakest links in any system. Even considering hardware or software, both are created by humans; hence, many 'failures' can be attributed to human error. In human spaceflight, as well as in other missions, we can find differences regarding how human error is perceived, based on the project failure or success. For example, if an operator (astronaut, flight controller, ground support personnel) fails to follow all the required procedural steps based on his/her judgment during a decision point, and this action does not cause any accidents, then the action may not be considered a human error. In fact, a human action is not considered to constitute a human error until it causes an accident, or a failure with a negative effect. Thus, in current practice, human error or failure is measured by the outcome, not the action itself.

26.4.2 One-Sixth-G Gravity

In gross terms, a structure will have six times the weight-bearing capacity on the Moon as on the Earth. Or, to support a certain loading condition, one-sixth the load-bearing strength is required on the Moon as on the Earth. In order to maximize the utility of concepts developed for lunar structural design, mass-based rather than weight-based criteria should be the approach of lunar structural engineers.

26.4.3 Internal Air Pressurization

The lunar structure will be a life-supporting closed environment. It will be a pressurized enclosed volume with an internal pressure of 6.9×10^4 to 10.3×10^4 Pa (10.0–14.9 psi). The enclosure structure must contain this pressure and must be designed to be 'failsafe' against catastrophic and other decompression incidents caused by accidents and natural impacts. Internal pressurization offers challenges to all lunar structures, but especially the inflatable concept.

The inflatable concept provides a volumetric advantage. It can be folded into a small volume for transport, and when it arrives at its destination it can be inflated to typical volumes such as spheres and cylinders, or to complex combinations of volumes. The site whereupon the structure is inflated needs to be prepared by removal of debris and rocks and smoothed out. A base of some sort needs to be placed on the lunar surface in case it is not part of the inflatable structure. Upon inflation, the structure needs to be rigidized using a variety of techniques, for example, UV, a temperature gradient, or its walls can be filled with a rigidizing foam. Cadogan et al. (1999) are pioneers in these concepts and technologies.

26.4.4 Shielding

A prime design consideration is that the structure be able to shield against the types of hazards found on the lunar surface: continuous solar/cosmic radiation, meteorite impacts, and extreme variations in temperature and radiation. In the likely situation that a layer of regolith is placed atop the structure for shielding, the added weight would only partially balance the forces on the structure (in the range of 10–20%) due to the internal pressurization mentioned above. In addition to general shielding, special radiation shelters will be needed during periods of increased solar activity. Shielding against micrometeorite impacts is done by providing dense and heavy materials, in this case compacted regolith, to absorb the kinetic energy. Lunar rocks would be more effective than regolith because of their fracture toughness, but rocks are more difficult to obtain and much more difficult to place atop surface structures. Some suggest that for shielding purposes alone, it is better to design and place human-rated structures underground. This may be so, but it is then necessary to factor in the added costs and difficulties of subsurface work.

26.4.5 Vacuum

A hard vacuum surrounds the Moon. This will preclude the use of certain materials that may not be chemically or molecularly stable under such conditions. This is an issue for continued research. Construction in a vacuum has several problems. One would be the possibility of out-gassing of oil, vapors and lubricants from pneumatic systems. Hydraulic systems are not used in space for this reason. The outgassing is detrimental to astronomical mirrors, solar panels and any other moving machine parts because they tend to cause dust particles to forms pods.

Another problem is that surface-to-surface contact becomes much more abrasive in the absence of an air layer. The increase in dynamic friction would cause fusion at the interfaces, for example, a drill bit fusing with the lunar rock. This is obviously aggravated by the fact that the vacuum is a bad conductor of heat. The increase in abrasiveness at interfaces also increases wear-and-tear on all moving parts; for example, railways and wheels.

Blasting in a vacuum is another interesting problem to consider. When the explosive in a blast hole is fired, it is transformed into a gas, the pressure of which may sometimes exceed 100,000 terrestrial atmospheres. How this would affect the area around the blast on the Moon and the impact of ejecta resulting from the blast is difficult to predict. Keeping in mind that a particle set in motion from the firing of a rocket from a lander could theoretically travel halfway around the Moon, the effects of surface blasting on the Moon would be something to be concerned about.

26.4.6 Dust

The lunar surface has a layer of fine particles that are easily disturbed and placed into suspension. These particles cling to all surfaces and pose serious challenges for construction equipment, air locks, and all exposed surfaces. Lunar dust consists of pulverized regolith and appears to be charged. The charge may be from the fractured crystalline structure of the material, or it may be of a surficial nature, for example, charged particles from the solar wind attaching themselves to the dust particles.

26.4.7 Ease of Construction

The remoteness of the lunar site, in conjunction with the high costs associated with launches from Earth, suggests that lunar structures be designed for ease of construction so that the extravehicular activity of the astronaut construction team is minimized. More likely is the use of robotic construction teams. Construction components must be practical and, in a sense, modular, in order to minimize local fabrication for initial structural outposts.

26.5 Structures of the Moon Base

26.5.1 First, Second and Third Generation Habitats

We envisage the first human habitats on the Moon to be pioneering settlements with the initial goals of survival and exploration. Once they are assured of survival, efforts will be expended to support infrastructure and growth. Self-sufficiency will be critical, but not viable, during the first two generations of structures. First generation structures are those preassembled from Earth, with second generation structures being those brought from Earth as components, assembled on the Moon, perhaps using some local materials. Third generation structures are those whose materials have been locally sourced. See our artist's visualizations showing all three structures. The first generation is a space ship, the second a Bigelow inflatable habitat, and the third a dome. Other structures shown partly underground, under a layer of printed regolith over structures transported, can be considered a combination of these (Fig. 26.5).

There are several dimensions to the design of structures for surface habitation. First are the engineering design constraints for construction, strength, volume, daily and emergency radiation and micrometeoroid shielding (as the Moon has no atmosphere, each rock coming from space will crash on the surface). Second are the constraints for psychological wellbeing, for example, architectural and aesthetic livability considerations, such as color, configuration and personal space. Here we

Fig. 26.5 Bigelow Aerospace inflatable habitat. The dome is an artist's impression of the future habitat. The umbrella-shaped structure on the left is a recent NASA fission reactor. Credits: Bigelow Aerospace, NASA, Martin Vaněk. Visualization by Martin Vaněk, AD13

wanted to show in our visualizations that living in the dome would be visibly more comfortable than underground.

Structural classes include rigid cylindrical and ellipsoidal concepts, inflatable concepts, multi-component cylinders encased in truss frameworks, and local material concepts using building blocks of regolith or processed regolith. We expect that eventually habitable structures will be mostly buried, with no more than access points exposed to the surface. Additive manufacturing will play a big role during the construction of the Moon Base, initially to support 3D printing of tools and parts and later to complete new structures, including all of the mechanical and perhaps also electrical conduits and systems.

Science and exploration will enable commerce and defense as they play interrelated roles. Planetary defense sites can be identified as part of the exploration part, which will also allow for identification of sites for scientific outposts and possible sources of raw materials for human settlements. We propose a multipurpose lunar settlement, knowing that a settlement mission may require a large investment. From

Fig. 26.6 Multipurpose lunar base. Dome with vegetation with a thin water layer shielding against radiation. Service rover, street lights, batteries and other utilities. Credits: visualization by Martin Vaněk, AD13

a system of systems perspective, this is designing for affordability that meets multiple objectives, hence designing with various capabilities in mind. Human survival is the primary focus for which human systems integration processes are needed to ensure human capabilities and limitations are integrated into the design of the lunar base throughout the lifecycle. To accomplish this, we need an understanding of the environmental factors of the Moon and how they affect human physiology and psychology, and the behavior of materials and structural dimensions and classes required to build lunar habitats from first, second or eventually third generations (Fig. 26.6).

26.5.2 The Cursed Governmentally Driven Moon Base efforts

Even with the recent boom of popularity of the Moon as the next destination in space, the reality is that there are no realistic plans from any government to visit our satellite again. All new proposals are exceptionally similar to those from the past, and so suffer from the same problems. The old space community is still trapped in the mission designs of the 1970s and 1980s, partly by the heartfelt nostalgia of the engineers who saw these designs as kids, and partly by a denial of the enabling benefits that commercialization would bring to their designs. Most official approaches to create a base on the Moon are centered on the three core ideas of old space:

- The technology is too complicated, thus only governments can manage it.
- The technology is too expensive, thus only international cooperation can afford it.
- The technology is not practical, thus only science can justify its development.

This view is the direct heritage from Earth's history with space: the failure of the Russian Moon program and its constant budget issues afterwards, the cancellation of all lunar programs in the US after Apollo 17 (including the Apollo Applications Program and the Space Exploration Initiative) and finally the ISS era, which defined the current view of space projects as inevitably decade-long and over budget. This was also a time when the need for collaboration among superpowers was apparent for any space development. This thinking can be seen in the main projects of the day: the ESA Moon Village and the Chinese and Russian proposals.

Both China and Russia represent perfect examples of the old-space thinking when it comes to lunar missions. Both countries are taking the traditional approach with a space program funded exclusively by the government, centralized through a space agency and aimed at fulfilling national goals. Although Russia has been talking about lunar missions for decades and is expecting to launch the Luna 25 lander in 2019 (Sharkov 2018), it is China who is making the best progress towards the Moon. The Chang'e 3 lander in 2013 was Earth's first lunar lander since 1976, and as such was met with great fanfare. It was the first step towards China's lunar program, which, very much in the style of the 1950s and 1960s, is taking incremental steps towards the Moon, with a landing projected on the far side in 2018 (Chang'e 4), a sample return mission in 2019 (Chang'e 5), and an eventual crewed landing planned for 2036.

The official purpose of these missions is scientific, but the 100%-governmental budget and the great lengths of the state's publicity of each mission show that the real intent is national pride and technology development. This makes them essentially a 21st century Apollo remake, but without the actual driver of the Western-like liberal democracies. The key characteristics of space developments in our era are extremely high budgets and absurdly long development times. Given its nationalistic goals and dependence on national and international politics, this approach from the previous century is highly unlikely to lead to a thriving human presence on Moon. Most of the designed systems and knowledge gathered for the mission will disappear once the Moon is not part of the country's goals, as the funding disappears—if it ever first appears.

26.5.3 The Fresh Wind by Decentralization: Moon Village Becomes Global

Opposite to the traditional approaches, the ESA Moon Village presents a fresh idea among the current efforts, even if still just an unofficial one. The concept was proposed by ESA's Director General Johann-Dietrich Wörner in 2016. It is unclear at the time of writing whether Moon Village is going to be driven by ESA or the Moon Village Association, or what is going to happen in the upcoming months. The idea encompasses the establishment of a lunar station based on international collaboration and commercial participation, and also coordinated through a complex network of relationships between each player rather than through a centralized committee. This focus on the involvement of multiple actors and the openness to commercial participation sets the project apart from any traditional lunar aspirations; however, it still lacks significant visible funding. The decentralized approach should avoid the problems that ISS faced during its development, with constant redesigns and changes of requirements, and let the lunar environment and the basics of economics dictate the flow of the project.

This idea somewhat materialized in 2017 with the Moon Village Association, a think-tank-like organization aiming to 'promote international discussions and formulation of plans to foster the implementation of the Moon Village.'[2] With the diversification of the role of the involved actors, projects like the Moon Village could gradually be realized by small entities with lunar ambitions. Companies like iSpace, who recently raised $90 M to deliver two small landers to the Moon, are gearing up to set up a payload delivery service to the Moon's surface (Nakamura and Furukawa 2017). Similar efforts show how a small engineering team can trigger interest among larger players. The Berlin-based company PTScientist, which has a launch slot on a Falcon 9 to launch two Audi rovers to the Moon in 2019, managed to sign up Nokia and Vodafone to relay their rovers' HD stream from the Moon to Earth. Vodafone pledged to set up 4G relays on the surface for this purpose, reducing the rovers' power requirements and creating the first off-Earth cellphone network (VodafoneGroup 2018). This effort might be leveraged by other lunar delivery companies in the future, such as Astrobotic, which could start sending more payloads to the same area, further increasing the interest of the Vodafone lunar network in a continuous cycle. However, the current efforts to create the Moon Village do not seem to be leveraging this opportunity.

There are no formal plans to implement the Moon Village project, and the concept is still clearly based on the idea that the main goal for space exploration should be science and that governments are needed to take the lead forward. Mr. Wörner himself made this very clear in his original publication as he left commercial activities in a secondary role, stating 'you might see not only scientific and technological activities, but also activities based on exploiting resources or even tourism.'[3]

---

[2] See https://moonvillageassociation.org/about/.

[3] See https://www.esa.int/About_Us/Ministerial_Council_2016/Moon_Village.

With a goal clearly based on noncritical science and commercial activities, the Moon Village will have to wait for a government to plan its implementation before it becomes a reality.

26.5.4 Other Moon Base Efforts

Despite the still debatable level of commitment from national space agencies and governments, numerous realistic plans have been developed in recent years. The International Lunar Research Park (ILRP) (Pittman et al. 2016) and the Evolvable Lunar Architecture (ELA) (Hall and Miller 2016) are both economically feasible for a space agency and present feasible timeframes for implementation that do not require a decades-long development process. This contrast with the governmental approach brings attention to the main difference between realistic and unrealistic plans: the leveraging of commercialization as a key element of the mission design.

Commercialization allows for a quick development with low budgets of space systems (Utrilla 2017) and lets the mission sponsor partially externalize the risk of development by having an investment partner that takes on most of the sustained costs of operations. The elimination of an unconstrained government budget and the imposition of a commercial timeline rationalize and simplify requirements, as was the case with Falcon 9, thus making the resultant mission design a more realistic approach in most cases. Jeff Bezos has become more open recently, explicitly stating that he is going to begin Moon settlement in 2023 (Kettley 2018). Meanwhile, Elon Musk is still primarily focusing on Mars (Musk 2017). It is common sense that with a growing micro-economy on the Moon, Musk's payload of 150 t and announced cost of $7 M would be a game changer (Fig. 26.7).

A part of NASA seems to have taken good note of this idea and is pushing for a lunar version of the extremely successful Commercial Orbital Transportation Systems (COTS) program, which helped NASA develop SpaceX's Falcon 9 and Dragon, and Orbital ATK's Antares and Cygnus, bringing the US back into play in the world scene of commercial launchers. The Lunar-COTS (LCOTS) presents a development scheme leading to a human presence on the Moon sustained by commercial partners, where NASA leverages commercial capabilities to provide the key resources needed on the lunar surface while also reducing costs for missions farther into deep space (Zuniga et al. 2015). This approach of commercial partnership not only reduces costs and schedules, but it also allows a natural participation of small players, be it countries or companies. Given the current *pensée unique* regarding the need for international collaboration, this is yet another point in favor of creating a commercial-based approach towards lunar development.

If implemented, the Lunar-COTS program would allow the US to reach the Moon within a timeframe similar to that of the ELA study (10–12 years (Zuniga et al. 2015)), making it the fastest Moon program since Apollo, and adding a sustainable aspect that permits such endeavors in the current era. This timeframe is

Fig. 26.7 Landed Big Falcon Rocket (BFR) in the Multipurpose lunar base. Credits: engineering design: SpaceX. Visualization by Martin Vaněk, AD13

similar to those from the ILRP and ELA and must be compared with China's schedule (20 years to human landing), NASA's SLS development (16 years to first flight of main launcher, counting from the start of development of Ares V in 2004 to the expected SLS launch on 2020), and the complete lack of a schedule for the Moon Village. The comparison of proposed budgets leaves the traditional approach in no better position, and the inclusion of the untold benefits of developing a world-leading high-tech commercial industry alongside the space program completes the case for the commercialization of the Moon.

The only barriers preventing us from permanently living in our satellite are the established core ideas of old space. NASA hast started taking note of this (Foust 2018a), but we might need a small nudge to finally persuade countries to consider creating a base on the Moon. Planetary defense and the *Multipurpose Lunar Base* project might be it.

26.5.5 *Privately Funded Lunar Programs*

As noted above, the focus of lunar exploration is shifting from public-funded to private initiatives. The latter are focused on two key aspects. Many efforts plan to use lunar resources to support commercial and government space activities. The earliest use of space resources would be using lunar volatiles as fuel for GEO-based communications satellites. Indeed, this potential application is one reason that the nation of Luxembourg, with its strong investment in communications satellites, started a space-resources initiative in 2015.[4] A second key, nontraditional approach to the Moon comes from private philanthropic groups and individuals interested in establishing a lunar human settlement. As noted above, the wealthiest man on Earth, Jeff Bezos, has focused his efforts on such a settlement. In 2014, a group of lunar experts and high-net-worth Silicon Valley entrepreneurs met at a large investment house, Draper, Fischer, and Jurvetson, to consider how such a privately-funded endeavor could be done They concluded that an initial lunar settlement could be completed within a decade for less than $5 billion (Jurvetson 2014). Initial efforts to begin a settlement by 2023 are now underway (Kettley 2018).

It is likely that large-scale private investments in the Moon and lunar infrastructure will emerge in the next decade. This would enable the development of an infrastructure that could be used to mount a major laser-focused approach for both propulsion into the Solar System and beyond, and as an effective planetary defense method by the second half of the 21st Century. This idea was publicly proposed by Pete Worden at Asteroid Day 2018. It differs significantly to how the US perceives planetary defense policy, which completely omits the laser mitigation method (The White House 2018).

26.6 Main Issues of Building the Multipurpose Laser Facility on the Moon

As can be expected, an ambitious project like a lunar multipurpose laser facility is plagued with engineering challenges and complications. However, none of them constitute a dealbreaker that would render the project completely unfeasible, and most of them can be dealt with using current technology and techniques. Technically, the main engineering challenges are related to the power source, residual heat, beam divergence, target accuracy and laser power. Although this is nothing compared to the challenge of building such a facility on the Moon in the first place, it is important to address these issues early on to ensure that the project is feasible. Other constraints, such as timing due to the lunar orbit and general practicality of the system, are also interesting to discuss and will give the complete picture of the usability of the system (Figs. 26.8 and 26.9).

---

[4] See http://www.spaceresources.public.lu/en.html.

Fig. 26.8 A laser array surrounded by radiators for heat dissipation. Bottom view. Credits: engineering design of the laser array: Philip Lubin. Visualization and the radiators by Martin Vaněk, AD13

26.6.1 Beam Divergence

Even in the vacuum of space, any directed energy system will have to face the divergence of the laser beam due to diffraction effects. The large differences between the operational scales of the system (meter-wide laser aperture and target spot, planetary-scale distances) make divergence the largest issue for our system. A smaller target spot or a larger targeting distance would require a lower divergence, in turn needing a shorter wavelength for the laser. This can quickly run away from our current laser-building capacity. A single 10-m-aperture laser attempting to target a meter-scale spot on an asteroid barely 0.05 AU away (a stretch in interplanetary space, meaning the target would already be within terminal approach) would require wavelengths in the range of hard X-rays or gamma rays, completely killing the concept. There is no way around diffraction itself, but the focusing constraints can be relaxed by increasing the output of the laser system. This can be achieved using the Vorontsov Phased Array principle, as proposed by Lubin: small, modular units make up the power output of the laser, allowing for smaller lasers to be used and a for a larger target spot. This converts the physical constraint (diffraction) into

Fig. 26.9 A laser array surrounded by radiators for heat dissipation. View from above. Credits: engineering design of the laser array: Philip Lubin. Visualization and the radiators by Martin Vaněk, AD13

an engineering constraint: a given laser power would be able to target asteroids with a given target spot for a certain range of distances. This operational envelop could easily be expanded by adding more lasers to the modular system, and this would eventually reach true interplanetary reach (1 AU) for array sizes of tens of kilometers (Fig. 26.10).

26.6.2 Target Accuracy

Once the beam issues are solved, targeting it accurately to the asteroid is the next big step. There are two problems here. First, knowing where the asteroid is, and second, pointing the laser towards the asteroid's position with meter-level precision

Fig. 26.10 Laser emitters in detail. Credits: engineering design: Philip Lubin. Visualization by Martin Vaněk, AD13

in order to hit the intended part of the asteroid. The first issue can be solved, as the same lasers forming the directed energy array can be used as a LIDAR (Riley et al. 2014). LIDAR is a well understood and broadly used technology in space missions and has been used in the sub-planetary scale for decades to measure the Moon's distance to Earth with cm accuracy. Active illumination can also help scan the sky for targets, providing a use for the array whenever no active threat mitigation is going on. The challenge begins once the asteroid's position is fully determined. Hitting a meter-scale spot at a considerable fraction of an AU would need pointing accuracies of nanoradians. This can be achieved from orbit, as proven by the Hubble telescope, but will certainly be more challenging while sitting on a celestial body rotating at a rate of microradians per second. This is an issue that needs to be further investigated, but there is no reason to think that our targeting objective cannot be achieved with a LIDAR-based feedback system and the seismic stillness of the Moon. A potential workaround is to increase the size of the target spot, which would increase power requirements but also reduce divergence constraints, increasing in turn the possible target distance. In any case, pointing is just an engineering challenge, and not a dealbreaker for the system (Fig. 26.11).

26.6.3 Laser Powering and Heat

The deflection capacity of the system will be directly determined by the total power output of the system, since the thrust generated will be directly proportional to the applied power, as shown theoretically by Phipps et al. (2010) for any form of laser

Fig. 26.11 Laser emitters in detail. Credits: engineering design: Philip Lubin. Visualization by Martin Vaněk, AD13

ablation, and experimentally by (Gibbings et al. n.d.) for asteroid material. Since beam divergence and accuracy will be independent constraints based on the laser wavelength and the design of the pointing system, the single variable defining the operational envelop of the laser will be its power output. For a given laser design (pointing and wavelength), the power will determine how much thrust can be applied to any target asteroid, with the final deflection being a function of the asteroid's initial mass and the total time the thrust is applied. Being a modular, scalable system, the operational envelop can be extended by increasing the system's power. Access to an adequate power source is thus essential, and will determine the usability of the whole system.

With the constraint of placing the lasers on the far side of the Moon, the obvious solution would be to sit close to the lunar poles: close enough to have access to the permanent sunlight spots in the poles, but far enough to guarantee no line of sight of Earth is ever possible. Solar power could then be used for the initial operation of the system, but dedicated nuclear power plants would likely be necessary in the long term, when the system scales up to km-size diameters. Other latitudes are also possible, such as near the equator of the Moon, using nuclear power plants from the beginning, but any latitude would require two laser systems to be implemented to maximize sky coverage (a single position would always cover only half the sky not blocked by the Moon), making the poles a location that solves two problems at once (power access and coverage). Whatever the final location, the large power requirements of the system would cause large amounts of heat that would need to be dissipated. The deployment of large radiator fields would solve the issue, assuming they can be effectively shielded from the heat reflected from the lunar surface to operate during the lunar day, or that suitable shadowed locations are found. NASA's recent developments on fission reactors show promising results for nuclear power operations on the Moon, including issues like heat dissipation and reactor design. Three installations—two close to the poles and one on the equator—might be ideal from the political, power and coverage perspectives. To this end, developing reliable nuclear reactors early on seems to be a must for the feasibility of the planetary defense base.

26.6.4 Other Concerns Related to the Moon Base Laser Facility

Once the main engineering issues are addressed, there are other concerns that might make the system completely useless. The first and most worrying one is the matter of timing. If the system is sitting on the far side of the Moon, its visibility range is determined by the month-long lunar orbit around the Earth. This would mean that the system might have to wait up to half a month to start the deflection process on a target detected on the wrong side of the sky. This would be unacceptable for a system that initially can only target asteroid on terminal trajectory. The main solution is to have two sites operationally ready at any given time: two locations at lunar latitudes of +82° and −82° would see about 95.6% of the total sky at any given time, minimizing the chances of a target appearing with short notice from the 4% of the sky that is uncovered. This fraction of uncovered sky can be minimized by optimizing the latitude of both locations and would be nearly zero for locations close to the poles. The only constraint is how far the lasers must sit from the poles to have no visibility of Earth at any given time. The issue of bad timing is also reduced using the laser system for sky-scanning with active illumination: active scanning would increase the number of asteroids detected and deepen our knowledge of their precise orbits, eventually minimizing the chances that a target is just discovered during its terminal approach. We argue that the problem of detections with low warning time is not an excuse to use the last resort deflection methods, such as the nuclear armed HAMMER concept. The objective should be to deepen our knowledge of Solar System asteroids first, because then politically sensitive technologies will become irrational options.

Despite its workarounds, the bad-timing issue is a key drawback of the system. Other alternatives, such as kinetic impactors could avoid all these issues. An always-ready kinetic impactor, not so different in design from modern intercontinental ballistic missiles (ICBMs), could be launched with no given notice to any target discovered in terminal approach. The operational envelop of an impactor-based system is only constrained by the size of the target and the number of impactors launched and could be easily extended by just maintaining a fleet of these impactors ready to go. The impactors would have to be completely nuclear-free to ensure they are not used as ICBMs in terrestrial war, but otherwise the system is easy to implement and far cheaper than a lunar base. However, any rocket designed to target an asteroid at speeds of tens of km per second could easily be used as an anti-satellite weapon, and any nation stockpiling them would have a fleet of ICBMs ready to go by just changing the payload from a dummy lead block to a similar-sized nuclear device. The political issues are then obvious (for example, a fleet of kinetic impactors might easily allow anyone to avoid any antiproliferation treaties and stockpile nukes), and the potential to stockpile enough impactors to deflect anything on terminal approach would be greatly reduced. A laser system on the Moon avoids political issues, requires novel governance perspectives without the interference of realist thinking, and could respond to most short-term threats and potentially eliminate the risk of any target being discovered in short notice through active illumination. Moreover, the installations on poles would have a chance to deflect even comets coming from above or below the ecliptic.

If the laser's usability is limited by its total size, and initially it would only be able to target asteroids close by, its alternative use as an active-illumination scanner increases its usability as a planetary defense system. This is because it would reduce the chances of undetected threats coming in, as happened with the Chelyabinsk meteor. During the time it takes for the system to reach a suitable size to tackle long-term or distant threats, one-off missions to asteroids might do the trick, with a smaller system operating in a similar fashion from a spacecraft that would apply laser deflection to an asteroid from just a few kilometers away. However, if the laser system managed to reach the required size, it would be the only one capable of targeting any type of threat, with enough warning both in time and distance to make it an ideal planetary defense system.

By this stage, the laser would also be used to propel deep-space and interplanetary spacecraft on a regular basis, and to actively scout the whole Solar System to provide the most accurate picture of our surroundings to date. The multipurpose laser system might be the most challenging planetary defense option to build, but it is the only one capable of becoming the ideal system to protect Earth while propelling human knowledge and capabilities to the stars.

26.7 Multipurpose Lunar Base: The Ultimate Planetary Defense and the Gateway to the Universe (Fig. 26.12)

26.7.1 Laser Propulsion, Planetary Defense and Lunar activities

The idea for sailing in space was proposed by Johannes Kepler to Galileo in 1610 from Prague:

> It is not improbable, I must point out, that there are inhabitants not only on the moon but on Jupiter too, or (as was delightfully remarked at a recent gathering of certain philosophers) that those areas are now being unveiled for the first time. But as soon as somebody demonstrates the art of flying, settlers from our species of man will not be lacking. Who

Fig. 26.12 Artist depiction of the Multipurpose lunar base. Credits: visualizations by Martin Vaněk, AD13

> would once have thought that the crossing of the wide ocean was calmer and safer than of the narrow Adriatic Sea, Baltic Sea, or English Channel? Given ships or sails adapted to the breezes of heaven, there will be those who will not shrink from even that vast expanse. Therefore, for the sake of those who, as it were, will presently be on hand to attempt this voyage, let us establish the astronomy, Galileo, you of Jupiter, and me of the Moon. (Kepler 1611)

The ideas promulgated in this remarkable letter are now coming to pass. Particularly significant are accelerating efforts to use and settle the Moon and using laser light-sails, the latter being developed for interstellar probes through the Breakthrough StarShot project.[5] The laser field shown in the illustration annex is a depiction of a real engineering design by Philip Lubin (2016).

When men first set foot on the Moon in 1969, many thought there would soon be large-scale human activity on the Moon, including human settlements. However, Government space programs were little interested in human expansion into the Solar System, concentrating instead on Earth-focused space activities. Even the International Space Station (ISS), ostensibly about human exploration of space, is Earth-focused. Indeed, the primary question of human exploration is whether humans can live and reproduce in Lunar or Mars gravities (1/6th and 1/3rd Earth gravity, respectively). But the human-rated centrifuge that would have simulated Lunar and Mars gravities was never built for the ISS. Yet, recent space agency discussions have to some degree turned to lunar topics.

The aforementioned Moon Village concept envisages an international cooperative science facility on the Moon. But to date, no substantive funds have been expended nor programs developed for this concept. Similarly, in 2017 the Trump

[5] See https://breakthroughinitiatives.org/initiative/3.

Fig. 26.13 Servicing rover. Credits: visualization and design by Martin Vaněk, AD13

Administration directed NASA to return humans to the Moon and plan for human Mars missions (Cofield 2017). However, the only NASA lunar mission, the Resource Prospector Mission, was cancelled in May, 2018 (Foust 2018b). Other space agencies, in particular those of China and India, although planning ambitious lunar science missions, have not definitively announced large-scale lunar programs. The lunar focus is shifting, however, to private space activities funded by high-net-worth individuals, such as Amazon's Jeff Bezos. As mentioned, he is not disclosing too much, but was quite clear at the ISDC 2018 conference in Los Angeles that there are two options for him: to build the Moon base or to run out of money, adding that the latter is less probable (Fig. 26.13).[6]

26.7.2 *Laser Light-Sails*

The idea for light-sails owes its origin to Johannes Kepler, who observed that comet tails point away from the sun, suggesting these are caused by the same. Pyotr Lebedev showed that light can produce pressure at the turn of the 20th century

[6] Heard in person during the conference by Nikola Schmidt.

(Lebedev 1901). Light sails work much like the wind does to impart momentum to a sail. In particular, light photones carry both momentum and energy. When a photon bounces off a light sail, it deposits a small quantum of momentum. This is the source of movement for the sail and any attached payload. Robert Forward first proposed laser-propelled light-sails for interstellar flight in the 1970s (Forward 1984). In 2015, under the support of NASA's Innovative and Advanced Concepts (NIAC) program, Phillip Lubin showed that with rapid advances in low-cost lasers and miniaturization of spacecraft, it would be feasible to send gram-class spacecraft to the nearest stars at 10% light speed (Lubin 2016).

On April 12, 2016, the 55th anniversary of Yuri Gagarin's historic first human spaceflight, Yuri Milner and Stephen Hawking announced Breakthrough StarShot, whose objective is to send nanoprobes to the nearest star. The current objectives differ from Lubin's original proposal in that the laser would be based on Earth rather than space, and the objective is to send the gram-class probes at 20% light speed. The program envisages spending $100 million over the next 5–7 years to develop technologies in three key areas: low-cost laser systems that can be developed to enable a 100 GW laser at kilometer scales and a cost of about $10 billion; a light-sail/nanocraft combination that could be accelerated in a few minutes at tens of thousands of gravities and survive the 20-plus year interstellar flight; and laser communications technology that can effectively transmit planetary images across 4.3 light years from the nearest star system, Alpha Centauri.

If the research program is successful, a billion-dollar class prototype could be built in the next 10–15 years with the full-scale system constructed, probably in Chile's Atacama Desert, within the next few decades. The nearest star system to our Sun is the Alpha Centauri star system consisting of three stars, two of them similar to the Sun and a third being a much smaller star known as a 'Red Dwarf.' The star's location in the sky is far south of the equator at –61° south delination. It is thus only visible in the Southern hemisphere. We would need a laser in the Southern hemisphere to propel a probe there, in preferably as high and dry a site as possible to get a high energy blaser beam through the atrmosphere. At 5000-m altitude, the Atacama is above about a third of the Earth's atmosphere, has great weather (no rain in decades) and a good infrastructure. It is an ideal site for a ground-based StarShot laser system. However, satisfactory sites may also exist in New Zealand. This program if successful would represent a major public-private partnership. The early high-risk phases would be privately financed, with the final system paid for in partnership with publicly funded space agencies.

While the initial system would be based on Earth, a final system might best be placed on the lunar far-side, both for operational reasons and to ensure that it does not raise concerns about interference with Earth-focused space operations or appearance as a weapon. Phillip Lubin pointed out in 2016 that a system of this size, properly based and configured, could be used to deflect potentially hazardous asteroids and comets (Lubin et al. 2016). Such a facility for both interstellar and Solar System propulsion and planetary defense may be a key component in humanity's future lunar infrastructure and provide a building block for untested global governance systems (Fig. 26.14).

26.7.3 The Shape of the Proposed Laser Facility in Case of Being on the Moon

Out of the many potential options to deflect potentially hazardous objects for planetary defense, the one with the best applicability on the Moon is probably a *Directed Energy System* (DES), such as the Lunar Sentinel proposed by Thangavelu and Vasmate (2016). This system would deflect potentially hazardous objects based on surface material vaporization or could disrupt them using resonant frequencies to mitigate the damage caused by the impact. A directed energy system can work at planetary distances and can deflect hazardous objects in terminal approach, with barely days or weeks of warning (Thangavelu and McVicker 2015). Most other proposed methods, such as kinetic impactors, nuclear deflectors, gravity tractors or deflection by alteration of surface albedo, require a spacecraft in close proximity to the asteroid (be it via rendezvous or direct impact) that can be stored on and launched from Earth, negating any associated benefit from a lunar base. These systems also require either expendable spacecraft, such as HAMMER (Barbee et al. 2018), or

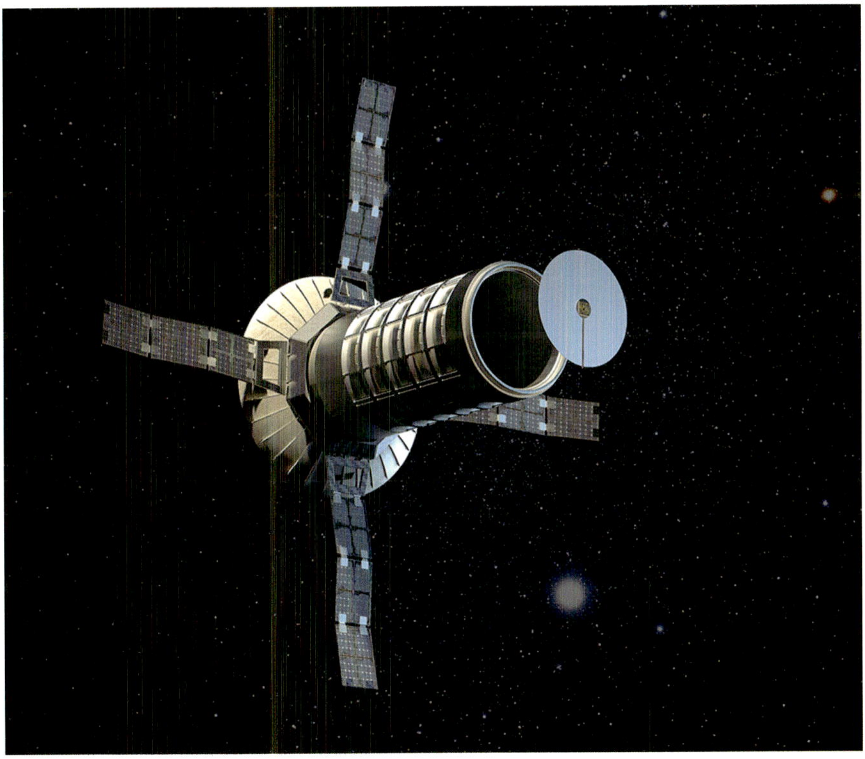

Fig. 26.14 Container of nanoprobes releasing them for laser acceleration to Proxima Centuari. Design partly based on inspiration from Breakthrough Initiatives. Credits: visualization by Martin Vaněk, AD13

complex and expensive rendezvous maneuvers and long-term operations (a few decades), reducing their effectiveness as defense systems. Directed energy systems provide the most viable solution for both long- and short-term threats while adding the benefits of versatility to operate from close proximity to Earth, reusability, and potentially targeting several targets at once. This, along with its multiple other applications, makes it the ideal option for a permanent planetary defense facility.

While a DES facility could potentially be placed anywhere on cislunar space, including Earth, there are a number of political problems involved. A laser system that could deflect an asteroid would also be a powerful anti-satellite weapon. Also, if placed in orbit in cislunar space, it could be targeted at Earth with extreme precision, a dangerous enough prospect to cause conflict through its mere existence. The solution to these problems is building the facility on the only place of near-Earth space that cannot target Earth's satellites or ground facilities: the far side of the Moon. With about 82% of the surface of the far side being invisible from Earth, the risk of the laser facility being used as a weapon against our planet or its infrastructure all but disappears. There are other advantages that the Moon brings beyond its elimination of immediate political risk. The lack of atmosphere prevents any loss of power or dispersion on the way of the beam beyond that of the interplanetary environment, and if placed on the poles it could have access to nearly constant solar power. While a large facility might require a nuclear reactor for its power supply, initial operations and testing could begin with a small laser array powered by the constant flux of solar energy on the lunar poles.

This concept is not new. Thangavelu and Lubin have proposed viable concepts for in-space laser defense systems, the first one on the lunar far side and the second from an orbital facility. The beauty of these proposals is that they show that a laser facility powerful enough to deflect most potentially hazardous objects is feasible within current technology. A GW-class laser from the lunar DES could be formed by a combination of kW-class lasers combined in a phased array, using modular beam combining and adaptive optics to ensure the optical quality of the combined beam. This concept is similar, being scalable to essentially any range of power desired. Current Ytterbium laser amplifiers can yield an energy efficiency of 35%, with a mass per unit power of around 5 kg/kW. This is expected to improve in the near future to reach 50% energy efficiency and a 1 kg/kW mass ratio, reducing the costs of deploying the lasers on the lunar surface and reducing its power requirements (Lubin et al. 2016). Similar systems have already been tested in military environments and in form factors small enough to be deployed in mobile systems and that can be adapted for use in space.

The main benefit of a directed energy system over other planetary defense options is its broad range of alternative commercial applications. The lasers could be used for prospecting of asteroids via active illumination, and active manipulation of asteroids' orbits could be directly used for asteroid mining by changing the orbits of mining targets to come closer to Earth or the Moon. This would make mining targets more accessible for exploitation, so long as the system was powerful enough and capable of firing continuously without significant shutdowns for heat dissipation. The potential to change orbits of comets could even make these great sources of water for in-space use.

The large power of the laser array could also be used to propel solar-sail spacecraft to relativistic speeds: a 40–70 GW laser array could accelerate a wafer-scale, 1-m-sail spacecraft to 26% of the speed of light in 10 min, allowing it to reach Alpha Centauri in a mere 20 years, as proposed by Breakthrough StarShot (Fig. 26.15). Larger spacecraft could be accelerated to lower speeds appropriate for exploration of the Deep Solar System, e.g. a 10,000-kg spacecraft could be accelerated to 1000 km/s, being able to reach the planetoid Sedna in around 4 months. The use of the array as an acceleration system could also be scaled down to allow propellanless injection into interplanetary trajectories of large spacecraft, towards Mars or the asteroids, further assisting human expansion into the Inner System.

The large number of advantages of a lunar DES facility raise the legitimate question of whether these concepts are not too far in the future, given the limited mass that we are currently capable of putting in Earth orbit. Mass is indeed the main limiting factor for the feasibility of the idea, as for the large majority of space missions, but it is important to consider the current state of the orbital launch industry and its ongoing development. Ariane 5, which was the most affordable rocket until

Fig. 26.15 Field of laser arrays prepared to fire and accelerate nanoprobes to Proxima Centauri. Credits: concept: Philip Lubin. Visualization by Martin Vaněk, AD13

2012, has been capable of putting 16 tons in low Earth orbit for \$160 M–\$200 M. Since 1998, however, the latest iteration of SpaceX Falcon 9 has been able to launch 23 tons to LEO for \$62 M in expendable mode, barely a quarter of the cost of Ariane 5. If SpaceX reaches its promised objectives for the Big Falcon Rocket, planned for the first Mars cargo trip in 2022, the company will be able to launch 150 tons for \$7 M, about 0.4% of the cost of Ariane 5. This would mean a cost reduction of nearly 33% per year for the second and third decades of the 21st century, a testament to the revolutionary times that we are experiencing in the space industry. This large reduction in the cost of launches would potentially make the mass of the facility a secondary concern, leaving the cost of putting 1 kg on the Moon at around \$5000. We are nearing the time where the only limiting factor for humanity's expansion into space is our will to do so.

Nevertheless, having the laser facility for Breakthrough StarShot in Atacama Desert already funded by Yuri Milner, moving it completely to the Moon in case of political tensions caused by building such an extremely powerful laser, while challenging, would still be within the realm of possibility.

26.8 Conclusion: Earth Defense Is Not National Defense

As clearly shown above, planetary defense capability can be built on a true civilian scientific project sending probes to the stars. This is one of the critical ideas of our volume. We want to soundly argue that taking nationally driven defense policies and applying them on the planetary defense scale is not only erroneous, but dangerous. Engaging a cosmopolitan outlook is not wishful thinking but can be embraced by visionary projects that already have private funding secured. We believe that this small exercise undertaken from Chaps. 1 to 26 of this volume has demonstrated that visionary thinking is not wishful thinking. All these ideas are underpinned by a solid political theory, clear science and precise engineering. We look forward to the reactions that await us.

In this chapter, our aim was to take the normative stigma of nuclear weapons and use it as an opportunity to argue for an alternative political system in the context of the Moon settlement realities. The developed suggestions for research of security communities points us to the concept of communities of fate, free of dominating national outlooks, historical realism and based on actual actors, practices and knowledge. We showed that living on the Moon is difficult, but not impossible, and that a future lunar base can have many possible applications. A multipurpose lunar base built by, say, a multinational global cosmopolitan village is meant to be the first serious settlement by mankind on another celestial body; however, as has been demonstrated, it is highly probable that the next steps in the moon regolith will be funded from private sources. The inclusion of private and commercial actors in cooperation with the state ones can help to overcome the mentioned old ideas about Moon colonization of governmental exclusivity, costs and effectivity. It can allow for a stable regulatory, protective and internationally legal environment, enabling long-term human presence on the Moon and overcoming the limitations posed by national Moon efforts.

We cannot imagine a better idea than Breakthrough Starshot, because if the required laser to propel nanoprobes to the star is able to deliver the capability to deflect an asteroid, we will achieve exactly the kind of security-plus we have been talking about throughout this book. Security can be positive and provide opportunities to flourish and enable the establishment of the Moon as the gateway to the universe. Approaching Moon settlements with the multi-stakeholder concept of communities of fate can achieve that.

Acknowledgement This study was supported by the grant awarded by the Technological Agency of the Czech Republic, project TL01000181: *"A multidisciplinary analysis of planetary defense from asteroids as the key national policy ensuring further flourishing and prosperity of humankind both on Earth and in Space,"* and co-funded by the Institute of Political Studies, Faculty of Social Sciences, Charles University in Prague.

References

Barbee, B. W., Syal, M. B., Dearborn, D., Gisler, G., Greenaugh, K., Howley, K. M., et al. (2018). Options and uncertainties in planetary defense: Mission planning and vehicle design for flexible response. *Acta Astronautica, 143*(August 2017), 37–61. doi:https://doi.org/10.1016/j.actaastro.2017.10.021

Barnett, A. (1998). Summary: Security Communities

Cadogan, D., Stein, J., & Grahne, M. (1999). Inflatable composite habitat structures for lunar and mars exploration. *Acta Astronautica, 44*(7–12), 399–406

Cofield, C. (2017). President Trump Directs NASA to Return to the Moon, Then Aim for Mars. space.com. https://www.space.com/39050-trump-directs-nasa-humans-to-moon.html. Accessed 12 July 2018

Deutsch, K. (1957). Political Community and the North Atlantic Area. *The European Union*

Ditrych, O. (2014). Security community: A future for a troubled concept? *International Relations, 28*(3), 350–366. doi:https://doi.org/10.1177/0047117814545952

Forward, R. L. (1984). Roundtrip interstellar travel using laser-pushed lightsails. *Journal of Spacecraft and Rockets, 21*(2), 187–195

Foust, J. (2018a). NASA authorization bill increases emphasis on commercial partnerships. *Spacenews*. http://spacenews.com/nasa-authorization-bill-increases-emphasis-on-commercial-partnerships/. Accessed 23 April 2018

Foust, J. (2018b). NASA argues Resource Prospector no longer fit into agency's lunar exploration plans. spacenews.com. https://spacenews.com/nasa-argues-resource-prospector-no-longer-fit-into-agencys-lunar-exploration-plans/. Accessed 12 July 2018

Gibbings, A., Vasile, M., Hopkin, J.-M., Burns, D., & Watson, I. (n.d.). *LASER BEES. A Concept for Asteroid Deflection & Hazard Mitigation*. http://www.planetary.org/explore/projects/laser-bees/Gibbings-Laser-Bees-201202.pdf

Hall, A., & Miller, C. (2016). A Summary of the Economic Assessment and Systems Analysis of an Evolvable Lunar Architecture That Leverages Commercial Space Capabilities and Public–Private Partnerships. *New Space, 4*(1), 4–6. doi:https://doi.org/10.1089/space.2015.0037

Held, D. (2010). *Cosmopolitanism: Ideals and Realities*. Polity

Inglis, D. (2014). Cosmopolitans and cosmopolitanism: Between and beyond sociology and political philosophy. *Journal of Sociology, 50*(2), 99–114. doi:https://doi.org/10.1177/1440783312438788

Jurvetson, S. (2014). Moon Base Alpha — Strategies for Low Cost Lunar Settlement Workshop at DFJ. Presentation on August 23, 2018. Flicker.com - *a report from a presentation*. https://www.flickr.com/photos/jurvetson/30929394984. Accessed 10 July 2018

Kepler, J. (1611). *Ioannis Kepleri Mathematici Caesarei Dissertatio, cum nuncio sidereo, nuper ad mortales misso a Galilaeo Galilaeo, Mathematico Patauino*. Francofurti: Apud D. Zachariam Palthenium

Kessler, O., & Guillaume, X. (2012). Everyday practices of international relations: People in organizations. *Journal of International Relations and Development, 15*(1), 110–120. doi:https://doi.org/10.1057/jird.2011.29

Kettley, S. (2018). Moon colony: Amazon chief Jeff Bezos targets moon for Blue Origin lunar colony by 2023. *express.co.uk*. https://www.express.co.uk/news/science/984907/Moon-colony-Jeff-Bezos-Amazon-Blue-Origin-NASA. Accessed 10 July 2018

Lebedev, P. (1901). *Untersuchungen über die Druckkräfte des Lichtes*. Annalen der Physik

Lubin, P. (2016). A roadmap to interstellar flight. *JBIS - Journal of the British Interplanetary Society, 69*(2–3), 40–72

Lubin, P., Hughes, G. B., Eskenazi, M., Kosmo, K., Johansson, I. E., Griswold, J., et al. (2016). Directed energy missions for planetary defense. *Advances in Space Research, 58*(6), 1093–1116. doi:https://doi.org/10.1016/j.asr.2016.05.021

Musk, E. (2017). Making Humans a Multi-Planetary Species. *New Space, 5*(2), 46–61. doi:https://doi.org/10.1089/space.2017.29009.emu

Nakamura, Y., & Furukawa, Y. (2017). Space Startup Raises $90 Million to Start Moon Ads by 2020 - Bloomberg Businessweek Middle East. *Bloomberg Businessweek*. http://businessweekme.com/2017/12/14/space-startup-raises-90-million-start-moon-ads-2020/. Accessed 23 April 2018

Nyman, J. (2016). What is the value of security? Contextualising the negative/positive debate. *Review of International Studies, 42*(05), 821–839. doi:https://doi.org/10.1017/S0260210516000140

Phipps, C., Birkan, M., Bohn, W., Eckel, H.-A., Horisawa, H., Lippert, T., et al. (2010). Review: Laser-Ablation Propulsion. *Journal of Propulsion and Power, 26*(4), 609–637. doi:https://doi.org/10.2514/1.43733

Pittman, R. B., Harper, L. D., Newfield, M. E., & Rasky, D. J. (2016). Lunar Station: The Next Logical Step in Space Development. *New Space, 4*(1), 7–14. doi:https://doi.org/10.1089/space.2015.0031

Riley, J., Lubin, P., Hughes, G. B., O'Neill, H., Meinhold, P., Suen, J., et al. (2014). Directed energy active illumination for near-Earth object detection. *Proceedings of SPIE 9226, Nanophotonics and Macrophotonics for Space Environments VIII, 9226*, 922606-922606–15. doi:10.1117/12.2061355

Sharkov, D. (2018). Russia Will Go to the Moon Next Year With First Lander in Decades. *Newsweek*. http://www.newsweek.com/russia-will-go-moon-next-year-first-lander-decades-790446. Accessed 23 April 2018

Sherwood, B. (2005). Lunar architecture and urbanism. *SAE transactions*, 354–361

Silva-Martinez, J. (2016). Human systems integration: process to help minimize human errors, a systems engineering perspective for human space exploration missions. *REACH, 2*, 8–23

Thangavelu, M., & McVicker, J. M. (2015). QBOLT - Directed Energy System Concepts for Asteroid Threat Mitigation. In *4th IAA Planetary Defense Conference – PDC 2015* (pp. 13–17). Frescati, Italy

Thangavelu, M., & Vasmate, V. (2016). LUNAR SENTINEL : Planetary Defense from the Moon. *AIAA SPACE2*, 5475

The White House. (2018). *National Near-Earth Object Preparedness Strategy and Action Plan*. http://www.whitehouse.gov/ostp. Accessed 6 July 2018

Utrilla, C. M.E (2017). Establishing a framework for studying the emerging cislunar economy. *Acta Astronautica, 141*, 209–218. doi:https://doi.org/10.1016/j.actaastro.2017.10.005

VodafoneGroup. (2018). Vodafone and Nokia to create first 4G network on Moon. http://www.vodafone.com/content/index/media/vodafone-group-releases/2018/vodafone-and-nokia-to-create-first-4g-network-on-moon.html#. Accessed 23 April 2018

Winner, L. (2003). Do artifacts have politics? *Technology and the Future, 109*(1), 148–164. doi:https://doi.org/10.2307/20024652

Zuniga, A. F., Rasky, D., & Pittman, R. B. (2015). Lunar COTS: An Economical and Sustainable Approach to Reaching Mars. *AIAA Space 2015 Conference and Exposition*, 1–24. doi:https://doi.org/10.2514/6.2015-4408

Zürn, M. (2018). A Theory of Contested Global Governance, *9*(1), 138–145. doi:https://doi.org/10.1111/1758-5899.12521

Chapter 27
Will Planetary Defense Change a Realist World?

James J. Wirtz

27.1 Introduction

Until recently, the subject of planetary defense straddled the line between science and science fiction. In 2002, I included a discussion of the topic in the first edition of the textbook, *Strategy in the Contemporary World*, as an entertaining diversion for undergraduates. After all, I could not pass up the opportunity to write the following words in a serious scholarly work: "every year thousands of extraterrestrial visitors arrive to stay on planet earth" (Wirtz 2002). At the time, I relied on a single source on the subject: a paper written in 1996 by a group of Air Force officers who depicted a *fictional* scenario unfolding 30 years in the future (Urias et al. 1996).[1] *Tempus fugit*: Their vision of the future is upon us, arriving a few years early to boot.

The chapters in our volume demonstrate that planetary defense is no longer the stuff of science fiction. Instead, it is a highly developed scientific and engineering endeavor that has begun to characterize the threat posed by the impact of celestial bodies on the Earth-Moon system in general, and planet Earth specifically. As Clemens Rumpf notes in Chap. 12, a good deal of progress has been made in term of "rationalizing the threat," the first and maybe most important step in building planetary defenses. Most asteroids larger than 1 km in diameter have been identified and assessed, which is important since impacts involving bodies of this size can produce a global extinction event. Nevertheless, smaller objects also can wreck enormous damage. Although it is only about 500 m in diameter, asteroid 101955 Bennu would generate a force of approximately 1200 MT of TNT if it hits Earth, a low-probability event that might occur between 2175 and 2199. The bad news is

[1] In a disclaimer at the outset of the paper, they noted that "at this time, Planetary Defense is not an assigned or approved mission for the United Stated Department of Defense or the United States Air Force."

J. J. Wirtz (✉)
Naval Postgraduate School, Monterey, CA, USA
e-mail: Jwirtz@nps.edu

N. Schmidt (ed.), *Planetary Defense*, Space and Society,
https://doi.org/10.1007/978-3-030-01000-3_27

that only 1% of the near-Earth objects in the 100-m size range have been identified; smaller bodies are far more numerous than larger ones, and even small objects can cause enormous death and destruction if they hit a populated area. In 2013, for instance, the Chelyabinsk meteor, which was approximately 20 m wide, exploded while entering the Earth's atmosphere with a force of about 500 kt of TNT.

27.2 Towards Planetary Defense

The first step in constructing a planetary defense is thus well underway: risk assessments are continuously refined as new data about actual objects are gathered by astronomers. This leads to the observation that the time has arrived to take the next major step in planetary defense by facilitating, synchronizing or coordinating private and state-sponsored efforts to search for relatively small asteroids.

Step two involves building the warning function of planetary defense so that we can identify real threats with enough lead time to take action. As Peter Veres and Nikola Schmidt report in Chap. 4, there already is an extensive global effort to search for small objects undertaken by the community of professional and amateur astronomers. The problem is that private, state-sponsored and international efforts to synchronize these activities are underdeveloped. One would think that standardization and digitization of observation reports in formats easily accessible to researchers might yield immediate dividends. In Chaps. 5 and 6, Peter Veres also notes that developing better instrumentation to conduct space surveys might be the best place to spend the marginal "planetary defense dollar." Better survey equipment and techniques would not only help refine threat estimates but would also identify targets before they crash into Earth. The second step of planetary defense—working to build situational awareness—has started. Individuals, academic institutions and state agencies have begun to tackle the problem, albeit somewhat haphazardly, and agreement seems to be emerging that better coordination and equipment can accelerate progress. In 2016, for example, the National Aeronautics and Space Administration (NASA) established the Planetary Defense Coordination Office to, among other things, better organize the search effort. Step one in planetary defense identifies what to look for; step two is about doing the looking.

Step three involves taking action once the threat is identified, which at this time seems primarily focused on deflecting an object from its trajectory so that it does not impact the Earth. The fundamental parameters of this problem are straightforward: the smaller the object and the longer the available lead time, the less force is necessary to deflect it from its Earth-bound trajectory. Nahum Melamed in Chap. 8 even discusses a computer application funded by NASA that provides a good approximation of what type of force would have to be applied to shift the orbit of a range of targets. At present, we do not possess the technology to divert large objects; mercifully, as current risk assessments suggest, they pose a minimal threat. With enough warning, however, nuclear weapons at least in theory could be used to divert or even break apart more modest threats. If all goes as planned, in 2024 NASA will actually assess the kinetic impactor method to divert an asteroid in the appropriately named Double Asteroid Redirection Test. The idea is to use a small spacecraft to alter the path of the non-threatening asteroid Didymos B, which is actually a smaller satellite of Didymos A. Didymos B is still about ten times the size of the Chelyabinsk meteor.

A successful Double Asteroid Redirection Test would forever transform planetary defense from science fiction into science. Planetary defense capabilities would be rudimentary, but they would be a reality. With a little bit of luck, we might actually enjoy a few hundred years to improve our survey and deflection systems before we have to put our planetary defenses into action to deflect a significant threat.

27.3 Progress Without Politics?

What is remarkable about the current evolution and trajectory of planetary defense is how much progress has been made at relatively little cost and with little political fanfare or concern. National efforts mostly, but not exclusively, conducted in the United States (note the NASA-FEMA planetary defense plan released in June 2018 (The White House 2018)), have proceeded apace with technological advances, improved research methodologies and new findings.

A global scientific community is emerging that eagerly incorporates new equipment and research into the planetary defense enterprise. The fact that a redirection test is actually scheduled for 2024, combined with the rapid development of the science and technology behind that test, is bound to give the casual observer a good dose of future shock, so to speak. Planetary defense is not necessarily expanding the limits of technology and the boundaries of science, but the epistemic community working on the problem appears to be quickly and efficiently incorporating technological advances and research findings into the global defense enterprise. It benefits from an outburst of enthusiastic support from amateur astronomers and the academic community. It also is helped along by the fact that government officials view planetary defense as a good way to spend the "budget dust" left behind on ledger sheets by larger programs.

In terms of theory, scholars do not spend a lot of time formulating explanations of highly successful endeavors that proceed smoothly, with much international participation, at low cost, while generating little political interest or concern. It is possible that such phenomena are actually commonplace but are more easily recognized in hindsight when achievements become socially, militarily or politically salient. For instance, a few scientists transformed facets of chemistry into nuclear physics in the late 19th century; nevertheless, this development went largely unnoticed at the time (Bernstein 2008). Others have suggested that when problems are modest and solutions are within reach, collaboration becomes more likely because absolute gains clearly outweigh relative gains or the high costs of policy synchronization (Dinar and Dinar 2017). The truth is that even the most ardent functionalists, who posited that cooperation between scientists and policy experts could occur to solve shared problems despite political barriers, would probably be surprised by the success of this epistemic community.

Given the likelihood that we might have hundreds or even thousands of years before a true crisis, current trends suggest that the Earth might possess a robust planetary defense before it actually ends up in the crosshairs. Indeed, because policymakers (and most rational observers for that matter) lack information about long-term outcomes, they generally base their estimates of the future on trends (Gartner 1999). Current trends thus bode well for planetary defense, despite the fact that it is not politically salient anywhere on the planet. If politicians and scientists alike approach all problems in the same way—apolitical assessment, development of optimization strategies for resource allocation, building situational awareness to refine assessments, launching increasingly ambitious remedial experimentation with maturing technology and know-how—it would not be long before most of the ills troubling mankind would be a thing of the past. The significant phenomenon outlined by our contributors might indeed be a new type of social interaction that animates a scientific and engineering enterprise. Like functionalists from the last century, our contributors suggest that technical cooperation could spark political compromise, cooperation and evolution—the Planetary Council mentioned in Chap. 21 is a case in point.

Defense enthusiasts energized by visions of advanced technologies might bemoan a lack of progress and demand greater international commitment to planetary defense. Enthusiasm is admirable, but such calls are misguided given the fact that planetary defenses are already keeping pace with advancing technological and situational awareness. It also appears that there is virtually no political support for bearing the significant costs that would be entailed in accelerating progress, especially because of the potentially long timeframes involved in developing robust defenses against medium-sized asteroids. We probably have hundreds or even thousands of years before we confront a real moment of truth; current trends suggest that we will have a solution in place when needed, and we probably will have the capability and situational awareness needed to deflect a Chelyabinsk-sized meteor before mid-century. There is nothing broken about the planetary defense enterprise—planetary defense activism runs the risk of disrupting the constructive social and political setting that facilitates progress by actually drawing popular attention to the enterprise. There are no votes in planetary defense.

27.4 Planetary Defense and the Emergence of Cosmopolitanism

Several of our contributors, however, see a somewhat different political problem lurking behind the planetary defense success story. In their minds, national governments, international law, international regimes and the general state of international collaboration cannot cope with a planetary defense crisis, i.e. a need to take action to divert an asteroid. They are uncomfortable with those who will probably be at the controls of our initial planetary defense (i.e. the United States). They wonder whether or not Washington "will take the shot" to save others, especially if the United States and its partners are unaffected by the looming asteroid impact or whether they will be willing to take the risk of missing or only deflecting the asteroid partially so it crashes elsewhere. If placed in the wrong hands, they believe that the planetary defense apparatus could be used to coerce those who are not at its controls. Also troubling is the fact that nuclear weapons offer the only practical way at present to alter the trajectory of likely targets.

Planetary defense actually supplies governments with a good reason to keep a few "bombs in the basement," especially some extremely high-yield devices. National control of planetary defenses thus helps to perpetuate the presence of nuclear weapons in military arsenals. In their minds, political acrimony, social turmoil and even war might be produced by one nation's decision to activate or not activate planetary defenses that will probably involve nuclear weapons. Even a successful defense could produce debris (a common problem encountered by existing air and missile defenses); rubble pile asteroids can separate into dozens of small bodies, creating an asteroid shower and potential mayhem. For those who champion a *cosmopolitan approach* to planetary defense, the technological and scientific progress enjoyed by the defense effort is actually outpacing the evolution of the political concepts and institutions needed to govern those defenses.

It is impossible to comment extensively in this chapter on various streams of cosmopolitanism; the interested reader has already been treated to such an exposé in Chaps. 20, 21, 24 and 26. Nevertheless, three facets of the concept are especially relevant to planetary defense. The first is the notion that all living creatures on the Earth are literally in the same boat. Therefore, the political, social, and ethnic concepts and boundaries that divide us are irrelevant, especially in the case of existential threats. Second, only a social and political construct that reflects our shared fate—cosmopolitanism—can effectively govern and employ a planetary defense. For example, in a cosmopolitan world, a nuclear planetary defense would not constitute a political or military threat because there would be no impulse or reason to employ nuclear weapons other than deflecting an asteroid. Nuclear weapons would not be perceived as a threat to international order in a cosmopolitan world. Cosmopolitanism thus eliminates the morally questionable action of increasing the risk of one existential threat (i.e. nuclear holocaust) to reduce the risk of another existential threat (i.e. impact of a large asteroid). Third, our contributors are not entirely consistent when it comes to the ontology of cosmopolitanism. Sometimes it is depicted as a necessary prerequisite for planetary defense; in other words, cosmopolitan institutions need to be created through deliberate political actions alongside the construction of planetary defenses. At other times, the emergence of cosmopolitan outlooks and institutions is depicted as a spontaneous and universal response to a growing human presence in space as new frames of reference and concepts alter common political perceptions of relevant social, ethnic or political divisions on the planet. They sometimes take a neo-functionalist view of the possible rise of cosmopolitanism, where technical and scientific cooperation will produce positive political consequences. Put somewhat differently, if enough people see the Earth from outer space, not only rocket scientists will recognize that we are all in the same boat.

Even if realists were aware of the supposed relationship between planetary defense and cosmopolitanism, they would probably suggest that the cosmopolitan project is rather fanciful. Two types of evidence would support this position. First, the US government, albeit with the support of a small group of international enthusiasts and state partners, is largely responsible for the current planetary defense effort. These efforts have virtually no political salience and are generally met with incredulity by average people, who would rather spend money fixing potholes instead of "shooting down asteroids." In a practical sense, planetary defense lacks the necessary political salience to gain significant funding, let alone alter notions of nationalism and international relations. Planetary defense is evolving quickly within the confines of the current nation-state system. The idea that cosmopolitanism is a necessary condition for effective planetary defense is not supported by the evidence supplied by recent history or obvious technological and political trends. The normative call for a cosmopolitan planetary defense is perfectly logical on both scientific and moral grounds, but it is impossible to see how it would lead to any practical political results. Instead, excessive cosmopolitan rhetoric, by highlighting the global nature of the defense enterprise, could easily lead to a backlash against the modest expenditures involved; there are always potholes to fix at home.

There also is absolutely no evidence that any government sees planetary defense as a means to gain international leverage over friend or foe alike. Indeed, scenarios whereby states manipulate defenses to achieve terrestrial political objectives seem farfetched and are based on capabilities and know-how that will not be available for quite some time. There are simply too many uncontrollable factors posed by the threat that render it unsuitable as a practical weapon or instrument of coercion. Militaries prefer to use weapons with known and controllable effects; with this observation in mind, threatening to take a pot shot at an asteroid to intimidate an opponent would probably be viewed as harebrained by professional soldiers. In any event, there are easier ways to destroy a city, or a country for that matter, than altering the course of an asteroid.

Those who see the asteroid threat-planetary defense nexus as a vector towards cosmopolitanism also fail to recognize that we face a basket of existential threats that have done little to alter the primacy of the nation-state. Phil Torres recently provided a survey of potentially existential problems that are the products of human ingenuity: nuclear weapons, genetic engineering, self-replicating nanotechnology, artificial intelligence, and (runaway) global warming (Torres 2016). Nevertheless, these relatively more familiar and in some cases more politically salient threats have done little more than foster improved international collaboration within the nation-state system and for the most part actually exacerbate international tensions. In other words, the fact that a threat faces the entire planet more or less equally is not sufficient to produce a cosmopolitan outlook or system. By contrast, planetary defense, which responds not to a manmade threat but to a natural disaster, seems to have generated the most cosmopolitan-like response among the scientific and policy communities. This response has done nothing to change the dominant political concepts and practices on the planet; nevertheless, this response, and the creation of rudimentary planetary defenses, has emerged in a world dominated by nation-states.

27.5 Conclusion

In several important ways, the planetary defense narrative highlights the growing obsolescence of the nation-state and the possibility of a cosmopolitan approach to world politics. For instance, the language of planetary defense puts us all in the same boat. The plural pronouns used throughout this chapter accurately reflect that planetary defense is defense of all the Earth's residents, regardless of commonly accepted political, ethnic and social divisions. Klaatu might object, but defense against asteroids is indeed cosmopolitan because it protects *our* planet and *our* species from an existential threat. Today, nation-states are the dominant actors in the defense effort, but that cannot alter the fact that planetary defense is cosmopolitan in nature. The fact that this incongruity seems to have had no political or practical impact especially on the rapid progress made by the defense enterprise is certain, but it is impossible to estimate how it might shape the evolution of world politics over the long term.

The planetary defense "movement" itself is highly cosmopolitan in the sense that it exists as a sort of spontaneous, worldwide scientific and engineering activity under no national flag. It also is a remarkable scientific endeavor that attacks a problem in an apolitical manner, relying more on brainpower than lavish government spending to get results. Moreover, all interested parties—individuals, institutions and states—are invited to the party, and the latest scientific findings appear to be readily available. In fact, it is so cosmopolitan that it lacks requisite synchronization and organization, a shortcoming that apparently is now being addressed by tapping the resources of nation-states, especially the US space program. The planet's defenses are not just cosmopolitan in terms of their function; they are cosmopolitan in terms of their animating philosophy, processes and development.

Those who see planetary defense as a clear path to cosmopolitan governance of planetary defense and who demand action to bring this situation about fail to understand that such rhetoric damages their normative project. Cosmopolitan rhetoric can give planetary defense the patina of global welfare, potentially slowing progress by politicizing the defense endeavor. Instead, it might be better to let a cosmopolitan state of mind emerge naturally and gradually as advances in space and planetary activity occur over the coming decades. A successful intercept of an incoming asteroid, colonization of the Moon and Mars or discovery of life or the remnants of past life somewhere else in the Solar System will cast dominant political concepts and institutions back on Earth in a new light.

This is not the first time that observers have suggested that scientific advances will produce positive political changes on the Earth, helping to end the risks created by the nation-state and international relations, but given the acceleration of change globally, trends should emerge soon. Maybe it's time to write a fictional scenario set around 2050: it could be entitled *Cosmopolity Planet Earth*.

References

Bernstein, J. (2008). *Nuclear Weapons: What you need to Know*. New York: Cambridge University Press.

Dinar, S., & Dinar, A. (2017). *International Water Scarcity and Variability: Managing Resource Use Across Political Boundaries*. Oakland: University of California Press.

Gartner, S. S. (1999). *Strategic Assessment in War*. New Haven: Yale University Press.

The White House. (2018). *National Near-Earth Object Preparedness Strategy and Action Plan*. http://www.whitehouse.gov/ostp. Accessed 6 July 2018

Torres, P. (2016). *The End: What Science and Religion Tell Us about the Apocalypse*. Durham: Pitchstone.

Urias, J. M., DeAngelis, I. M., Ahern, D. A., Caszatt, J. S., Fenimore III, G. W., & Wadzinski, M. J. (1996). *Planetary Defense: Catastrophic Health Insurance for Planet Earth*. Paper Prepared for the Air Force 2025 Project.

Wirtz, J. J. (2002). A New Agenda for Security and Strategy. In J. Baylis, J. Wirtz, E. Cohen, & C. Gray (Eds.), *Strategy in the Contemporary world: An Introduction to Strategic Studies* (p. 323). New York: Oxford University Press

Index

N. Schmidt (ed.), *Planetary Defense*, Space and Society,
https://doi.org/10.1007/978-3-030-01000-3

O

P

GPSR Compliance

The European Union's (EU) General Product Safety Regulation (GPSR) is a set of rules that requires consumer products to be safe and our obligations to ensure this.

If you have any concerns about our products, you can contact us on ProductSafety@springernature.com

In case Publisher is established outside the EU, the EU authorized representative is:

Springer Nature Customer Service Center GmbH
Europaplatz 3
69115 Heidelberg, Germany

Batch number: 10756373

Printed by Printforce, the Netherlands

BÜYÜK YAKINSAMA

Bilgi Teknolojisi ve Yeni Küreselleşme

RICHARD BALDWIN

Çeviri: Kemal Akın

BÜYÜK YAKINSAMA
Bilgi Teknolojisi ve Yeni Küreselleşme

RICHARD BALDWIN

Çeviri: Kemal Akın
Çeviri Editörü: Dila Balcı

The Great Convergence
Information Technology and the New Globalization

Genel Yayın Numarası: 348
ISBN: 978-605-2294-65-9
1. Basım, Eylül 2021

EFLATUN Basım Dağıtım Yayıncılık Danışmanlık Yatırım ve Tic. Ltd. Şti.©2021
Efil©2021

Sertifika No: 45550

Genel Yayın Yönetmeni: Serenay Dıraz
Grafik Tasarım: Hatice Kübra Demirel

Baskı ve Cilt:
Ayrıntı Basım Yayım ve Matbaacılık Hizmetleri Sanayi Ticaret Anonim Şirketi
Saray Mahallesi 126.Cadde No:20 Kazan / Ankara
Tel: 0312 394 55 90 . E-mail: info@ayrintibasimevi.com.tr
Sertifika Nu.: 49599

EFİL YAYINEVİ
EFLATUN Basım Dağıtım Yayıncılık Danışmanlık Yatırım ve Tic. Ltd. Şti.
Bağcılar Mahallesi Şemsettin Günaltay Caddesi 283. Sokak Ata Apartmanı
No: 9/7 06670 Çankaya/Ankara Türkiye
Tel : (+90) 312 442 52 10
GSM : (+90) 530 108 99 76 Faks : (+90) 312 442 52 12

www.efilyayinevi.com • instagram.com/efilyayinevi • twitter.com/efilyayinevi